JUNOS Enterprise Switching

JUNOS Enterprise Switching

Harry Reynolds and Doug Marschke

O'REILLY®

Beijing · Cambridge · Farnham · Köln · Sebastopol · Tokyo

JUNOS Enterprise Switching

by Harry Reynolds and Doug Marschke

Published by O'Reilly Media, Inc., 1005 Gravenstein Highway North, Sebastopol, CA 95472.

O'Reilly books may be purchased for educational, business, or sales promotional use. Online editions are also available for most titles (*http://my.safaribooksonline.com*). For more information, contact our corporate/institutional sales department: 800-998-9938 or *corporate@oreilly.com*.

Editor: Mike Loukides
Production Editor: Sarah Schneider
Copyeditor: Audrey Doyle
Proofreader: Sada Preisch

Indexer: Lucie Haskins
Cover Designer: Karen Montgomery
Interior Designer: David Futato
Illustrator: Robert Romano

Printing History:

July 2009:   First Edition.

ISBN: 978-0-596-15397-7

[LSI]   [2011-02-11]

1297276114

Table of Contents

Foreword

Charles Darwin once said, "I have called this principle, by which each slight variation, if useful, is preserved, by the term of Natural Selection." This principle of evolution applies to business as well as to nature. Individuals, companies, and industries evolve and compete with one another in preparation for the future.

Back in 1998, a group of engineers decided to split from the mainstream and form a new company, Juniper Networks. That company has evolved to create and acquire many products over the last decade. While Darwin might have viewed this as an evolutionary step in products, we view it as a competitive step.

Humankind differs from the rest of the animal kingdom not only because our instincts go beyond the primal survival urge (a.k.a. Darwin's "survival of the fittest"), but also because of our intellect and capacity to improve. When one group splits off from another, it does not necessarily sound the death knell for the original group. More likely, the two groups will start taking steps to outdo one another. In other words, they compete.

Those who are not directly involved in that competition can reap the benefits of the intellectual sparring. We have observed many such competitive moves within the routing industry over the years. Now, we are entering another such arena within the switching side of our industry.

Evolution of the Bridging World

More than 20 years ago, Radia Perlman developed an interesting algorithm while working for Digital Equipment Corporation. At the time, it was meant to solve an irritating little problem brought on by the forced evolution of data networks. Although networking in 1985 was significantly different than it is today, it's amazing how some of the basic building blocks have survived.

This book discusses some of those building blocks, both historically and with reference to the present-day network infrastructure. For example, the Spanning Tree Protocol, which was based on Perlman's algorithm, is like one of the amino acids from which today's switching was created. (I've always despised biology, but it seems to fit here!)

You'll also learn how to improve on features we've always believed to be necessities "just because they are." This book helps you understand the evolution—or stagnation—of networking to date, and then shows you how to see past it and unlock your networking potential in a few short steps.

What Is the Big Deal About Switching Anyway?

Interestingly, Layer 2 networking in general, and Ethernet internetworking in particular, are technologies vastly overlooked by many people in the current networking world.

Ethernet has become a commodity. We see many a WAN link today delivered as an "Ethernet handoff," which tells the tale of how common and simple it has become. And with it has emerged a number of supporting technologies and design mediums.

Many people see this area of networking as "voodoo magic." That term implies it can be good as well as bad, but it doesn't explain why! Without waxing philosophical or launching a religious debate in this foreword, I'll leave it at this: you may learn things in this book that you knew only bits and pieces of before.

As we expand our networks to include IP telephony, data centers, or *really* large Layer 2 networks, we should pay attention to the details of many technologies that we have taken for granted over the years.

Even with as much Layer 2 experience as I have, serving as technical editor for this book has opened my eyes on several topics. It has spurred some entertaining debates, and perhaps it will do the same for you!

How This Book Will Help You (a.k.a. What's in It for Me?)

I don't think better people could have been found to write this book. Doug and Harry not only have an incredible set of technical talents—which I've come to know over the years—but they also have a great sense of style. Although those virtues alone make them qualified, their track record of success in teaching, training, and consulting has also been put to use for your edification.

Too many technical books jump off the Boring Board into the abyss of Techie Soup. While this book covers plenty of deep technical matters, I think you will find its approach refreshing and compelling.

Now, I can't guarantee that it will lead to any evolutionary process in your own life. But I can certainly say that after you read it, you'll have some eye-opening conversations (at least with yourself!). This book will help you quickly acquire knowledge surrounding the topics presented, and it will add to your successful implementation of those technologies.

One thing I've gained from being part of this book's evolutionary process is the burning desire to dive in and get my hands on these switches to see what they can do. While that may not mean much to some of you, anyone who has perused my personal lab on the Internet (*http://smorris.uber-geek.net/lab.htm*) should understand.

Evolution is progress. The antithesis of that is Congress.... No, sorry, wrong dissertation. The antithesis of that is stagnancy. We should always strive to find new ways to improve our conditions. Whether you apply that to your personal life or to networking, it has the ring of truth.

Darwin said, "In the long history of humankind (and animal kind, too) those who learned to collaborate and improvise most effectively have prevailed." Let us go forth and collaborate on and improvise our internetworking.

At the very least, as you go through this book and pay attention to the details, you'll feel like Saturday Night Live's Church Lady: "Well.... Isn't that special!"

—Scott Morris
Four-time CCIE and JNCIE-M, and a Juniper Networks Certified Instructor
June 1, 2009

Preface

It wasn't but a day or so after we finished *JUNOS Enterprise Routing (http://oreilly.com/catalog/9780596514426/)*, the companion book to this one, that our editor started pressuring us to start writing again. Writing a book may seem easy on the surface; how hard could it have been to write some words on these pages? But in reality, writing a book is a time-consuming and collaborative effort that affects everyone close to you. Also, at the time of the request, the EX Series was a new product line, and although we had extensive experience in the switching space, we did not have many customer deployments of the EX Series switches.

We were really impressed when the first EX units arrived for us to put in the rack that we reserved for this book. As JUNOS "experts," we expected a lot, and we were impressed at the ease of configuration and stability that the JUNOS software provided to the EX. Although the performance and configuration of the EX switches are slightly different from those of a JUNOS router, the EX platform fit our needs, and hopefully it will fit yours, too.

The world of Ethernet switching changed when Juniper Networks got involved. At the time of this writing, the EX Series switches are already making significant inroads into the enterprise marketplace.

In this book, we have extended the same approach and writing style we used in *JUNOS Enterprise Routing* so that you, the reader, can have a set of books that work together. We did not write these books with the idea that you'd read one before the other, so some content overlap may exist. Most people will want to work up the OSI stack and read this book before the routing book; still, for many reasons, the routing book came out first, and we kept this in mind as we organized this book.

This book does double duty. It is both a field guide and a certification study guide. Readers who are interested in attaining a Juniper Networks certification level are wise to note that we discuss and cover topics that are relevant to the official exams (hint, hint), and at the end of each chapter we provide a list of examination topics covered, as well as a series of review questions that allow you to test your comprehension. Since the EX does both switching and routing, you should be able to answer a majority of enterprise certification questions by reading and understanding the content of all of the chapters in *both* books.

We wrote this book to serve as a field guide for any time you work on your EX. We present a lot of tutorials and samples, with lots of actual command output. We like to think that the detailed theoretical coverage we provide goes well beyond any certification exam, to give you something that can't be tested: the ability to get things to work the right way the first time. When plan A fails, this material provides the steps needed to monitor network operation and quickly identify and resolve the root cause of malfunctions. We believe this level of coverage extends the life of both this book and its predecessor, far beyond any short-term certification goals.

Some of our chapters tend to be on the longer side; we're sorry about that. Or maybe we should say "You're welcome" in keeping with the "more is better" philosophy. Just dog-ear the pages, write notes in the margins, and perhaps even update the topology illustrations with something more akin to your own fiefdom. We hope you'll make this copy your own.

What Is JUNOS Enterprise Switching?

The idea of switched networks is definitely not a new concept. LANs and Ethernet networks have been around since the 1970s, and some of you reading this book may actually remember the days of vampire taps and Thicknet cabling! Like the Atari 2600, those stories are great to relate and are actually quite amazing in their own technology light, but thankfully, technology advances. Now a LAN is synonymous with Ethernet; people know that if they plug their laptop's network port into the wall, they will wind up on an Ethernet network.

Switching had historically been a fairly simple process, which allowed bridges to outperform routers of the same era. After the MAC address table was built, a simple filtering and forwarding process was all that was needed. *Intelligence* and *switches* were never used in the same sentence. If forwarding intelligence was needed, it was slow, and it was often reserved for higher OSI layers.

However, as technology improved, intelligence was no longer a compromise for speed. That was the premise of Juniper Networks' "speed without compromise" routers that brought Juniper into the networking space against the big giant, Cisco—a dangerous place littered with the remnants of all the other companies that had failed before. Some may argue about how successful Juniper has been, but nobody can argue that Juniper will remain a player for many years to come, as evidenced by the large deployed base of its products and the much-loved JUNOS software that is its foundation.

As technology evolved, so did Juniper, as did its historic focus on routing within service provider networks. Juniper set its sights squarely on the enterprise, a market that was still being dominated by Cisco, and it saw a place where it could leverage the experience and reliability of JUNOS. The first "Layer-2-aware" device that Juniper Networks released was the MX Series. Although it was well received, the MX platforms were targeted at service providers and were really seen as routers with dense Ethernet connectivity but that also happened to offer some Layer 2 services. The first true Ethernet LAN switch is the EX Series, the basis of this book.

Will Juniper displace Cisco in the enterprise switching market? Only time will tell. So far, the EX platforms have been well received in the industry (and don't forget, they have JUNOS going for them). The purpose of this book is to provide details on the product's capabilities, and to show examples of its deployment and operation in support of those features. This book is not designed to sell boxes, or to convince you to use one vendor over another. We believe that after reading the material you will be able to make an informed decision as to what products work best for your needs. In all probability, the Juniper EX was not even an option the last time you shopped for LAN switching gear. We think you will want to keep the EX in mind the next time.

So, come join us in the exciting new world of enterprise switching and JUNOS on the new EX Series of high-performance LAN switches. With all the excitement in the air it seems like 1998 and the release of Juniper's first router, the M40, all over again.

The Juniper Networks Technical Certification Program (JNTCP)

This book is an official study guide for the JNTCP Enterprise tracks. For the most current information on Juniper Networks' Enterprise certification tracks, visit the JNTCP website at *http://www.juniper.net/certification*. We've included lots of questions at the end of each chapter to help with your cert plans.

How to Use This Book

We are assuming a certain level of knowledge from the reader. This is important because if you and the book don't match that assumption level, you'll feel off-track right from the get-go. So, we are assuming you have knowledge of the following:

Basic networking
> This book is about LANs and LAN interconnection. The introduction chapter provides a good overview of networks, TCP/IP, and LAN interconnection. However, it's assumed that the reader has basic familiarity with data communications and LANs; otherwise, it may be a fast ride.

Computers
> What can we say? We assume the reader owns at least one and is familiar with some of its networking operations, such as web browser use.

What's in This Book?

The ultimate purpose of this book is to be the single most complete source for working knowledge related to Juniper Networks enterprise switching on the EX platform. Although you won't find much focus on actual packet formats and fields, topics for which there is already plentiful coverage on the Internet and in bookstores, you will find information on how to effectively deploy JUNOS switching technology in your network.

Here's a short summary of the chapters and what you'll find inside:

Chapter 1, *LAN and Internetworking Overview*
> Here you will find a detailed, condensed, and somewhat irreverent breakdown of how we got to where we are; the popularity of Ethernet, TCP/IP, and the demise of OSI; and basic LAN interconnect terminology and concepts. Nothing earth-shattering, but all good information and things we think you should be familiar with before jumping into the more specific chapters that follow.

Chapter 2, *EX Platform Overview*
> This chapter details EX platform capabilities and options. This includes several *day in the life of a frame* processing examples designed to help you understand internal operations so that you can better appreciate the platform's performance and capabilities. How many PoE ports are available? What uplink options are possible? What is the name of the management interface? Find out the answers to these questions, and many more, here.

Chapter 3, *Initial Configuration and Maintenance*
> This chapter begins by examining the steps you need to take once you have your brand-new, shiny switch in your hands. It starts with a factory-default configuration and builds the initial configuration required for deployment. This includes any basic interface configuration that would be necessary.

Chapter 4, *EX Virtual Chassis*
> The EX virtual chassis (VC) is an exciting and important capability that provides virtualization of a multichassis stack. This capability provides many performance and reliability enhancements, such as the support of redundant routing engines.

Chapter 5, *Virtual LANs and Trunking*
> This chapter details the concept of VLAN tagging, and how to configure VLANs and trunking in a mixed JUNOS and IOS environment. Here we handle tricky things such as native versus default VLANs, and current best practices for multi-vendor interoperability.

Chapter 6, *Spanning Tree Protocol*
> This chapter dives into the world of Spanning Tree Protocol (STP), providing first an overview of the protocol and then the implementation specifics. The original STP is examined as well as the updated standards for Rapid Spanning Tree Protocol

(RSTP) and Multiple Spanning Tree Protocol (MSTP). Lastly, we describe Redundant Trunk Groups at the conclusion of the chapter.

Chapter 7, *Routing on the EX*

The EX is a switch that runs JUNOS, and can therefore route pretty darned well. This chapter details the Routed VLAN Interface construct and how to perform inter-VLAN routing on an EX. We also cover some basic JUNOS routing concepts along the way.

Chapter 8, *Routing Policy and Firewall Filters*

Routing policy and firewall filters are features that are on every box that runs JUNOS software and, quite honestly, are most often found on routers. However, they have their place on switches as well, and we cover them in this chapter. Those who have already read *JUNOS Enterprise Routing* can probably skim this chapter, as the policy is the same and only the firewall filters and policers are different on the switches.

Chapter 9, *Port Security and Access Control*

What good is a high-performance LAN if all it does is allow you to get hacked more quickly? This chapter details port-level security features such as MAC limits, DHCP snooping, and ARP inspection, in addition to support for IEEE 802.1X access control. Learn how to harden your switched network's parameters here, and how to perform fault isolation when things do not go as planned.

Chapter 10, *IP Telephony*

In this day and age, IP telephony is the deployment normal, finally displacing the traditional PBX. This chapter looks at some of the tools that are now available for VoIP, such as Power over Ethernet and LLDP-MED. It contains case studies on solving problems when IP phones support these features, and when they do not.

Chapter 11, *High Availability*

The chapter provides an overview of the High Availability (HA) features offered on the EX Series at the time of this writing. These features include hardware redundancy, Graceful Routing Engine Switchover, Graceful Restart, Bidirectional Forwarding Detection, Virtual Router Redundancy Protocol, and aggregated Ethernet.

In addition, you can also use this book to attain one of the Juniper Networks certification levels related to the Enterprise program. To that end, each chapter in the book includes a set of review questions and exam topics covered in the chapter, all of it designed to get you thinking about what you've just read and digested. If you're not in the certification mode, the questions will provide a mechanism for critical thinking, potentially prompting you to locate other resources to further your knowledge.

Topology of This Book

Figure P-1 displays the topology of the book that appears beginning in Chapter 2. It consists of six EX Series switches running JUNOS 9.1 (in a few cases, a newer software version is used for reasons that we explain in the related material), two Cisco 3550 switches running IOS Release 12.2(44)SE3, one J Series router, one Cisco 2600 router, some IP phones, and various hosts that are really just Cisco 2500 routers that are used to generate some pings.

The book uses both Layer 2 and Layer 3 topologies, but the physical connectivity stays the same. Also, Chapter 10 uses an additional Cisco switch, Cisco router, and IP phones. Similarly, Chapter 9 uses a different test bed and test topology due to the need for a supported client and authentication server.

Once again, you might recognize that the devices' hostnames relate to different distilled liquids. Recall that in the routing book, the names were based on brewed libations. As before, we chose the names due to their international appeal and preservation of the crops from which they are created.

Conventions Used in This Book

The following typographical conventions are used in this book:

Italic
> Indicates new terms, URLs, email addresses, filenames, file extensions, pathnames, directories, and Unix utilities

`Constant width`
> Indicates commands, options, switches, variables, attributes, keys, functions, types, classes, namespaces, methods, modules, properties, parameters, values, objects, events, event handlers, XML tags, HTML tags, macros, the contents of files, and the output from commands

`Constant width bold`
> Shows commands and other text that should be typed literally by the user

`Constant width italic`
> Shows text that should be replaced with user-supplied values

 This icon signifies a tip, suggestion, or general note.

 This icon indicates a warning or caution.

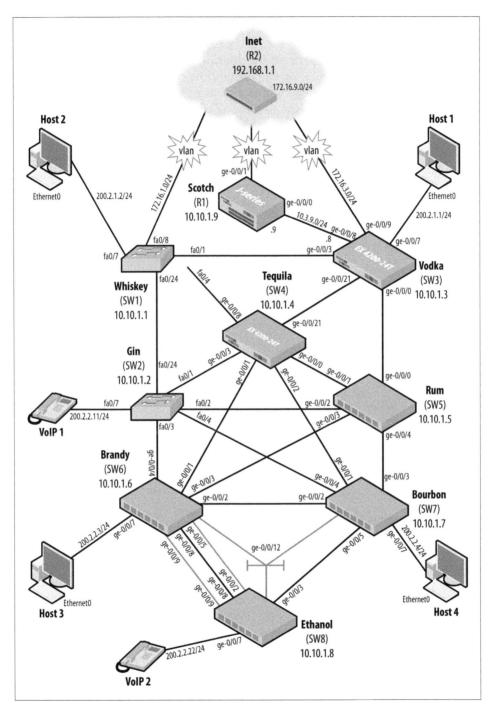

Figure P-1. A typical Layer 2 topology

Using Code Examples

This book is here to help you get your job done. In general, you may use the code in this book in your own configuration and documentation. You do not need to contact us for permission unless you're reproducing a significant portion of the material. For example, deploying a network based on actual configurations from this book does not require permission. Selling or distributing a CD-ROM of examples from this book does require permission. Answering a question by citing this book and quoting example code does not require permission. Incorporating a significant amount of sample configurations or operational output from this book into your product's documentation does require permission.

We appreciate, but do not require, attribution. An attribution usually includes the title, author, publisher, and ISBN. For example: "*JUNOS Enterprise Switching*, by Harry Reynolds and Doug Marschke. Copyright 2009 Harry Reynolds and Doug Marschke, 978-0-596-15397-7."

If you feel your use of code examples falls outside fair use or the permission given here, feel free to contact us at *permissions@oreilly.com*.

Safari® Books Online

Safari When you see a Safari® Books Online icon on the cover of your favorite technology book, that means the book is available online through the O'Reilly Network Safari Bookshelf.

Safari offers a solution that's better than e-books. It's a virtual library that lets you easily search thousands of top tech books, cut and paste code samples, download chapters, and find quick answers when you need the most accurate, current information. Try it for free at *http://my.safaribooksonline.com*.

Comments and Questions

Please address comments and questions concerning this book to the publisher:

O'Reilly Media, Inc.
1005 Gravenstein Highway North
Sebastopol, CA 95472
800-998-9938 (in the United States or Canada)
707-829-0515 (international or local)
707-829-0104 (fax)

We have a web page for this book, where we list errata, examples, and any additional information. You can access this page at:

http://www.oreilly.com/catalog/9780596153977

or:

http://www.proteus.net

To comment or ask technical questions about this book, send email to:

bookquestions@oreilly.com

For more information about our books, conferences, Resource Centers, and the O'Reilly Network, see our website at:

http://www.oreilly.com

About Scott Morris, Lead Tech Reviewer

Scott Morris is a four-time CCIE (#4713) and JNCIE-M (#153), as well as a Juniper Networks Certified Instructor (JNCI) for M Series and J Series routers. Scott is currently a senior instructor for Internetwork Expert, Inc., as well as a consultant for various service providers around the world. He has more than 23 years of experience in all aspects of the networking industry and often tech-reviews networking books for Cisco Press and McGraw-Hill. When not attending events around the globe, he is often at home with his lovely wife and two beautiful daughters. He hasn't yet taught them binary, although that will likely happen. His home basement sports a four-rack networking lab that generates so much heat it requires its own air-conditioning unit—take a peek at *http://smorris.uber-geek.net/lab.htm*.

Acknowledgments

The authors would like to gratefully and enthusiastically acknowledge the work of many professionals who assisted us in developing the material for this book. Although our names are printed on the book as authors, in reality no author works alone. There have been many people whose contributions have made this book possible and others who have assisted us with their technical accuracy, typographical excellence, and editorial inspiration.

Many thanks are owed to the official technical editor of this material, Scott Morris. Despite our lack of planning, he was able to adjust to our schedule and return the chapters in a timely manner. He used his tremendous experience not only to fix technical errors, but also to encourage us to add topics we had overlooked.

We would also like to acknowledge Juniper Networks in general, and in particular Chris Spain and Michael Banic, for the assistance provided on various fronts and for providing much of the gear we needed to write this book. We also would like to thank Proteus Networks for housing this gear at no cost in its data center, as well as Chris Heffner for providing gear support.

And last but not least, we give special thanks to Patrick Ames for his assistance and constant annoyance to keep the book going in the right direction. His thick skin after bouts of anger from the authors is legendary and much appreciated.

From Doug Marschke

First, I would like to personally thank Harry Reynolds for all of his help and encouragement while writing this book. This is really his book, and I was just along for the ride. Despite my obvious missed deadlines, he was patient and did not yell at me too much throughout this process. Due to the timing of starting a new company and changing the face of my bar, this book was difficult not from lack of motivation but from lack of time. To that end, I must personally thank my business partners at Proteus Networks, Joe Soricelli and Chris Heffner, and my business partner in the Taco Shop at Underdogs, Nick Fasanella. They essentially picked up my slack as I devoted my time to completing this book. And finally, but most importantly, I must thank Becca Morris for her love and support that never seem to waver despite how much I mess up. The homemade brownies and cookies also helped to sugar-infuse my late nights, and they tasted really good.

From Harry Reynolds

I would (again) like to thank my wife, Anita, and two lovely daughters, Christina and Marissa, for understanding and accommodating my desire to engage in this project. Also, special thanks to my manager at Juniper Networks, Sreedhevi Sankar, for her understanding and support. I really appreciate her willingness to accommodate the occasional glitch in my "day-job" schedule as needed to make this book happen. Also, thanks to the Juniper EX team, and to PLM, for providing us with gear and technical support during the effort.

## CHAPTER 1

LAN and Internetworking Overview

Over the years, the importance of data communication has closely paralleled that of computing. In fact, most individuals are completely unaware of just how often data communications networks affect and enable our daily lives. Network technologies are ubiquitous to the point where a modern car has more processing power than a 1960s-era mainframe, along with several communications networks that control everything about the vehicle's operation, from its chipped key ignition security, engine fuel/air mixture, and transmission shift points to the sensing and reporting of low tire pressure.

The modern world is hooked on computing, and in this addiction data networks are like a drug in the sense that most PC users find their machines rather dull and bordering on boring when their Internet access is down.

In fact, given the easy access we all have to nearly the sum of man's knowledge (typically for free), the Internet and its World Wide Web make activities such as writing technology books seem rather old hat. But the amount of background information regarding general data communications concepts, local area network (LAN) technologies, and the TCP/IP protocol suite, which makes all of humankind's knowledge available and easy to find, is still an invaluable library. It's the library that enables the digital libraries. And so we start right here.

The primary goal of this book is to document the application and use of Juniper Networks EX switches in a number of LAN and interoperation scenarios. The coverage has already jeopardized our editor's budgeted page count, which leaves us little room given that we wrote this chapter last and we are a bit tired and in dire need of sunshine. Therefore, we make no attempt in this chapter to re-create the "complete history of networking and LAN technologies" wheel.

Instead, this chapter's goal is to provide an extremely targeted review of networking and LAN history, including internetworking principles related to LAN interconnect. Aside from an irreverent take on history for which the authors apologize beforehand (did we mention we saved this chapter for last?), what follows is also an extremely focused targeting of key principles that you should understand before moving on to the remainder of this book.

There aren't a lot of pages, but it's a fun, informative, and action-packed ride; trust us, both you and the trees are better for it.

The topics covered in this chapter include:

- Networking and OSI overview
- Ethernet technologies
- The TCP/IP protocol suite
- LAN interconnection

What Is a Network?

A network can be defined as two or more entities with something to say that is not already known to the intended recipients, and a channel or medium over which to convey this information. Simple enough, right?

Network technologies, much like fashions, seem to flare in popularity and then quietly fade away in favor of the next thing. At one point in the dark past of networking, users were compelled to source their network gear from a single vendor, oftentimes the same vendor that provided the data processing equipment. This was due to a lack of open standards that resulted in vendor-proprietary solutions for both the hardware and networking protocols.

Although good for the vendor, a single source for anything is generally bad for economics, and in some cases it can also hamper innovation and performance; after all, if a vendor has you locked into its solution, there may be little motivation for the vendor to spend money on research and development in an effort to improve the basic technology. Nope, users wanted to be able to select from best-of-breed solutions, ones that optimize on performance or price, while still enjoying end-to-end interoperability.

What Is a Protocol?

The term *protocol* is bandied about all the time in this industry. The essence of a protocol is simply a mutually agreed upon set of rules and procedures that in turn govern the behavior of the participants to help ensure orderly operation. Thus, the English language is a protocol, as are the rules of Parliament. In both cases, a violation of protocol can cause confusion or malfunction.

Data communications networks are based on layered protocols that provide a divide-and-conquer approach to the daunting task of getting complex information from one source to some remote host over error-prone and unreliable communications paths, all the while providing authentication, authorization, and in some cases actual translation services to compensate for differences in local data representation.

Enter the Open Systems Interconnection (OSI) model, which we detail in the next section.

The OSI Model

The OSI model, and the International Organization for Standardization (ISO) suite of protocols that were originally based on the model, failed to see much adoption. As evidence, consider GOSIP. The Government Open Systems Interconnection Profile was first published in 1990, and essentially stated that all U.S. government communications networks must be OSI-compliant for consideration in networking bids. This was a big deal, and in theory it sounded the death knell for vendor-proprietary solutions. It was to include the U.S. Department of Defense's (DoD) ARPANET (Advanced Research Projects Agency Network) protocols; e.g., TCP/IP, which at the time *was* the de facto multivendor interoperability solution. Given this level of backing, it is hard to understand how OSI/ISO could fail. The answers are multifaceted:

The OSI protocols were slow to market/produced no products
> The best ideas in the world are not very useful if they have no tangible manifestation in reality. Many of the official OSI protocols were never fully implemented and most were never deployed in production networks. Ironically, the OSI layers that did have products tended to function at or below Layer 4, which is just where existing technologies (i.e., TCP/IP) already existed and could be used as models. Stated differently, TCP/IP has a Network layer (IP), and so did the OSI model; it was called Connectionless Network Service (CLNS). The TCP stack does not have a true Session or Presentation layer, and it's in these very upper layers where it seems the ISO bit off more than it could shove into a Layer 3 packet, so to speak.

The OSI protocols were overtly complex and suffered from a slow development velocity
> The OSI protocols attempted to go above and beyond existing network functionality. In effect, it was a protocol for the world's current *and* future needs. This was a fine aspiration, but trying to solve every known or projected issue, all at once and in a worldwide forum, was just too hard. The resulting standards were too complicated or too incomplete to implement, especially when TCP/IP was already working.

The IETF is too practical, and far, far too nimble
> The Internet Engineering Task Force (IETF), which produces Internet RFCs and drafts, uses a guiding principle known as "rough consensus and running code" (*http://www.ietf.org/tao.html*). The ability to move forward with working solutions without being bogged down in international law and geopolitics means that IETF standards significantly outpace their international counterparts, and are typically backed by a working implementation to boot! In contrast, the ITU/ CCITT, which produced key ISO standards such as X.25 and B-ISDN in I.361, would meet every four years to make updates and solve problems. In theory, the world was supposed to patiently await their collective wisdom; in reality, IP simply ran the whole process over and never looked back to see what that bump in the road even was.

Basically, you could summarize all of these reasons as "The world already had a workable set of interoperability protocols known as TCP/IP, and the cost of waiting for *official* standards, which in the end always seemed to lack parity with the latest IP offerings anyway, was simply too jagged a pill to swallow."

So, why are we (seemingly) wasting your invaluable time with a discussion of a grand failure? The answer is because although the OSI protocol stack itself failed, the related reference model lives on as a common way of expressing what role some networking device performs. The OSI model sought to partition the challenges of end-to-end communications among dissimilar machines into a layered approach, in which the protocol options, roles, and responsibilities for each layer were clearly defined.

Figure 1-1 shows the venerable OSI model in all its seven-layer glory, along with some selected protocol options for each layer. Note that upper layers were the least well defined, and few saw any production network use.

Application	ISO 9040/ 9041 VT	ISO 8831/ 8832 JTM	ISO 8571/ 8572 FTAM	ISO 9595/ 9596 CMIP
Presentation	ISO 8823/CCITT X.226 Connection-Oriented Presentation Protocol			
Session	ISO 8327/CCITT X.225 Connection-Oriented Session Protocol			
Transport	ISO 8073/CCITT X.224 Connection-Oriented Transport Protocol (TP 1–4)			
Network	ISO 8473 Connectionless Network Service		ISO 8208/CCITT X.25 Packet Level Protocol	
Link	ISO 8802.2 / ISO 9314-2 FDDI, ISO 8802.3 CSMA/CD, ISO 8802.4 Token Bus, ISO 8802.5 Token Ring		ISO 7776 CCITT X.25 LAP/LAPB	ISO 7809 HDLC
Physical	Various Media Options from CCITT, EIA, IEEE, etc.			

Figure 1-1. The OSI Reference Model

Key points about the model are:

- Each layer interacts with a peer layer, which may be at the end of the link or at the actual receiver. In practical context, this generally means that a given layer adds some protocol header, and maybe a trailer, which is then acted upon and removed by the remote peer layer.

- Some layers have a link-level scope whereas others have an end-to-end scope. A communications path can contain numerous independent links, but no matter how far-flung its constituent links are, it relies on a single Transport layer entity that exists only in the endpoints.

- Each layer provides a service to the layer above it, and receives services from the layers below.

- There is general modularity that provides options and the ability to "mix and match" specific technologies at a given layer. The specifics of each layer are opaque to those above and below it, as only service semantics are defined between the layers. This means that any LAN technology could be used to provide Link layer services to the Network layer, and in fact the ISO Connectionless Network Layer Service (CNLS) Network layer protocol could operate over Token Ring (802.5), CSMA/CD (802.3), and even Token Bus (802.4).* It should be noted that each such LAN technology typically came with its one slew of Layer 1 options, such as coaxial cable, twinaxial cable, or unshielded twisted pair (UTP).

Layer functions

As noted previously, a layered model works by using a divide-and-conquer approach, with each layer chipping in to do its part. The main function of each layer is as follows:

Physical layer
> The Physical layer is where the bits meet the road, so to speak. All communications systems require one. Layer 1 places bits onto the transmission medium for transmit, and pulls them off for receipt. It cares not what those bits mean, but some Physical layers have framing and/or Forward Error Correction (FEC) that allows them to detect problems and in some cases act better than they really are. Bits are the Protocol Data Units (PDUs) sent at Layer 1. EIA-232, SONET, V.32bis modems, and 1000Base-T are examples of Layer 1 technologies.

Link layer
> The Link layer deals with frames. It adds a header and trailer to frame upper-layer traffic, and generally provides link-by-link error detection. Some Link layers also provide error correction and multipoint addressing, as well as multiprotocol support through a type indication. Frames are the PDUs sent at Layer 2. Frame Relay and HDLC, as well as LAN MAC frames, are examples of Layer 2 technologies.

Network layer
> The Network layer is the first end-to-end layer. It can be said that a Network layer packet passes pretty much as it was sent, all the way to the remote endpoint. The

---

* Note that IEEE 802.x standards are prefixed with an additional "8" when adopted by the relevant OSI entity, thus ISO 8802.3 is equivalent to IEEE 802.3.

Network layer identifies endpoints (not the next Link layer hop), and may provide error detection/correction, protocol identification, fragmentation support, and a Type of Service (ToS) indication. Packets (or datagrams) are the PDUs sent at Layer 3. IP is a Network layer, as is X.25.

 X.25 is technically a Layer 3 packet protocol. When used to support IP it functions more as a link between two entities. When sending IP over X.25 it can be said that there are two Network layers, but in this context IP is seen as the real Layer 3 as its endpoints may lie beyond the endpoints of the X.25 connection. Much of the same is true of other connection-oriented technologies, such as the Public Switched Telephone Network (PSTN) in analog form and its digital cousin, the Integrated Services Digital Network (ISDN).

Transport layer

The Transport layer deals with end-to-end error control and the identification of the related application process through ports (sockets). This layer may also perform connection establishment, sequencing, flow control, and congestion avoidance. The term *Service Data Unit* (SDU) is often used to describe what is sent or received when dealing with Layer 4 or higher. In the TCP/IP model, segments are the PDUs sent at Layer 4.

Session layer

The Session layer deals with session establishment, synchronization, and recovery. Given that we are now in the realm of things that never really happened, it's hard to say what this means. TCP/IP has no official Session layer, but protocols such as Fault-Tolerant Overlay Protocol (FTOP) have a user sign-in phase.

Presentation layer

The Presentation layer deals with application-specific semantics and syntax. In theory, this layer can convert to some machine- and application-independent format—say, ASN.1—upon transmission, and then back into the desired format upon reception. This is a pretty tough row to hoe, and seems much like a protocol converter in a layer. Again, there is no real example to give, other than that TCP/IP does use ASN.1 or HTML/XML data formats to help promote communications between dissimilar machines.

Application layer

The Application layer is not the application. It's the application's interface or API into the communication's stack. This is akin to a Windows or Unix socket in the TCP/IP context.

Figure 1-2 illustrates key concepts regarding layered protocol operation.

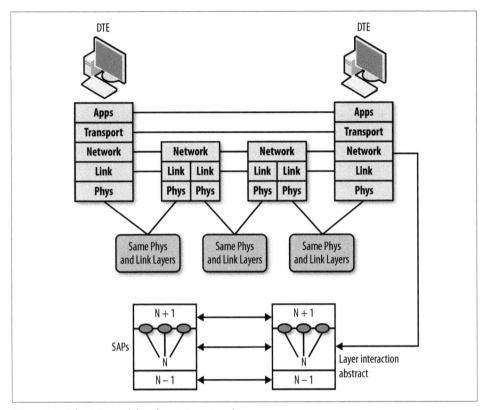

Figure 1-2. The OSI model and an internetwork

Figure 1-2 shows two communicating pieces of data terminal equipment (DTE) along with what appears to be a router-based form of interconnection, given the presence of Layer 3 in the intermediate nodes. The lower portion details generic layer interaction. Note that each layer communicates with a peer layer, which may be at the other end of the link, or at the far endpoint depending on the layer's scope. Each layer accepts service requests from the layer above it, and in turn makes requests of the layer below.

The Network layer is the first end-to-end layer. As such, it's technology-independent, meaning the same Network layer packet that is sent is pretty much the one received (minus the obligatory Time to Live [TTL] decrement designed to protect against routing loops). In contrast, Layers 1 and 2 vary by network technology type. Stations that communicate directly (i.e., those that share a link) must use compatible network technology. Stated differently, if the DTE on the left is running 10Base-T and is using Ethernet v2 framing, the first hop data circuit-terminating equipment (DCE)/router must be compatible. It may operate at 100 Mbps, given that repeaters and bridges can adapt Ethernet speeds, but it must support IP in Ethernet v2 encapsulation for communications to succeed across that link. If the connection is direct, the speed and duplex as well as other physical parameters must also match. The link between the two routers is a different link, and therefore does not have to be compatible with either DTE, as this link is not used for direct DTE communications.

Network Types and Communication Modes

There are many different kinds of network technologies and methods for communicating information among some set of stations. Generally speaking, a network is classified as being a wide area, local area, or metropolitan area technology (i.e., WAN, LAN, MAN). In some cases, the same technology can be used in all three environments. At one point this was the promise of ISDN, and then B-ISDN (ATM), but now it seems to be the domain of Ethernet, which holds more convergence promise than any other Link layer protocol.

Simply put, a LAN describes a set of nodes that communicate over a high-speed shared medium in a geographically confined area. A WAN can span the globe, tends to operate at lower speeds, and is often point-to-point (P-to-P) rather than multiaccess. Some multipoint WAN technologies are still in use; chief among them are Frame Relay and, in less developed parts of the world, the old standby, X.25.

Communication modes

The exchange of information between endpoints can occur in one of several ways:

Point-to-point
> As its name implies, this mode involves two endpoints, one as a source and the other as the recipient. Many WAN technologies are P-to-P. Modern P-to-P technologies are full duplex (FD), which means that both ends can send and receive simultaneously.

Multipoint
> Multipoint topologies are often associated with WANs. Historically, a multipoint technology describes a hub-and-spoke (sometimes called *star*) arrangement whereby a central site can send to all remote sites at the same time, but each remote site is allowed to send back only to the central site, à la IBM's polled Synchronous Data Link Control (SDLC) protocol. Multipoint also refers to virtual circuit technologies such as Frame Relay and ATM that allow a single physical interface to be used to send to multiple destinations by using the correct circuit identifier (i.e., a Frame Relay DLCI or ATM VCI).

Broadcast
> A broadcast network uses shared media or some replication function to allow a single transmission to be seen simultaneously by all attached receivers. LANs always operate in a broadcast manner, making this one of their defining characteristics. Note that the use of switches (bridges) or routers isolates the broadcast domain, a technique used for both performance and security reasons.

Broadcast networks can operate in simplex, half-duplex (HD), or FD mode based on specifics.

Non-Broadcast Multiple Access (NBMA)
An NBMA network is a form of virtual-circuit-based topology that does not permit true broadcast, but by virtue of having a virtual circuit to every other endpoint, an NBMA network can emulate broadcast functionality by sending the same message multiple times over each of its locally defined virtual circuits.

NBMA networks can operate in simplex, HD, or FD mode based on specifics. Practically all modern network technologies are now FD anyway, but some distribution systems are inherently simplex.

So, Where Did We LANd?

After the dust settled on the past 40 years of data networking, we may be so bold as to say that a few key trends have emerged:

- OSI is dead, and we can only hope it's resting peacefully, as the chip—nay, PDU—that it bears upon its shoulder would make for one nasty ghost.

- IP is the dominant convergence technology that serves as the basis for everything from interactive data to email to telemedicine, virtual reality, and even old-school services such as telephony and many television distribution systems. There appears to be no serious threat to this venerable workhorse on the near horizon, except maybe its younger sibling, IPv6, which is creeping into more and more networks each day.

- Ethernet rules the LANs, and most MANs, and is also being seen in long-haul WANs as part of Layer 2 virtual private network (VPN) services (that typically ride over IP-enabled Multiprotocol Label Switching, or MPLS, networks), or as native Ethernet as part of a PBB or long-haul SONET/SDH transport. Ethernet keeps getting faster (40 Gigabit Ethernet is now available) and cheaper, and updates such as Operational Administration and Maintenance (OAM) continue to extend its reach by providing it with some SONET-like maintenance and alarm-reporting capabilities. Ethernet is built into every PC, and virtually all broadband access is based on this Ethernet connection, attaching it to the DSL or cable modem used to access the service.

The next section focuses on modern Ethernet technologies as a primer for what is to come in the rest of this book. Today the term *LAN switching* is assumed to mean Ethernet. For those with a penchant for obscurity and a bit too much money and time on their hands, we hear you can find Token Ring Media Access Units (MAUs) on eBay at bargain basement prices these days; just be sure to get the high-speed 16 Mbps version. Running at 4 Mbps would look bad given that Ethernet made the 10 Mbps leap back in 1980, when the Digital Intel Xerox (DIX) consortium published the ESPEC v1 specification. There was even a (very) short-lived 32 Mbps "FD" Token Ring MAU available.

Ethernet Technologies

Ethernet has been around for so long, and is now so widely used, that a complete overview would easily fill its own book. The goal here is to stay concise and to convey only key points to clear up areas that are known to cause confusion.

A Brief Look Back

The philosopher George Santayana one stated, "Those who do not remember their past are condemned to repeat their mistakes." It is to him that we dedicate this section. Well, him and everyone else who fell prey to buying into now-obsolete market failures such as ATM, Token Ring, FDDI, Token Bus, ARCnet, TCNS, or SMDS. All sought to solve the needs of high-speed communications over a shared medium, and all are now only footnotes in history—a history that is penned by the victorious Ethernet.

Ethernet v2 is a de facto standard first published by the Digital, Intel, and Xerox vendor alliance. It was based on a prototype satellite communications network called ALOHAnet. When Bob Metcalf later adapted the technology to run over coaxial cable in the early 1970s, the term *Ether* was used to pay homage to its original use of electromagnetic radiation through the vacuum of space, whereby the alleged media was the mythical *luminiferous ether*, a substance that the ancient Greeks believed conducted the planets through their orbits.

When used for minicomputers, LANs were a novelty. Enter the IBM PC in the early 1980s, and suddenly LANs are a hot commodity. The official standards bodies could not stand by and watch a vendor consortium do all the work. As a result, the IEEE 802 committee was formed to standardize LANs. The committee initially met in February 1980, hence the 80-2 Committee.

What was Ethernet became IEEE 802.3, which then branched off into various medium-specific standards such as 10Base-5, 10Base-2, 10Base-T, 100Base-T, 1000Base-TX, and so on. In the IEEE terminology, the number represents the bit rate, the term *Base* indicates the baseband (digital) signaling (there is a 10Broad36 spec for analog use over cable), and the last value identifies a medium, either indirectly via maximum segment length or by type. For example, the "5" in 10Base-5 means 500-meter cable length, which in turn indicates a thick coaxial cable medium, whereas the "T" in 10Base-T stands for UTP.

Despite the blessing by an official standards body, the irony is that the most common usage of Ethernet is, in fact, actually Ethernet v2 and *not 802.3*. The next section details the differences so that you can speak the truth when describing your network.

Ethernet or 802.3, That Is the Question

According to the OSI model, LANs operate at Layer 1 and Layer 2. Hence, they are considered a Link-layer technology. In addition, a LAN's Link layer is broken into two parts, or sublayers: the Media Access Control (MAC) and Logical Link Control (LLC) components.

The goal here was noble. LANs should be able to provide a common service and interface to the upper layer (the Network layer), regardless of the fact that each LAN technology has a unique MAC sublayer that functions to provide orderly access to the shared medium. This makes some sense, in that a collision is unique to Ethernet LANs, so why should the IP layer know or care at all about one, as the same IP is also run over Frame Relay, which is a collision-free technology.

In contrast, Ethernet specifies only a MAC layer, as shown in Figure 1-3.

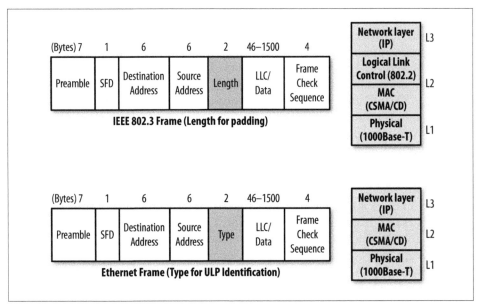

Figure 1-3. Ethernet versus 802.3

Although there is a notable difference in the MAC layer, given that Ethernet uses a Type code and 802.3 redefines the field as a Length value, the real answer to the proverbial question of Ethernet or 802.3 lies in the absence, or presence, of the Length field and 802.2/LLC. If the frame has a Type field, it's Ethernet, plain and simple.

No magic is needed to differentiate between these two frame types because Type codes are selected to not conflict with valid IEEE Length values. Thus, any value less than 0×0600 is interpreted as a Length, and a value greater than $0 \times 05DC$ is seen as an EtherType. In case you are doing the conversion, $0 \times 05DC$ in hex symbolizes 1,500 bytes in decimal. The need to preserve this compatibility with Ethernet is one reason the IEEE 802.3 standard was never updated to support *jumbo* frames, which are frames larger than 1,500 bytes.

An Ethernet frame identifies the upper layer protocol using an EtherType, where the value 0×0800 indicates IP. Because there is no length indication padding, which is needed for Ethernet as the smallest amount of user data that can be sent is 46 bytes, it has to be performed by the upper layers.

This interlayer dependency, though long since accommodated by IP stacks, was seen as an egregious violation of the principle of layer independence. As such, the IEEE opted for a Length field at the MAC layer. This meant the MAC layer could now do its own padding, which is cool and all, but there was still a need to identify what the heck was in the frame. Enter LLC, which, from a practical perspective, pretty much functions to replicate Type code functionality, except now using three bytes rather than two, and you get to use fancy-sounding terms such as *Service Access Point* (SAP), which in the end simply identifies the upper layer. Note that another form of LLC is defined, LLC type 2, that provides a connection-oriented, reliable traffic exchange à la the connection-oriented balanced exchange procedures defined in Link Access Procedure, Balanced (LAPB). LAPB is used at Layer 2 of the X.25 model; in LANs, LLC type 2 uses a Set Asynchronous Balanced Mode Extended (SABME) initiation command to set up extended mode (modulus 128) sequencing, however. This LLC mode was used only in Token Ring/Bus networks, and then typically only for protocols such as the Systems Network Architecture (SNA) protocol. Need we say more? A third type of LLC, type 3, was never implemented. It provided a connectionless mode with acknowledgments.

We already stated that IP is the Network layer protocol of choice in the modern world. Yes, there are standards that define how an IP datagram can be encapsulated in Ethernet, 802.3/LLC, or 802.3 with LLC, combined with the official standards-supported method of escape back to the use of an EtherType via the Subnetwork Access layer Protocol (SNAP), which, as indicated, accommodates the use of the original EtherType codes, except now conveniently buried within a SNAP header, which itself is embedded within an LLC header. Talk about wheels within wheels....

Although this is good to know, it must be stressed that Juniper Networks switching and routing gear running JUNOS supports *only* Ethernet-based encapsulations for IP. JUNOS is known to interoperate with all networking gear of consequence, and is found in most large IP networks and in the backbones of virtually all Tier 1 service providers. It would seem that lack of support for IP over IEEE 802.3 is not an issue for anyone, and this is the point that is made. Ironically, the only thing that uses 802.3 encapsulation in JUNOS is Intermediate System to Intermediate System Level 1 (IS-IS), which is an OSI-based routing protocol that was originally intended to support OSI's CLNS routing. The irony is that for most JUNOS gear, actual CLNS routing is not supported; IS-IS is used to build, and this is the rub, *IP* routing tables! I'm not sure why, but I can't help but smirk as I write this, being both an IP bigot and not myopic enough to fail to see the bold IP writing on the wall, so to speak.

These days the practical truth is that IEEE 802.3 defines updates to the Physical layer standards and capabilities, and IP makes use of the faster speeds while choosing the relative comfort of an Ethernet-based frame; the fiber-optic medium or wire cares not one bit about the Type or Length field, so all is well. Unless you are specifically discussing LLC and the use of SAP, or MAC layer padding, the two terms are pretty much interchangeable, and are often used that way. This near synonymous nature is what often leads to confusion about the true differences, however.

The MAC Layer

There are many physical varieties of Ethernet, but generally speaking they all share the same MAC frame format and protocol. This is one of Ethernet's greatest strengths. Its specification was based on bit times, not rates, making it easy to ramp up the speed (typically by an order of magnitude), while leaving the rest untouched.

It is worth noting here that the MAC we refer to is the entire Media Access Control layer of networking. A MAC address, which we all know and love, is merely a portion of this. All media have both PHY (physical) and MAC characteristics. The Ethernet MAC defines frame structure as well as the CSMA/CD shared media access procedure.

There are a few MAC characteristics that bear some extra attention, so let's get cracking.

CSMA/CD

In its original form, Ethernet was based on a shared (not switched as is now the norm) medium and the use of a single baseband bit rate. This meant that only one node could actively use the cable at any time. Rather than bother with passing tokens or other shenanigans, Ethernet's inventors opted for an opportunistic-based MAC called Carrier Sense Multiple Access with Collision Detection, or CSMA/CD. Sounds fancy, but humans do this all the time. We, after all, also share a medium and emit sound energy in the same band, which means there really should be only one person speaking at any time for maximum productivity.

In CSMA/CD, a station that wants to speak first listens to sense whether another station is active. If so, it waits. This is the Carrier Sense part. As quiet for one is quiet for all, there is nothing to prevent multiple stations from seizing the opportunity (a case of *carpe medium*, to use yet another pun), and this would be the Multiple Access part. When this occurs all the messages are corrupted, so the stations involved should detect the collision so that they can start an exponentially increasing back-off timer and try again later.

The Ethernet MAC algorithm can be summarized as "It's nice to share, and if you have something to say don't wait for an invitation; if at first you don't succeed, try and try again, 16 consecutive times, and then give up," as there is likely a cable fault (unterminated) that is causing the station's own energy to reflect back (as a standing wave), which in turn activates the collision detection circuitry at every transmission attempt.

Opportunistic Versus Deterministic MAC

Token-passing proponents were always fond of raising the fact that, in theory, an Ethernet LAN could fail to convey any useful information due to repeated occurrences of collisions, which simply waste bandwidth and time. This is true, in theory, much like there is a certain probability that a pot of water will freeze over a flame. Yes, it could happen, but no, it never has.

By specification, an Ethernet LAN is permitted no more than 1,024 MAC entities. Each time there is a consecutive collision, a station picks a wait time, at random, from a pool that increases exponentially, until it hits 1,024 (the number of permitted nodes), and then truncates at 1,024 for counts 11–15.

Although it's true that some forms of Ethernet can have collisions, and that collisions result in resource waste, there has never been a confirmed case of a properly functioning Ethernet, with the supported number of MAC entities (1,024), ever having 16 consecutive collisions. Further, studies have shown that, on average, the time needed to successfully access the medium is lower for Ethernet than for a Token Ring LAN, where the latter wastes resources waiting around for the next token on an otherwise idle LAN.

Stated differently, with Ethernet you *might* have to wait, whereas with Token Ring you *know* you have to wait, but when it's your turn access is guaranteed. Think of taking the freeway versus taking a train or plane for a 50-mile trip. The freeway may be congested, slowing your progress; or it may not be, but you are allowed to enter the flow as soon as you can wedge your car in. A plane, on the other hand, generally leaves at some fixed time, one that I have always found myself waiting for (and all too often, even that fixed time was less than deterministic, in my humble opinion). So, I would choose the opportunistic system any day, and this choice is paralleled in LAN technology with the demise of all things token-based.

The widespread use of FD Ethernet, enabled by switches, has eliminated the potential for collisions, which simply made a good thing all the better.

The shift away from shared media. The shift away from shared coax, to a hub-and-spoke (star-based) topology that was based on twisted pair, was a monumental point for Ethernet. Although no one could have imagined it at the time, advances in technology would shift these *multiport repeaters* into a bridging or switching role. Unlike a repeater, both of these devices terminate collision domains. Because a UTP link can have only two stations (P-to-P), and because of a separate transit and receive pair (or frequency), there was no reason to run HD anymore.

This is significant, because in addition to doubling potential throughput by allowing simultaneous send and receive this also eliminated the potential for collisions.

The ability to use preexisting UTP, a medium originally intended for analog telephone use at 20 KHz, to build a high-performance and highly survivable LAN was too much for the market to resist. Not having to deal with buying and installing coaxial cable was advantage enough. However, the shift to P-to-P links also eliminated the single point of failure associated with shared media. In theory, one cable break with Thick Ethernet, or one jabbering station that won't shut up, could bring down the entire LAN. Now the repeater or switch port simply partitions to isolate the malfunctioning link/node to prevent such network-wide disruption.

The shift from a bus to a star topology was followed closely by a bump in speed from 10 Mbps to 100 Mbps via the 100Base-TX standard. Ethernet has been the dominant LAN ever since, and is now, for all intents and practical purposes, the only LAN the world appears to need, or want, for that matter.

MAC addressing

One part of the MAC layer specification is the shared media access method; the other is the frame structure. An important part of the MAC frame is the MAC address. Ethernet LANs use a 48-bit-long MAC address that is non-hierarchical or flat. Figure 1-4 details the structure of a MAC address. The Most Significant Byte (MSB) of the MAC address is sent first and is the high-order address octet.

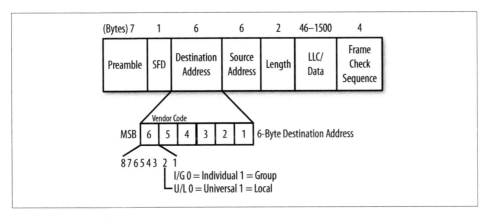

Figure 1-4. MAC address structure

The MAC address contains flags to indicate whether it's a group (multicast) or an individual (unicast) address, and whether the address is universally administered (via the IEEE) or is a locally assigned value. The broadcast address is a special form of multicast that uses all 1s to indicate it is intended for every station on the LAN.

The network interface card (NIC) or Ethernet port's burned-in address (BIA) is presumed to be globally unique by virtue of a managed vendor-ID space. Vendors apply for one or more blocks of MAC addresses, as identified by the first 24 bits, and are then responsible for distributing MAC chips with unique addresses using the remaining 24 bits.

The address space is said to be flat in that all 48 bits are needed to identify a station. There is no concept of information hiding or aggregation of multiple MACs into a single super MAC. A switch must know and learn (which means *store*) the complete 48-bit MAC address for every active station in the LAN. This is a significant point, as it impacts scalability. There is a reason that the worldwide Internet is not based on Layer 2 switching. No switch on Earth could learn and store a 48-bit address for every one of the more than *1.8 billion* machines projected to form the Internet of 2010! (Source: *http://www.clickz.com/stats/web_worldwide.*)

In contrast, routers operate at Layer 3, where the addresses that are used support hierarchical structuring. This allows a router to summarize (or hide) information, which in turn allows it to scale far beyond the scope of any Layer 2 device. Consider the common case of a single default route, which summarizes every possible IP address (more than 4 billion are possible) into a single table entry. A core router must be able to reach every host on the planet without relying on a default route. IP address hierarchy currently permits this feat with a table that contains only about 280,000 entries as this is written. Far better efficiency could be had if IP addresses were originally allocated in a manner that better accommodates summarization—a mistake that will, ostensibly, not be repeated with IPv6 address allocation.

Ethernet Standards Wrap-Up

Ethernet technologies continue to evolve as speeds ramp up and new functionality, such as OAM, continues to breathe life into this venerable workhorse. Table 1-1 summarizes the characteristics of widely used Ethernet Physical layer standards.

Table 1-1. Key Ethernet standards

Standard	Speed/mode	Topology	Medium	Segment length	Comments
10Base-5 DIX 1980, 802.3 1983	10 Mbps/HD	Bus	Thick coax, RG-8/U	500 m	The original, vampire taps and all
10Base-2 802.3A 1985	10 Mbps/HD	Bus	Thin coax, RG-58	185 m	Bye-bye taps, hello BNC T connectors
FOIRL 802.3D 1987	10 Mbps/FD	Star	Two fibers	1,000 m	The beginning of star-wired buses; P-to-P allows FD
10Base-T 802.3I 1990	10 Mbps/FD	Star	Two pairs Category 3 UTP or better	100 m	No more coax or expensive fiber media; the beginning of the end for other LANs
10Base-FL 802.3J, 1993	10 Mbps/FD	Star	Two multimode (MM) fibers, 62.5/125 µm	2,000 m	Updated FOIRL specification

Standard	Speed/mode	Topology	Medium	Segment length	Comments
100Base-TX 802.3U, 1995	100 Mbps/FD	Star	Two pairs Category 5 UTP or better	100 m	So much for FDDI and its expensive optics
100Base-T4, 802.3U, 1995	100 Mbps/HD	Star	Four pairs Category 3 UTP or better	100 m	Uses eight wires, allows use of installed CAT-3 for 100 Mbps
100Base-T2, 802.3Y, 1997	100 Mbps/FD	Star	Two pairs Category 3 UTP or better	100 m	So much for 100Base-T4; same speed on half as many wires and FD
1000Base-LX, 802.3Z, 1998	1,000 Mbps (GE)/ FD	Star	Two single-mode (SM) fibers, 10 µm	5 KM/5,000 M	The first Gigabit Ethernet (GE) flavor
			Two MM fibers, 62.5/125 µm	550 m	
1000Base-SX, 802.3Z, 1998	1,000 Mbps (GE)/ FD	Star	Two MM fibers, 62.5/125 µm, 50/100 µm	220 m 550 m	A popular fiber flavor with wide deployment
1000Base-LH (SX) (non-standard)	1,000 Mbps (GE)	Star	Two SM fibers, 10 µm	10–70 km	Non-standard flavor of 1000Base-LX with better optics
1000Base-T, 802.3AB, 1999	1,000 Mbps (GE)/ FD	Star	Four pairs Category 5 UTP or better	100 m	GE over UTP copper, albeit using all eight wires with echo cancellation for FD
10GBase-R, 802.AE, 2002	10 Gbps (10 GE)/ FD	Star	LR, SM 9/125 µm	10 km	First 10 GE over fiber, both long and short reach
			SR, MM 62.5/125 µm	26 m	
10GBase-T, 802.3AN, 2006	10 Gbps (10 GE)/ FD	Star	Four pairs UTP	55–100 m	Had to happen; 10 GE on UTP, needs CAT-6a for max distance

Note that the wide variety of Physical layer media options for Gigabit Ethernet (GE) has resulted in the concept of Small Form-factor Pluggable (SFP) optics. The optics name is somewhat of a misnomer here, as copper-based SFPs are also available. Note that 10 GE SFPs are called XFPs, and provide the same function.

Although none too cheap, the ability to mix and match switch or NIC ports to the physical layer du jour by simply inserting the desired module is a big advantage. XFP support is especially important with 10 GE, given that there are at least 10 different Physical layer standards specified! The original version of this technology was referred to as a GE Interface Converter (GBIC). The newer SFPs have been further reduced in size and are sometimes called mini GBICs.

Currently the IEEE is working on standards for the next Ethernet speed overhaul, specifically 40GbE and 100GbE.

A word on auto-negotiation

Table 1-1 makes it clear that there is no shortage of Ethernet flavors to choose from. With so many options, finding the set of mutual capabilities that yields the highest performance between any two pairs of nodes can be daunting. Automatic selection of the best level of compatibility is the motivation behind auto-negotiation.

The current best practice is to use auto-negotiation rather than to hardcode parameters. Over the years, the standards have matured enough to work reliably, and manually setting these parameters has been found to be error-prone. Figure 1-5 shows the operating mode priority, which always selects the *best* mode that is *mutually* supported, as well as a table showing the outcome for various combinations of Ethernet auto-negotiation pairings.

The key takeaway is that pairing one end set for auto-negotiation with another end that is hardcoded is almost always a bad thing. The result is often a *duplex mismatch*, which can be a very nasty thing, as in many cases the result is significantly diminished performance that *may not be detected* and therefore will be allowed to remain in place, causing long-term service degradation. The issue is that the HD end senses the remote end's FD operation as a collision, resulting in needless back-offs and retransmission attempts.

Auto-negotiation has been defined in several IEEE standards, and is optional for most flavors of Ethernet. The protocol was updated and made mandatory for 1000Base-T as part of the GE 802.3a specification. Auto-negotiation is mandatory for normal 1000Base-T operation due to the need to determine the Master/Slave timing role for each end's Physical layer; this function is unique to 1000Base-T, and is determined during the auto-negotiation process.

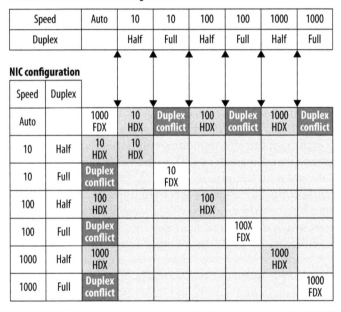

Figure 1-5. Ethernet auto-negotiation

Ethernet Technology Summary

Ethernet won the LAN battle. Today when the three-letter acronym (TLA) LAN is used, you can bet it regards some flavor of that tireless workhorse known as Ethernet. Although there was a time when a LAN switch needed to support Ethernet and Token Ring/FDDI ports, that time has passed. Yes, it has passed much like the proverbial token, on into that great sunset that awaits all mortal beings, be they LAN, WAN, or (hu)MAN (pun intended).

Juniper Networks' EX switches are Ethernet-based. The information in the next section prepares the reader for upcoming deployment labs, which by matter of modern practicality are strictly Ethernet-based.

The TCP/IP Suite

Having arrived here, you likely agree that Ethernet technology is the bee's knees, so to speak, and is the be-all and end-all of all your local area networking needs. You have installed your shiny new EX switch, and you now have all those network ports just waiting for something to do.

And then it strikes you. All that revved-up LAN infrastructure is little more than a doorstop without upper-layer protocols to drive useful data over it. A LAN, with its Physical and Data Link layer services, provides you with a grand communications potential, but you can realize this potential only when there are applications written to operate over that technology.

There was a time when users were forced to choose one proprietary protocol suite over another; or as needs often dictated, when they ran multiple protocol suites to serve various user communities. Netware and its IPX/SPX had great file and print sharing, while Banyan Vines had a really cool directory service. Unix workstations and servers typically supported engineering communities with their Network File System (NFS), remote r commands, and related TCP/IP-based networking support. And who can forget AppleTalk and its DDP, often used for the graphic artist, or IBM's SNA/SAA, often found transporting the business's accounting and financial applications. Oh, and then along came Microsoft with its NetBIOS/NetBUI-enabled Windows for Workgroups solutions, which often found their way into ad hoc networks when they were not trying to compete with Novell. In the typical case, this litany of protocols would run on a multiprotocol backbone in a "ships in the night" fashion, passing each other by and never even being aware of the close encounters. Few hosts were bilingual, so most user communities shared a cable and little else. Protocol converters (remember those?) and application gateways were big business when users in different communities needed to interoperate.

Enter OSI

At one point, the grand saviors to this mess were—you guessed it—the OSI's ISO models and related ITU/CCITT standards. The plan was to have the world switch to a standards-based open system, which would then promote wide-ranging connectivity and, just as important, would once and for all end all the multiprotocol babble the world of networking was mired in. Too bad OSI imploded.

What was left was pretty much what folks started with, which is to say some level of international standards that were actually in use and were intended to form the foundation of the larger OSI solution. This left people with circuit technologies such as dial-up modems, ISDN, and leased lines, and a few packet-based options such as LANs and X.25 for WAN use. There was one new thing, however, and that was the idea of what a widely adopted set of open standards could do. By replacing multiple stacks with one, costs could be reduced, support was simplified, and there was greater potential for interoperability.

Given that the OSI model was similar to the preexisting TCP/IP suite, that TCP/IP was known to work, and that it in fact did more every day than OSI accomplished in its whole life, a candidate came into focus. The fact that TCP/IP was well understood and widely deployed, was already supported on most operating systems, and was generally made available for free caused that candidate to again stand out. Once the U.S. government's mandate of using only OSI protocols was exempted in the case of IP, until such time that equivalent functionality was available in OSI protocols, the writing was clear.

The world had its communications savior, and its name was Internet Protocol. Thus began the mantra of *IP over everything, and everything over IP.*

Exit OSI, Enter IP

As we ranted earlier, while all the OSI work was moving forward at its own snail-like pace, the fine folks at the IETF, the entity that generates Internet RFCs and drafts, kept on doing what was needed to solve actual issues and to create new functionality in the TCP/IP suite of protocols. A number of factors added up in IP's favor, including the following:

- The open nature of TCP/IP made the technology and its related specifications freely available. In contrast, OSI specifications cost money, and so did the resulting products, of which there were few.

- TCP/IP was derived from early work on the DoD's ARPANET, which went online in 1969! It was one of the first routed/packet-switched protocols ever implemented. TCP/IP's original purpose was to avoid issues with the connection-oriented services of the day, which had a tendency to be disrupted every time some battle happened and bombs tore up phone lines. The TCP stack was intended to provide robust and resilient communications in battlefield conditions, and was known to work in this regard with a long field-proven history. I mean, who doesn't want his Internet surfing at Amazon or his iTunes downloading to be robust and reliable, despite the near-battlefield conditions that comprise a modern distributed denial of service (DDoS) attack on large portions of the Internet's infrastructure?

- Because the U.S. government contracts were historically based on TCP/IP, and then later when OSI was mandated by virtue of the exception, for lack of OSI functionality, all OSs and products of any consequence had TCP/IP protocol support. If it was not already built in and simply waiting to be activated, you could certainly find a TCP stack for a given OS far more easily, and for far less money, than some OSI stack that would at best let you use a subset of the functionality on a far more limited set of machines.

- Sir Timothy John Berners-Lee invented the World Wide Web (WWW) *application*, and suddenly every normal (non-computer-geek) person, and quite literally his or her mother, could see a use for the Internet that they would be willing to *pay* for. Things always seem to get real popular real quick whenever there's money to be made.

- As much as this dates me, I can remember the pre-WWW Internet. There was FTP from a command line, and there were command-line search tools such as the FTP site indexer Archie, which in short time begat the updated (and Gopher-based) search capabilities of Jughead and, ultimately, Veronica. Yep, back in the day we would walk barefoot through the snow to our 2,400 bps dial-up modems, and we would be happy for the chance to use a command-line FTP client to download some plain-text copy of an RFC. Nope, we did not need images, fancy multimedia, a GUI point-and-click interface, or the ability to view the contents of our shopping carts. Nope, back in the day, when I went shopping I had an actual shopping cart in my hands, and I could look at its contents without any fancy click buttons (I say in my best cranky old-man voice).

- After the web browser and HTTP/HTML, the Internet and its underlying TCP/IP enabler became "the WWW." It still irks me to hear news reports of "The WWW is under attack due to <insert latest cause here>," then in reality it's the Internet that is being attacked and the stupid WWW application is but one of many that are affected. No one but us aged geeks seems to care that during these outages we're unable to use the finger application to determine how many cans of soda are left in some science lab's vending machine! The Internet is more than just the WWW.

When the preceding points were factored, the choice was clear. The world already had a multivendor interoperable protocol, it was known to work, and money could now be made from it. TCP/IP became the de facto set of open standards while the official ones faded into that great night.

Vive IP!

The IP Stack, in a Nutshell

Once again we find ourselves broaching a subject that is *extremely* well documented, and found in numerous places. The wheel is not re-created here, as trees are far too valuable. Once again we try to stick to the facts and distill the most important and widely used set of protocols down to a few paragraphs. Wish me luck....

Figure 1-6 depicts the TCP/IP stack, along with some selected applications.

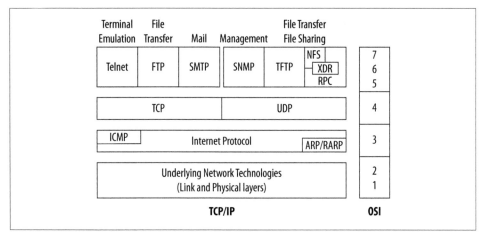

Figure 1-6. The Internet Protocol stack

Figure 1-6 also shows the good old OSI model alongside the IP stack. After all, being able to compare non-OSI things to its well-delineated layers is about the only useful thing left of that grand effort.

The network that lies beneath

As is so often the case, things begin in the physical realm, where the bits meet the media, so to speak. In the IP stack, both the Physical and the Link layers are combined into a single underlying Network layer. IP, which lives at Layer 3, is well shielded from the incredibly long list of supported technologies. For our purposes, we can place Ethernet technology here, allowing the Physical layer to be 10Base-2, GE over optical fiber, or whatever the current flavor of Ethernet happens to be. Likewise, the Link layer becomes CSMA/CD and the Ethernet MAC frame structure. For some flavors of Ethernet the MAC layer can dispense with collision detection (CD) and operate FD with no need for carrier sensing. In other modes both are still needed, as well as the binary exponential back-off algorithm used to recover from collisions.

To help drive home the significance of layering and the beauty of layer independence, consider that in all of these cases the same IP entity operates in the same manner, albeit sending less, or more, as the physical link speed dictates. All the details are handled by the related layers, such that IP simply sees a datagram transmission and reception service.

ARP me, Amadeus

Moving up, you next encounter Address Resolution Protocol (ARP). ARP is a critical component of IP's independence from the underlying technology. Some technologies do not use link-level addressing, or are inherently P-to-P, and therefore do not require any dynamic binding of a destination IP address to a Link layer hardware address. In these cases, ARP is simply not used. Multipoint links, such as a LAN, represent a different story, as shown in Figure 1-7.

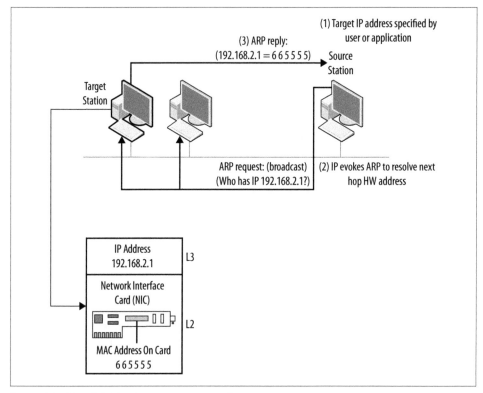

Figure 1-7. IP and ARP—they taste great together

Figure 1-7 shows two IP stations that share an Ethernet link. At step 1, the user or her application specified an IP address as the destination for some session.†

The IP layer forms the resulting packet, but before it can be handed to the MAC layer for transmission, the correct destination MAC address must be specified. IP looks for a match in its ARP cache, where recent responses are stored. Here, no match is located, forcing it to evoke ARP's services, shown as step 2 in the figure. The ARP request is broadcast at the MAC layer to get around the "cart before the horse" issue of needing to talk to a station to learn its MAC address *before* you actually know its MAC address.

In this example, the target is alive, and it sees the ARP request is intended for its local IP address. Its ARP reply contains its Ethernet MAC address, which completes the process. Usually this reply is sent as unicast back to the requester, but an unsolicited

† In most cases, a Domain Name System (DNS) name is specified, but the result is an IP address, so we can skip the IP-to-domain name binding complexity for now.

ARP reply can be broadcast to prepopulate ARP tables in what is known as Gratuitous ARP.

Layer 3 to Layer 2 address resolution is a must when there are multiple possible destinations on a single medium. "Close" does not count in the networking world.

IP, freely

Next up the chain is Internet Protocol (IP) itself. IP provides a datagram service to its upper layers. The term *datagram* implies a connectionless mode of operation and a resulting unreliable transport service. IP supports:

- Identification of source and destination network addresses
- A header checksum to detect corruption of its own header (not payload)
- A ToS (Type of Service) indication
- A precedence value to influence the probability of discard during congestion
- Fields to support and control fragmentation (and later reassembly) of large datagrams to accommodate dissimilar maximum transmission units (MTUs)
- A TTL field to limit effects of routing loops
- A Protocol field to identify the owner (the upper layer protocol) that is responsible for the datagram's payload
- A Length field to accommodate padding, which is needed for Ethernet
- Options, which when present alter packet handling (i.e., source routing and record route)

A lot of functions, to be sure, but again note that IP *does not* provide connection setup, flow control, payload error detection, or a simple discard response (with no retransmission) in the event of IP header errors, or the inability to process for any reason (e.g., lack of buffers due to congestion). IP leaves those functions to the users of its service: the upper layers. Some applications—for example, broadcast media—may not care much about error correction as they are negatively impacted by any retransmission attempt, whereas other applications—say, e-commerce—will be quite concerned with error control, or at least one would hope. IP provides the same services to both and lets the upper layers sort things out.

Note that some underlying network technologies—for example, X.25 or a Token Ring LAN running 802.2 LLC type 2—provide their own error detection *and* retransmission-based correction, which is another good reason to leave IP streamlined and built for routing. X.25 was the only Layer 2 protocol to also offer error correction. At least until DOCSIS came along!

When links are error-prone, relying on end-to-end retransmissions, as is the case with a protocol such as Frame Relay, it quickly results in extreme throughput degradation, even when only tens of such links are involved. For this reason, error-correcting Link layers are still used in support of IP when error-prone transmission links are used.

To carry the previous example of IP and underlying network technology independence a bit further, consider that an IP packet can easily cross tens of links between endpoints. Consider this traceroute from Juniper Networks headquarters in Sunnyvale, California, to the official government website for the Central Asian nation of Uzbekistan:

```
bash-3.2$ traceroute www.gov.uz
traceroute to www.gov.uz (195.158.5.137), 30 hops max, 40 byte packets
 1  mrc2-core1-3.jnpr.net (172.24.28.2)  0.541 ms  0.517 ms  0.735 ms
 2  172.24.19.33 (172.24.19.33)  0.447 ms  0.438 ms  0.421 ms
 3  172.24.230.90 (172.24.230.90)  1.287 ms  1.274 ms  1.497 ms
 4  ns-egress-fw-vrrp.jnpr.net (172.24.254.6)  1.213 ms  1.196 ms  1.403 ms
 5  66.129.224.34 (66.129.224.34)  1.926 ms  1.911 ms  1.893 ms
 6  POS2-1.GW5.SJC2.ALTER.NET (208.214.142.9)  1.873 ms  1.675 ms  1.659 ms
 7  161.ATM4-0.XR2.SJC2.ALTER.NET (152.63.48.82)  2.325 ms  2.311 ms  2.293 ms
 8  0.so-1-0-0.XL2.SJC2.ALTER.NET (152.63.56.141)  2.391 ms  2.376 ms  2.357 ms
 9  0.ge-3-0-0.XT2.SCL2.ALTER.NET (152.63.49.110)  3.363 ms  3.324 ms  3.555 ms
10  sl-crs2-sj-0-1-0-1.sprintlink.net (144.232.9.1)  4.744 ms  4.733 ms  4.710 ms
11  sl-crs1-rly-0-4-2-0.sprintlink.net (144.232.20.187)  74.735 ms  70.007 ms
73.500 ms
12  sl-crs1-dc-0-8-0-0.sprintlink.net (144.232.19.213)  69.516 ms sl-crs1-dc-0-12-
2-0.sprintlink.net (144.232.19.223)  69.208 ms sl-crs1-dc-0-8-0-0.sprintlink.net
(144.232.19.213)  69.481 ms
13  sl-bb20-par-1-0-0.sprintlink.net (144.232.19.147)  159.589 ms  158.893 ms
159.475 ms
14  sl-bb21-fra-13-0-0.sprintlink.net (213.206.129.66)  159.159 ms  159.999 ms
159.981 ms
15  sl-gw10-fra-15-0-0.sprintlink.net (217.147.96.42)  174.860 ms  174.846 ms
174.828 ms
16  sl-MTU-I-278357-0.sprintlink.net (217.151.254.134)  160.784 ms  161.635 ms
160.751 ms
17  bor-cr01-po3.spb.stream-internet.net (195.34.53.101)  220.050 ms  220.278 ms
219.633 ms
18  m9-cr01-po4.msk.stream-internet.net (195.34.53.125)  210.818 ms  211.339 ms
210.590 ms
19  m9-cr02-po1.msk.stream-internet.net (195.34.59.54)  213.698 ms  213.475 ms
214.951 ms
20  synterra-m9.msk.stream-internet.net (195.34.38.38)  213.879 ms  213.869 ms
214.622 ms
21  83.229.225.243 (83.229.225.243)  230.080 ms  229.762 ms  229.750 ms
22  83.229.243.98 (83.229.243.98)  259.745 ms  260.105 ms  259.709 ms
23  195.69.188.148 (195.69.188.148)  282.930 ms  282.916 ms  283.508 ms
24  195.69.188.2 (195.69.188.2)  276.545 ms  276.534 ms  276.887 ms
25  84.54.64.66 (84.54.64.66)  277.306 ms  278.145 ms  278.347 ms
26  firewall.uzpak.uz (195.158.0.155)  283.783 ms  283.323 ms  282.876 ms
27  ta144-p86.uzpak.uz (195.158.10.181)  278.743 ms  276.536 ms  277.146 ms
28  195.158.4.42 (195.158.4.42)  276.802 ms  277.392 ms  276.766 ms
29  195.158.5.137 (195.158.5.137)  284.062 ms !X  284.042 ms !X  283.711 ms !X
```

Here, the results show that a fair number of hops are needed to reach the target site. As an upside, the presence of `.uz` domains near the target indicates that at least the website resides in a faraway and exotic land, as opposed to being hosted in some Silicon Valley-based company.

A key aspect of IP internetworking, and of routers in general, is that on *each* such link a *completely different* type of network and transmission technology can be used. The packet's first hop might be over a modern LAN in sunny California, whereas a subsequent link may involve a jaunt in a Frame Relay frame transported via a trans-Atlantic SONET link that runs Point-to-Point Protocol (PPP). And later yet, as the IP packet nears its final hop and, likely, middle age, given that its TTL field has been decremented at each hop, the packet could jump into an X.25 packet, where it's well protected for its arduous journey over an error-prone analog leased line the recipient uses for Internet access.

In summary, IP is a lot like the U.S. postal service, or any non-registered mail system for that matter. Your envelope indicates who the letter is from and where it's going, using hierarchical addressing that permits information hiding, whereby more and more of the address becomes significant as the letter winds its way to the destination. Here, all the stuff up to and including the street address is like the IP address, and the recipient's name is like a protocol identifier. You can have multiple protocols at the same IP, just as multiple people can dwell at the same address. You can pay more for airmail, or save with bulk, which is a reasonable analogy to IP's ToS indication. There is some weight/size limit, and if in excess of this limit the package may have to be split into smaller parts (fragmentation). Although no doubt diligent in the face of snow and all that, regular mail is a connectionless service that is only a best effort. You do not need permission to send someone a letter (no connection establishment), and simply dropping a letter in the mailbox is no guarantee that it will be delivered successfully, but in most cases it will. All of this sounds exactly like what IP does, except that IP deals with packets and not packages.

IP addressing

IP addressing is a topic that is well covered in numerous other places. Technically, this being an Ethernet-based LAN switching book and all, it could be argued that anything above Layer 2 is beyond the scope of this book.

Yes, this could be said. But to do so would ignore the truth that virtually all new networks are IP-based, and each day some multiprotocol LAN gets closer to pure IP nirvana as one more of its legacy protocols is decommissioned in favor of IP transport. As such, a thorough understanding of IP addressing is critical for anyone dealing with LAN switching in the modern context. So, the packet stops here, to use yet another poor pun, and we pause to take a very condensed tour of what matters in IP addressing. It's only 32 bits. How bad can it be?

Hierarchical. This is nothing but a fancy way of saying an IP address has more than one part. Here, we mean it has both a *network* and a *host* portion. This is significant, and to a large degree it is what allows a router to scale to a worldwide Internet while a bridge would melt under the "load" of all those MACs.

The takeaway: routers route to networks, not hosts. The host portion of IP is of concern only to the last hop router when it attempts direct delivery. In fact, because routing is based on a *longest match*, it's likely that remote routers are even ignoring parts of the network portion because of *supernetting*, which is also called *route summarization*. In the end, the router only needs to direct the packet out a sane interface, one that gets it one step closer to its destination on a path that is not a loop. You do not need to examine 30 bits of network address to do this; trust me.

In fact, some non-core routers may use a default route, which is the ultimate in information hiding. The 0/0 route by definition says the router should try to match *zero bits*. Matching against nothing always succeeds, and is always the least specific match possible, but it's a match nonetheless. Using a default means that in effect, the packet is routed not by virtue of matching its destination address particulars, but quite the opposite: by virtue of matching *none*. Hence a low-end router can still forward to each of the 4 billion possible IPv4 addresses with a single default route entry, along with its directly connected network. If the router has two egress interfaces, two default routes in the form of a 0/1 and a 128/1, each pointing to a different interface, of course, this provides pretty decent load balancing to all possible destination IPs. Not bad, huh? I'd like to see you do that with a bridge.

Classless is the norm (or, how we learned to subnet). When first envisioned, IP addresses were class-based. Figure 1-8 shows the original IP address class breakdown along with a binary-to-hex-to-dotted decimal conversion example.

Figure 1-8 has a lot of information, and all of it is important. The left side shows the original IP address class breakdown. The address class is determined by the setting of the high-order bits. For example, a Class B address always begins with a 10 pattern, as shown. Behind this plan was the perception that computers were special-purpose machines and that the Internet would remain an academic/military community, so the early Internet architects saw a need for a few very large networks (Class A), a good number of medium-size networks (Class B), and a larger number of small networks (Class C). The figure shows that for each class, there is some number of supported networks, and each such network in turn supports some number of host computers. The box on the upper right shows the net effect in the form of how many networks are available in each class (126 for Class A), along with how many hosts each such network class supports. The math is a function of 2 to the power of the address space (7 and 24 for Class A networks and hosts, respectively), and then subtracting 2 for the combinations of all 0s and all 1s, which are generally reserved for indicating *this* and *all*, respectively.

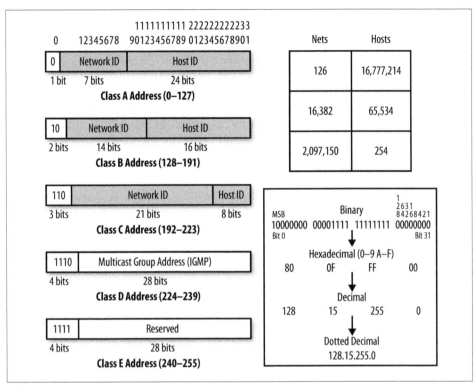

Figure 1-8. "Classy" IP addressing

To be effective with IP addressing, you must understand binary, hexadecimal, and the more human-friendly dotted decimal format, as all are needed at various times when working with IP. Remember when working with hexadecimal that you break each byte into two nibbles of four bits each. The resulting values therefore range from 0 to 15, but in hex 10–15 are coded as the letters A–F, respectively. In contrast, when working on the decimal value, all eight bits of the byte are grouped to yield a value from 0 to 255.

The dashed box on the lower right of the figure provides an example of this conversion process. This busy little box also shows the IP network byte order, which has the most significant bit of the MSB sent first, from left to right. Stated again, bit 0 is the most significant of the 32 bits that make up the IP address, and it's the first bit sent. The low-order octet also shows a power-of-10 breakdown for each of the eight bits in the octet. In this example all are set to 0, hence the value shown in Figure 1-8. A setting of 11000000 codes (128 + 64), or 192, as it has the bits for both 128 and 64 set. In hex this would be a C, given that 1100 0000 codes (8 + 4) = 12.

The class-based scheme was a fine plan, but as things worked out, classful addressing is far less than ideal. The updated IPv6 protocol has no such concept, and in all practicality, neither do *modern* IPv4 networks. The problem is basic inefficiency. It's great that a single Class A network can support more than 16 million machines and all, but that many machines on one logical subnet is preposterous (from a performance and reliability design perspective). Heck, even a single Class C, with its support for 254 hosts, is generally wasted in a routed environment.

LAN-based routed networks tend to have tens, rather than hundreds, let alone thousands, of machines. In the end, what people wanted was *more networks*, each with *fewer hosts*. As noted, the issue is that a recipient of a single Class A network would be hard-pressed to go back to his regional numbering authority to ask for yet more network numbers, when it was shown that he was using only a small fraction of the host space available in the Class A allocation he already had.

 So, what does it mean that nearly the first command everyone enters on an IOS-based router is `ip classless`, which provides support for IP subnetting (and supernetting)? No such command is needed in JUNOS, so that's one less command for you to type.

Classless IP routing simply means that for each IP address (prefix) there is an associated network mask. In contrast, with classful routing the address's class is used to derive a *presumed* network mask. Having an explicit mask allows the user to define what portion of the 32-bit address identifies the network; once the network portion is known the remainder is considered to be host addresses.

Subnetting is the process of extending the mask, making it longer so as to extend network numbering into the host field. Thus, more networks are gained, at the cost of fewer hosts on each network. *Supernetting* is the opposite, and creates fewer networks by reducing network mask length. Supernetting is an important concept behind Classless Inter-Domain Routing (CIDR), which is an effort to summarize networks into fewer routing entries, wherever possible. This is done to try to keep the size of Internet routing tables from growing at a pace that outstrips computer processing power, a real threat that at one point genuinely jeopardized global Internet stability!

Figure 1-9 shows classless IP routing at work.

An often misunderstood aspect of CIDR is the fact that *different network masks* are used, at different places, to route the *same packet!* The network mask does not have to be the same length, except for the collection of hosts that attach to the same logical IP subnet/network. As a result, a core router may use a default (class-based) network mask to direct traffic to a customer's network attachment point. In Figure 1-9, a Class B address is assigned, resulting in a /16 mask, which is also represented in the pre-CIDR notation format of 255.255.0.0. Within the customer's network, the single Class B address has been subnetted, in this example to provide some 254 additional subnets through a /24 (or 255.255.255.0) network mask.

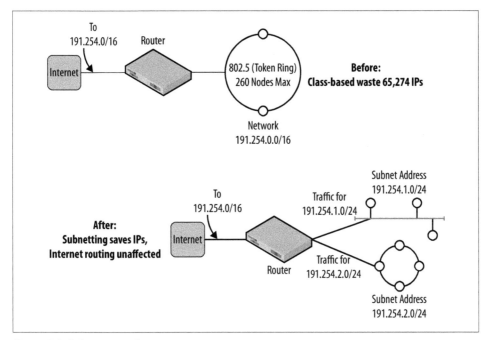

Figure 1-9. Subnetting and supernetting

Subnetting and supernetting are now old news in IP. Their widespread use is doing much to forestall the predicted demise of available IP addressing space, as well as the meltdown of core routers due to impractical routing table size, several times over.

VLSM and Discontiguous Subnets

Variable Length Subnet Masking (VLSM) refers to the ability to assign network masks of varying lengths to different portions of a network to maximize the proverbial bang for the IP addressing buck. For example, /30s, /31s, or even /32s might be used on P-to-P links, whereas /26 might be assigned to a LAN. There is no magic here, as IP routers always route to the most specific, or longest, match. The key is in having a routing protocol that supports the conveyance of a network mask along with the IP prefix. Older protocols such as Routing Information Protocol (RIP) v1, or Cisco's IGRP, lack this capability, which forces you to use the same mask length each time a given (major/class-based) address is assigned. This is because the lack of network masks in routing updates forces the local router to assume that major/classful prefixes that match a local address assignment must use the same mask length as that assigned to the interface on which it was received. When VLSM is used, these assumptions are not valid and routing problems can surface.

Problems with discontiguous subnets are also related to routing protocols that do not convey a network mask. The issue is when some network address—say, a Class B 172.16—is subnetted on two routers that are separated by a link with a different major network (not a 172.16) address. In this case, protocols that do not support a mask perform auto-summarization to the classful network, resulting in both ends sending and receiving the same 172.16/16 update. The result is the loss of subnet routing for the discontiguous subnets. Using a routing protocol such as Open Shortest Path First (OSPF), IS-IS, or RIP v2, solves both issues through inclusion of an explicit network mask along with each network prefix.

ICMP, the bad news protocol

Moving up the stack in Figure 1-6 we next hit the Internet Control Message Protocol (ICMP). ICMP is classified as a sublayer. ICMP is an official part of IP, but is itself *encapsulated inside* IP and is therefore shown above it. ICMP is often used to report errors when handling IP datagrams, hence the not-so-funny title of this section. Common errors are Destination Unreachable, TTL Expired, Options Handling Issue, and Fragmentation Needed But Not Permitted. ICMP messages can also be used to provide information such as reporting a timestamp or the local link's subnet mask. ICMP is the mechanism behind the echo request and response functionality affectionately referred to as *ping*.

UDP, multiplexing, and not much else

User Datagram Protocol (UDP) provides a best-effort connectionless Transport layer service. Recall that Layer 4 is the first end-to-end layer, and is therefore processed only by the destination machine. Given that IP is also a best-effort protocol, it can be said that UDP does not add much in the way of reliability. UDP offers no error correction or flow control; it does provide error detection (with silent discard) against the UDP header and payload.

UDP's most important function is the notion of *ports*. The port abstract is similar to a Unix socket, and provides multiplexing among multiple processes that each share the same IP address. Recall that the IP layer's Protocol field identified the owner of its payload, which may be UDP. IP then hands its payload to the UDP process, where the first step is error detection. If all is well, the UDP header is stripped and the destination port is used to direct the packet's payload to the appropriate process. Port values below 1023 are standardized for use by well-known (server) processes. Clients pick their port at random, selecting some unused value in the ephemeral range of 1,024 to 65,535. In most cases, services that can use multiple transport protocols, that is, either TCP or UDP, use the same port values; the Protocol field at the IP layer ensures there is no ambiguity in such cases.

The connectionless nature of UDP makes it well suited to point-to-multipoint applications (multicast) and short-lived transactional services such as DNS queries.

TCP, a transport for all seasons

Transmission Control Protocol (TCP) provides reliable, connection-oriented services. TCP supports ports for the same reasons as UDP, but in addition, TCP has:

- Connection setup, maintenance, and teardown phases that ensure that both ends agree regarding connection state. Traffic can be sent only when connected.
- Flow control, which prevents data loss due to lack of a buffer in the connection endpoints (not at the IP layer).
- Error detection based on a header/payload checksum, as well as through sequenced exchanges. This provides detection for corrupted data, in addition to lost or duplicated data, the latter being conditions that often occur given the datagram operation of the underlying IP.
- Retransmission-based error correction based on sophisticated congestion avoidance and recovery mechanisms that attempt to optimize communications among endpoints with greatly dissimilar processing capabilities, and to intelligently monitor and adapt to current end-to-end transmission delays.

Given that everything from Layer 3 down is often switched in datagram fashion (connectionless), a method of operation that's officially classified as *unreliable*, it's obvious how important TCP is to the world. When data integrity matters, and when you need to move a lot of information, TCP is likely your protocol of choice. The connection-oriented nature of TCP means that in some cases, more traffic is sent to set up and tear down a connection than is actually sent over the connection, and that a given TCP connection can connect only two endpoints.

What's this Internet thing for again, eh, sonny?

In the IP suite, applications are found directly above the Transport layer; there is no discrete Presentation or Application layer, but some IP applications provide these types of services. Some applications—for example, ICMP, or OSPF routing—make direct use of IP. Other routing options, such as RIP and Border Gateway Protocol (BGP), make use of UDP or TCP, respectively. More end-user-focused applications such as Telnet allow terminal emulation, or file transfer via FTP.

And then there is HTTP, the grand enabler of the modern Internet. Combined with the HTML specification, this is the only application that most people will ever use (to the extent that the Internet has become synonymous with the WWW, much to the irritation of this author). The Internet existed long before the WWW, and was quite useful to those with some level of *.clue*. The WWW allowed the great unwashed masses to rush in and make immediate productive and commercial use of the Internet. Although this killer app single-handedly ended the old geeks' club that was the academic- and research-focused Internet, it also did a lot to boost router sales, which for me is reason enough to welcome HTTP into the IP suite.

From email to telemedicine, e-commerce to games, you can bet your last packet there's an IP application written to support it.

IP encapsulation example

Figure 1-10 shows the TCP/IP stack at work, with an example of IP encapsulation within an Ethernet frame.

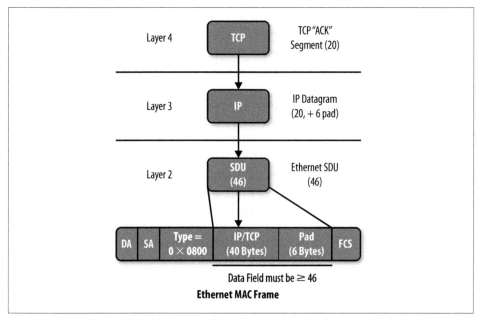

Figure 1-10. TCP/IP-over-Ethernet encapsulation example

This example begins with a TCP acknowledgment segment that needs to be sent. Although user data can be piggybacked onto such an ACK segment, this assumes that some user data is pending, and that's obviously not always the case; the lack of user data does not exempt the TCP entity from having to ACK traffic received from the remote end. If we assume no TCP options (in many cases, options such as a maximum segment size or a timestamp are present), then a 20-byte TCP segment, the minimum size of its header, is passed to the IP layer. Along with the data are internal semantics (primitives in OSI-speak) that convey variables such as the destination IP address, and special ToS values, and so forth.

The IP layer accepts its duty and builds the needed header. Again, assuming no options, that's another 20 bytes, for a total of 40 when the TCP header is also factored. Ethernet has maintained the need for a minimum frame size, which relates to ensuring reliable collision detection as a function of a frame's minimum transmission time versus the maximum allowed propagation delay; basically, the station should still be sending by the time its signal has propagated to the far end and any resulting collision has had time to make it back. The result is a need for four bytes of padding, which is added by IP and accounted for via the Length field. Any data outside the datagram's total length is assumed to be padding and is discarded by the far end. As is so often the case, the PDU rolls downhill only to darken Ethernet's door. The service request also tells Ethernet to set the Type field to 0x0800, IP's EtherType, and in our case, we can assume a successful ARP cache hit so that the next hop's MAC address is also passed along.

Direct Versus Indirect Delivery

IP is all about routing. One of the most basic aspects of IP datagram forwarding is a routing decision in the form of whether the destination address is on the sending machine's local subnet. If it is, direct delivery is performed and the ARP and subsequent packet are sent out over the interface with that direct route.

When the target subnet does not match a local subnet, indirect delivery is needed. This simply means that one or more intermediate stations will need to forward the packet on the local station's behalf. Forwarding other people's traffic is what routers are all about, so here the next hop would be to a device with at least two network connections, and with a willingness to forward traffic between those interfaces—in other words, a router. Usually an end station uses a default route to direct all non-local traffic to its default gateway (router). From there the packet typically picks up more intelligent forwarding that is based on least-cost routes that are dynamically learned via routing protocols that operate between the routers.

An important point about indirect delivery is that a packet that's intended for a remote machine is sent to the MAC address of the local subnet's default router. The router takes notice, strips the frame, and performs a longest-match lookup against the destination address, only to find that, alas, once again it's not the intended recipient (routers get lonely, too). After stiffening its lip and decrementing the TTL, the *same* IP packet is then *reframed* and sent out of a different interface to the next forwarding hop, where the process repeats until either the packet arrives at the target host (in which case you have a /32 match, which is as long as it gets, baby), or the packet's TTL expires and it's ignominiously discarded, with nary but an ICMP (TTL expired) error message that suffices as its death knell.

Ethernet constructs its frame, populates the destination MAC and the Type field with the value provided by IP along with the service request, and goes about the dirty work of successfully placing the frame upon the wire. At the remote end, a reversal of this process occurs, ending with the remote TCP receiving its ACK and flushing its retransmit buffer of the related data, resting in what it knows is a job well done.

Internet Protocol Summary

IPv4: it made the Internet what it is today, and what it will be tomorrow. It's the OSI *that worked*, and it's here to stay, so we deal with it. Each time the protocol is predicted to have met its natural limit, due to a lack of addresses, a need for class of service (CoS),

VPNs, encryptions, or something else, some bright engineers find a good workaround. For example, Network Address Translation/Port Address Translation (NAT/PAT) has done much to extend IP's useful life by allowing use of a private network addressing space *within* a private network, which is then translated and effectively hidden behind a lesser number of *real* IP addresses. As another example, IP Security (IPSec) was originally planned to be inherent to IPv6 but was backported to its predecessor, providing one less compelling reason to change what is still working.

With all of that said, IPv6 is making headway into today's networks. Many mobile devices are IPv6-enabled, and believe it or not, we are heading into a world where we will be surrounded by IP-addressable entities, be it your refrigerator, washing machine, or cable TV box. IPv6, with its 128-bit addressing space, combined with what we learned from IPv4 address allocation mistakes, promises that future generations will be free from having to worry about the Internet running out of addresses every few years. This is good, as it seems they will have plenty more to worry about, but that is another story and one best not told here.

IP is the convergence technology of choice in today's networks. It rides over every type of transport, and if it can be digitized, it likely rides inside IP. *IP over everything and everything over IP*. Learn it. Live it. Love it.

LAN Interconnection

Or: Repeaters, Bridges, and Routers, Oh My!

This section focuses on terminology and technology surrounding *modern* LAN interconnection. Given the qualifier, you can safely presume there will be no discussion of Source Route Bridging (SRB), translational bridging, or multiprotocol routing. When you're bridging in a purely Ethernet environment there is no need to translate, and SRB is but a part of the dark Token Ring past that surfaces only to haunt humanity in the occasional bad dream.

Figure 1-11 shows the relationship of LAN interconnect devices to the OSI model.

Once again the OSI model shows its remaining utility. One simple figure makes it clear that repeaters operate at Layer 1, bridges at Layer 2, and routers at Layer 3. This means that repeaters spit bits, bridges frames, and routers packets. You cannot get to a packet without first dealing with the frame in which it was encapsulated. Thus, Figure 1-11 shows that routers process the Link and Physical layers. However, unlike a repeater or bridge, a router terminates both in its bid to obtain access to the inner Network layer packets, which are reframed and sent on the egress link.

Some key characteristics of each device are described in the following subsections.

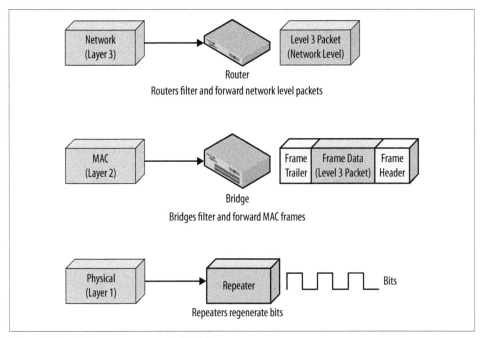

Figure 1-11. LAN interconnect and the OSI

Repeaters

As noted, a repeater is a Physical layer device that regenerates the signal to compensate for transmission losses and distortion. There was a time when repeaters were very common, as bridges and routers had not yet been invented, or were very slow *store-and-forward*-style boxes that were expensive in terms of both money and performance impact.

10Base-T and its related hubs were initially multiport repeaters. Such a repeater was sometimes call a *bus in a box*, given that it extended the collision domain and provided no filtering. Besides extending distance, repeaters also improved reliability through their partitioning functions. This feature would logically disconnect a segment that had a continuous carrier, a condition that could otherwise lock out the entire LAN.

Repeaters were associated with the handy 5 4 3 rule of thumb that indicates the maximum size of an Ethernet CD should consist of no more than five segments interconnected by four repeaters, of which only three of those segments can be multipoint; the remaining two are supposed to be repeater-only P-to-P links. These limits were intended to prevent issues with repeaters robbing bits from frames to the point where they would be deemed corrupted, and the added delays of longer transmission paths that impacted CD as a function of minimum transmission time. Repeaters propagate collisions, and therefore prevent FD operation.

The widespread use, and therefore low cost, of Ethernet switching has made the repeater a thing of the past. Who wants a bus in a box when you can get a switch in a box for the same price?

Bridges

A bridge is a device that operates at Layer 2 and provides a filtering and forwarding function. Historically, bridges were store and forward, which meant the entire frame had to be received before it could be processed; the result was high latency and low packets per second (pps) rates. Modern bridges operate in a cut-through mode, which means they can begin processing and even forwarding a frame while it's still being received.

At one point, the higher performance of cut-through was something to brag about, so vendors called such devices *switches*. Advances in silicon and design mean that now everything is switched, making the term *switched* somewhat ambiguous. Because routers and other devices can also operate in cut-through mode, it's best to qualify such statements with the layer in question, as in "I have a Layer 3 switch" when referring to a router.

Ethernet bridging is transparent. This means that end stations are not aware of the bridge, and therefore they take no special steps to go from their local segment to a remote one. The bridge listens to all traffic, learns the source MAC addresses, and then filters and forwards based on the destination MAC address. The result is that known unicast traffic is only sent out of the port that leads to that device. Other stations do not see this traffic, which allows them to make their own transmissions at the same time, thereby increasing overall throughput.

Bridges used to be expensive; historically, they tended to offer only a few ports, and only some offered WAN interface support to offer *remote bridging*. Bridges terminate collision domains, which is what makes FD Ethernet, and remote bridging, possible. Because multiple stations can simultaneously transmit on different bridged segments, a bridge can also dramatically improve performance. In theory, a two-port bridge takes one big segment from 10 Mbps to two smaller ones, *each* with 10 Mbps.

Protocol-agnostic

Bridges operate at Layer 2. A bridge can support Ethernet, or Token Ring, or both in some manner or another. There is no such thing as an IP bridge. IP is inside a frame, and bridges do not make it that far. In similar fashion, there is no such thing as an Ethernet router. Ethernet, with its flat MAC addressing, is not routable and never will be. Figure 1-12 shows a bridge in operation.

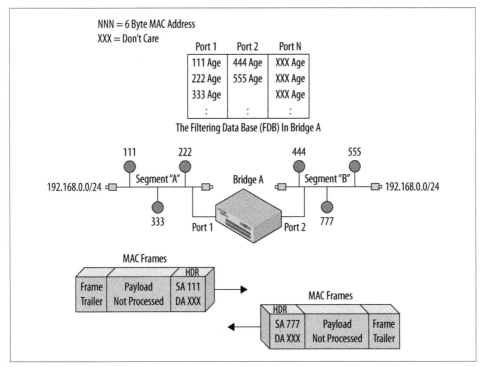

Figure 1-12. Bridging

Key in Figure 1-12 is the conveyance of entire frames, based on a MAC address database. Note that the payload can be any Upper Layer Protocol (ULP) and is not processed or inspected in the act of bridging. Also of note is that being transparent, the end stations are not aware of the bridge. As such, all stations believe they are on the same network from a Layer 3 perspective. This is shown by virtue of the *same* IP network 192.168.0.0/24 appearing on *both sides* of the bridge.

Loops are bad, really, really bad

Given that the bridge bases its forwarding decisions on the MAC address, and that unknown MACs are flooded, or sent over all links except the link on which the frame was received, there is a definite issue with redundant/parallel links in a bridge's environment. If both links are forwarding, such flooded traffic repeatedly ping-pongs back and forth. This is bad enough at Layer 3, but given that Layer 2 has no TTL mechanism to limit the number of iterations, the entire Layer 2 network can crash as a result of a *broadcast storm.*

Given the fatal result of a loop, it's clear that transparent bridging, by its nature, must operate across a set of links that are known to be loop-free. To ensure that this was always the case a specific protocol known as *Spanning Tree* was invented. Spanning Tree Protocol (STP) operates between bridges and uses a number of parameters to guarantee that only one forwarding path is operational between all endpoints (we cover STP in detail in Chapter 6). STP is optional, but most bridges use it. STP simply blocks redundant links. If there is no redundancy there is nothing for STP to do, and it's therefore not needed.

Bridge processing in detail. This section describes the basic operation of a transparent bridge. We cover most of these processing actions in more depth in subsequent chapters. Our goal here is to provide you with a big-picture overview to ensure that you keep sight of the forest despite all those trees.

Listening

An STP-capable bridge first listens to determine the presence of other STP entities, and to determine whether, based on their parameters, the local port should be blocked or should transition to the learning state.

Blocking

Based on the exchange of STP messages the port may be blocked. In the blocked state it does not learn, or forward. It continues to listen for STP messages to catch the case where a change in topology allows the local port to become unblocked.

Learning

After leaving the listening or blocking state, a bridge port spends some time cramming for its big test by learning as many MAC addresses as it can. Bridges learn by promiscuously monitoring all traffic, and inspecting the Source MAC (SMAC) of each frame it sees. The goal of the learning delay is to prevent the flooding that occurs when a bridge has to forward an unknown MAC address. Learning is an ongoing process, and is also used to catch MAC moves, which is when an SMAC is found to ingress on a different interface.

Filtering and forwarding

After building its MAC address table through promiscuous monitoring of all traffic, the bridge then filters and forwards traffic based on the Destination MAC (DMAC) address. The bridge inspects the DMAC of each frame it sees. When the DMAC is associated with a learned SMAC entry on the same port, the frame is filtered, as the target station should already have a copy. When the DMAC of the received frame is associated with a learned SMAC on a different port, the frame is forwarded out of that port only. When the DMAC is unknown, the bridge has to *flood* the frame, which is the act of sending a copy out of *all ports*, *except* the ingress. Ongoing flooding can impact network performance. The assumption is that as a result of the flood, the target station will reply, which then allows its SMAC to be learned so that subsequent traffic can be filtered and forwarded. This is often referred as *transparent bridging*.

Books on bridging usually walk through the transparent bridging and source route bridging models. In today's networks, switches generally implement transparent bridging, as source route bridging was usually reserved for Token Ring networks, which you will have a hard time finding these days.

So much for the 80/20 rule. The rule of thumb used to be that a well-designed LAN had 80% of its traffic local to that segment/CD, and the other 20% switched to a remote location. The goal was to avoid the painful performance hit associated with bridges back in the day.

A modern star-wired Ethernet uses a switch with P-to-P-oriented link types to form a star-wired LAN. The effect is a shift to a *0/100* rule, and the switched rather than repeated nature means that each station is offered dedicated throughput based on its access speed. Thus, a 24-port GE switch can, in theory, offer each attached device a *dedicated* FD GE transmission link, and the aggregate throughput of the switch could be 48 Gbps! (That's 24 Gig × 2 for FD.) I'm not trying to sound cranky, but back before the Great War we were happy with a *shared* 10 Mbps of *HD* throughput, and that wasn't even guaranteed, given that collisions could bring that to zero if a large number of stations were attached.

Routers

Folks started shifting from repeaters to bridges as the technology matured. A few problems began to surface with large bridged internetworks. Some of these were:

- No tolerance for loops and poor use of redundancy as only one path can be active.
- No TTL mechanism, making loops fatal and causing a need for loop restriction.
- No traceroute or ping. Troubleshooting a Layer 2 network is difficult, as things either work end to end, or they don't. Transparency is great, until something is broken.
- Not well suited to dissimilar network types/no fragmentation support.
- No explicit CoS mechanism (until 802.1p), and then not finely grained.
- No IP layer processing, and therefore no firewall filters or IP services such as IPSec or NAT.
- Poor performance when there's a lot of Broadcast, Unknown, or Multicast (BUM) traffic, flooded over all ports, much like a (slow) repeater.

Routers operate at Layer 3 and were seen as a solution to the aforementioned issues. The fact that routers are specific to each Network layer protocol created the need for Multiprotocol Routing (MPR) when more than one Network layer protocol was in use. Some protocol stacks were unroutable, as they either had no network layer, or in the case of SNA were connection-oriented and therefore had endpoint addressing present only during connection setup.

Figure 1-13 shows a router at work.

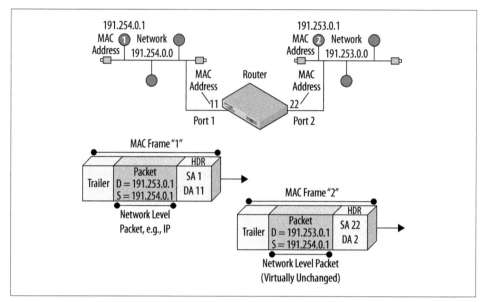

Figure 1-13. Routing

Key in Figure 1-13 is the fact that the same network packet is passed through the router, and that the router attaches to different networks. The Link layer is used to get packets to and from the router; at ingress the frame is stripped, and at egress a new one is built. The IP packet is processed by the router, but that packet flows all the way to the end-point. The only change at the IP layer is a decremented TTL field, and the resulting updated header checksum.

Note that in the routed model the frame's MAC layer always identifies a *next hop* on the *local link*. As a result, the frame on the first hop identifies the router's MAC address, rather than that of the target station. ARP is not routable, and as such a station never attempts to ARP a remote destination. Note that if Figure 1-13 were based on bridges rather than routers, the two stations would be on the *same network* by definition. As such, the ARP exchange would pass through the bridge and operate end to end between the IP endpoints.

Multi-Protocol Routing

As routers began to get faster, the benefits of routing began to take their toll on bridging. The mantra *Route when you can, bridge when you must* developed to show this senti-ment. In some cases, you had to bridge due to lack of routing support for that protocol on that router, or because the protocol itself was unroutable. The trend was to keep unroutable protocols local, to provide workgroup types of connectivity, and to only route across the WAN for backbone connectivity. Although this required more work to configure and set up, routers simply had too many advantages over bridges when all was said and done, especially at a large scale or in complex types of connectivity.

Routed Versus Routing

There is often confusion between a *routed* protocol such as IP and a *routing* protocol such as RIP. The former involves packets that transit routers based on a longest match against their destination address. In contrast, the routing protocol populates the router's routing table based on the dynamic exchange of routing protocol messages. Most IP routing protocols actually ride inside IP, but generally speaking they are not themselves routed in that they are sent only to the next neighboring router. Thus, a router uses its directly connected links to learn how to reach directly connected neighbors, and with this reachability a routing protocol then exchanges messages to convey knowledge about remote destinations. The routers select what they feel is the best path for each destination and install that entry in their routing table. You can still route IP without an IP routing protocol; you just need a single router such that everything is directly connected, or the administrative burden of trying to maintain full and logical connectivity using static routes.

One protocol to rule them all. MPR became a drag, and even more so with the addition of each new routable protocol to an existing backbone. Although technically feasible, the world's shift to IP transport did not come quickly enough to spare some from the resulting torment. Can anyone remember running Cisco's EIGRP to simultaneously support IP, IPCX, and DDP? Or how about the alternative that had you run IP RIP, IPX RIP, and AppleTalk's RTMP simultaneously to achieve the same thing (three separate routing tables, one for each protocol)?

The need to route AppleTalk is gone, and IBM's SAA is now based on IP transport. Their respective applications live on, but they're now provided via an IP infrastructure. The result is a renewed focus on all things IP, both in the routing and in the LAN switching fronts.

The migration to an IP-based world has led to some interesting side effects, such as bridges that are *IP-aware*. Such a device can inspect the IP payload of certain traffic to perform a service, such as validating an ARP request based on a previous Dynamic Host Configuration Protocol (DHCP) exchange. Although a violation of the layer principle, such capabilities are common for devices that can process at both Layer 2 and Layer 3, such as is the case with the Juniper EXs. You should avoid the temptation to refer to such a device as an *L Brouter*, however. The world already went there once, and it wasn't fun. Many set off on the quest to find the definitive definition of that mythical beast; most have yet to return.

LAN Interconnect Summary

The modern state of LAN interconnect is all about IP routing over wide areas and Ethernet-based switching in the campus or workgroup. Most IP switches are also capable of IP routing. This gives you the freedom to bridge when you want and to route when you want, using the same device and in some cases the same interfaces. Talk about the best of all worlds: high-performance switching with IP-enabled features in the workgroup, with a high-performance routed IP backbone for communications with remote groups.

Conclusion

The world has a long history with LANs and the resulting desire to interconnect LANs over both local and remote areas. Many a failed technology is strewn along the roadside, but we seem to have arrived at a good place.

IP is the internetworking protocol of choice. Ethernet is the LAN technology of choice. Bridging and routing allow you to interconnect at the Ethernet or IP layers, as your needs dictate.

This section provided a review of Ethernet and IP technology, as well as that of LAN interconnect, with more than a bit of cranky commentary along the way as to how we got there. You should now be well prepared to launch into the remainder of this book, which is focused on the use of Juniper Networks' EX LAN switches in a Layer 2 environment.

Chapter Review Questions

1. Which is true regarding 802.3?
    a. Uses a Length field and LLC
    b. Uses a Type field and LLC
    c. Does not use LLC
    d. Defines a Token Ring LAN
2. What is the first layer with end-to-end flow?
    a. Layer 1
    b. Layer 2
    c. Layer 3
    d. Layer 4
3. True or False: a bridge extends a LAN's collision domain.
4. What is the role of ARP?
    a. Permits Layer 2 and Layer 3 address independence
    b. Is used in Ethernet only
    c. ARP is not required in a bridged network
    d. All of the above

5. How many networks and how many hosts can you address with a 200.0.0.0/30 address?

   a. 1/2

   b. 2/2

   c. 10/10

   d. 200/200

6. What class is the IP address 128.69.0.0/24?

   a. Class B with a Class B mask

   b. Class B with a Class C mask

   c. Class A with a default mask

   d. Class C with a default mask

7. What type of media is common in a star-wired bus?

   a. Thick coax

   b. Thin coax

   c. UTP

   d. Fiber

   e. Both C and D

8. In the indirect delivery model, the source station sends an ARP for which of the following?

   a. The target host

   b. The next hop, which is a default gateway

   c. The next hop, which is a bridge

   d. None of the above; ARP is not needed for indirect communications

9. What protocol provides a reliable transfer service?

   a. IP

   b. ICMP

   c. UDP

   d. TCP

10. What type of error message does a *bridge* send when it needs to fragment but is unable to do so?

   a. An ICMP fragment needed error

   b. An ICMP destination unreachable error

   c. An IGMP fragment needed error

   d. None of the above; it does silent discard and cannot generate any errors

11. Which author wrote this chapter?
    a. Opinionated "Hot-Head Harry"
    b. Just-the-facts "Diplomatic Doug"

Chapter Review Answers

1. Answer: A. 802.3 LANs use both a length code to provide padding at the MAC layer, and LLC to identify the upper layer.

2. Answer: C. IP, a Network layer protocol, is the first with end-to-end scope. Although it is end to end, IP is acted upon at each routed hop. TCP, a Transport layer protocol, goes end to end and is generally processed only by the endpoints.

3. Answer: False. Unlike a repeater, a bridge terminates the CD.

4. Answer: A. ARP is used in all LANs and some WANs, as well. It's needed because of the independence between Layer 2 and Layer 3.

5. Answer: B. The address provided *is* a network address, so only one network can be addressed as 200.0.0.0/30. On that network you can have a host 1 and a host 2, given there are two host bits to work with. The combinations of all 0s and all 1s are generally not permitted. You could use VLSM to further subnet—say, to a /32, which would yield two networks with the same number of hosts, assuming there is support for /31 addressing on that device.

6. Answer: B. The 128 in the high-order octet indicates that this is a Class B address. Such an address has a default network mask of /16 or 255.255.0.0. This is a sub-netted Class B that is using a Class C mask. As the whole notion of classes is gone, it's best to just say this is a prefix with a 24-bit mask.

7. Answer: E. Coax cable was rarely used P-to-P, and was a true multipoint bus. UTP and fiber mandate P-to-P-type links, on the other hand.

8. Answer: B. ARP is always used for IP over LANs. In indirect delivery, the next hop is the router/default gateway, so that is the MAC address that is resolved via ARP.

9. Answer: D. Of the others listed, only UDP is a Transport layer, and it provides best-effort service.

10. Answer: D. This is a trick question, perhaps. Ethernet bridges are transparent. If it were to send an error message, it might blow its cover.

11. Answer: A. Yes, it was Harry. But he is old enough to have formed such opinions, being one of the few who actually spent time learning OSI and writing classes on Token Ring. If anyone is interested, he can sell you all of his OSI and Token Ring/SRB materials cheap.

## CHAPTER 2

EX Platform Overview

Juniper Networks' long-awaited entry into the Ethernet switching market began on March 31, 2008, with the release of the EX3200 and EX4200 Ethernet switching platforms. To date, Juniper Networks has been notable for its wide range of high-performance IP routing platforms that share a common JUNOS Software code base. Although Juniper's hardware and application-specific integrated circuit (ASIC) design prowess should not be underestimated, many consider JUNOS to be the real mojo behind the company's success. JUNOS software has been field-tested and proven robust in the largest service provider networks on the planet. Add in its unique usability features and the benefits of a single code train and you can begin to understand why JUNOS enjoys such brand loyalty among its users.

This chapter details the hardware design and general capability of the EX platforms, and introduces foundation concepts of JUNOS software and the associated command-line interface (CLI). A detailed discussion of Juniper's routing platforms and IP routing in general is beyond the scope of this book. Readers interested in these topics can consult product documentation, training materials, or any of the numerous books published on the subject, including this book's companion volume, *JUNOS Enterprise Routing*, by Doug Marschke and Harry Reynolds (O'Reilly). For information on general routing in a JUNOS software environment, check out the JNCIA-M, JNCIS-M, JNCIP-M, and JNCIE-M study guides, which are currently available from Juniper Networks in PDF format as a free download at *http://www.juniper.net/training/certification/books .html*.

To help put the EX product line into perspective, here is an overview of the current JUNOS-based routing and switching product lines:

J Series routers to include the J2300, J2320, J2350, J4350, and J6350
> The J Series routers are software-based platforms that offer predictable high performance and a variety of flexible interfaces that deliver secure, reliable network connectivity to remote, branch, and regional offices.

M Series routers to include the M7i, M10i, M40e, M120, and M320
> The M Series multiservice edge routing platforms provide advanced IP/Multiprotocol Label Switching (MPLS) edge routing services at scale. These ASIC-based

platforms offer a wide range of interfaces and service capabilities with throughput ranging from 5 to 320 Gbps.

T Series routers to include the T320, T640, TX Matrix, and T1600

The T Series core routing platforms offer ASIC-based forwarding performance, an extensible design, and numerous carrier-class reliability features. A single T640 can offer up to 320 Gbps of throughput (640 Gbps full duplex, or FD), and can be upgraded to a T1600 with 1.6 Tbps of capacity, or clustered as part of a TX matrix to scale up to 2.5 Tbps!

MX Series Ethernet routers to include the MX240, MX480, and MX960

The MX platforms are focused on Ethernet-centric services within carrier networks. The MX960 offers up to 960 Gbps of switching and routing capacity with up to 480 Gigabit Ethernet ports and 48 ports of 10 Gigabit Ethernet per system.

The EX3200, EX4200, and EX8200 series Ethernet switch portfolio from Juniper Networks represents a new era in networking. This family of high-performance, carrier-class networking solutions is designed to address evolving business requirements while enabling a secure, reliable network that's ideal for today's converged network deployments. EX technology supports low-cost fixed configurations, and grows as you need Virtual-Chassis options and a carrier-class Terabit Chassis model. The EX switches run the same JUNOS software train as found powering the world's largest routing systems, and with that come the ease of use and reliability for which Juniper has become famous.

 As a general note, the term *enterprise network* is used to describe a network that serves the communication needs of its owner, rather than the needs of the network owner's customers, such as is the case in a service provider's network. Although it's safe to assume that a typical service provider's network is both large and complex, such assumptions prove troublesome in the case of the enterprise network. Here we could be dealing with scales that range from a PC and printer interconnected via local cabling, all the way to a multinational entity whose network may exceed the node count and general complexity of a Tier 1 service provider's network.

Juniper currently positions the J Series and the M7i/M10i platforms as enterprise-level devices, but a large enterprise network could justify the performance of a T Series platform. In fact, the reverse is also true in that a traditional service provider network may find an appropriate need and use for platforms designated as low-end enterprise gear— for example, as part of a managed CPE service or in a performance monitoring or route-server role. In a similar fashion, the new EX Series of switches, both in the smallest fixed configuration or as part of a large Virtual Chassis (VC), are expected to be found in both provider and enterprise networks alike. Given the ever-increasing popularity of all things Ethernet, and the seemingly endless need for increased connectivity and applications with larger bandwidth demands, the future seems bright for high-performance switching gear.

So, whether you operate a modest enterprise or a gargantuan service provider network, the chances are good that sooner or later you will be exposed to Juniper's routing and switching equipment, and likely on several different hardware platforms. The saving grace is that JUNOS software is pulled from a single train of code with a common feature base that is largely shared across all platforms. So, regardless of your actual hardware platform, there is a single version of software code to load, and perhaps more importantly, to *learn*. Having a single code train has lots of *hidden* benefits, such as stability, ease of expandability, and lower operational costs. Once you get used to a consistent set of features and configuration syntax across a multitude of platforms, it's hard to go back. Especially when that single code base has more than 10 years of proven, rock-steady performance in the largest IP networks ever deployed.

The topics covered in this chapter include:

- EX platform overview and general capabilities
- Hardware architecture and packet flow
- JUNOS software overview
- CLI overview
- Other cool CLI features and capabilities

EX Hardware Overview

Juniper Networks entered the switching market with the release of the EX3200 and EX4200 platforms. Both switches offer up to forty-eight 10/100/1,000 Gigabit Ethernet ports and support for an optional 4× Gigabit Ethernet/2×10 Gigabit Ethernet uplink module. The EX3200 supports a field-replaceable power supply unit (PSU) and a fan tray with a single blower. You can also use an optional remote power supply (RPS) to provide redundant power.

The EX3200 is typically deployed in the access layer or for small-scale LAN deployments, but still packs some serious forwarding performance: the 24-port models offer 88 Gbps of throughput and forward Layer 2 traffic at wire rate, on all ports, clocking in at a respectable 65 Mpps. The 48-port version bumps these numbers up to 136 Gbps and 101 Mpps!

Figure 2-1 shows the EX3200 chassis front and rear panel connectors.

The EX4200 platform adds carrier-class reliability and VC clustering capabilities, allowing it to scale up to 480 Gigabit Ethernet and twenty 10 Gigabit Ethernet ports, while consuming only 10 rack units (RUs) and providing the ease of management associated with a single network entity. Each standalone EX4200 supports redundant load-sharing PSUs and a multiblower fan module, both of which can be hot-swapped in the field. As with the EX3200, the remote PSU option is also supported on the EX4200.

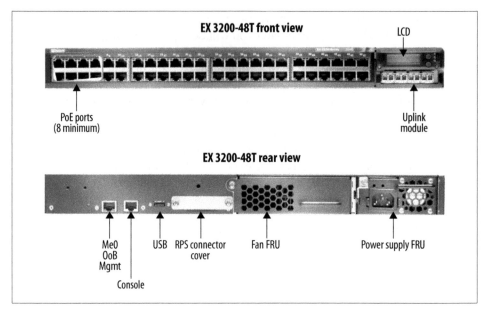

Figure 2-1. The EX3200 chassis

When part of a VC, each chassis is linked by a virtual backplane that supports a ring topology that allows survival of single-backbone cable faults. The VC backbone can operate at up to 128 Gbps (64 Gbps FD). In addition to the standalone redundancy features, a VC configuration supports redundant routing engines (REs) with 1:*N* redundancy. This is because any EX4200 chassis in a VC can become a backup (and therefore could be elected as the master) RE, such that in a three-chassis VC there will be a master and a backup RE, as well as one chassis functioning as a Line Card (LC) that is awaiting its chance to become the backup RE, in the event of a failure of the current master or backup RE.

When the platform redundancy and VC features are coupled with JUNOS software capabilities such as Graceful Routing Engine Switchover (GRES), In-Service Software Upgrades (ISSU), Graceful Restart (GR), and Non-Stop Routing (NSR), it's easy to see that the availability of a VC (or redundant 8200 series) can approach "five 9s." EX support for High Availability (HA) features in the JUNOS Software 9.2 release, along with general design principles, is detailed in Chapter 11. Complete details on EX VC technology are provided in Chapter 4.

Several significant JUNOS HA features are not supported for EX switches in the 9.2 release. For example, NSR is not supported, and by extension neither is ISSU. Currently, EX HA software features include GR for Layer 3 routing protocols (Border Gateway Protocol [BGP], Open Shortest Path First [OSPF], and Intermediate System to Intermediate System [IS-IS]) only. GRES is supported in a VC or with an EX8000 series with redundant REs.

The EX4200 is often deployed in the distribution layer of large-scale LAN designs. In this capacity, it terminates multiple feeds from access layer switches and aggregates them onto high-speed uplinks for transmission into the core. The EX4200 provides the same wire-rate throughput and switching capacity as the EX3200, but also includes two 54 Gbps Virtual Chassis Port (VCP) ports for switching over a VC. When in a VC, each member switch is still capable of local switching at wire speed. The result is that a 10-member VC in which *all* traffic is *locally* switched provides a total switching capacity of 1.36 Tbps (10 × 136 Gbps), with an aggregate throughput of 1.01 billion packets per second!

Figure 2-2 shows the EX4200 chassis front and rear panel connectors.

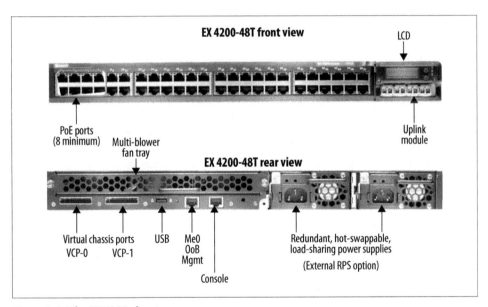

Figure 2-2. The EX4200 chassis

Both fixed-configuration switches offer high port density, full or partial Power over Ethernet (PoE), a rich set of Layer 2 bridging capabilities, robust IP routing, and hardware-based Layer 2/Layer 3 security features built into the base software license. When coupled with its industry-leading wire-rate forwarding performance and the proven track record of JUNOS software, it would seem that Juniper was serious about its entry into the Ethernet switching market.

It's worth noting that both the EX3200 and EX4200 chassis support the same uplink and PSU part numbers, which greatly simplifies sparing. It's suggested that when PoE is used anywhere in the network, all spare PSUs should be sized according to the largest need. You can always install a 930-watt supply in a chassis that does not require that much power, with no ill effects. Later, when the failed supply is repaired, the higher-capacity PSU can be returned to the spare pool. In contrast, inserting a power supply with insufficient wattage results in deactivation of PoE on as many ports as needed to remain within the switch's available power budget.

You can mix AC and DC PSUs on the EX4200. However, currently the DC PSU cannot provide power to PoE ports. This capability may change in the future and is based on the belief that DC power is typically used within a service provider's network, or in data centers, where PoE for end-user devices is not generally required.

The EX8200 Series

Not wanting to rest on its laurels, Juniper soon followed this one-two EX punch with the high-performance, fully redundant, chassis-based EX8200 series. The 8200 series represents some serious switching iron designed for the most intense data center and high-capacity backbone environments. The EX8200 is expected to ship with JUNOS Software 9.4.

The 8208 offers eight 200 Gbps I/O slots in a 14-RU footprint, while the 8216 doubles that number in only 21 RUs of space. Note that in FD terms, each I/O slot represents 100 Gbps of FD capacity. As such, both models are 100 Gigabit Ethernet ready, with the EX8208 supporting 384/64 1 GE/10 GE ports and the EX8216 sporting up to 768/128 1 GE/10 GE ports. The 8200 series switches are normally deployed in the core of a large-scale LAN deployment. The 8200 series EX switches are based on a 1.2 GHz PowerPC platform and ship with 1 GB of flash memory that can be upgraded to 4 GB.

Figure 2-3 shows the front of an EX8216 switch.

The EX8208 has eight dedicated LC slots, two switch fabric/RE slots, and one switch fabric-only slot on the front panel. The EX8208 uses a single fan tray that provides side-to-side airflow, and six power supply bays located on the front of the chassis base. The base model ships with two PSUs, one switch fabric/RE module, and one fabric-only module. A fully redundant configuration adds a second switch fabric/RE module, and fully loads all six power supply bays.

The EX8216 has 16 dedicated LC slots, 2 route engine slots on the front of the chassis, and 8 switch fabric slots at the rear of the chassis. The 8216 incorporates two fan trays (side-to-side airflow), and six power supply bays at the front base of the chassis.

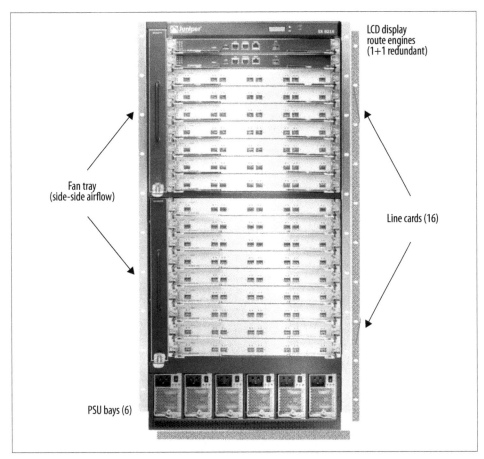

Figure 2-3. The EX8216 chassis

The EX8200 series switches offer fully redundant REs, switching fabric, power supplies, and cooling to maximize reliability and uptime. And, as already mentioned, all EX switching platforms run the same JUNOS software to provide consistent configuration and capabilities across the entire switching line, and given the single JUNOS software train, EX switches are also consistent with much of the routing line as well!

In the initial offering, the EX8200 series does not provide PoE options. This is in keeping with its targeted market of data centers and service provider networks where PoE is typically not required. Future versions may support PoE as customer needs dictate.

Figure 2-4 provides a typical campus network that is based on current best practices for both scalability and reliability.

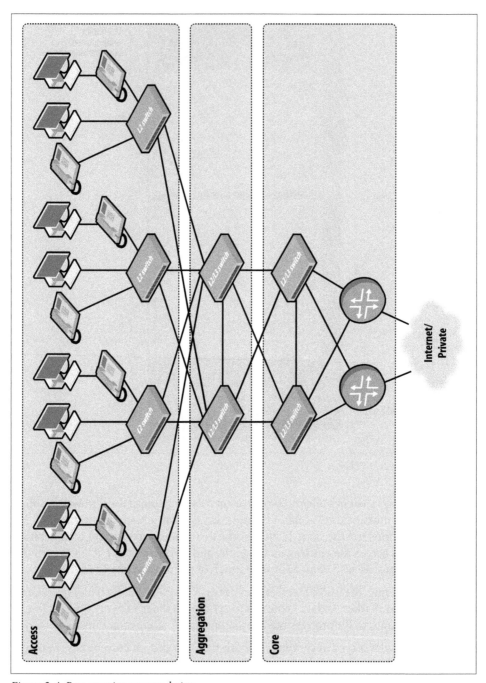

Figure 2-4. Best practice campus design

The important thing to notice in Figure 2-4 is the presence of three distinct network partitions in the form of access, distribution, and core. Within each partition, specific hardware and protocols are deployed to optimize performance and reliability, while also reducing costs.

The access layer is often based on a Layer 2 solution using low-end switches with limited redundancy. Layer 2 access control, user authentication, flow-based rate limiting/policing, and class of service (CoS) functionality are typically performed at the network edge.

The distribution layer provides intra-access switching and aggregation of traffic into the core. Distribution devices often integrate Layer 2 and Layer 3 functionality, providing switching for the access layer and routing toward the core. This is a key point given that routed networks, although more complicated, have many advantages over a pure Layer 2 design; this design keeps the edge simple and adds Layer 3 complexities where maximum benefit is achieved. The distribution layer is typically tasked with Layer 3 security and services, that is, firewall filters and Network Address Translation (NAT), as well as CoS-aware high-speed switching between the access and core layers.

The core layer is typically based on high-speed Layer 3 routing between distribution layer devices, and to external network attachments. In most designs, the core layer is not responsible for securing or policing individual flows, as these functions are best suited to edge layer devices. Core devices tend to act on aggregate bundles, or traffic classes, with regard to policing or CoS-related processing actions.

Both the core and distribution layer devices need to have HA given their critical role in the network. Generally speaking, network HA is influenced by hardware and software reliability/redundancy within each node, and the presence of redundant network connections between these nodes.

Separate Control and Forwarding: It's a Good Thing

As with other Juniper Networks routers, EX platforms share the same design philosophy of a clean separation between the control and forwarding planes. Such a design provides protection from unanticipated loads in either plane, and also accommodates technology-specific solutions for the equally difficult but orthogonally opposed problems associated with running modern, complex signaling and routing protocols (control plane), while simultaneously forwarding and oftentimes touching (i.e., altering a packet's CoS marking) large numbers of packets (data plane).

EX platforms facilitate this divide in software through a mix of built-in rate limiting and access control lists (ACLs), an approach that has proven successful in the J Series product line. This virtual separation helps to ensure that devices continue to function and are reachable for corrective actions, even during abnormal levels of control or data plane activity, whether the result of a network malfunction or configuration error, or due to an intentional denial of service (DoS) attack. Figure 2-5 shows the general design of Juniper Networks routers and switches.

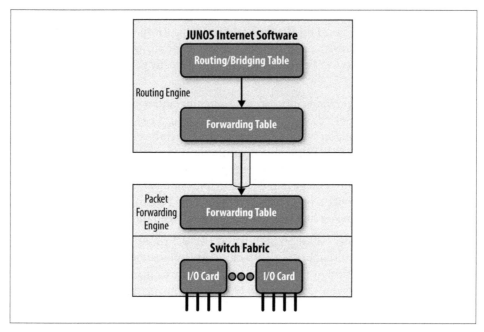

Figure 2-5. The separation of control and forwarding

As with Juniper routing platforms, in the EX Series the control plane is instantiated by an RE running JUNOS software, while the forwarding plane is an ASIC-based entity referred to as the EX Packet Forwarding Engine (EX-PFE).

Figure 2-5 shows how the RE runs JUNOS software, which provides the CLI and general management access and troubleshooting tools (ping, traceroute, etc.), and maintains the master copy of the routing/switching table, which is built through static configuration or by dynamic Layer 2/Layer 3 bridging/routing protocols. The RE also maintains and monitors the PFE, and keeps its copy of the routing/switching table current as network conditions change.

Meanwhile, the PFE lives to forward frames (or packets when operating in Layer 3 mode) as fast as it can. It faithfully accepts the next hop forwarding instructions and performs the required binary searches to quickly locate the Media Access Control (MAC) address or longest-match IP address, so packets may be quickly dispatched toward their networking fate. In keeping with the "Performance Without Compromise" slogan touted for Juniper's routers, the EX switches implement security and CoS-related services in PFE hardware at wire speeds.

This design means that the RE is never directly involved in packet forwarding (i.e., there is no process switching), which ensures that sufficient resources are available for actual control functions such as processing a Spanning Tree Protocol (STP) or OSPF routing update. Traffic that is received by the PFE, but that needs to be processed by the RE, is termed *exception* traffic, in that it must be sent to the RE for additional processing. Exception traffic is subjected to rate-limiting and access-control features that ensure that the RE and its communications link to the PFE are not overrun during periods of excessive exception traffic, such as might result from a DoS attack, or by unintended address learning churn resulting from a Layer 2 loop.

As a practical example, consider the ability to issue the equivalent of a `debug all` command in a production network, without actually degrading transit and local processing performance!

On some vendors' gear, such a command could easily lead to additional network disruption stemming from the local device's inability to maintain forwarding rates, given the added processing burden imposed by packet debug. Adding a new problem never makes troubleshooting the original issue any easier, and is why the user is often warned *against* using such debug commands in production routers. A tool that you cannot risk using due to unpredictable impact to network operation is a tool not worth owning. This situation does not arise on Juniper switches, which is what the separation of control and forwarding is all about, and why no such admonitions exist in the EX documentation regarding `traceoptions`, which is equivalent to `debug` in the JUNOS world.

Just because you can does not mean you should. Best practice dictates that no device should be unnecessarily burdened with superfluous processing. Besides making sense, this helps to guard against the *camel's back* syndrome, in which each added processing burden has no ill effect, until the camel's back finally breaks, spilling packets all over the desert sand.

Although you can trace all protocol activity in a production network given JUNOS safeguards and design characteristics, this does not mean you should leave such tracing in effect after it's no longer needed. The rule of thumb here is to jump in, set up tracing, figure out what is wrong, fix it, confirm the fix, and then remove the now unneeded tracing.

EX Hardware: The Numbers

The previous section provided a general description and overview of the EX switching line. This section summarizes the system feeds, speeds, and general capabilities of each EX model.

We begin with Table 2-1, which details supported port types, counts, and related power draw with and without PoE for the EX3200 and EX4200 switches.

Table 2-1. EX3200 and EX4200 ports and power draw

Switch	Ports	Port type	PoE ports	Max power (with PoE)	Uplink option
EX3200	24	10/100/1000Base-T	8	190 (320) W	4-port Gigabit Ethernet (GbE) (SFP)
					2-port 10 GbE (XFP)
	24	10/100/1000Base-T	24	190 (600) W	4-port GbE (SFP)
					2-port 10 GbE (XFP)
	48	10/100/1000Base-T	8	190 (320) W	4-port GbE (SFP)
					2-port 10 GbE (XFP)
	48	10/100/1000Base-T	48	190 (930) W	4-port GbE (SFP)
					2-port 10 GbE (XFP)
EX4200	24	10/100/1000Base-T	8	190 (320) W	4-port GbE (SFP)
					2-port 10 GbE (XFP)
	24	10/100/1000Base-T	24	190 (600) W	4-port GbE (SFP)
					2-port 10 GbE (XFP)
	24	100Base-FX/ 1000Base-X	N/A	190 (no PoE)	4-port GbE (SFP)
					2-port 10 GbE (XFP)
	48	10/100/1000Base-T	8	190 (320) W	4-port GbE (SFP)
					2-port 10 GbE (XFP)
	48	10/100/1000Base-T	48	190 (930) W	4-port GbE (SFP)
					2-port 10 GbE (XFP)

The key takeaway from Table 2-1 is that the EX3200 and EX4200 switches come in 24- and 48-port varieties, and support either partial or full PoE. The number of PoE ports in turn determines the wattage demands placed on the power supply. Table 2-1 assumes a worst-case maximum PoE power draw of 15.4 watts per PoE port.

> The EX3200 and EX4200 support the same power supply options. However, the number of PoE ports is a fixed parameter that cannot be increased with a larger power supply. This means there is little point in permanently installing a 930-watt power supply in an eight-PoE port chassis, given that the additional power budget does not increase the number of PoE ports, which remains fixed at eight.
>
> EX switches disable PoE ports when the pool of available power is insufficient. This means that installing a smaller power supply in a full PoE chassis will reduce the total number of PoE-enabled ports.

The EX4200 has two rear-panel 64 Gbps (FD) VCP connectors used to form VC clusters with 128 Gbps of interchassis bandwidth. The EX4200 also offers an all-SFP-based (all

Small Form-factor Pluggable-based) 100/1,000 Mbps Ethernet chassis option intended for environments with preexisting fiber, or when maximum distances are required. On the EX4200, the uplink module can also be used to form a VC over extended distances.

Here are the performance details for supported optics:

Gigabit Ethernet SFP

- 1,000Base-T SFP (copper)
- 100 FX: 1310 nm, 2 km
- SX: 850 nm, 220 m on 62.5 µ FDDI-grade fiber; 550 m on 50 µ multimode fiber
- LX: 1,310 nm, 550 m on multimode fiber; 10 km on single-mode fiber
- LH: 1,550 nm, 70–80 km on single-mode fiber

10 Gigabit Ethernet Small Form-Factor Pluggable (XFP)

- SR: 850 nm, 300 m reach
- LR: 1310 nm, 10 km
- ER: 1,550 nm, 40 km
- ZR: 1,550 nm, 80 km

As with other Juniper products, third-party optics modules *should* work, but these will not be supported in the event of problems. This is in contrast to some vendors that intentionally disable a port when their branded (and often highly priced) optics module is not correctly sensed.

Table 2-2 details the port breakdown for the 8200 series switches.

Table 2-2. EX8200 speeds and feeds

Switch	200 Gbps slots	Module type	Max ports	Type
EX8208	8, 1.6 Tbps	48-port 10/100/1000Base-T	384	RJ-45 (copper)
		48-port 100Base-FX/1000Base-X	384	SFP
		8-port 10 GbE	64	SFP+
EX8216	16, 3.2 Tbps	48-port 10/100/1000Base-T	768	RJ-45 (copper)
		48-port 100Base-FX/1000Base-X	768	SFP
		8-port 10 GbE	128	SFP+

In the initial release, the EX8200 series does not provide PoE options, but this functionality may be added at a later time.

Table 2-3 details hardware redundancy features for the EX switches and indicates whether a component is considered a Field Replaceable Unit (FRU). When FRU is not indicated, the entire chassis must be replaced or repaired to regain functionality.

Table 2-3. EX Series redundancy features

Component	EX3200	EX4200	EX8200
RE	No.	Yes, when in a VC.	Yes, as an FRU.
Switch fabric	No.	Yes, when part of a VC. A VC ring can survive a single VC trunk failure.	Yes, N+1 supported as an FRU.
PSU	Yes, when remote PSU option is in effect. PSU is an FRU.	Yes, redundant hot-swappable PSUs as an FRU; also supports remote PSU option.	Yes, redundant hot-swappable PSUs as an FRU.
Fan	No. Single-blower fan tray is an FRU.	No, but FRU fan tray has three blowers and should survive failure of one blower up to ambient temperature of 40°C/104°F.	Yes, as an FRU. Each fan tray has three blowers.

Both the EX8208 and EX8216 can hold up to six AC power supplies. The EX8208 switch offers 3+3 redundancy at maximum power consumption, meaning that each of the three active PSUs can have a spare. The EX8216 offers 5+1 redundancy at maximum power consumption, which is to say that five active PSUs are backed up by a single spare unit.

For all EX chassis types, the failure of a fan module will generate a chassis alarm, but does not result in operational impact until an excessive temperature threshold is reached. At this point, the switches will power themselves down to avoid ASIC damage. In the case of the 8200, the fan can operate at variable speeds, and will ramp up its speed in an effort to avoid an over-temp-induced shutdown.

Use the CLI to Monitor Environment

Although we have not yet officially covered the CLI and its usage, it's appropriate to point out that you can display the current chassis environment and the yellow/red alarm temperature thresholds using CLI commands, as shown here:

```
lab@Rum> show chassis environment
Class Item                      Status    Measurement
Power FPC 0 Power Supply 0       OK
      FPC 0 Power Supply 1       Absent
Temp  FPC 0 CPU                  OK        39 degrees C / 102 degrees F
      FPC 0 EX-PFE1              OK        42 degrees C / 107 degrees F
      FPC 0 GEPHY Front Left     OK        29 degrees C / 84 degrees F
      FPC 0 GEPHY Front Right    OK        28 degrees C / 82 degrees F
      FPC 0 Uplink Conn          OK        29 degrees C / 84 degrees F
Fans  FPC 0 Fan 1                OK        Spinning at normal speed

lab@Rum> show chassis temperature-thresholds
                        Fan speed      Yellow alarm      Red alarm
Item                  Normal  High  Normal  Bad fan  Normal  Bad fan
FPC 0 CPU                 60    70      80       70      95       85
FPC 0 EX-PFE1             60    70      80       70      95       85
FPC 0 GEPHY Front Left    60    70      80       70      95       85
FPC 0 GEPHY Front Right   60    70      80       70      95       85
FPC 0 Uplink Conn         60    70      80       70      95       85
```

Table 2-4 summarizes important hardware capabilities and general platform scaling limits. These numbers can change as the platforms evolve; it's always best to check the latest documentation.

Table 2-4. EX capabilities and scaling limits

Feature	EX3200/EX4200 series	EX8200 series	Comment
Throughput/PPS	24 ports: 88 Gbps/65 Mpps 48 ports: 1.6 Gbps/101 Mpps	8208: 1.26 Tbps/960 Mpps 8216: 2.56 Tbps/1,920 Mpps	Wire rate on all ports, overbooking possible on VC trunk ports. Throughput cited for the 8200 series is based on 10 GE ports.
Jumbo frames	Up to 9,216 bytes	Up to 9,216 bytes	Enabled per port.
Queues per scheduler	8	8	Queuing supported per port (not per Interface Logical Unit [ifl]. Policing can be used to shape at the virtual LAN (VLAN) level.
Port and VLAN/inet policers	512/512	1,024/1,024	Port/VLAN level and Internet family (Layer 3) policers to limit interface bandwidth; ingress only.
MAC address table	32,000	64,000	Stores learned MAC addresses.
IPv4 unicast/ multicast routes	16,000/2,000	512,000/256,000	Forwarding table entries for IPv4, routing table is limited by RAM.
IPv6 unicast/ multicast routes	4,000/512,000	128,000/64,000	Forwarding table entries for IPv6, routing table is limited by RAM.
Security ACLs	14,000	64,000	Firewall filters for Layer 2 or Layer 3.
Generic Routing Encapsulation (GRE) tunnels initiated/ terminated	2,000	2,000	Used for IP tunneling and remote port mirroring.
Port mirroring sessions	One local, one remote	Seven local, seven remote	Switched Port Analyzer (SPAN)-like function for wiretap/traffic analysis.
Processor/flash	600/1 GHz, 512/1 GB	1.2 GHz/1 GB (upgrade-able to 4 GB)	PowerPC-based, processor-based.
LAG groups/ members	32/8, 64/8	256/12	On an EX4200 VC, a link aggregation group (LAG) can span member switches.

EX Feature Support

As with all JUNOS software-based products, you can expect rapid feature velocity for the EX Series, based on the standard three-month JUNOS release cycle. The tables in this section characterize initial EX Series Layer 2 and Layer 3 feature support. You should always check the current software release documentation set for the latest in features and functionality.

Note that at the time of this writing, EX switches support a soft licensing model for *advanced features*, as described shortly. Licenses are sold per chassis, and only two licenses are required in a VC or redundant RE 8200 series system (i.e., a license is needed on both the active and backup REs, but not on LCs). The soft model results in warnings displayed/logged at commit time, and comments that a license is needed when viewing the related portion of the configuration. Despite the warnings, the licensed feature is expected to work with no restrictions, other than the side effects of your guilty conscience and lack of JTAC support on that feature, should any issues arise later. In the initial release, a single license is available that unlocks all licensed features.

Layer 2 features

Table 2-5 summarizes base Layer 2 functionality currently supported by the EX Series. Subsequent chapters detail what all of this means, and provide configuration and troubleshooting examples. So, for now, consider this a heads-up as to what is coming down the road for you in this book.

Table 2-5. Layer 2 base feature support

Feature	EX platform	Comment
Spanning tree	All	Standard, multiple instance, and rapid STP supported (802.1D, 802.1s, 802.1w)
Redundant Trunk Group (RTG)	All	Provides primary/backup port redundancy with no STP
802.1Q	All	VLAN tagging, support for 4,094 user-assignable VLANs per EX-PFE
802.1X	All	Standards framework for Layer 2 access security; supports port, multi, and VLAN assignment modes
802.3ad, 802.3X flow control, and link aggregation maximum groups/members	3200, 4200, 8200	Link aggregation, dynamic via LACP or static: EX3200 32/8 EX4200 64/8 EX8200 64/12
Layer 2 multicast	All	Internet Group Management Protocol (IGMP) snooping support allows switches to reduce multicast flooding when there are no group members
Dynamic ARP inspection (DAI) and Dynamic Host Configuration Protocol (DHCP) snooping	All	Monitors DHCP exchanges to prevent spoofing, using this information to prevent Address Resolution Protocol (ARP) spoofing by enforcing DHCP IP-MAC binding to ARP exchanges
Generic VLAN Registration Protocol (GVRP)	All	Protocol used to manage VLAN across switches
LLDP, LLDP-MED	All	Link Layer Discovery Protocol, Media Endpoint Discovery; part of unified communication to auto-sense devices (e.g., IP phones)

Layer 3 and general system features

Table 2-6 summarizes general EX system capabilities and Layer 3 feature support. As before, subsequent chapters detail what all of this means, so for now consider this another heads-up. Note that some features are considered advanced and require a license for legitimate usage. Items marked with an * were not included in the initial 9.0 release. Many of the Layer 3 features are JUNOS-enabled, which helps to drive home a key benefit of the single code train model.

Table 2-6. Layer 3 base and advanced feature support

Feature	Platform	Base/advanced	Comment
Basic IPv4 routing: Static, RIP v1/v2, OSPF v2	All	Base	Significant routing support in base feature set
Policy-based routing*	All	Base	Leverages JUNOS policy and firewall to provide policy rather than longest-match DA-based routing
VRRP, BFD	All	Base	HA features
IPv4 multicast—IGMP, PIM SM, PIM DM	All	Base	Software (RE)-based tunnel services for Protocol Independent Multicast in Sparse mode (PIM SM) and Dense mode (PIM DM)Protocol Independent Multicast in Sparse mode (PIM SM encapsulation and de-encapsulation functions
Non-Stop Routing (NSR)*	EX4200 in a VC, EX8200 series	Base	Allows failover to a redundant RE with stateful protocol replication, no control or data plane disturbance
Graceful Restart (GR)	All	Base	Allows an RE reboot or routing daemon restart without incurring loss in the data plane; protocol sessions are disrupted
In-Service Software Updates (ISSU)*	EX4200 in a VC, EX8200	Base	Allows the upgrade of a dual RE machine without incurring a control or data plane hit; based on NSR functionality
Secure device management via J-Web, CLI	All	Base	SSH and OpenSSL support
Device management interface (DMI)	All	Base	Accommodates integration with third-party network management
Centralized management	All	Base	Leverages the Network and Security Manager product to centralize configuration management
BGP routing	All	Advanced	World-class BGP, on a switch!
IS-IS routing	All	Advanced	Not too common in today's enterprise; IS-IS in support of IPv4 routing, not CNLS/CNLP (OSI)
MPLS*	All	Advanced	Not too common in today's enterprise; traffic engineering support
IPv6 routing*	All	Advanced	Unicast and multicast support, OSPF v3/IS-IS
Enhanced GRE tunnel support*	All	Advanced	Support for more than seven GRE tunnels

EX Hardware Summary

This section provided a general overview of the EX3200, EX4200, and EX8200 hardware and software capabilities. The EX3200 has limited hardware redundancy and does not support clustering in a VC. This makes it a good choice for access layer deployments in non-mission-critical environments. The EX4200 provides redundancy enhancements, most notably the ability to cluster up to 10 EX4200s into a single VC. A VC is highly reliable in that it supports redundant REs, PSUs, and VC switch fabric using a ring topology. The 8200 chassis-based switches offer redundant REs, switch fabric, PSUs, and in the case of the EX8216, redundant fan modules.

The high-performance EX hardware is backed up with an impressive range of Layer 2 and Layer 3 base features, perhaps the best of which is the fact that EX hardware runs the venerable JUNOS software, which brings numerous reliability, performance, and stability advantages to the game.

The next section explores the hardware architecture of the EX3200 and EX4200, and provides a packet walk-through showing the processing steps involved with packet switching/routing.

EX Series Architecture

This section provides a high-level overview of typical EX Series platforms to prepare the reader for upcoming detailed discussions of packet flow in this chapter, as well as in Chapters 4 and 11.

The EX-PFE ASIC

The heart of the EX platform is the EX-PFE ASIC. This Juniper-developed chip offers extreme network and VCP/fabric flexibility, which allows it to be used in a number of different modes within a particular switch model. As with other Juniper designs, this PFE mechanism can be replicated as needed to provide increased switch capacity through a multichip PFE complex. For example, an EX3200-24 is based on a single PFE ASIC, whereas the EX4200-48 uses three. Meanwhile, a single 8200 series LC uses four PFE chips to support 8 × 10 GE front panel ports, along with the necessary switch fabric connections. Each EX-PFE chip can forward more than 102 Mpps at wire rate, whether all Layer 2, all Layer 3, or any mix.

EX-PFE effectively offers some 88 Gbps of switching/communications capacity, which can be carved up and delegated to 1 × GE, 10 × GE, or VCP/fabric ports as each hardware application requires. For example, each EX-PFE can provide 24 × 1 Gigabit ports + 2 VCP ports, or alternatively can provide 24 × 1 Gigabit + 1 × 10 Gigabit + 1 VCP port, or in yet another application, 2 × 10 Gigabit + 2 VCP ports.

We highlight the flexibility of the EX-PFE ASIC in the next section, where you can see it operating in a number of the aforementioned modes.

EX3200 Architecture

Our hardware architecture discussion begins with the EX3200 switch, as the rest of the EX Series shares many common aspects of its design. Figure 2-6 illustrates the primary processing stages of an EX3200-48.

The bottom-left portion of Figure 2-6 shows where the CPU components are housed. The 8 MB boot flash gets the JUNOS kernel going so that the JUNOS image, which is stored on the 1 GB main flash, can load into and run from the 512 MB of RAM memory. The 600 MHz PowerPC-based CPU is protected by non-user-configurable ACL/rate limiting, and has interconnections to other components, such as the LCD status display and EX-PFE ASICs for control and communications purposes. The built-in and non-user-alterable ACL function is provided by the JUNOS kernel and includes connection number and connection rate limits, in addition to traffic policers that ensure network control is not locked out during periods of protocol churn, or during a DoS attack that bombards the router with ping or other types of exception traffic.

The EX3200 supports 512 MB of DRAM, and a rear panel USB that can be used to expand flash memory. None of the built-in memory is considered upgradeable, and is therefore not classified as an FRU. The CPU also drives the 1 Gbps em0 Out of Band (OoB) management interface, in addition to the EIA-232-based console.

The 48-port EX3200 uses two EX-PFE chips in the arrangement shown; because the EX3200 does not support a VC, each EX-PFE chip is tasked with providing 24 × 1 GE, + 1 × 10 GE, + 1 fabric port. In contrast, the 24-port version of the EX3200 makes use of a single EX-PFE chip configured in a 24 × 1 GE + 2 × 10 GE arrangement.

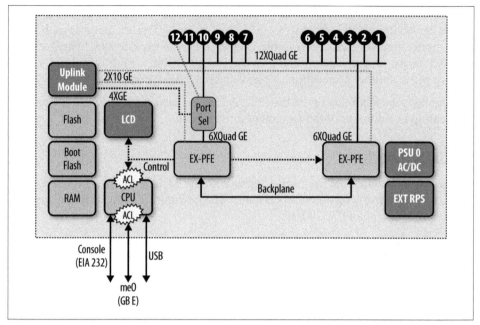

Figure 2-6. The EX3200-48 block diagram

In both the 24- and 48-port versions, the four highest-numbered Gigabit Ethernet ports (i.e., ge-0/0/20–ge-0/0/23 on the 24-port model) are shared between the corresponding network ports and the optional 4 × 1 GE uplink module. Built-in logic senses when a 4 × 1 GE uplink is installed and automatically disables the corresponding network interface ports. Use of the 2 × 10 GE uplink module option allows you to regain use of the high-order 4 GE ports. An EX3200-24 contains a single EX-PFE ASIC, and the EX3200-48 uses two PFE chips. As detailed in the next section, the EX4200s make use of an extra EX-PFE to eliminate this restriction.

Figure 2-6 also shows the EX3200's single PSU, which can be either AC or DC, in addition to support for an optional RPS.

EX4200 Architecture

The hardware architecture of the VC-capable EX4200 is based on the same EX-PFE technology found in the lower-end EX3200, but using more PFE ASICs. The EX4200-24 uses two EX-PFEs (the 24-port EX3200 uses only one), and the EX4200-48 makes use of three PFE chips. The extra PFE capability is used to drive the EX4200's rear-panel VCP ports, and allows simultaneous use of all uplink and front-panel GE ports. Figure 2-7 illustrates the major hardware components of an EX4200-48.

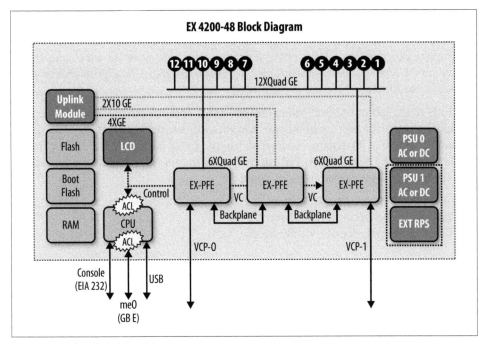

Figure 2-7. The EX4200 architecture

The EX4200 shares many of the same design characteristics found on the EX3200, so the focus here is on the delta. The primary difference is the presence of three EX-PFEs, and how the work is divided among them. In this switch, each PFE provides some aspect of front-panel network connectivity, as well as some portion of the VCP/switch fabric functionality. For example, the first and last PFEs each drive one of the 24 × 1 GE arrays, and each provides one 64 Gbps (FD) VCP port for VC use. The first PFE also provides one of the 10 GE uplink ports. Meanwhile, the second PFE drives two switch fabric ports (128 Gbps FD), provides one 10 GE uplink module, and also drives the 4 × 1 GE uplink ports. Due to this arrangement, the 4 × 1 GE uplink ports and the highest-numbered 1 GE ports are no longer shared, and can therefore be used at the same time.

The EX4200 is based on 1 GB of DRAM and a 1 GHz PowerPC processor. Like the EX3200, it's equipped with 8 MB/1 GB of boot/main flash, as well as support for a rear-panel USB flash extension; the built-in memory is not upgradeable. Figure 2-7 also shows the EX4200's support for redundant PSUs or an optional RPS.

Front-panel LEDs

EX switches provide two LEDs that indicate each port's operational and traffic status. A complete description of each possible state is available at *http://www.juniper.net/ techpubs/en_US/release-independent/junos/topics/reference/specifications/network-port -leds-ex-series.html*.

The LED to the left of the port indicates link status and link activity, and the LED to the right indicates the port's administrative status, duplex mode, PoE status, and speed, depending on the LCD display mode. You use the Enter button on the LCD panel to toggle between the ADM, DPX, PoE, and SPD indicators. The default mode displays port speed using a specific number of blinks:

> One blink per second—10 Mbps
> Two blinks per second—100 Mbps
> Three blinks per second—1,000 Mbps

A Day in the Life of a Packet

This section explores the main areas of EX-PFE processing as a frame (or packet) is received, processed, switched (or routed), and ultimately dispatched toward its final destination. The PFE processing steps vary based on whether the EX is operating in Layer 2 mode as a *switch*, or in Layer 3 mode as a *router*. Both modes are analyzed.

Layer 2 switching

Figure 2-8 shows the processing stages for traffic received on an interface that is configured for Layer 2 operation. This is to say that a given logical unit is configured with `family bridging` rather than `family inet`.

We begin with the reception of an Ethernet frame at step 1. This header extraction stage strips the frame's header and writes the original packet into a shared-memory-based switch fabric. As with other Juniper PFE designs, transit traffic is written into the shared memory switch fabric once. As a result of the switching/routing process, that same frame may then be read from that memory location multiple times at egress, as needed for multicast replication functions. Meanwhile, what's actually being processed in subsequent stages is a *notification* message, the essence of which is all that header good stuff, which in addition to Layer 2 also includes the Layer 3 IP/IPv6 header, as well as Layer 4 port info, assuming, of course, that IP is actually present in the frame. This information accommodates the various Layer 2/Layer 3 security checks, switching/ routing, and CoS functions that may occur in subsequent processing stages.

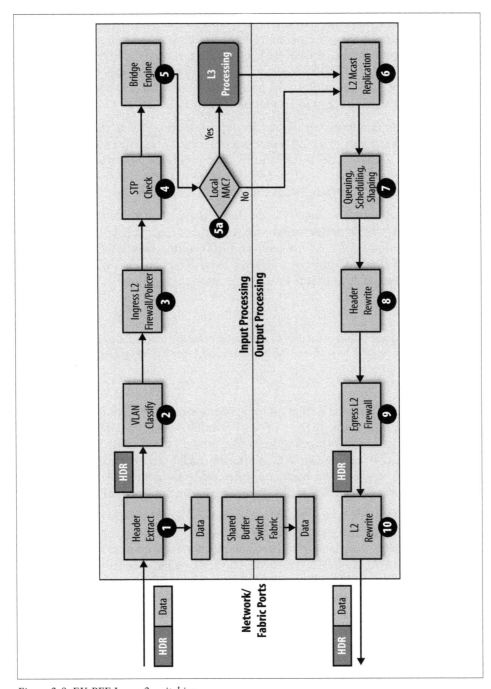

Figure 2-8. EX-PFE Layer 2 switching

The VLAN is extracted at step 2 by the VLAN classification stage. This is accomplished using an explicit header tag, the incoming interface index, or the source MAC address, depending on the operating mode of the logical interface. Each EX-PFE ASIC can index a total of 4,096 VLANs, but VLAN assignment is limited to 1 to 4,094 inclusive. This is because VLAN ID 0 is considered the untagged, or native, VLAN, whereas VLAN ID 4095 (0×FFF) is reserved for implementation-specific functions.

Depending on the configuration and results of previous processing, the packet is now subjected to either a port or a VLAN-level firewall filter. These filters can match on a variety of Layer 2/Layer 3 fields for security, telemetry tracking (counting), policing, forwarding class assignment, and so forth. Layer 3-related firewall checks are performed in the Layer 3 processing stages, which we'll discuss shortly.

We cover firewall filters and security in detail in Chapter 9. Note that *JUNOS firewall filter* is analogous to *ACL* in Cisco terminology. Currently, each EX-PFE ASIC supports up to 2,000 port-level, 2,000 VLAN-level, and 1,000 router-based ACLs, for a total of 5,000 ingress ACL-related functions. Because these filters are executed in hardware, there is usually very little impact to forwarding rate with or without filtering actions in effect.

 Policing is supported only in the ingress direction on the EX platform. This applies to port-level, VLAN-level, and Layer 3 (IPv4) policers.

If the packet is not dropped during firewall processing, step 4 kicks in to subject the packet to an STP state check. In this case, transit traffic is dropped if the corresponding port is not found in the *forwarding* state. We provide details on STP operation in Chapter 6. Note that well-known MAC addresses, such as those used for STP control messages, are recognized as such and are redirected to the control plane as exception traffic when received on an STP-enabled port that is in a non-forwarding state.

Step 5 deals with MAC address learning functions. Here the packet's source MAC address is matched against the list of learned MACs by the bridging engine. If a match is found, the timeout is updated and the incoming interface is compared to the existing state (in case the MAC address has moved). The bridging function redirects the packet to the control plane for MAC learning when no entry is found, or when the interface association needs to be refreshed in the case of a MAC move. At step 5a, the packet's destination MAC address is compared to the switches' own MAC address; if a match is found the traffic is redirected to the Layer 3/Route stage, which is detailed next. Traffic that is not addressed to the switches' MAC is subjected to the Layer 2 switching function, which begins our tale of output processing stages.

Note that currently each EX-PFE chip can retain 24,000 learned MAC addresses. MAC addresses that have not been learned, or that had to be forgotten due to exhaustion of MAC learning table size or age-out, results in flooding of that traffic out all ports in the same bridging instance. The large number of MACs that can be learned by the EX-PFE prevents the inefficiencies of ongoing flooding due to insufficient MAC table size.

Output processing: Layer 2 switching

The first step in the Layer 2 switching function's output process is shown in step 6 of Figure 2-8. This stage replicates Layer 2 traffic that needs to be sent out multiple logical interfaces, such as is the case in a broadcast/multicast Destination MAC (DMAC) address, or for a unicast DMAC that has not yet been learned (e.g., a Source MAC [SMAC] address that has not been seen on that VLAN/bridging instance as the source of any traffic). In these cases, the rules of Layer 2 switching state that the frame must be flooded out to all ports that are in a forwarding state, except, of course, the port on which the packet arrived. Note that what is actually replicated here is a pointer to the copy of the packet that was written into memory way back at step 1.

At step 7, the egress traffic is subjected to CoS processing, which includes queue selection (a total of eight queues are supported per port), priority scheduling, and drop prioritization. Note that per-VLAN (logical interface level) queuing is not supported. And, as you likely surmised, EX Series CoS is covered in detail in Chapter 10, but it's covered much more broadly in our companion book, *JUNOS Enterprise Routing*.

Step 8 of output processing involves *marker rewrite*, also known as *header rewrite*. This stage is often associated with Layer 3 functionality involving IP Precedence/DiffServ Type of Service (ToS) field remarking, but can also be used for Layer 2 rewrite involving the 802.1p priority bits.

Step 9 provides Layer 2 egress firewall filter functionality. Currently, only VLAN-based (and router/Layer 3, as applicable) filter actions are supported at egress. Port-based egress filters are currently not supported, nor is egress policing of any type. A total of 2,000 egress filters (both Layer 2 VLAN-based and Layer 3) can be defined per EX-PFE chip. Layer 2 frames can be dropped, counted, or policed as they transit this stage.

The final egress stage re-forms the Layer 2 frame, which may have had its header remarked at step 8; recalculates the CRC; and transmits the frame onto the medium. Note that as a result of the Layer 2 replication actions at step 6 (replication can also occur in the Layer 3 processing flow) there can be a one-to-many relationship between ingress and egress traffic.

Layer 3 routing

Figure 2-9 details the processing stages for traffic that is subjected to Layer 3 processing. Layer 3 processing is performed on traffic received on a Layer 3-enabled logical interface, or for traffic received on a Layer 2 interface, but for which the destination MAC address matched the address owned by the switch. The example shown is based on the latter case, but the primary concepts and packet processing are similar for packets received on Layer 3 interfaces. Each EX-PFE supports up to 1,000 Layer 3 interfaces, which can operate as a native IP interface or as a Routed VLAN Interface (RVI).

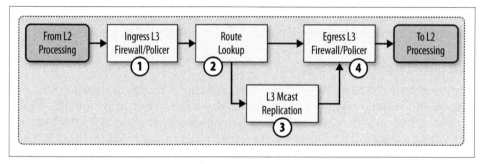

Figure 2-9. EX-PFE Layer 3 processing stages

Things begin in Figure 2-9 at step 1, where the packet (well, technically, its header fields, along with some other magic sauce in the guise of a notification message) is subjected to router/Layer 3 firewall filters. These ACLs operate at the IP and Transport layers (TCP/UDP), and provide similar security, policy, ingress policing, and accounting functionality as those available at Layer 2.

The actual route lookup is performed at step 2, unless, of course, the packet fails to pass muster and is dropped back at step 1. Here, a longest-match lookup is performed against the destination IP address to determine the outgoing interface/next hop or egress firewall action that should be performed. When the destination IP address represents a multicast group address, additional processing is done at step 3 to facilitate IP packet replication and selection of the list of outgoing interfaces to which copies of the packet should be sent. Packets whose longest match is against a locally owned IP address, or that were sent to a multicast group for which the local switch is an active member, are shunted out of the PFE for processing by the RE. These may be locally destined traffic, such as a ping or Telnet session, or a routing update, such as RIP or OSPF.

At step 4, packets whose lookup resulted in a firewall filter next hop are so processed, or else they are handed back to the Layer 2 engine, where they're subjected to Layer 2 egress processing and ultimately transmitted, assuming they are not dropped by the CoS stage or filtered by an egress Layer 2 filter action.

EX Series Architecture Summary

This section provided a detailed description of the hardware capabilities of the EX3200, EX4200, and EX8200 series switches. The varying levels of redundancy and switching capacities, and the ability to grow a VC in an incremental manner as needs dictate, mean that virtually any organization can find an EX platform that meets its design needs.

The next section shifts the focus from EX hardware to the JUNOS software that provides the brains for all that silicon brawn. Given that the same JUNOS software runs on all of the company's J Series, MX Series, M Series, and T Series platforms, there could be worse things to learn.

JUNOS Software Overview

To quote from *JUNOS Enterprise Routing*:

> JUNOS software is cool. It just is. The designers of JUNOS software put tremendous thought into making a stable, robust, and scalable operating system that would be a positive for the router. They were able to learn from previous vendors' mistakes, and create an OS that other companies will forever use as their model.

JUNOS software is cool, and the fact that the same JUNOS software that runs on the EX platform runs on the largest router that Juniper makes, the TX Matrix, is *really* cool.

JUNOS software is a modular operating system that promotes stability and reliability. The modularization is achieved through the use of software daemons that run in protected memory space. The stability stems from well-written and tested code; choosing a well-known, open source, and stable kernel of FreeBSD on which to build JUNOS did not hurt, of course.

The kernel and its functions are normally not directly visible to the user, but many features of FreeBSD have been ported to the command line of JUNOS, and many that have not are still available, albeit at the shell prompt. The kernel supports the various user processes (daemons), maintains the forwarding table synchronization between the RE and the PFE, and provides interface-related routing functions such as performing ARP and installing local and direct routes in the routing table (RT).

Riding on top of the kernel are the software processes that provide JUNOS services and the user interfaces needed to configure, monitor, and troubleshoot these services. These include the CLI, interface control, routing, address learning, and so forth. Figure 2-10 illustrates the modular architecture of JUNOS software, listing some of the user space processes that run on top of the kernel. Note that you can always obtain a complete list of these processes by issuing a show system processes command. You can also restart most daemon processes from the CLI using the restart command, as shown here:

```
lab@Rum> restart ?
Possible completions:
  802.1x-protocol-daemon  Port based Network Access Control
  adaptive-services     Adaptive services process
  audit-process         Audit process
  cfm                   Connectivity fault management process
  chassis-control       Chassis control process
  chassis-manager       Chassis Manager
  class-of-service      Class-of-service process
  dhcp                  Dynamic Host Configuration Protocol process
  dynamic-flow-capture  Dynamic flow capture service
  ecc-error-logging      ECC parity errors logging process
  ethernet-link-fault-management  Ethernet OAM Link-Fault-Management process
```

```
ethernet-switching      Ethernet Switching Process
event-processing        Event processing process
firewall                Firewall process
general-authentication-service  General authentication process
gracefully              Gracefully restart the process
immediately             Immediately restart (SIGKILL) the process
interface-control       Interface control process
ip-demux                Demux Interface Daemon
ipsec-key-management    IPSec Key Management daemon
l2-learning             Layer 2 address flooding and learning process
lacp                    Link Aggregation Control Protocol process
license-service         Feature license management process
link-management         Link management process
lldpd-service           Link Layer Discovery Protocol
mib-process             Management Information Base II process
mountd-service          Service for NFS mounts requests
mspd                    Multiservice Daemon
nfsd-service            Remote NFS server
pgcp-service            Packet gateway service process
pgm                     Pragmatic General Multicast process
ppp                     PPP process
pppoe                   Point-to-Point Protocol over Ethernet process
redundancy-interface-process  Redundancy interface management process
remote-operations       Remote operations process
routing                 Routing protocol process
sdk-service             SDK Service Daemon
service-deployment      Service Deployment System (SDX) process
sflow-service           Flow Sampling (Sflow) Daemon
snmp                    Simple Network Management Protocol process
soft                    Soft reset (SIGHUP) the process
virtual-chassis-control  Virtual Chassis Control Protocol
vrrp                    Virtual Router Redundancy Protocol process
web-management          Web management process
lab@Rum> restart
```

A description of each process's function is beyond the scope of this book. You can find some additional information on JUNOS processes in the technical documentation at *http://www.juniper.net/techpubs/software/nog/nog-baseline/frameset.htm*. Note that a large chunk of the documentation is bundled with the software running on your switch. A later section demonstrates use of the embedded documentation.

Each software process is fully independent, so a failure of one process does not affect the others. For example, Figure 2-10 shows how the Simple Network Management Protocol (SNMP) process pulls information from the interface, chassis, and routing processes. If the SNMP process fails for any reason, it affects only that process and not the others, and in most cases can be restarted and returned to service. This is a major shift from other routing vendors that operated monolithic code in which one change in the interface code could affect just about anything, seemingly without rhyme or reason.

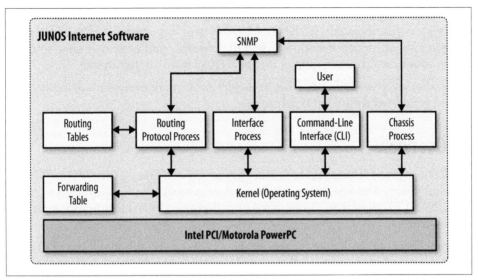

Figure 2-10. The JUNOS software architecture

As previously noted, the J Series, M Series, MX Series, and TX routing platforms all run JUNOS, which is built from a common code base. Note, however, that having the same code base is not the same as running the same binary image. The EX platforms run an EX-specific binary image that is built from the common JUNOS code base. Although some features are platform-dependent, JUNOS is JUNOS, and the general rule is a common configuration and operational mode syntax, and a high degree of feature parity across all platforms. The first time you need to upgrade the software on your EX switch, this point and its immediate benefits are driven home. There's no need for a fancy feature calculator to (hopefully) find an image with the desired feature that actually runs on your hardware. Nope; just find the latest EX release (there are four major releases each year), download it, load it, and go play with the new feature.

The major difference in the EX image versus the image running on, say, an M Series router, is the successful porting to a RISC-based PowerPC platform, and the inclusion of Layer 2-specific functions performed by the Ethernet switching daemon (eswd), such as Layer 2 address learning and STP. Inclusion of the LLDP daemon provides a Cisco Discovery Protocol (CDP)-like functionality, which also includes the LLDP-MED extensions used to support IP telephony devices.

JUNOS Software Summary

This section provided an overview of the JUNOS software architecture, detailing how the modular software process runs atop a hardened BSD-based kernel on a variety of hardware platforms. The key takeaway here is the common JUNOS base that is shared across a huge portion of the company's product line and that has been proven stable in the largest service provider networks for more than 10 years.

Although the EX may be the new platform on the block, it inherits the benefits and capabilities of the legendary JUNOS that happily runs some of the world's largest routing systems in most of the world's largest IP networks. And now, that same software sits on your desktop, in the form of that new EX switch that just arrived.

Well, what are you waiting for? Are you chicken? C'mon, everyone's doing it.... It's time to get started with JUNOS software!

CLI Overview

Hardcore engineers always prefer a CLI as opposed to a fancy-pants GUI web-based frontend thingy. The odd thing is that even novice users often quickly learn to prefer the JUNOS CLI, given that it's one of the most user-friendly and feature-rich interfaces ever invented.

The remainder of this chapter focuses on the CLI, but gives an honorable nod to some slick user-friendly features as well. Generally speaking, the GUI makes things so easy, as long as you know what you want and why it's used, it's a cinch to click your way through it. Using the CLI provides direct access to all of the features of JUNOS, and it takes much less time and page space than dealing with GUI screens. As such, biting a bit of the CLI bullet really is the best way to learn JUNOS.

J-Web and EZSetup

When Juniper decided to reach into the Enterprise it understood that it would be encountering a new audience, one that is often familiar with the IOS CLI but in many cases may be brand new to networking, with little to no CLI experience. To address this concern, a web-based GUI interface was developed. The J-Web interface was first released on the J Series products, and was updated and enhanced to support the EX Series as well.

Figure 2-11 shows the main screen for the EX J-Web interface.

Yes, the J-Web interface is pretty slick.

J-Web supports both HTTP and HTTPS protocols, provides clean operational status displays and configuration (which includes wizards that help get you up and running), and performs common maintenance functions such as software upgrades. The built-in DHCP server found in a factory-default configuration makes it easy to attach a PC for quick access to the web-based EZSetup feature. Once configured, J-Web provides ongoing configuration and system management capabilities, including really cool performance dashboard displays.

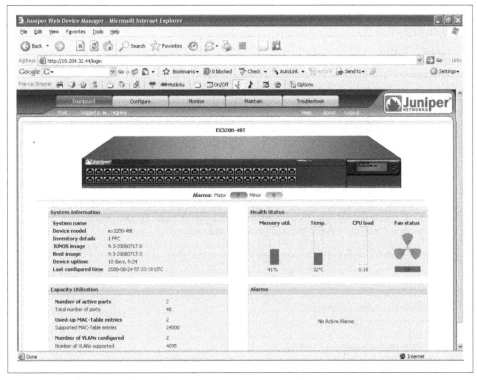

Figure 2-11. The EX J-Web interface

EZSetup

The EZSetup feature, as its name suggests, is designed to get a brand-new switch up and running by prompting for commonly needed configuration data (see Chapter 3 for more details on the EZSetup feature). The web version of EZSetup is presented when a factory-default configuration is in place. The EZSetup feature is also available via the CLI, but only when logged into a root shell, and again a true factory-default configuration must be in effect.

> After loading a factory-default configuration with the `load factory-default` configuration mode command, you must assign a `root` password before you can `commit` (committing actually places the configuration into effect, as described later in this chapter). Once the modification is committed, the result is no longer a true factory-default configuration, which in turn blocks shell-based evocation and suppresses the J-Web prompt for the same.

The following code shows EZSetup starting from a root shell on a true factory-default configuration:

```
root@% ezsetup

*****************************************************************************
* EZSetup wizard                                                           *
*                                                                          *
* Use the EZSetup wizard to configure the identity of the switch.          *
* Once you complete EZSetup, the switch can be accessed over the network.  *
*                                                                          *
* To exit the EZSetup wizard press CTRL+C.                                 *
*                                                                          *
* In the wizard, default values are provided for some options.             *
* Press ENTER key to accept the default values.                           *
*                                                                          *
* Prompts that contain [Optional] denotes that the option is not mandatory. *
* Press ENTER key to ingore the option and continue.                      *
*                                                                          *
*****************************************************************************

EZSetup Initializing..done.

Initial Setup Configuration
--------------------------

Enter System hostname [Optional]:
```

CLI Operational Modes and General Features

The CLI has two modes: *operational* and *configuration*. Operational mode is where you troubleshoot and monitor the router hardware and software, as well as its network connectivity and protocol operation. Configuration mode is where the actual configuration statements for interfaces, routing protocols, and everything else, for that matter, are specified.

 Every command that can be run in operational mode can also be used in configuration mode with the additional keyword run. For example, if the command show route is issued in operational mode, it can be issued as run show route in configuration mode. This is similar to IOS's do command, except it works all the time and in all configuration mode contexts.

Operational mode

When a non-root user first enters the router via Telnet, SSH, or direct console access, the user will see a login prompt. After entering the correct username and password, the user will be placed directly into operational mode. Operational mode is designated by the > (chevron) character at the router prompt of username@hostname. As shown in the following code, user lab logs into a switch called Rum:

```
login: lab
Password:

--- JUNOS 9.0R2.10 built 2008-03-06 10:31:45 UTC
lab@Rum>
```

An exception to being automatically placed into operational mode is when logging in as the root user. In this case, the user is placed directly into a shell (designated by the percent [%] sign) and must start the CLI process manually:

```
Rum (ttyd0)

login: root
Password:

--- JUNOS 9.0R2.10 built 2008-03-06 10:31:45 UTC
root@Rum% cli
root@Rum>
```

Most of the commands that you will run in operational mode are show commands, which allow you to gather information about the switches' hardware, software, and general protocol operation. The ping, traceroute, telnet, and ssh utilities are also available in operational mode, as are clear commands that are used to reset counters, learn MAC/ARP entries, or reset protocol adjacencies.

In addition, there are some very JUNOS-specific operational mode commands such as request, restart, and test. The request commands perform systemwide functions such as rebooting, upgrading, and shutting down the box. The restart commands are similar to the Unix-style kill commands, allowing you to restart various software processes. The test commands allow verification of saved configuration files, proactive testing of policies, and interface-related test functions that vary by hardware type.

 The restart commands should be used with caution! Network disruption can easily result from restarting the chassis control, routing, or switching processes, for example. Generally, you restart a process under guidance from technical support, when fault analysis indicates that the process is hung or generally misbehaving.

The next section examines some additional CLI features that you will not want to live without.

Command completion

The command completion feature saves you lots of time and energy, and it provides syntax checking as you type. Gone are the days when you type a command on a line and after you press Enter the command is either invalid or not supported on that version of software. Any error or ambiguity will be detected early, and the switch will present a list of valid completions for the current command. You can disable command completion on a per-login basis by modifying the CLI environment with an operational mode set cli command:

```
lab@Rum> set cli ?
Possible completions:
  complete-on-space    Set whether typing space completes current word
  directory            Set working directory
  idle-timeout         Set maximum idle time before login session ends
  prompt               Set CLI command prompt string
  restart-on-upgrade   Set whether CLI prompts to restart after software upgrade
  screen-length        Set number of lines on screen
  screen-width         Set number of characters on a line
  terminal             Set terminal type
  timestamp            Timestamp CLI output
```

But a good reason to do so has not yet been noted.

You can evoke command completion by using *either* the space bar or the Tab key. Note that the Tab key *also* completes user-assigned *variables* such as interface names, IP addresses, firewall filters, and filenames. For example, you use the show configuration firewall command to view that portion of the configuration from an operational mode prompt. The example makes use of both space and tab completion for commands and variables, respectively, and also uses the family and filter switches to display a subset of the firewall stanza in the form of a specific filter named test_L3_filter:

```
root@Rum> sh<space>ow conf<space>iguration fire<space>wall family i<space>net
filter ?
Possible completions:
  <filter-name>        Filter name
  test_L3_filter       Filter name
root@Rum> show configuration firewall family inet filter t<Tab>est_L3_filter
<enter>
term 1 {
    from {
        protocol icmp;
    }
    then count icmp_counter;
}
term 2 {
    then accept;
}
```

Notice that the space bar is used until a variable is reached, at which time the Tab key is used to auto-complete the user variable for the filter name of test_L3_filter. The example also shows the CLI's context-sensitive help function, which is evoked with the ? key. The context-sensitive help function provides valid completions for both standard commands and user-defined variables. Pretty slick, eh?

In the previous example, the syntax checker went word by word each time the space bar or Tab key was pressed, and the minimum characters were typed to avoid ambiguity. When a command is ambiguous, the CLI states the issue and lists the possible completions:

```
root@Rum> show e
              ^
'e' is ambiguous.
Possible completions:
  esis                 Show end system-to-intermediate system information
  ethernet-switching   Show Ethernet-switching information
  event-options        Show event-options information
root@Rum> show e
```

Here, the command show e cannot be completed as typed, so the CLI prompts the user for further clarification as to which value, ethernet-switching or event-options, should be displayed. The CLI then repaints the ambiguous command, patiently awaiting disambiguation.

Emacs keys. The CLI is based on Emacs-style keystrokes. This allows you to quickly position the cursor, or edit the command line, using keystroke shortcuts. Here are some useful Emacs keystrokes:

Ctrl-b

Moves the cursor back one character. The left arrow can also be used for this purpose when set to vt100.

Ctrl-f

Moves the cursor forward one character. The right arrow can also be used for this purpose when set to vt100.

Ctrl-a

Moves the cursor to the beginning of the command line.

Ctrl-e

Moves the cursor to the end of the command line.

Ctrl-k

Deletes all words from the cursor to the end of the line.

Ctrl-w

Deletes an entire word to the left of the cursor.

Ctrl-x

Deletes or clears the entire line.

Ctrl-l

Redraws the current line.

Ctrl-p

Scrolls backward through the previously typed commands. The up arrow can also be used for this purpose when set to vt100.

Ctrl-n

Scrolls forward through the previously typed commands. The down arrow can also be used for this purpose when set to vt100.

Ctrl-r

Searches the previous CLI history for a search string.

Ctrl-u
Erases the current line.

As noted, when the terminal type is set to vt100, you can use the up and down/sideways arrow keys for previous/next command buffer recall and cursor positioning, respectively.

 The JUNOS CLI maintains a separate command history for operational versus configuration mode commands.

The pipe. Another important CLI feature is support for piping the output of any command, whether configuration or operational, to a set of processing functions such as count, match, and so forth.

The actual arguments supported by the pipe (|) function are constrained based on the command's context as either configuration or operational mode. In the following example, the operational mode show bgp summary command is shown being piped to the display function, but the only supported argument is xml. Again, this is because of the operational mode context:

```
root@Rum> show bgp summary | display ?
Possible completions:
  xml                    Show output as XML tags
```

The same general show command syntax is now issued in configuration mode context. And the result is a different set of options for the display argument:

```
[edit]
root@Rum# show | display ?
Possible completions:
  changed         Tag changes with junos:changed attribute (XML only)
  commit-scripts  Show data after commit scripts have been applied
  detail          Show configuration data detail
  inheritance     Show inherited configuration data and source group
  omit            Omit configuration statements with the 'omit' option
  set             Show 'set' commands that create configuration
  xml             Show output as XML tags
```

Note that the set argument is now available for the display function. By the way, the purpose of display set is to convert the curly-brace-delimited configuration into the sequence of set commands that created the configuration; this is a *very* useful pipe function. Note that the full configuration hierarchy is displayed for each set command displayed:

```
[edit interfaces ge-0/0/4]
lab@Rum# show | display set
set interfaces ge-0/0/4 unit 0 family inet filter input test
set interfaces ge-0/0/4 unit 0 family inet address 10.5.7.5/24
set interfaces ge-0/0/4 unit 0 family inet6
```

```
[edit interfaces ge-0/0/4]
lab@Rum#
```

The pipe command allows the display buffer to be massaged and displayed in various ways. The most common applications of the pipe function are detailed here:

count

Count the output lines:

```
root@Rum> show interfaces terse | count
Count: 67 lines
```

display

Show additional data; for example, XML tags, set commands, or details about what the config means:

```
[edit interfaces ge-0/0/4]
lab@Rum# show | display detail
##
## Logical unit number
## range: 0 .. 16385
##
unit 0 {
    ##
    ## family: Protocol family
    ## constraint: Can't configure protocol family with encapsulation ppp-over-
ether-over-atm-llc
    ## constraint: Can't configure protocol family with encapsulation ppp-over-
ether
    ##
    ##
    ## inet: IPv4 parameters
    ## alias: inet4
    ## constraint: family inet is not supported on encapsulation frame-relay-ppp
    ##
. . .
```

except

Omit matching lines from the output:

```
root@Rum> show interfaces terse | except ge-
Interface               Admin Link Proto    Local            Remote
bme0                    up    up
bme0.32768              up    up   inet      128.0.0.1/2
                                             128.0.0.16/2
                                             128.0.0.32/2
                                   tnp       0x10
dsc                     up    up
gre                     up    up
ipip                    up    up
lo0                     up    up
lo0.0                   up    up   inet      10.10.1.5        --> 0/0
lsi                     up    up
me0                     up    down
me0.0                   up    down
```

```
mtun                        up    up
pimd                        up    up
pime                        up    up
tap                         up    up
vlan                        up    up
```

find

Begin the output at the specified match:

```
root@Rum> show interfaces ge-0/0/0 extensive | find alarm
Active alarms  : LINK
Active defects : LINK
MAC statistics:                        Receive          Transmit
  Total octets                            0                 0
  Total packets                           0                 0
  Unicast packets                         0                 0
  Broadcast packets                       0                 0
  Multicast packets                       0                 0
  CRC/Align errors                        0                 0
  FIFO errors                             0                 0
  MAC control frames                      0                 0
  MAC pause frames                        0                 0
  Oversized frames                        0
  Jabber frames                           0
  Fragment frames                         0
  VLAN tagged frames                      0
  Code violations                         0
Filter statistics:
  Input packet count                      0
  . . .
```

match

Display only the lines matching the specified string:

```
root@Rum> show log messages | match fail
Mar 11 11:50:34  Rum chas[517]: cm_pid_get_process: stat of /proc/520 failed
Mar 11 11:50:34  Rum chas[517]: failed SIGSTOP to pfem 520 - kill failed
. . .
```

no-more

Do not paginate the output at a more prompt:

```
root@Rum> show system statistics arp | no-more
arp:
        0 datagrams received
        0 ARP requests received
        0 ARP replys received
        0 resolution requests received
        0 unrestricted proxy requests
        0 received proxy requests
        0 proxy requests not proxied
        0 with bogus interface
        0 with incorrect length
        0 for non-IP protocol
        0 with unsupported op code
```

```
                    0 with bad protocol address length
                    . . .
```

save

Save the output to a file in the user's home (or specified) directory:

```
[edit]
root@Rum# show interfaces | save rum_interfaces_config
Wrote 157 lines of output to 'rum_interfaces_config'
```

Multiple pipe commands are treated as a logical AND, meaning the output must match *both* of the commands listed. Consider this example, which combines the count and match arguments to count how many interfaces are currently down:

```
root@Rum> show interfaces terse | match down | count
Count: 50 lines
```

Ouch, a lot of down interfaces, to be sure. Don't be disheartened. It's still early in the book.

Here we show multiple incarnations of the same pipe argument type (match), to provide additional filtering. Here, matching messages must have both match criteria to pass the filter:

```
root@Rum> show log messages | match fail | match kill
Mar 11 11:50:34  Rum chas[517]: failed SIGSTOP to pfem 520 - kill failed
Mar 11 11:50:34  Rum chas[517]: failed SIGCONT to pfem 520 - kill failed
Mar 11 11:50:34  Rum chas[517]: failed SIGHUP to pfem 520 - kill failed
Mar 11 13:30:12  Rum chas[516]: failed SIGSTOP to pfem 519 - kill failed
. . .
```

Pipe commands are not limited to a logical AND, however, as a logical OR operation can also be performed. This is done by wrapping the string in quotation marks and using the OR operator. Here we parse the logfile for messages having either match term. Note that CLI-based matches are *not* case-sensitive:

```
root@Rum> show log messages | match "(fail|error)"
Mar 11 11:50:34  Rum chas[517]: PFELC:Error connecting master kernel,4294967295
Mar 11 11:50:34  Rum chas[517]: cm_pid_get_process: stat of /proc/520 failed
Mar 11 11:50:34  Rum chas[517]: failed SIGSTOP to pfem 520 - kill failed
Mar 11 11:50:34  Rum chas[517]: cm_pid_get_process: stat of /proc/520 failed
Mar 11 11:50:34  Rum chas[517]: failed SIGCONT to pfem 520 - kill failed
. . . .
```

Configuration Mode

In the previous section, you were exposed to the CLI's operational mode. To actually configure a JUNOS device, you must enter the CLI's configuration mode by entering configure at the operational mode prompt. When in configuration mode, the router prompt changes to the hash (#) symbol, and the user's position in the configuration hierarchy is displayed. Here, the top of the configuration is indicated by the [edit] prompt:

```
root@Rum> configure
Entering configuration mode

[edit]
root@Rum#
```

By default, multiple users can enter the router and make changes at the same time. To avoid any issues that may arise, you can use the configure exclusive or configure private option. The former allows only a single user to configure the router, and the latter allows multiple users to configure *different* pieces of the configuration simultaneously. The private mode grants users their own copy of the current candidate configuration, while ensuring that each user's changes are unique. This means that only the changes made by a given user are activated when that user performs a commit. If two users attempt to make the same change, such as adding an IP address to the *same* interface, the change is rejected and both users must exit configuration mode and begin again to resolve the conflict.

In contrast, with the exclusive option no other users can make changes to the configuration except the single user that entered configuration mode. In both the private and exclusive cases, the CLI forces the users to either commit or discard their changes to avoid database conflicts with another user, then enter configuration mode later:

```
[edit]
lab@Rum# quit
The configuration has been changed but not committed
warning: Auto rollback on exiting 'configure exclusive'
Discard uncommitted changes? [yes,no] (yes) yes

warning: discarding uncommitted changes
Exiting configuration mode

lab@Rum>
```

A normal configuration user can exit with unsaved changes; this is considered bad form as others may not know how to quickly determine the nature of the changes that were left in place (and were not yet activated with a commit):

```
lab@Rum> configure
Entering configuration mode

[edit]
lab@Rum# delete interfaces

[edit]
lab@Rum# quit
The configuration has been changed but not committed
Exit with uncommitted changes? [yes,no] (yes) yes

Exiting configuration mode

lab@Rum>
```

While in configuration mode, the `set` command is used to place new values into the *candidate* configuration. For example, to enable a Telnet server on the box, use a `set` command:

```
[edit]
root@Rum# set system services telnet

[edit]
root@Rum#
```

A `delete` is the opposite of `set`, and it is used to remove portions of the configuration. `delete` acts on the *most specific* item specified; if you are not careful, you can remove more than you intended. Here, omission of the keyword `telnet` results in deletion of *all* statements at the [`edit system services`] hierarchy, which can be a far sight more than just Telnet!

```
[edit]
root@Rum# delete system services telnet

[edit]
root@Rum#
```

The JUNOS configuration is based on an object-oriented hierarchy. The top of the hierarchy is called `edit`, and each configuration stanza is placed in a corresponding subhierarchy. This is roughly analogous to a PC filesystem, in which *C:* is the root and *C:\data* is a directory under the root.

The configuration hierarchies that live directly below [`edit`] can be viewed with a `set ?` command:

```
[edit]
root@Rum# set ?
Possible completions:
> access                 Network access configuration
> access-profile         Access profile for this instance
> accounting-options     Accounting data configuration
> active-probe
+ apply-groups           Groups from which to inherit configuration data
> chassis                Chassis configuration
> class-of-service       Class-of-service configuration
> dialer
> ethernet-switching-options  Ethernet-switching configuration options
> event-options          Event processing configuration
> firewall               Define a firewall configuration
> forwarding-options     Configure options to control packet forwarding
> groups                 Configuration groups
> interfaces             Interface configuration
> multicast-snooping-options  Multicast snooping option configuration
> poe                    Power-over-Ethernet options
> policy-options         Routing policy option configuration
> protocols              Routing protocol configuration
> routing-instances      Routing instance configuration
> routing-options        Protocol-independent routing option configuration
> security               Security configuration
```

```
 > snmp                  Simple Network Management Protocol configuration
 > system                System parameters
 > vlans                 VLAN configuration
 [edit]
```

Entries beginning with a > have their own subhierarchies, and so on, and so forth. We will explore most of these hierarchies at some point in this book. Figure 2-12 shows a partial tree of the configuration hierarchy.

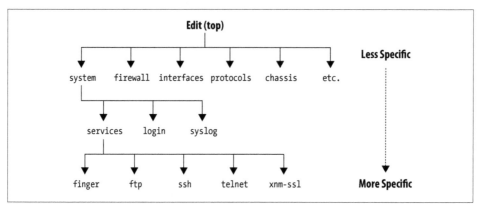

Figure 2-12. A portion of the JUNOS configuration hierarchy

Figure 2-12 shows the top of the hierarchy, edit, and some of the next-level hierarchies that lie directly below it, such as system, interfaces, and protocols. The system path is further expanded to show that it has sublevels, one of which, the services hierarchy, is in turn expanded to show some of the specific services available. The takeaway here is that a configuration-related command intended to alter a system service will have system services in its fully qualified command path, that is, the path needed when at the [edit] root.

Although you do not need to memorize the tree structure, it is important to understand the hierarchical structure and how it relates to configuration mode commands. For example, the save command saves information from the *current* hierarchy down. Therefore, the outcome of a save command when issued at the [edit] hierarchy is far different from when it's issued at a lower hierarchy, such as [edit protocols ospf area 0]. In the former case, the entire configuration is saved, whereas in the latter only the configuration relating to OSPF area 0 is saved.

As mentioned previously, the delete command is the opposite of set. Usually the command is used to remove a single line, but it can be used to remove an entire hierarchy.

Be Careful Out There

JUNOS provides the operator with a long length of rope. Use caution to make sure it does not hang you. For example, issuing a delete protocols, when you meant to delete only one interface from area 0 of OSPF—say, with a `delete protocols ospf area 0 interface ge-0/0/0` command—does not result in any complaints, or even a sanity check in the form of an "are you sure, enter y/n" dialog. Nope, the whole protocols stanza is simply gone. Moving on to commit such a change could be disastrous. A similar condition is true when clearing protocol neighbors in operational mode, where a failure to identify which neighbor assumes you wish to clear them all.

There's not much more to be said, other than to be careful. And remember that `commit confirmed` and/or `rollback 1` are good tricks to keep handy when making configuration changes, as both function to restore a previous configuration that, for whatever reason, is now deemed better than the current one. Note that if you experience a "mulligan moment" and have not yet committed the mistake, `rollback 0` is what you need; this restores a fresh copy of the candidate configuration, which again matches the active configuration. Just be careful next time.

Navigating the configuration hierarchy

Configuration commands such as `set` and `delete` can be issued from the top, or root level, using a fully qualified path. Alternatively, you can park at a configuration sub-hierarchy and issue relative commands. The `edit` command functions like a change directory (`cd`) command in that it allows you to position yourself at a lower hierarchy. Here, the `edit` function is used to park at the [`edit system services`] hierarchy:

```
[edit]
root@Rum# edit system services

[edit system services]
root@Rum# set web-management http
```

Note how the CLI banner changes to reflect the user's new position in the hierarchy. A *relative* `set web-management http` statement is then issued to enable the HTTP web management service (J-Web). The results are confirmed with a `show` command, which again is relative to the current hierarchy:

```
[edit system services]
root@Rum# show
telnet;
web-management {
    http;
}
```

Although using the `edit` command is not necessary, it does allow the user to issue shorter (relative) `set` statements, compared to issuing the same statements from the top level. Just like choosing a color for a new car, you can choose how you want to configure the router, as long as the desired result is achieved. In general terms, most users opt to park at a subhierarchy when many `set` commands are needed. Then again, more experienced users avail themselves of space completion and command recall, and can often issue multiple `set` statements quite rapidly; recall that Ctrl-w deletes the last word, so an up arrow/Ctrl-p followed by a Ctrl-w rapidly prepares the user to enter the next statement at the same hierarchy.

Once you are in a certain directory, there are multiple ways to navigate the directory tree using the up, top, and exit commands. The up command moves you up one level in the directory tree or multiple levels if a numerical value is given after the command:

```
[edit system services]
root@Rum# up

[edit system]
root@Rum# edit services

[edit system services]
root@Rum# up 2

[edit]
root@Rum#
```

You can use the top command at any hierarchy to move up to the root level of the configuration tree. This command has the added functionality of allowing multiple configuration statements after you issue the command, such as top edit or top set. Here the user moves from the [edit system services] hierarchy to the top, and then back down to the [edit protocols ospf area 0] hierarchy in one fell swoop:

```
[edit system services]
root@Rum# top edit protocols ospf area 0

[edit protocols ospf area 0.0.0.0]
root@Rum#
```

It's always nice to be able to see your work while you're still working on it. It's also nice to be able to focus only on the areas that happen to concern you at that moment. Use the show command to view all or just selected parts of the configuration. Here the operator uses show along with a description of which configuration hierarchy should be displayed:

```
[edit]
root@Rum# show system services
telnet;
web-management {
    http;
}

[edit]
root@Rum# show interfaces ge-0/0/0
unit 0 {
    family ethernet-switching;
}
```

Yes, the JUNOS CLI is cool. There, I said it again.

This section introduced you to the configuration hierarchy and the basic commands used to alter a configuration by adding or deleting statements. The next section focuses on how to activate or deactivate those changes, as operational outcome should dictate.

Active and candidate configurations, commits, and rollbacks

JUNOS uses a candidate configuration model, in which all changes are performed against a *copy* of the configuration that was current at the time the user entered configuration mode. As a result, the changes do not take effect until the candidate configuration is made active through a commit process. During this process, some sanity checking is performed against the candidate configuration.

When problems are detected, one or more warnings may be displayed (while the new configuration is still activated), or a commit failure can occur. When a commit failure is declared, the candidate configuration is preserved and users are free to make additional edits to address the error so that they can try the commit again. In some cases, a new error may be displayed when the last error is corrected because the configuration parser is allowed to proceed further into the candidate configuration.

Pay attention to any commit-time warnings printed on the terminal, and ensure that no warnings are being placed into the logfile (*/var/log/messages*) at the time of commit. This is especially true if things do not appear to be working as they should with a new configuration.

JUNOS strives to be flexible and will permit some hardware-dependent configuration on a platform that is currently lacking the hardware needed for the feature to work. This is a feature, not a bug, and as such a warning may not occur on the user's terminal, but often a log message is generated.

Here, a commit error is declared when a policer is applied in an output direction, which is not currently supported on the EX platform. Note the use of commit check, which performs *only* the sanity check and does not activate changes regardless of success:

```
[edit]
lab@Rum# commit check
[edit interfaces ge-0/0/4 unit 0 family inet]
  'filter'
    Referenced filter 'test' can not be used as policer not supported on egress
[edit]
  'interfaces'
    error parsing interfaces object
error: configuration check-out failed
```

When a successful commit does occur, the candidate configuration becomes the active configuration, and the previous active configuration is archived into a file called *juniper.conf.1.gz*, which is a binary file compressed form. During this process, the previous *juniper.conf.1.gz* is renamed *juniper.conf.2.gz*, and so on, such that a total of the previous 50 (0–49) active configurations are retained on the switch.

The first of these is the current active configuration; it is called *juniper.conf.gz* and is stored along with the previous three configurations at */config*. The remaining 46 previous configurations are stored at */var/db/config/*.

You access a previous configuration by issuing a `rollback n` command. In this case, the *n* refers to a rollback index, which can be 0 (restore current active config, that is, discard changes and start over) through 49. A `rollback` command causes the current *candidate* configuration to be replaced by the previous configuration identified, causing it to become the current candidate configuration. It must be stressed that the newly restored configuration is the candidate *until* you actually perform a commit. Thus, when you have just committed a configuration, and you realize a serious mistake was made, the quickest recovery mechanism is typically a `configure`, `rollback 1`, `commit` sequence, which assumes you left configuration mode after committing the mistake—say, by issuing a `commit and-quit`, which exits from configuration mode after a successful commit.

As mentioned earlier, you can perform a `commit check` at any time to run the sanity-check routine against the candidate configuration, without actually placing any changes into effect. One convenient use of this feature is to build some new configuration, in stages, without actually committing the new configuration until all changes are complete and you are in a maintenance window. For example, you enter configuration mode and load a saved configuration file to make some edits. Before wrapping up, you issue `commit check` to confirm there are no blatant errors, and then save the file for the next editing session. You then perform a `rollback 0` to regain a fresh candidate configuration copy, and exit configuration mode with no warnings given that no changes were ever made.

Commit confirmed.   This section is for anyone who has ever been roused from sound slumber to go on-site to recover a box that can no longer be remotely contacted because of a configuration error. It's equally for those who have experienced the sinking feeling of pressing the Enter key on such a configuration error, only to realize too late the error of your ways, as all contact is lost to an important piece of networking gear that happens to be located in another country!

The `confirmed` option to the `commit` command requires that another `commit` be issued within some period of time, the default being 10 minutes. If no such commit is sensed, the box automatically performs a `rollback 1`, `commit` sequence, in effect activating the previous configuration. This process is demonstrated in the following code, where failure to commit within one minute results in automatic restoration (and activation) of the previous configuration:

```
[edit]
lab@Rum# set system host-name Mur

[edit]
lab@Rum# commit confirmed 1
commit confirmed will be automatically rolled back in 1 minutes unless confirmed
commit complete
```

```
lab@Rum# quit
Exiting configuration mode

# commit confirmed will be rolled back in 0 minutes
. . .
Broadcast Message from root@Rum
        (no tty) at 15:28 UTC...

Commit was not confirmed; automatic rollback complete.

lab@Rum> configure
Entering configuration mode

[edit]
lab@Rum# show system host-name
host-name Rum;
```

 You have to disconnect and log back into the router to actually see a hostname change take effect with JUNOS software. In this example, the user does not disconnect, which explains why the system continues to display the old hostname.

Confirmed commits are a real lifesaver and should always be considered when performing remote operations.

We just covered some very important JUNOS capabilities, so a bit of review is in order. When you enter configuration mode, a copy of the current configuration is forked off to become the candidate configuration. No changes in switch operation occur until you activate all changes in the candidate through the commit process, which activates the candidate configuration.

At this time, the previous active configuration becomes accessible as rollback 1. At the second commit, the precious active configuration becomes rollback 1 and the previous rollback 1 becomes rollback 2, and so on. Figure 2-13 illustrates this process.

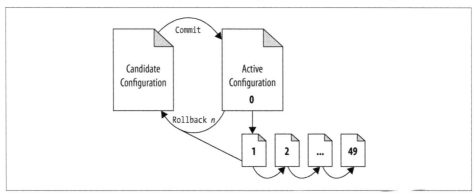

Figure 2-13. Commits, rollbacks, and saved configurations

Using `rollback 0`, which *overwrites* the current candidate by recopying the active configuration, is a common case that is often missed by those new to JUNOS. As an example, imagine that user `lab` logs into a switch, makes a few changes, and then rudely logs out without committing the changes:

```
lab@Rum> configure
Entering configuration mode

[edit]
lab@Rum# set protocols rstp interface all edge

[edit]
lab@Rum# quit
The configuration has been changed but not committed
Exit with uncommitted changes? [yes,no] (yes) yes

Exiting configuration mode

lab@Rum> exit

Rum (ttyu0)

login:
```

Now, a new user logs into the router and upon entering configuration mode is warned of the stale changes:

```
root@Rum% cli
root@Rum> configure
Entering configuration mode
The configuration has been changed but not committed

[edit]
root@Rum#
```

Fortunately, JUNOS offers a few ways out of this predicament. In this example, the pipe function is used to quickly compare the candidate and active configurations. This allows the user to make an intelligent decision as to what she should do next:

```
[edit]
root@Rum# show | compare
[edit protocols]
+    rstp {
+        interface all {
+            edge;
+        }
+    }
```

The output clearly shows the configuration delta, and makes it obvious that committing this change could have a significant impact on the network's operation. Although you could parse the logs to determine who made the change, and attempt to track them down for clarity, you could also perform a `rollback 0` to replace the adulterated candidate with a fresh copy of the active configuration:

```
[edit]
root@Rum# rollback 0
load complete

[edit]
root@Rum# show | compare

[edit]
root@Rum#
```

The (lack of) output shows the earlier change has been eliminated, leaving you free to make your edits without having to worry about unintended consequences stemming from someone else's changes. Note that JUNOS is very flexible. You could have used the save command to save a copy of the candidate before you did the rollback so that the uncommitted changes would not be lost should someone later confess to having made them. Alternatively, you could mandate use of exclusive configurations, which, as noted previously, prevents users from exiting configuration mode with uncommitted changes in the first place.

Loading and saving configurations

When in configuration mode, you use the save command to save all or part of a candidate configuration to local or network storage. To save an active rather than candidate configuration, issue a show configuration command and pipe the results to save when in operational mode, or perform a rollback 0 if in configuration mode to ensure that the active and candidate configurations are in sync.

Recall that save begins from the *current* hierarchy. To save the entire candidate configuration, make sure you are positioned at the top of the configuration hierarchy. In the following example, only eight lines of the configuration are saved, and all are related to OSPF area 0:

```
[edit protocols ospf area 0.0.0.0]
lab@Rum# save test
Wrote 8 lines of configuration to 'test'
```

In the next example, the addition of top ensures that the *whole* configuration is captured, resulting in 86 lines being written:

```
[edit protocols ospf area 0.0.0.0]
lab@Rum# top save test
Wrote 86 lines of configuration to 'test'

[edit protocols ospf area 0.0.0.0]
lab@Rum#
```

Sometimes it is not desirable to save the entire configuration; to save some portion of a configuration, simply navigate into the desired hierarchy. For instance, if every router in your network has the same system login information, you may want to save only that portion to load into other routers later, so a save is issued at the [edit system login] hierarchy:

```
[edit system login]
lab@Rum# save only_system_login
Wrote 12 lines of configuration to 'only_system_login'
```

The automatic system archival feature eliminates the need to manually save and then transfer your configuration files (using the `file copy` command); the feature can be set to archive upon each commit, or at preset intervals. The CLI's help feature is used to display some of the archival options:

```
lab@Rum# set system archival configuration ?
Possible completions:
+ apply-groups          Groups from which to inherit configuration data
+ apply-groups-except   Don't inherit configuration data from these groups
> archive-sites         List of archive destinations
  transfer-interval     Frequency at which file transfer happens (minutes)
  transfer-on-commit    Transfer after each commit
```

As with the **save** command in general, configuration snapshots can be transferred to a remote FTP or secure copy (SCP) server based on the URL details specified under the **archival-sites** hierarchy. The configuration shown here results in an FTP-based transfer at each commit. The switch will authenticate to the server as user **ftp** with password **ftp**, and the file will be uploaded to the */config/junos/Rum directory*:

```
archival {
    configuration {
        transfer-on-commit;
        archive-sites {
            "ftp://ftp:ftp@foo.bar/config/junos/Rum";
        }
```

Here, a manual load and network-based file copy operation is demonstrated. Note the use of **run** for the operational mode `file` command:

```
[edit]
lab@Rum# save rum_base_config
Wrote 124 lines of configuration to 'rum_base_config'

[edit]
lab@Rum# run file copy rum_base_config
ftp://instructor:training1@172.16.69.254/juniper/rum_config
ftp://instructor:training1@172.16.69.254/junip100% of 2386  B 1803 kBps

[edit]
lab@Rum#
```

The **load** command is the opposite of **save**. There are several important variations of this command; all require that you issue a **commit** before the change is placed into effect:

```
[edit]
lab@Rum# load ?
Possible completions:
  factory-default    Override existing configuration with factory default
  merge              Merge contents with existing configuration
  override           Override existing configuration
  patch              Load patch file into configuration
```

```
    replace               Replace configuration data
    set                   Execute set of commands on existing configuration
    update                Update existing configuration
[edit]
```

Most of the load options support an argument to control where the information that is loaded actually comes from (e.g., a terminal buffer, a local file, or a network resource).

Use the override option to completely replace the current candidate configuration with another one. Here, the contents of a file called *junos_is_cool*, which is located in the user's home directory, overrides the current candidate:

```
[edit]
lab@Rum# load override junos_is_cool
load complete
```

Use the merge option when you wish to *add to*, rather than *replace*, the candidate configuration. For instance, here you *add* the system login configuration that was saved previously to the current candidate, rather than replace it:

```
[edit]
lab@Rum# load merge only_system_login
load complete
```

Using a terminal buffer to cut and paste configuration data between boxes can be a real timesaver. There are several ways to cut and paste into a JUNOS configuration. The most common method is to use the terminal option to paste updated configuration data. The terminal data can override, or merge, with the existing candidate, as desired:

```
lab@Rum# load merge terminal
[Type ^D at a new line to end input]
system {
services {
        ftp;
        ssh;
        telnet;
    }
}
^D
load complete
```

Although quite useful, this method can take some practice. The proper number of levels and braces must always be present to avoid an error, because the terminal command assumes the entire top-level hierarchy is known and all delimiting braces are present.

This is shown in the following code, where the user is at the edit hierarchy and is trying to paste in data that relates to the [edit systems services] hierarchy. The problem is that the loaded data is missing the system portion of the hierarchy, resulting in errors:

```
[edit]
lab@Rum# load merge terminal
[Type ^D at a new line to end input]
services {
    ftp;
  terminal:2:(7) syntax error: ftp
```

```
[edit services]
  'ftp;'
     syntax error
ssh;
   telnet;
}
[edit]
  'services'
    warning: statement has no contents; ignored
load complete (1 errors)
```

It's times like these that using the **relative** switch saves the day. In this case, you first navigate to the desired configuration hierarchy, and then load the data, which is now understood to be *relative* to the current working location in the configuration:

```
[edit]
lab@Rum# edit system

[edit system]
lab@Rum# load merge terminal relative
[Type ^D at a new line to end input]
services {
    ftp;
    ssh;
    telnet;
}
^D
load complete
```

In summary, **load override** is used when you want to do a wholesale configuration file swap, and **load merge** adds the new information to the existing configuration. By using the **update** or **replace** option you can further control which portions of a configuration will be added to or replaced, respectively.

Yet another option is to load a series of **set** commands, as they would have been typed in configuration mode with the **load set** option:

```
lab@Rum# load set terminal
[Type ^D at a new line to end input]
set system services ftp
set system services ssh
set system services telnet
^D

load complete
```

Recall that a configuration mode **show** command (or a show configuration from operational mode) can have the results piped to **set**, to generate the **set** statements.

The JUNOS CLI Summary

This section introduced you to the JUNOS software CLI, and gave an honorable nod to the GUI-based J-Web interface, for those so inclined.

If you are new to Juniper, you have to admit its CLI is pretty slick, with features such as space-based command completion, tab-based variable completion, context-sensitive help, candidate configurations with basic sanity checking, rollbacks, confirmed commits... the list just goes on and on.

The next section builds on this foundation with a look at additional CLI and JUNOS functionality.

Advanced CLI and Other Cool Stuff

There are lots of other fantastic configuration options that can be used, and an explanation of all of them would require another book. The JUNOS documentation contains many timesaving tips, and *JUNOS Cookbook (http://oreilly.com/catalog/9780596100148/)* by Aviva Garrett (O'Reilly) is a great resource, too. To whet your appetite, here are three JUNOS software CLI tips.

SOS

If JUNOS is causing you problems, just ask it for help. You can do this in a few ways. Most people are instinctively aware of the option to use the question mark (?) to display possible command completions:

```
[edit]
lab@Rum# set protocols rstp ?
Possible completions:
  <[Enter]>              Execute this command
+ apply-groups           Groups from which to inherit configuration data
+ apply-groups-except    Don't inherit configuration data from these groups
  bridge-priority        Priority of the bridge (in increments of 4k - 0,4k,8k,..60k)
  disable                Disable STP
  forward-delay          Time spent in listening or learning state (4..30 seconds)
  hello-time             Time interval between configuration BPDUs (1..10 seconds)
> interface
  max-age                Maximum age of received protocol bpdu (6..40 seconds)
> traceoptions           Tracing options for debugging protocol operation
  |                      Pipe through a command
[edit]
lab@Rum# set protocols rstp
```

Note that the > character indicates a directory that contains subdirectories, and the + indicates a command that takes multiple arguments; no symbol means the command takes a single argument or is in fact the end statement of a command.

The help command is a secret resource of which few are aware. Sometimes a small piece of a command is remembered but not the full statement; help apropos can aid in finding the remaining syntax by searching through large portions of the JUNOS documentation for string matches. For example, let's see whether we can jog our memory on forward-delay:

```
[edit]
lab@Rum# help apropos forward-delay
set logical-routers <name> protocols stp forward-delay <forward-delay>
    Time spent in listening or learning state
set logical-routers <name> protocols rstp forward-delay <forward-delay>
    Time spent in listening or learning state
set logical-routers <name> protocols mstp forward-delay <forward-delay>
    Time spent in listening or learning state
set logical-routers <name> protocols vstp forward-delay <forward-delay>
    Time spent in listening or learning state
set protocols stp forward-delay <forward-delay>
    Time spent in listening or learning state
set protocols rstp forward-delay <forward-delay>
    Time spent in listening or learning state
set protocols mstp forward-delay <forward-delay>
    Time spent in listening or learning state
set protocols vstp forward-delay <forward-delay>
    Time spent in listening or learning state
```

Here, the results remind us that this is an STP parameter controlling time spent listening and learning, and we also see at which configuration hierarchies the statement can be applied. Even more information can be provided by issuing the help topic or help reference command. The former will display general usage guidelines for that command:

```
[edit]
lab@Rum# help topic stp ?
Possible completions:
  bridge-priority     Priority to become root bridge or LAN's designated bridge
  configuration-name  MSTP region name recorded in MSTP BPDUs
  cost                Link cost for determining designated bridge and port
  edge                Interface is edge port until BPDU is received
  extended-system-id  Bridge identifier for different routing instances
  force-version       Spanning tree version is original IEEE 803.1D STP
  forward-delay       Time STP bridge port remains in listening, learning state
  hello-time          Interval for root bridge sending configuration BPDUs
  interface           Interface participating in MSTP or RSTP instance
  interface-msti      Logical interface in Multiple Spanning Tree instance
  max-age             Maximum expected interval between hello BPDUs
  max-hops            Maximum number of hops to forward BPDU in MSTP region
  mode                Link mode to identify point-to-point links
  msti                Multiple Spanning Tree instance configuration
  priority            Priority of interface to become root port
  protocols           Protocol is Multiple STP or Rapid STP
  revision-level      Revision number of MSTP configuration
  trace-example       Sample tracing configuration for STP
  traceoptions        Trace options for Spanning Tree Protocols
  vlan                Single or range of VLAN IDs associated with MSTI
  vstp                VLAN Spanning Tree Protocol instance configuration
  vstp-requirements   Requirements, limitations for VLAN Spanning Tree Protocol
[edit]
lab@Rum# help topic stp forward-delay
                        Configuring the Forwarding Delay
```

The forwarding delay timer specifies the length of time an STP bridge port remains in the listening and learning states before transitioning to the forwarding state. Setting the interval too short could cause unnecessary spanning-tree reconvergence. Before changing this parameter, you should have a thorough understanding of STP.
To configure the forwarding delay timer, include the following statement:
  forward-delay seconds;
You can configure this statement at the following hierarchy levels:
  * [edit protocols protocol-name]
  * [edit routing-instances routing-instance-name protocols protocol-name]
  * [edit protocols vstp vlan vlan-id]
  * [edit routing-instances routing-instance-name protocols vstp vlan
        vlan-id]

After learning what a certain command actually accomplished and when it should be used, we can view the syntax and possible options with the help reference command:

```
[edit]
lab@Rum# help reference stp forward-delay
forward-delay

    Syntax

            forward-delay seconds;

    Hierarchy Level

        [edit protocols protocol-name],
        [edit routing-instances routing-instance-name protocols protocol-name],
        [edit protocols vstp vlan vlan-id],
        [edit routing-instances routing-instance-name protocols vstp vlan
        vlan-id]

    Release Information

        Statement introduced in JUNOS Release 8.4.

    Description

        Specify the length of time an STP bridge port remains in the listening
        and learning states before transitioning to the forwarding state.

    Options

        seconds--(Optional) Number of seconds the bridge port remains in the
        listening and learning states.
            Range: 4 through 30
            Default: 15 seconds

    Usage Guidelines

        See "Configuring the Forwarding Delay".

    Required Privilege Level
```

```
routing--To view this statement in the configuration.
routing-control
```

Scheduled Commits and Wildcards

Most changes that need to be made on a router can be done at only certain times, referred to as a *maintenance window*. Because these windows are often at the most inconvenient times for those who have to make them, changes represented by commit can actually be scheduled (comments can also be added):

```
[edit]
lab@Rum# commit at 19:00 comment "testing commit at feature"
configuration check succeeds
commit at will be executed at 2008-03-19 19:00:00 UTC
Exiting configuration mode
```

When a commit is pending, the database is locked, preventing other users from making any modifications:

```
lab@Rum> configure
Entering configuration mode
Users currently editing the configuration:
  lab terminal u0 (pid 3443) on since 2008-03-19 15:28:38 UTC
      commit-at

[edit]
lab@Rum# set system host-name foo
error: configuration database locked by:
  lab terminal u0 (pid 3443) on since 2008-03-19 15:28:38 UTC
      commit at
```

If the system needs to be unlocked before the specified time, an operational mode clear command can stop the timed action:

```
lab@Rum# run clear system commit
Pending commit cleared

[edit]
lab@Rum#
```

Wildcards and regular expressions

Wildcard and regex support are two features that operate together to simplify working with large configurations. You can use wildcards in operational and configuration modes, to display only certain values or to replace/delete matching strings, respectively. In this example, wildcards and a regex match are used to delete a range of matching interfaces:

```
[edit]
lab@Rum# wildcard delete interfaces ge-0/0/"[3-5]"
  matched: ge-0/0/3
  matched: ge-0/0/4
Delete 2 objects? [yes,no] (no) yes
```

Note the use of quotes around the regex itself, and the handy confirmation dialog that helps guard against the inevitable regex mistake. In this example, an operational mode command lists the status of only matching interfaces:

```
[edit]
lab@Rum# run show interfaces terse ge-0/0/"[3-4]"*
Interface              Admin Link Proto   Local               Remote
ge-0/0/3               up    up
ge-0/0/3.0             up    up   eth-switch
ge-0/0/4               up    up
ge-0/0/4.0             up    up   eth-switch
```

Lastly, common configuration changes can be made in one fell swoop with the `replace` command. Any string can replace any other string, with a string defined from a range of one character to any POSIX 1003.2 expression. The `replace` function is especially useful when you need to update an IP or MAC address that is referenced in various places, such as firewall filters, policies, protocols, and so on.

In this example, we update all instances of "200" with "201", which in this context refers to a VLAN tag:

```
[edit]
lab@Rum# replace pattern ^200$ with 201
```

The regex anchor characters ^ and $ are used to match only user variables that begin and end with exactly 200, which helps to avoid trouncing on unintentional matches such as might occur with an IP address containing the value 200. The CLI `compare` function is evoked against its default target, which is `rollback 0`, and the output confirms that only the VLAN ID has changed, while the IP address remains untouched due to the diligent use of regex matching:

```
[edit]
lab@Rum# show | compare
[edit interfaces ge-0/0/0 unit 0]
-    vlan-id 200;
-    family inet {
-        address 10.3.5.5/24;
-        address 200.0.0.1/24;
-    }
+    vlan-id 201;
+    family inet {
+        address 10.3.5.5/24;
+        address 200.0.0.1/24;
+    }
```

Copying, Renaming, and Inserting

The CLI also allows you to copy portions of the configuration, to rename one value to another, and to insert a string into an existing list of values. The latter is particularly useful for order-sensitive processing such as with ACLs, where you can insert a new address at the desired location without having to clear the entire list and rebuild it from scratch.

Here, a Layer 3 interface is copied and the `rename` function is used to provide it with a unique IP address:

```
[edit interfaces]
lab@Rum# copy ge-0/0/4 to ge-0/0/5

[edit interfaces]
lab@Rum# edit ge-0/0/5 unit 0

[edit interfaces ge-0/0/5 unit 0]
lab@Rum# show
family inet {
    filter {
        input test;
        output test;
    }
    address 10.5.7.5/24;
}

[edit interfaces ge-0/0/5 unit 0]
lab@Rum# rename family inet address 10.5.7.5/24 to address 10.9.9.10/24

[edit interfaces ge-0/0/5 unit 0]
lab@Rum# show
family inet {
    filter {
        input test;
        output test;
    }
    address 10.9.9.10/24;
}
```

Conclusion

This chapter discussed EX platform models and hardware capabilities as well as typical market applications. We also provided an overview of the initial Layer 2 and Layer 3 features, as well as a quick tour and operational guide to the JUNOS software that runs them.

The CLI is one of the most flexible and user-friendly frontends in the entire industry, allowing expert status to be achieved in record time. As your familiarity with the CLI increases, you will discover even more features. Now that the groundwork has been established, the rest of the book will dive into the specific EX switching and routing scenarios that will afford you much CLI practice. Yes, much indeed.

Chapter Review Questions

1. Which of the following Juniper Networks switches can support a VC?
   a. EX4200
   b. EX2300
   c. EX3200
   d. EX8200

2. Which hardware component provides the user CLI/management interface?
   a. Packet Forwarding Engine
   b. Route processor
   c. System control board
   d. Routing engine

3. True or false: adding a higher-wattage power supply can increase the number of PoE ports from the value the chassis initially ships with.

4. Which commands save the entire configuration? (Choose two.)
   a. `[edit]`
      `save config`
   b. `[edit protocols]`
      `save config`
   c. `save running configuration`
   d. `[edit]`
      `show | save config`

5. Which pipe argument is used to find every occurrence of the word *error* in some file?
   a. `match error`
   b. `find error`
   c. `search error`
   d. `except error`

6. What is the default password to enter configuration mode on the router?
   a. `juniper`
   b. `enable`
   c. None, a configuration password is not supported
   d. `root`

7. Which configuration mode command is issued to navigate to the `[edit protocols ospf]` directory?
   a. `cd protocols ospf`
   b. `edit protocols ospt`

    c. `cd /edit/protocols/ospf`

    d. `dir protocols ospf`

8. Which configuration mode command activates configuration changes?

    a. `apply`

    b. `copy`

    c. `save`

    d. `commit`

9. What is the top level of the configuration tree called?

    a. *C:/*

    b. *top*

    c. *edit*

    d. *root*

10. After performing a commit, which CLI command returns to the previously activated configuration?

    a. `rollback 1`

    b. `rollback 0`

    c. `rollback active`

    d. `rollback previous`

11. Which command is used to obtain detailed information about configuring OSPF area ranges?

    a. `ospf ?`

    b. `help topic ospf area-range`

    c. `help reference ospf area-range`

    d. `man ospf`

Chapter Review Answers

1. Answer: A. Only the EX4200 can be used as part of a VC.

2. Answer: D. The routing engine is the component in the router that controls all management functions, including commands that would be used to debug the router.

3. Answer: False. Installing a PSU with additional amperage does not activate any additional PoE ports. PoE port count is determined by part number only. However, installing an insufficient PSU will result in deactivation of PoE ports to bring the PoE load to within that PSU's budget.

4. Answer: A and D. The **save** command saves the configuration from the current hierarchy down. Displaying the configuration and piping it to the CLI's **save** function has the same effect.

5. Answer: A. The pipe command **match** will find every occurrence of a string in the output of the command. The **find** command will locate the first occurrence of the string, **search** is an invalid option, and **hold** will hold text without exiting the **-More--** prompt.

6. Answer: C. There is no password to enter configuration mode. Users are allowed into configuration mode based on access privileges.

7. Answer: B. Recall that in JUNOS the **edit** statement functions like a **cd** (change directory) command. While there is no **cd** command, you use the **edit** statement to park at some hierarchy. Once so positioned, you enter configuration statements that are qualified by the current hierarchy, which means that only relevant syntax is displayed with the CLI **help** function. In contrast, when at the top (**[edit]**) of the configuration hierarchy, you can modify anything in the configuration but must issue fully qualified statements to remove any ambiguities.

8. Answer: D. To activate changes in the switch's configuration, issue a **commit** command. Of the remaining options, **copy** and **save** are valid CLI commands but are used for configuration management.

9. Answer: C. When at the top level of the configuration tree, the CLI banner will display the **[edit]** prompt.

10. Answer: A. The first archive is stored in **rollback 1**. **rollback 0** is used to copy the active configuration to the candidate configuration; the other options are not valid **rollback** commands.

11. Answer: C. **help topic** commands provide a functional overview and **help reference** commands detail configuration options.

## CHAPTER 3

Initial Configuration and Maintenance

Now that we have introduced you to JUNOS, it is time to get started with initial switch configuration. This chapter walks through the configuration and hardware status checks typically performed when installing a brand-new EX switch. A new switch is set to a factory-default configuration, and the goal of this chapter is to teach you how to configure the switch with the most common settings needed to get the chassis up and running with network management and user accounts. The topics covered include:

- Factory-default configuration and EZSetup
- Initial configuration using the command-line interface (CLI)
- Secondary configuration using the CLI
- EX interfaces
- Chassis monitoring and network tools

JUNOS is a flexible operating system and there are several ways to skin the initial configuration cat, which is not to say that any animals were harmed during production of this material. We begin with a quick overview of the EZSetup Wizard, but the focus rapidly shifts to the CLI, where the real work can be quickly done.

This chapter walks you through the addition of a new Juniper Networks EX4200 switch, in this case named Tequila, using both EZSetup and the CLI, and demonstrates chassis status monitoring and validation of initial configuration services and operation. The new switch is being integrated into a legacy switched network that comprises two Cisco Catalyst 3550 switches named Whiskey and Gin. Figure 3-1 shows the planned topology expansion.

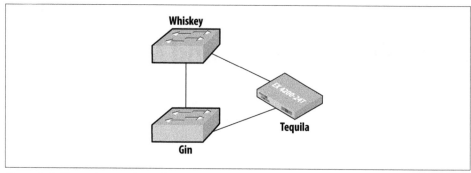

Whiskey

Gin

Tequila

Figure 3-1. Planned network topology expansion

The Factory-Default Configuration and EZSetup

As we discussed previously, the EX4200 switches are rack-, desk-, or wall-mountable. In our case, the switch is rack-mounted using the mounting brackets and rubber feet supplied with the new switch. This is primarily a spinal type of effort, and we will not document it further here. Details on rack mounting, and other installation options, are available in the user documentation.

Factory-Default Configuration

When the switch is installed in the rack, it contains a factory-default configuration. A factory-default configuration is not blank; it contains the following default behavior and settings:

- Enables Layer 2 switching mode on all ports
- Configures system logging
- Enables Link Layer Discovery Protocol (LLDP) and Rapid Spanning Tree Protocol (RSTP) on all interfaces
- Enables Power over Ethernet (PoE) on all supported ports
- Enables Internet Group Management Protocol (IGMP) snooping
- Resets the LCD menu and any Virtual Chassis (VC) configuration on commit

The following code displays the factory-default configuration. Note that only the root user can log in when the default configuration is active and no password is set:

```
root@% root
root@% OS 9.2R1.9 built 2008-08-05 07:25:22 UTC
root@% cli
root> configure
Entering configuration mode
[edit]
root# show
## Last changed: 2008-08-05 11:13:01 UTC
```

```
version 9.2R1.9;system {
    syslog {
        user * {
            any emergency;
        }
        file messages {
            any notice;
            authorization info;
        }
        file interactive-commands {
            interactive-commands any;
        }
    }
    commit {
        factory-settings {
            reset-chassis-lcd-menu;
            reset-virtual-chassis-configuration;
        }
    }
    ## Warning: missing mandatory statement(s): 'root-authentication'
}
interfaces {
    ge-0/0/0 {
        unit 0 {
            family ethernet-switching;
        }
    }
    ge-0/0/1 {
        unit 0 {
            family ethernet-switching;
        }
    }
    ge-0/0/2 {
        unit 0 {
            family ethernet-switching;
        }
    }
    ge-0/0/3 {
        unit 0 {
            family ethernet-switching;
        }
    }
    ge-0/0/4 {
        unit 0 {
            family ethernet-switching;
        }
    }
    ge-0/0/5 {
        unit 0 {
            family ethernet-switching;
        }
    }
    ge-0/0/6 {
        unit 0 {
            family ethernet-switching;
```

```
        }
    }
    ge-0/0/7 {
        unit 0 {
            family ethernet-switching;
        }
    }
    ge-0/0/8 {
        unit 0 {
            family ethernet-switching;
        }
    }
    ge-0/0/9 {
        unit 0 {
            family ethernet-switching;
        }
    }
    ge-0/0/10 {
        unit 0 {
            family ethernet-switching;
        }
    }
    ge-0/0/11 {
        unit 0 {
            family ethernet-switching;
        }
    }
    ge-0/0/12 {
        unit 0 {
            family ethernet-switching;
        }
    }
    ge-0/0/13 {
        unit 0 {
            family ethernet-switching;
        }
    }
    ge-0/0/14 {
        unit 0 {
            family ethernet-switching;
        }
    }
    ge-0/0/15 {
        unit 0 {
            family ethernet-switching;
        }
    }
    ge-0/0/16 {
        unit 0 {
            family ethernet-switching;
        }
    }
    ge-0/0/17 {
        unit 0 {
            family ethernet-switching;
```

```
        }
    }
    ge-0/0/18 {
        unit 0 {
            family ethernet-switching;
        }
    }
    ge-0/0/19 {
        unit 0 {
            family ethernet-switching;
        }
    }
    ge-0/0/20 {
        unit 0 {
            family ethernet-switching;
        }
    }
    ge-0/0/21 {
        unit 0 {
            family ethernet-switching;
        }
    }
    ge-0/0/22 {
        unit 0 {
            family ethernet-switching;
        }
    }
    ge-0/0/23 {
        unit 0 {
            family ethernet-switching;
        }
    }
    ge-0/1/0 {
        unit 0 {
            family ethernet-switching;
        }
    }
    xe-0/1/0 {
        unit 0 {
            family ethernet-switching;
        }
    }
    ge-0/1/1 {
        unit 0 {
            family ethernet-switching;
        }
    }
    xe-0/1/1 {
        unit 0 {
            family ethernet-switching;
        }
    }
    ge-0/1/2 {
        unit 0 {
            family ethernet-switching;
```

```
                }
            }
            ge-0/1/3 {
                unit 0 {
                    family ethernet-switching;
                }
            }
        }
        protocols {
            igmp-snooping {
                vlan all;
            }
            lldp {
                interface all;
            }
            lldp-med {
                interface all;
            }
            rstp;
        }
        poe {
            interface all;
        }
```

 Notice the existence of the ge-0/1/1 and xe-0/1/1 uplink interfaces in
the factory-default configuration. In JUNOS, interfaces are allowed to
be configured even if they do not physically exist. The factory-default
configuration takes into account that the uplink module could be either
1G or 10G, and the missing physical module will simply be ignored.

The attempt to commit with a factory default fails due to lack of a root authentication
setting. This is expected:

```
[edit]
root commit
[edit]
  'system'
    Missing mandatory statement: 'root-authentication'
error: commit failed: (missing statements)

[edit]
root# quit
Exiting configuration mode

root> exit

root@%
```

To return a switch to the factory-default configuration, issue the `load factory-default` command in configuration mode or navigate the LCD menu by clicking the Menu button and choosing "Restore to factory default." Note that after the configuration is loaded, it still cannot be committed due to lack of a root password. This safety measure ensures that this important password is set before you install the router in your production network.

However, because such a setting alters the factory default, once the setting is committed the EZSetup feature is no longer available. EZSetup runs only with a true, unaltered factory default. See the warning at the end of the next section for more details.

EZSetup

The EZSetup feature is designed to walk you through the most common configuration settings to get your switch up and running quickly. The functionality is similar to the initial configuration dialog seen on IOS-based devices. After logging in as root you have the choice of entering the router's operational mode by typing **cli** or **ezsetup**. You can also access EZSetup through the LCD menu by choosing Initial Setup or via J-Web by connecting to either the ge-0/0/0 or the me0 interface and directing your browser to 192.168.1.1.

 To access the CLI-based EZSetup feature, you must have a console connection to the switch. The switch is supplied with an RJ-45 to DB-9 EIA-232 cable for this purpose.

Once activated, EZSetup prompts you for the following configuration parameters:

- Hostname
- Telnet service
- SSH service
- Switch management
- Simple Network Management Protocol (SNMP)
- Time
- Time zone

You launch EZSetup from a root shell prompt when a factory-default config is active, as shown in the following code. In this example, various parameters such as hostname and Out of Band (OoB) management addresses are assigned:

```
root@% ezsetup
**********************************************************************
* EZSetup wizard                                                     *
```

```
*                                                                         *
* Use the EZSetup wizard to configure the identity of the switch.         *
* Once you complete EZSetup, the switch can be accessed over the network. *
*                                                                         *
* To exit the EZSetup wizard press CTRL+C.                                *
*                                                                         *
* In the wizard, default values are provided for some options.            *
* Press ENTER key to accept the default values.                          *
*                                                                         *
* Prompts that contain [Optional] denotes that the option is not mandatory. *
* Press ENTER key to ingore the option and continue.                     *
*                                                                         *
*************************************************************************

EZSetup Initializing..done.

Initial Setup Configuration
--------------------------


Enter System hostname [Optional]:Tequila

Enter new root password:
Re-enter the new password:

Enable Telnet service? [yes|no]. Default [no]:yes

Enable SSH service? [yes|no]. Default [yes]:

Switch Management
1: Configure in-band Management [interface vlan.XX]
2: Configure out-of-band management [me0.0]
Choose Option [1 or 2], default [1]:2
Configuring me0.0
Enter Management IP address [192.168.1.1]:172.16.69.4
Enter Subnetmask [255.255.255.0]:
Enter Gateway IP address:
Invalid IP address format, expected <aa.bb.cc.dd>
Enter Gateway IP address:172.16.69.9

Configure SNMP [yes|no], default [yes]:yes
SNMP Configuration
Contact information for administrator, Enter contact [Optional]:
Community name, Enter community [Optional]:underdogs
Physical location of system, Enter location [Optional]:

Enter System Time and Date YYYY:MM:DD:hh:mm:ss [Optional]:

Time Zone [Optional], Enter "yes" to choose Timezone from list:yes
Select Timezone
Africa/Abidjan
Africa/Accra
Africa/Addis_Ababa
Africa/Algiers
Africa/Asmera
```

```
Africa/Bamako
Africa/Bangui
Africa/Banjul
Africa/Bissau
Africa/Blantyre
Africa/Brazzaville
...
...
....

America/Los_Angeles
America/Louisville
America/Maceio
America/Managua

Press key n to stop and any other key to continue...

Press key n to stop and any other key to continue...
Enter Timezone:America/Los_Angeles

The input configuration parameters are

System Hostname:              Tequila
Root password:                ******
System Telnet Service:        yes
System SSH Service:           yes
Management IP Address:        172.16.69.4
SubnetMask:                   255.255.255.0
Gateway IP Address:           172.16.69.9
Out-of-band management:       me0.0
SNMP Community:               underdogs
SNMP Location:
SNMP Contact:
Time-Date:
Time-zone:                    America/Los_Angeles
Interfaces:


Commit the new configuration?
Choosing option "yes" will add new configuration to existing configuration.
Option "No" will allow user to come out of EZSetup wizard.

Choose option [yes|no], default [yes]:yes

Committing the new configuration, please wait.....
Commit success.
root@tequila% cli
```

The following code shows the updated configuration. Notice that the system was committed and the hostname was changed. The highlighted portions of the configuration show the configuration delta that resulted from answering the EZSetup prompts, and the rest of the configuration was inherited from the factory-default configuration:

```
root@tequila> show configuration
## Last commit: 2008-08-05 04:41:28 PDT by root
```

```
version 9.2R1.9;
system {
    host-name Tequila;
    time-zone America/Los_Angeles;
    root-authentication {
        encrypted-password bJOYvGXKxTdr6; ## SECRET-DATA
    }
    services {
        ssh;
        telnet;
        web-management {
            http;
        }
    }
    syslog {
        user * {
            any emergency;
        }
        file messages {
            any notice;
            authorization info;
        }
        file interactive-commands {
            interactive-commands any;
        }
    }
}
interfaces {
    ge-0/0/0 {
        unit 0 {
            family ethernet-switching;
        }
    }
    ge-0/0/1 {
        unit 0 {
            family ethernet-switching;
        }
    }
    ge-0/0/2 {
        unit 0 {
            family ethernet-switching;
        }
    }
    ge-0/0/3 {
        unit 0 {
            family ethernet-switching;
        }
    }
    ge-0/0/4 {
        unit 0 {
            family ethernet-switching;
        }
    }
    ge-0/0/5 {
        unit 0 {
```

```
                family ethernet-switching;
            }
    }
    ge-0/0/6 {
        unit 0 {
            family ethernet-switching;
        }
    }
    ge-0/0/7 {
        unit 0 {
            family ethernet-switching;
        }
    }
    ge-0/0/8 {
        unit 0 {
            family ethernet-switching;
        }
    }
    ge-0/0/9 {
        unit 0 {
            family ethernet-switching;
        }
    }
    ge-0/0/10 {
        unit 0 {
            family ethernet-switching;
        }
    }
    ge-0/0/11 {
        unit 0 {
            family ethernet-switching;
        }
    }
    ge-0/0/12 {
        unit 0 {
            family ethernet-switching;
        }
    }
    ge-0/0/13 {
        unit 0 {
            family ethernet-switching;
        }
    }
    ge-0/0/14 {
        unit 0 {
            family ethernet-switching;
        }
    }
    ge-0/0/15 {
        unit 0 {
            family ethernet-switching;
        }
    }
    ge-0/0/16 {
        unit 0 {
```

```
            family ethernet-switching;
        }
    }
    ge-0/0/17 {
        unit 0 {
            family ethernet-switching;
        }
    }
    ge-0/0/18 {
        unit 0 {
            family ethernet-switching;
        }
    }
    ge-0/0/19 {
        unit 0 {
            family ethernet-switching;
        }
    }
    ge-0/0/20 {
        unit 0 {
            family ethernet-switching;
        }
    }
    ge-0/0/21 {
        unit 0 {
            family ethernet-switching;
        }
    }
    ge-0/0/22 {
        unit 0 {
            family ethernet-switching;
        }
    }
    ge-0/0/23 {
        unit 0 {
            family ethernet-switching;
        }
    }
    ge-0/1/0 {
        unit 0 {
            family ethernet-switching;
        }
    }
    xe-0/1/0 {
        unit 0 {
            family ethernet-switching;
        }
    }
    ge-0/1/1 {
        unit 0 {
            family ethernet-switching;
        }
    }
    xe-0/1/1 {
        unit 0 {
```

```
                family ethernet-switching;
            }
        }
        ge-0/1/2 {
            unit 0 {
                family ethernet-switching;
            }
        }
        ge-0/1/3 {
            unit 0 {
                family ethernet-switching;
            }
        }
        me0 {
            unit 0 {
                family inet {
                    address 172.16.69.4/24;
                }
            }
        }
    }
    snmp {
        community underdogs {
            authorization read-only;
        }
    }
    routing-options {
        static {
            route 0.0.0.0/0 next-hop 172.16.69.9;
        }
    }
    protocols {
        igmp-snooping {
            vlan all;
        }
        lldp {
            interface all;
        }
        lldp-med {
            interface all;
        }
        rstp;
    }
    poe {
        interface all;
    }
```

 As of JUNOS 9.2R1.9, the only way to evoke EZSetup after committing any changes is to delete the files in the */config* directory from a root shell, and then reboot the switch to force the load of a *true* factory config. Using load factory-default does not suffice; a root password must be set before you can commit the loaded configuration, after which the config is no longer considered factory-default enough to prompt for EZSetup.

Factory-Default Configuration and EZSetup Summary

When you purchase a new switch, the system is loaded with a factory-default configuration that enables Layer 2 switching on all interfaces as well as protocols such as RSTP and LLDP. One way to get the initial installation into the router is to use the EZSetup feature. A second way is to configure each statement via CLI commands. We will examine this option next.

Initial Configuration Using the CLI

Although the EZSetup menu is, well, easy, it often does not contain all the parameters that are needed in a specific deployment. As a result, users often bypass EZSetup and configure the switch via the CLI. Juniper recommends that you configure the following parameters during initial configuration. Note that with the exception of the system's domain name, all parameters can be configured via EZSetup, when desired.

This section begins by showing the commands that you would type to offer the equivalent functionality to EZSetup. We won't go into a detailed explanation here, as we discuss each command and topic in detail later in this chapter.

Hostname
Sets the switch's name:

```
root# set system host-name Tequila
```

Root password
Sets the password for user root:

```
root# set system root-authentication plain-text-password
New password:
Retype new password:
```

Remote access protocols
Connects to the router without a physical connection:

```
root# set system services ssh
root# set system services telnet
```

Default VLAN management (EZSetup option only)
Uses the default virtual LAN (VLAN) or a new VLAN for management. If you select the New VLAN option, you are prompted to specify the VLAN name, VLAN ID, management IP address, and default gateway. Select the ports that must be part of this VLAN. See Chapter 5 for VLAN details.

OoB management IP
Uses the dedicated OoB management interface:

```
root# set interfaces me0 unit 0 family inet address 172.16.69.8/24
```

OoB default route
> Creates a default route for the OoB network and ensures, via the no-export tag, that this route is never redistributed into any protocols:
>
> ```
> root# set routing-options static route 0.0.0.0/0 next-hop 172.16.69.9 no-export
> ```

Domain name
> Sets the Domain Name System (DNS) entry:
>
> ```
> root# set system domain-name ietraining.net
> ```

SNMP contact and community information
> Adds contact information:
>
> ```
> root# set snmp community underdogs authorization read-only
> ```

System time and time zone
> Sets the time and time zone:
>
> ```
> root@Tequila> set date ?
> Possible completions:
>   <time>  New date and time (YYYYMMDDhhmm.ss)
>   ntp     Set system date and time using Network Time Protocol servers
> root# set system time-zone America/Los_Angeles
> ```

The candidate configuration is activated to put the initial configuration into effect. Recall that with JUNOS, "If you don't commit, it don't mean feces" (euphemism inserted):

```
[edit]
root# commit
commit complete
[edit]
root@Tequila#
```

Note that after the commit, the system prompt changes to reflect the newly assigned hostname. With the preliminary configuration committed, attention now shifts to additional configuration that is often needed as part of the installation of a new switch.

CLI Configuration Summary

This section showed how quickly you can get a JUNOS device up and running with a few fell strokes of the CLI interface. The next section builds upon this information with some secondary configuration items that you should consider in any install.

Secondary Configuration

After completing the initial configuration via either the CLI or EZSetup, you will likely need additional configuration, such as:

* Non-root-user accounts and privileges

- OoB management
- Additional remote access functionality
- Dynamic Host Configuration Protocol (DHCP) services

Customized User Accounts, Authentication, and Authorization

There are two types of users on a Juniper Networks system: non-root users and the root user. All users must be authenticated before they can access the switch, and when desired, various levels of authorization are possible to limit the scope of actions or commands available to users in each class. Recall that the root user is the only predefined user, and that root can log in only via the console port until SSH access is configured; the root user is not permitted to remotely access the router via Telnet for security reasons.

 Non-root users can telnet to the router and su to root when authorized if the SSH service is not running. This is not recommended, as Telnet sends in plain text.

You must set a root password before the switch will allow you to commit a modified factory-default configuration. As we showed previously, to set up a root password, issue the set root-authentication statement under the [edit system] level. There are several options:

```
root@Tequila# set system root-authentication ?
Possible completions:
+ apply-groups          Groups from which to inherit configuration
                        data
+ apply-groups-except   Don't inherit configuration data from these
                        groups
  encrypted-password    Encrypted password string t
  load-key-file         File (URL) containing one or more ssh keys
  plain-text-password   Prompt for plain text password (autoencrypted)
> ssh-dsa               Secure shell (ssh) DSA public key string
> ssh-rsa               Secure shell (ssh) RSA public key string
```

The password can be a plain-text password that is encrypted automatically in the configuration, an SSH key, or an encrypted string for copying and pasting between other configurations. In this case, a password of jncie123 is supplied:

```
root@Tequila# set system root-authentication plain-text-password
New password:
Retype new password:
```

When issuing a plain-text password, JUNOS enforces default requirements for password length and structure to help ensure that a strong password is used. The password must be between 6 and 128 characters and must contain one change of case or special character. You can modify these defaults under [edit system password].

Once the password is set, JUNOS automatically displays an encrypted value; there is no need for a password encryption service as in IOS, and only Message Digest 5 (MD5) is supported to prevent any reverse engineering attempts, as often occurs with Cisco type 7 password encryption:

```
[edit]
root@Tequila# show system root-authentication
encrypted-password "$1$G8r2A1XZ$NhOj8UP.Z3J/NGsWcBr9kO"; ## SECRET-DATA
```

The SECRET-DATA tag is used to prevent users who do not have authorization from even viewing the encrypted portion of the configuration; nice and tight, just the way it should be. The encrypted string plays an important role; it allows the root password to be copied to an additional switch without user knowledge of the actual password. Here the password is loaded into another new switch named Vodka:

```
root@Vodka# load merge terminal relative
[Type ^D at a new line to end input]
encrypted-password "$1$G8r2A1XZ$NhOj8UP.Z3J/NGsWcBr9kO"; ## SECRET-DATA
load complete

[edit system root-authentication]
root@Vodka# show
encrypted-password "$1$G8r2A1XZ$NhOj8UP.Z3J/NGsWcBr9kO"; ## SECRET-DATA
```

Now let's move on to non-root users. These users can be defined with local user passwords and permissions, or an external server such as RADIUS or TACACS can be used for authentication and authorization. Figure 3-2 demonstrates.

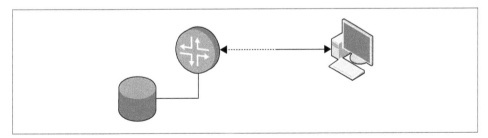

Figure 3-2. Server authentication and authorization

To use RADIUS or TACACS, two items must be configured: the server parameters and the authentication order. In JUNOS software, the authentication order determines the order in which the switch looks for a valid login attempt. For instance, the following configuration checks the authentication criteria against the TACACS+ server first, and if a communications (not an authentication) failure occurs, it then tries the next method, which is local authentication via the local password database:

```
root@tequila# set system authentication-order [tacplus password]
```

The authentication order has often been misunderstood and misconfigured in JUNOS. The most important point to remember is that the system will always try every method listed until success is achieved. So, if you want to make sure the local database is not consulted if the RADIUS or TACACS server returns a reject message, *do not* place the password keyword in the configuration. For example:

```
root@tequila# set system authentication-order tacplus
```

If you're betting on the fact that the local database will never be consulted, I suggest you stay away from Las Vegas. If the specified server is not reachable, the local database is *always* consulted. This default behavior is to protect *you* from being locked out of the switch if access to the authentication server is lost for any reason.

When creating a user or user template, three items need to be configured:

- Username
- Authorization
- Password

These three items could be on the switch, server, or a combination of both. Several levels of the authorization components can be configured, as shown in Figure 3-3.

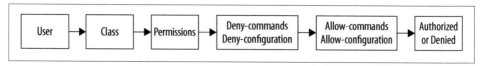

Figure 3-3. User authorization hierarchy

First, each user is applied to a class that defines the permissions for that user. The class can be predefined, as shown in Table 3-1, or user-created. A user's class determines the user's overall level of permissions. If the class permissions are a bit too broad or narrow, you can explicitly deny or allow additional commands that can be denied or allowed from the default permission bit setting. For instance, a class with the permission setting of reset allows a restart of all interfaces and processes. This may open up too many opportunities for user mistakes, so a tighter lockdown may be preferred, which you can specify via a set of deny commands with a regular expression. Similarly, you can explicitly choose to allow commands that would otherwise be denied by a given class with the allow keyword.

User authentication case study

In this section, we will provide a case study to illustrate some possible options associated with user definition.

Those familiar with *JUNOS Enterprise Routing*, by Doug Marschke and Harry Reynolds (O'Reilly), may find this case study a bit familiar. Despite the authors' apparent laziness, issuing a definition on a switch is no different from doing so on a router, and the selected case study represents a realistic scenario. JUNOS is, after all, JUNOS.

The case study sets the following requirements:

1. Define three local users, doug, harry, and lab, and provide them with maximum access.

2. Create a Network Operations Center (NOC) group consisting of 15 engineers. Each NOC engineer will have his own username, but will share the same permissions of read-only commands and maintenance commands for troubleshooting.

3. Create a design engineer group consisting of three engineers. This group will have full access to all CLI commands, except for the restart and request commands.

4. All users will be authenticated using a RADIUS server with a shared secret of ronkittle.

5. Authorization is defined on the local switch.

6. If the RADIUS server is down, only harry, lab, and doug may log in to the switch.

 One user that is not explored in this case study is the remote user. This is a user profile created for use on the switch when the authenticated user does not have a local switch profile, or the authenticated user's record in the authentication server specifies a local user. You can think of this as a default fallback account.

Each defined user must be associated with a login class, which assigns the permissions for the user. The login class can be one of the four default classes listed in Table 3-1, or a custom-defined class.

Table 3-1. Predefined JUNOS user classes

Class	Permissions
super-user	All
read-only	View
operator	Clear, Network, Reset, Trace, View
unauthorized	None

Users harry, lab, and doug require maximum access, so it makes sense to use a predefined JUNOS software class called super-user. Here we show the step-by-step process for harry only, as the process for users doug and lab is identical:

```
root@Tequila# set system login user harry class super-user authentication
plain-text-password
New password:
Retype new password:

[edit]
root@Tequila# show system login
user harry {
    class super-user;
    authentication {
        encrypted-password "$1$oOspqmHP$jlxUulOcAgPq3j88/7WQP/";
        ## SECRET-DATA
    }
}
```

Next, a group of 15 NOC engineers are defined. Since configuring 15 local users will be a pain to manage and tiresome to type, we will use a *user template*. A user template allows multiple users defined on the RADIUS server with unique passwords to be grouped to a single local Juniper user. Since a predefined class will not satisfy the authorization level for the NOC engineers of read-only commands and maintenance commands, we will define a custom class:

```
[edit system login]
root@Tequila# set class ops permissions [view maintenance trace]
```

 Refer to the access-privilege technical documentation to see each command that is allowed for every permission setting.

Next, we assign the user ops the new class, also called ops:

```
[edit system login]
root@Tequila# set user ops class ops

[edit system login]
root@Tequila# show class ops
permissions [ trace view maintenance ];
root@Tequila# show user ops
uid 2000;
class ops;
```

The RADIUS server will then have 15 users defined that all map to the same Juniper-local user of ops. For example, the configuration for 2 of the 15 users using a RADIUS server would be similar to the following:

```
becca   Auth-Type = Local, Password = "authorsgf"
        Service-Type = Login-User,
        Juniper-Local-User-Name = "ops"

calumet Auth-Type = Local, Password = "hometown!"
        Service-Type = Login-User,
        Juniper-Local-User-Name = "ops"
```

The design engineer group requirement will also use a template, but will make use of special allow and deny commands that we can also define in a class. If the permission bits that are set are too broad, we can deny individual commands within the permission settings. (And vice versa: if we need an additional command or set of commands that go beyond the permission setting, we can allow them.) These allow and deny statements could be a single command or a group of commands using regular expressions. They are also separated in allow or deny operational mode commands or configuration mode:

```
[edit system login]
root@Tequila# set class design ?
Possible completions:
  allow-commands        Regular expression for commands to allow explicitly
  allow-configuration   Regular expression for configure to allow explicitly
+ apply-groups          Groups from which to inherit configuration data
+ apply-groups-except   Don't inherit configuration data from these groups
  deny-commands         Regular expression for commands to deny explicitly
  deny-configuration    Regular expression for configure to deny explicitly
  idle-timeout          Maximum idle time before logout (minutes)
  login-alarms          Display system alarms when logging in
  login-tip             Display tip when logging in
+ permissions           Set of permitted operation categories
```

The design engineer's class will have the permission bits set to all, and all commands that start with r (request and restart) will be disallowed:

```
[edit system login]
root@Tequila# set class design permissions all

[edit system login]
root@Tequila# set class design deny-commands "^r.*$"

root@Tequila# set user design class design
```

 Although regular expressions are beyond the scope of this chapter, here's a quick list of common operators:

- . (any character)
- * (zero or more characters)
- ^ (start of string to which the regex is applied)
- $ (end of string to which the regex is applied)
- ? (zero or one character)

As mentioned, we can define users locally on the switch or on an external server such as RADIUS or TACACS. In this chapter's case study, we specified a RADIUS server earlier, in requirement 4. The RADIUS server's IP address and secret password are configured:

```
[edit system]
root@Tequila# set radius-server 10.20.130.5 secret ronkittle
```

For the system to use the RADIUS server, we must configure the authentication-order statement. This indicates which order of authentication method should be used; the default is the local switch database only. In this section of our case study, we must decide between the following configuration choices:

1. authentication-order [radius password]
2. authentication-order [radius]

In either configuration, the local database will be consulted if the RADIUS server is down, so the difference between the two options is evident when the RADIUS server returns a reject. This reject could be caused by a mistyped password or a username that is not defined in the RADIUS server. In option 1, the RADIUS server returns the reject and the local database will be consulted. Option 2 consults the local database only if the RADIUS server is unresponsive; processing stops if the server returns a reject message. The requirements state that the RADIUS server should always be used when available (as specified in option 1). If the RADIUS server is not available, users doug and harry will be allowed to log in using the local database since they are the only users with locally defined passwords on the switch. These users are also defined on the RADIUS server:

```
doug    Auth-Type = Local, Password = "reddawnrocks1"
        Service-Type = Login-User
```

Here is a complete system login configuration that meets all six of the criteria specified earlier:

```
[edit system]
root@Tequila# show
host-name Tequila;
authentication-order radius password;
ports {
    console type vt100;
}
root-authentication {
    encrypted-password "$1$85xXcov4$fLHtgMlqxRSg24zO8Kbe81"; ##
    SECRET-DATA
}
radius-server {
    10.20.130.5 secret "$9$KdgW87db24aUcydsg4Dj69AORSWLN24ZNd.5TFAt";
    ## SECRET-DATA
}
login {
    class design {
        permissions all;
        deny-commands "^r.*$";
    }
    class ops {
        permissions [ trace view maintenance ];
    }
    user design {
        uid 2004;
        class design;
```

```
                    user harry {
                        uid 2001;
                        class super-user;
                        authentication {
                            encrypted-password "$1$oOspqmHP$jlxUulOcAgPq3j88/7WQP/";
                            ## SECRET-DATA
                        }
                    }
                    user lab {
                        uid 2002;
                        class superuser;
                        authentication {
                            encrypted-password " $1$wD/I1Ybw$M7a/X51Gk36xaRs3XNRuQ1";
                            ## SECRET-DATA
                        }
                    }
                    user doug {
                        uid 2003;
                        class superuser;
                        authentication {
                            encrypted-password "$1$ocs3AXkS$JdlQW7z4ZIJblfFZD.fqH/";
                            ## SECRET-DATA
                        }
                    }
                    user ops {
                        uid 2000;
                        class ops;
                    }
                }
                services {
                    ftp;
                    ssh;
                    telnet;
                }
                syslog {
                    user * {
                        any emergency;
                    }
                    file messages {
                        any notice;
                        authorization info;
                    }
                    file interactive-commands {
                        interactive-commands any;
                    }
                }
```

Lastly, to verify that the user has the correct permissions, log in to the switch and issue a show cli authorization command:

```
design@Tequila> show cli authorization
Current user: 'design       ' class 'design'
Permissions:
    admin        -- Can view user accounts
    admin-control-- Can modify user accounts
```

```
clear        -- Can clear learned network information
configure    -- Can enter configuration mode
control      -- Can modify any configuration
edit         -- Can edit full files
field        -- Special for field (debug) support
floppy       -- Can read and write from the floppy
interface    -- Can view interface configuration
interface-control-- Can modify interface configuration
network      -- Can access the network
reset        -- Can reset/restart interfaces and daemons
routing      -- Can view routing configuration
routing-control-- Can modify routing configuration
shell        -- Can start a local shell
snmp         -- Can view SNMP configuration
snmp-control-- Can modify SNMP configuration
system       -- Can view system configuration
system-control-- Can modify system configuration
trace        -- Can view trace file settings
trace-control-- Can modify trace file settings
view         -- Can view current values and statistics
maintenance -- Can become the super-user
firewall     -- Can view firewall configuration
firewall-control-- Can modify firewall configuration
secret       -- Can view secret configuration
secret-control-- Can modify secret configuration
rollback     -- Can rollback to previous configurations
security     -- Can view security configuration
security-control-- Can modify security configuration
access       -- Can view access configuration
access-control-- Can modify access configuration
view-configuration-- Can view all configuration (not including
secrets)
Individual command authorization:
    Allow regular expression: none
    Deny regular expression: ^r.*$
    Allow configuration regular expression: none
    Deny configuration regular expression: none
```

Out of Band Network

The legacy network was deployed with an OoB management network that interconnects the routers and switches to allow management access, even during periods of
network malfunction that disrupt in-band network operations. Figure 3-4 shows the
OoB network details. The management network is a 172.16.69.0/24 subnet, and physical connections go through a VLAN-aware switch.

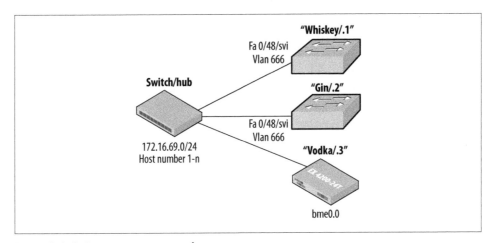

Figure 3-4. OoB management network

The Juniper devices support a separate physical port intended strictly for OoB use; the Cisco switches do not. To create an OoB network for these devices, the best practice is to dedicate a physical port that is bound to a management VLAN, which separates this interface from the rest of the interfaces and their transit network traffic. In our case, the final port of the Cisco switches (port 48) was chosen. A VLAN ID that is not used for any other purpose, 666, is chosen to tie the OoB port into a Switched Virtual Interface (SVI), which is used to support an IP-addressable interface without requiring a Layer 3 routing license. Also, it is a best practice to always change the OoB management VLAN on the Cisco devices, since by default all interfaces are part of that VLAN, which could open a large security risk.

 You cannot change the management VLAN on Catalyst 1900 and 2820 switches or on Catalyst 2900 switches with 4 MB of memory; VLAN 1 must be the management VLAN for these switches. You can change the management VLAN only on switches running Cisco IOS Release 12.0(5) XU.

Here's the relevant Cisco configuration from `Gin`:

```
interface FastEthernet0/48
 switchport access vlan 666
 switchport mode access
!
. . .
interface Vlan1
 no ip address
 shutdown
!
interface Vlan666
 ip address 172.16.69.2 255.255.255.0
```

The new Juniper switch `Tequila` simply needs to have the `me0` management interface configured with the proper IP address. If you are confused about what the heck that strange unit 0 piece is, please consult the interface configuration in "EX Interfaces" on page 143. For now, suffice it to say that it's a logical interface, which is akin to a subinterface on Cisco devices. However, on the Cisco side, the management interface must be a physical interface and not a subinterface:

```
lab@Tequila> show configuration interfaces me0
unit 0 {
    family inet {
        address 172.16.69.4/24;
    }
}
```

Remote Access

After the users are configured on the switch, we must decide what kind of remote access should be provided to the switch, assuming this was not configured via EZSetup or during the initial configuration. Remote access options include:

Finger

A protocol to get information about a user logged in to the switch. This protocol is no longer used and should never be enabled. For giggles, here is the `finger` output from the new switch:

```
% finger lab@172.16.69.4
[172.16.69.4]
Login: lab                        Name:
Directory: /var/home/lab          Shell: /usr/sbin/cli
On since Mon Sep 24 00:31 (UTC) on ttyd0, idle 0:01
No Mail.
No Plan.
%
```

FTP

Provides file transfer services. Although FTP is a widely used protocol, it transfers files in plain text, which can lead to security issues. When possible, you should use secure copy (SCP), which is enabled with the SSH service.

Rlogin

The Remote login protocol, which allows remote access to the JUNOS shell and CLI. This Unix utility has several security flaws and was used only in private environments. This utility is activated by a hidden command on the switch but, as discussed, should never be enabled on the switch.

A *hidden command* is a command that does not show up when you use ? in the CLI and does not auto-complete with the space bar. One of the most famous hidden commands in JUNOS software is `show version and haiku`. Try it yourself if you want to read some really bad poetry!

SSH

Allows communications over an encrypted tunnel. This ensures not only availability, but also data integrity and confidentiality. When SSH is enabled, this automatically enables SCP. The SSH option is available only in the domestic version of JUNOS due to government restrictions on encryption technology. Users on the list of permitted foreign countries can apply for the domestic version to obtain the SSH service.

Telnet

A common protocol, developed in 1969, for remotely managing a system. Telnet transits all data in clear text, so you should use SSH when possible.

Web management

Enables use of the J-Web GUI on the switch for management and configuration. These can be either encrypted or unencrypted Hypertext Transfer Protocol (HTTP/HTTPS) connections. Note that strong HTTPS encryption (128-bit) requires the domestic version of JUNOS and that a Secure Sockets Layer (SSL) security certificate be loaded.

JUNOScript server

Enables the switch to receive commands from a JUNOScript server via clear text or SSL connections.

Netconf

The Network Configuration protocol, which is defined in RFC 4741 and uses XML for configuration and messaging. Netconf is the Internet Engineering Task Force (IETF) standard created as a replacement for SNMP and is based on JUNOScript.

XML Tags

JUNOScript is a tool you can use to configure and manage the switch. JUNOS output and configuration contain XML tags that can be referenced by a JUNOScript client. Here is an example of a configuration and an operational command that displays the XML tags for each field:

```
root@Tequila> show system users | display xml
<rpc-reply xmlns:junos="http://xml.juniper.net/junos/9.2R1/junos">
    <system-users-information
xmlns="http://xml.juniper.net/junos/9.2R1/junos">
        <uptime-information>
            <date-time
junos:seconds="1217955562">4:59PM</date-time>
            <up-time junos:seconds="5287">1:28</up-time>
            <active-user-count junos:format="1
```

```
user">1</active-user-count>
              <load-average-1>0.15</load-average-1>
              <load-average-5>0.09</load-average-5>
              <load-average-15>0.07</load-average-15>
              <user-table>
                  <user-entry>
                      <user>lab</user>
                      <tty>u0</tty>
                      <from>-</from>
                      <login-time
junos:seconds="1217952963">4:16PM</login-time>
                      <idle-time junos:seconds="0">-</idle-
time>
                      <command>-cli (cli)</command>
                  </user-entry>
              </user-table>
          </uptime-information>
      </system-users-information>
      <cli>
          <banner></banner>
      </cli>
</rpc-reply>
lab@tequila> show configuration routing-options |display
xml
<rpc-reply xmlns:junos="http://xml.juniper.net/junos/9.2R1/junos">
      <configuration junos:commit-seconds="1217952936"
junos:commit-localtime="2008-08-05 16:15:36 UTC"
junos:commit-user="root">
              <protocols>
                  <lldp>
                      <interface>
                          <name>all</name>
                      </interface>
                  </lldp>
                  <stp>
                      <interface>
                          <name>all</name>
                          <cost>10</cost>
                      </interface>
                  </stp>
              </protocols>
      </configuration>
      <cli>
          <banner></banner>
      </cli>
</rpc-reply>
```

The most secure methods of remote access on the switch will be SSH and the transfer of files using SCP. To enable any service, simply set it under the [edit system services] level:

```
[edit system services]
root@Tequila# set ?
Possible completions:
+ apply-groups          Groups from which to inherit configuration
                        data
+ apply-groups-except   Don't inherit configuration data from these
```

```
                           groups
  > dhcp                   Configure DHCP server
  > finger                 Allow finger requests from remote systems
  > ftp                    Allow FTP file transfers
  > netconf                Allow NETCONF connections
  > service-deployment     Configuration for Service Deployment (SDXD)
                           management application
  > ssh                    Allow ssh access
  > telnet                 Allow telnet login
  > web-management         Web management configuration
  > xnm-clear-text         Allow clear text-based JUNOScript connections
  > xnm-ssl                Allow SSL-based JUNOScript connections
```

Each service typically has a variety of options, such as setting a maximum number of connections, rate-limiting the inbound connection attempts, and choosing supported protocol versions/options. In the case of the SSH service, we have the following options:

```
root@Tequila# set system services ssh ?
Possible completions:
  <[Enter]>                Execute this command
+ apply-groups             Groups from which to inherit configuration data
+ apply-groups-except      Don't inherit configuration data from these groups
  connection-limit         Maximum number of allowed connections (1..250)
+ protocol-version         Specify ssh protocol versions supported
  rate-limit               Maximum number of connections per minute (1..250)
  root-login               Configure root access via ssh
  |                        Pipe through a command
```

The following command enables SSH with the default parameters of 75 active sessions and 150 connection attempts per minute:

```
[edit]
root@Tequila# set system services ssh
```

Tequila then initiates an SSH session to Gin. The first connection to a given host needs to establish the RSA fingerprint for authentication by adding that host's public key to the user's list of host keys, which, by the way, are stored in the ~/.ssh directory:

```
lab@Tequila> ssh 172.16.69.2
The authenticity of host '172.16.69.2 (172.16.69.2)' can't be established.
RSA key fingerprint is 39:ef:fa:64:83:a0:40:48:f0:a5:ad:f5:0b:de:a9:7a.
Are you sure you want to continue connecting (yes/no)? yes
Warning: Permanently added '172.16.69.2' (RSA) to the list of known hosts.
lab@172.16.69.2's password:

Gin>
```

Once Gin is added to the list of known hosts, future sessions do not require identity confirmation:

```
root@Tequila> ssh 172.16.69.2
lab@172.16.69.1's password:
```

When SSH is enabled on the switch, it also automatically enables SCP to initiate secure file exchanges. You can upload or download files via SCP using variations of the file copy command. In this case, Tequila transfers a file called *test* to Gin. Tequila has already added Gin to its list of known hosts files:

```
root@Tequila> file copy test2 lab@172.16.69.2:test2.txt
lab@172.16.69.1's password:
test                                    100% 9480      9.3KB/s   00:00
```

Dynamic Host Configuration Protocol

DHCP is a protocol used by networked devices (clients) to obtain the parameters necessary for operation in an IP network. This protocol reduces system administration workload, allowing devices to be added to the network with little or no manual configuration. Without DHCP, every single client on the network would have to be configured manually.

Essentially, DHCP automates the assignment of IP addresses, subnet masks, default gateways, and other optional IP parameters such as DNS addresses. The protocol works on a client/server model in which the server assigns a client the IP information after seeing a request (Figure 3-5).

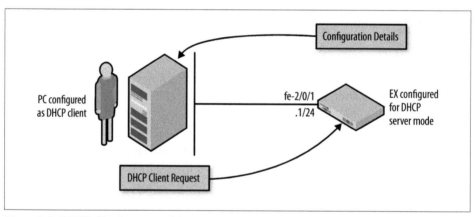

Figure 3-5. DHCP client/server model

DHCP consists of a four-step transfer process beginning with a broadcast DHCP discovery message from the client. The second step involves a DHCP offer message from the server to the client, which includes the IP address and mask, and DOCSIS-specific parameters. The client then sends a DHCP request to accept the offer it received from the server in the previous step. The DHCP server sends a DHCP response message and removes the now-allocated address from the DHCP scope (Figure 3-6).

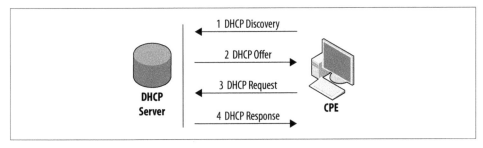

Figure 3-6. DHCP process

So far we have discussed DHCP requests that are on the same physical segment. DHCP requests from a client to a server are not restricted to the physical segment, LAN, or VLAN (discussed in Chapter 7). In this case, a relay agent is needed to pass requests between the client and server. This eliminates the need for a dedicated DHCP server in each LAN or VLAN environment (Figure 3-7).

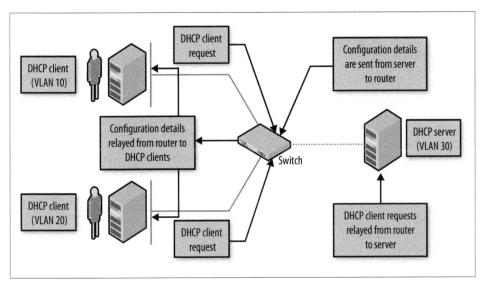

Figure 3-7. DHCP relay

DHCP server configuration in JUNOS

When configuring a DHCP server in JUNOS you can set the following parameters:

DHCP subnet
    To specify the subnet on which DHCP is configured

Address range
    To set the lowest address and highest address that can be used from the DHCP subnet

Exclude address(es)
To specify any addresses from the DHCP subnet that should not be used

Lease time
To set the time for which the allocated address is reserved

Server information
To set the IP address reported to the client, domain name, DNS servers, and domain searches

Field options
To set gateway routers and WINS servers

Boot options
To specify a boot server and file from which to get configuration information

Static bindings
To specify the static mapping for the client to use, such as an IP address, hostname, or client identifier

These options are displayed on Tequila and are configured under [edit system services dhcp]:

```
lab@Tequila# set system services dhcp ?
Possible completions:
+ apply-groups          Groups from which to inherit configuration data
+ apply-groups-except   Don't inherit configuration data from these groups
  boot-file             Boot filename advertised to clients
  boot-server           Boot server advertised to clients
  default-lease-time    Default lease time advertised to clients (seconds)
  domain-name           Domain name advertised to clients
> domain-search         Domain search list used to resolve hostnames
  maximum-lease-time    Maximum lease time advertised to clients (seconds)
> name-server           Domain name servers available to the client
  next-server           Next server that clients need to contact
> option                DHCP option
> pool                  DHCP address pool
> router                Routers advertised to clients
  server-identifier     DHCP server identifier advertised to clients
> static-binding        DHCP client's hardware address
> traceoptions          DHCP server trace options
> wins-server           NetBIOS name servers
```

For example, in the following code, Tequila is configured to allocate addresses out of the 172.16.69/24 range, starting at address 172.16.69.50 and ending at address 172.16.69.75, for a total of 25 addresses. It is also assigned a one-hour lease time and a default gateway of 172.16.69.4:

```
lab@Tequila# show system services dhcp
default-lease-time 3600;
router {
    172.16.69.4;
}
pool 172.16.69.0/24 {
    address-range low 172.16.69.50 high 172.16.69.75;
```

```
}

[edit]
lab@Tequila# commit and-quit
```

You can verify that the pool is configured properly by issuing the show system services dhcp pool and global commands:

```
lab@Tequila> show system services dhcp pool
Pool name        Low address     High address    Excluded addresses
172.16.69.0/24   172.16.69.50    172.16.69.75

lab@Tequila> show system services dhcp global
Global settings:
    BOOTP lease length          infinite

DHCP lease times:
    Default lease time          1 hour
    Minimum lease time          1 minute
    Maximum lease time          infinite

DHCP options:
    Name: router, Value: [ 172.16.69.4 ]
```

When the switch begins to allocate addresses, this can be viewed with the bind commands:

```
lab@tequila> show system services dhcp binding
IP Address      Hardware Address    Type     Lease expires at
172.16.69.50    00:a0:12:00:12:ab   dynamic  2008-09-27 13:01:45 PDT
172.16.69.51    00:a0:12:00:13:02   dynamic  2008-09-27 13:01:52 PDT
```

 Other useful commands include show system services dhcp statistics to display packet counters and show system services dhcp conflict to view any addresses that are found to be duplicated. Also, to see the actual DHCP messages, traceoptions (discussed in Chapter 8) can be configured and set to the log *fud*.

DHCP relay configuration in JUNOS

If the DHCP relay option is desired, this can also be configured on the EX Series switch.

 Since DHCP messages are broadcast and are not sent to a specific server or switch, the EX Series cannot function as both a DHCP server and a relay agent at the same time.

Relay options are configured under [edit forwarding options helpers bootp]. At first glance, the hierarchy may seem about as clear as mud, but breaking out each piece helps to clarify the situation:

forwarding options
>   Specifies the inherit forwarding of the DHCP packet

helpers
>   Describes that the switch is providing an aid to the network by sending the packet to a DHCP server

bootp
>   The original DHCP specification name (RFC 951)

Here are the most basic configuration parameters that are required:

Interface
>   Specifies the interface(s) that will have incoming DHCP packets that need to be relayed

Server
>   Configures the server IP address toward which to forward the DHCP request

The rest of the configuration is optional, including hop counts, wait times, Time to Live (TTL), and special relay options:

```
lab@Tequila# set forwarding-options helpers bootp ?
Possible completions:
+ apply-groups         Groups from which to inherit configuration data
+ apply-groups-except  Don't inherit configuration data from these groups
  client-response-ttl  IP time-to-live value to set in responses to client
  description          Text description of servers
> interface            Incoming BOOTP/DHCP request forwarding interface
configuration
  maximum-hop-count    Maximum number of hops per packet (1..16)
  minimum-wait-time    Minimum number of seconds before requests are forwarded
  relay-agent-option   Use DHCP Relay Agent option in relayed BOOTP/DHCP messages
> server               Server information
```

In the sample book topology the **relay** option is not used, but here is a quick example:

```
[edit forwarding-options helpers bootp]
user@host# show
description "Global DHCP Example";
server 172.18.24.38;
maximum-hop-count 4;
minimum-wait-time 1;
interface {
    ge-0/0/1
```

 To monitor **bootp** events configure **traceoptions** (as discussed in Chapter 8) and observe the */var/log/fud* logfile.

Secondary Configuration Summary

This was a big section, and it's a good time to take a step back and look at what we have accomplished. From a bare-bones basic configuration, we have added users, an OoB network, remote access, and possibly even DHCP. Now it is almost time to get into some networking and protocols, but first let's examine some basic interface and switch maintenance actions and procedures that may prove helpful at any time.

EX Interfaces

This section begins with some general JUNOS interface concepts before providing EX-specific interface configuration, operational analysis, and troubleshooting examples typical to an initial configuration.

Historically, Juniper Networks routers contained two major categories of interfaces: *permanent* and *transient*. Users cannot remove permanent interfaces, whereas they can move, change, and remove transient interfaces. Transient interfaces are typically based on modular interfaces, which in turn are based on a Flexible PIC Concentrator (FPC) and Physical Interface Card (PIC) model, in which FPCs and PICs can be moved about as desired in the system.

The EX3200 and EX4200 are fixed interface platforms, and therefore do not directly support the modular interface model. However, because EX switches run JUNOS software, it can still be said that some EX interfaces are permanent in regard to their intended function—say, as a network management interface, or tunnel services device—whereas others are seen as user or network interfaces, with a broader range of network-oriented functions. The following sections detail the JUNOS-based interface devices that often cause confusion among new users.

Permanent Interfaces

A permanent interface is any interface that is always present on the switch (it cannot be altered). These interfaces can be management interfaces such as OoB Ethernet, or a software-based pseudointerface such as a tunnel service device.

An EX4200 has three permanent management interfaces:

me0-
    Management Ethernet 0 is the OoB management Ethernet interface. This interface connects to the routing engine (RE) and is used for management services such as accessing Telnet/SSH, transferring files using FTP/SCP, and sending management messages such as syslog or SNMP traps to remote servers. The me0 interface is a *non-transit interface*, which means traffic cannot enter this interface and exit via a network interface, nor can it enter a network interface and exit through the management interface.

vme0-

Virtual Management Ethernet 0 serves the same roles as the physical me0, except in an EX4200 VC, where the virtual management interface automatically connects to the current master RE. Use of a vme0 is optional, but you cannot currently use both the vme0 and the me0 interfaces at the same time.

bme0-

This is an internal Gigabit Ethernet interface that links the RE to the various EX Packet Forwarding Engine (PFE) components, the specifics of which vary by EX model. There is never a need to directly configure this interface.

Many software pseudointerfaces are created automatically for use in a variety of functions:

lo0-

The loopback interface that ties to the switch itself and not to any one physical interface. A loopback interface is used to provide a stable address for management traffic and routing protocols.

pimd-

An RE-based Protocol Independent Multicast (PIM) de-encapsulation interface that allows a multicast rendezvous point (RP) to process PIM register messages.

pime-

An RE-based PIM encapsulation interface, used in multicast to create a unicast PIM register message to send to the RP.

ipip-

An RE-based IP-over-IP encapsulation interface used to create IP-in-IP tunnels.

gre-

An RE-based Generic Routing Encapsulation (GRE) interface used to create GRE tunnels.

dsc-

A discard interface that can be used to silently discard packets. This is often used to silently dump traffic associated with a distributed denial of service (DDoS) attack.

tap-

A virtual Ethernet interface historically used for monitoring on FreeBSD systems. This interface could be used to monitor discarded packets on a router but is no longer officially supported.

vlan-

A Routed VLAN Interface (RVI) used to internally route between VLANs.

vcp-

Virtual Chassis Port (VCP) interfaces, which exist on EX4200 models only.

Although used for transit traffic, the VCP protocol is not user-configurable and simply works, so these interfaces are included here, rather than in the network interface section. Note that using an uplink port as a VCE results in the creation of a vcp-255/1/n device, whereas the built-in VCPs are always vcp-0 and vcp-1.

Network Interfaces

Network interfaces, as the name implies, are the front-panel GE and 10 GE uplink ports used for Layer 2 and Layer 3 networking functions. Although Juniper routers support a wide variety of interface types, such as Ethernet, SONET, DS1/DS3, and ATM, among others, on EX switches it's pretty much Ethernet or bust, which is just fine given that they're intended to be used as Ethernet switches, after all.

Network interface naming

JUNOS interfaces follow a common naming convention—the interface media type abbreviation followed by three numbers that indicate the location of the actual interface. The general convention is MM-F/P/T, where:

MM = media type
F = chassis FPC slot number
P = PIC slot number
T = port number

EX switches use the same naming approach, which is handy for those already familiar with JUNOS. Just remember that in the EX context, the term *FPC* maps to a chassis number, and the PIC value is always 0 for network ports and 1 for uplink module ports, and things will go smoothly.

On the EX, the media type is either ge or xe, which denotes Gigabit Ethernet or 10G Ethernet, respectively. Note that GE interfaces are multirate-capable and can operate at 10 Mbps, 100 Mbps, or the full 1,000 Mbps (GE) rate. The chassis FPC number ranges from 0 to 9. It's always 0 on standalone EX3200s or EX4200s; non-zero values reflect a VC member number or the Line Card (LC) slot on an EX8200 series. The PIC slot number is 0 for all built-in network interfaces and 1 for uplink module ports. The port number ranges from 0 to 47, and reflects the actual front-panel port being used.

Logical units

Interfaces support the notion of a *logical unit*, which is analogous to a subinterface in IOS. A key difference is that in JUNOS, an operational network interface must have at least one logical unit, whereas in IOS, explicit use of a subinterface is optional. In some modes of operation, an interface can support multiple logical units, which can range from 0 to 16,385, with each such unit in turn supporting one or more protocol families and operating independently of other units that share the device.

The unit number follows the period (.) in an interface name. For example, ge-0/1/0.10 identifies a logical interface 10 on a GE interface housed in "FPC" slot 0, "PIC" slot 1, port 0.

Ethernet interfaces support multiple units through the use of VLAN tagging in JUNOS software. Note that an untagged Ethernet interface, which by definition can support only a single logical unit, must be assigned unit number 0. When VLAN tagging is enabled (for a Layer 3 interface), you can define multiple units, and each such unit is bound to a unique VLAN tag and one or more Layer 3 protocol families. There is no need for the unit number and VLAN tag to match, though this is the current best practice.

EX interfaces that are operating in Layer 2 don't support VLAN tagging, and therefore are limited to a single logical unit (0). EX interfaces are placed into Layer 2 mode when the ethernet-switching family is specified. Explicit tagging is not supported in this context because these interfaces are either trunks, which support traffic with various tags, or accesses, which generally send untagged traffic.

 In system log messages, a logical unit is known as an interface logical, or ifl, and the parent is an interface device, or ifd.

Interface Configuration

In JUNOS, network interfaces have two levels of configuration: physical properties and logical properties. *Physical properties* are tied to the entire physical port, whereas *logical properties* affect only that logical portion of the interface represented by unit numbers or channel numbers. When you configure an interface, you specify physical properties once and logical properties for each unit that is used.

Physical properties

You configure an interface's physical properties at the [edit interfaces <interface-name>] hierarchy. The CLI's ? function is used to display the options available in the 9.2 release:

```
{master}[edit interfaces ge-0/0/0]
lab@Vodkila# set ?
Possible completions:
  accounting-profile   Accounting profile name
+ apply-groups         Groups from which to inherit configuration data
+ apply-groups-except  Don't inherit configuration data from these groups
  description          Text description of interface
  disable              Disable this interface
> ether-options        Ethernet interface-specific options
  gratuitous-arp-reply Enable gratuitous ARP reply
> hold-time            Hold time for link up and link down
  mac                  Hardware MAC address
```

```
   mtu                     Maximum transmit packet size (256..9216)
   no-gratuitous-arp-reply Don't enable gratuitous ARP reply
   no-gratuitous-arp-request  Ignore gratuitous ARP request
   no-traps                Don't enable SNMP notifications on state changes
 > optics-options          Optics options
 > traceoptions            Interface trace options
   traps                   Enable SNMP notifications on state changes
 > unit                    Logical interface
   vlan-tagging            802.1q VLAN tagging support
{master}[edit interfaces ge-0/0/0]
```

Most of the keywords are self-explanatory, and we demonstrate virtually all of them at some point in this book. Some key physical properties include:

MAC

The mac keyword is used to alter the burned-in (Media Access Control [MAC]) address (BIA). There's rarely a need to change this, but for some testing purposes having known values can be helpful.

MTU

The maximum transmission unit (MTU) determines the maximum size frame that can be sent out of the interface. In JUNOS, the Ethernet MTU includes the frame's Source and Destination MAC addresses, and the type field, but not the 4-byte frame check sequence (FCS). The default MTU is 1,514, which permits a 1,500-byte IP payload. EX interfaces support an MTU from 256 all the way to a giant 9,216, but care should be taken to avoid MTU mismatches, which are particularly nasty to troubleshoot in a Layer 2 network.

 Cisco IOS does not include any Layer 2 overhead in its Ethernet MTU. Therefore, a Cisco Ethernet MTU of 1,500 is compatible with the default JUNOS Ethernet MTU of 1,514. Go figure.

Hold time

The hold-time parameter is used to configure a delay for the reaction to interface up or down events. By default, there is intentional delay, but sometimes it's desirable to ride out short-lived transitions, rather than generating SNMP alarms and destroying protocol adjacencies at each measurable blip.

Gratuitous ARP

The Gratuitous ARP statements control various Address Resolution Protocol (ARP) behaviors. The no-gratuitous-arp-request option disables the default behavior of responding to a Gratuitous ARP. A Gratuitous ARP request is used for duplicate IP address detection, and does not expect a reply, as there should be no duplicate IP addressing. From the receiving router's perspective, a Gratuitous ARP request is differentiated from a *real* ARP request targeted at the local router, because such an ARP contains a foreign Source MAC (SMAC) while both the target and source IP addresses match the local interface's IP. The gratuitous-arp-reply option enables the update of the switches' ARP cache based on the receipt of un-solicited (gratuitous) ARP replies. Use the no form of the keyword to return to the default behavior of ignoring such replies. A Gratuitous ARP response is usually sent in an attempt to update the ARP cache in remote nodes—for example, after a MAC address changes. From the local router's perspective, such a packet consists of an ARP reply using a broadcast Destination MAC (DMAC) address that contains the same (non-local) IP address as both the source and the target.

Traps

The traps and no-traps keywords enable or disable SNMP trap generation upon interface state transitions.

Ethernet-specific parameters are configured using the ether-options hierarchy. These options control auto-negotiation, or set speed and duplex settings or IEEE 802.3ad link aggregation. The ether-options are displayed in the following code:

```
{master}[edit interfaces ge-0/0/0]
lab@Vodkila# set ether-options ?
Possible completions:
> 802.3ad              IEEE 802.3ad
+ apply-groups         Groups from which to inherit configuration data
+ apply-groups-except  Don't inherit configuration data from these groups
  auto-negotiation     Enable auto-negotiation
  flow-control         Enable flow control
  link-mode            Link duplex
  no-auto-negotiation  Don't enable auto-negotiation
  no-flow-control      Don't enable flow control
> speed                Specify speed
{master}[edit interfaces ge-0/0/0]
lab@Vodkila# set ether-options
```

We cover link aggregation in Chapter 11. The flow-control keyword causes the EX to react to pause frames, which is the default. The use of pause is applicable only to 1,000 Mbps (GE) operation. The non-blocking nature of EX switches means they have no need to actually generate flow control, so this option simply allows the EX to reach to pause frames sent by a congested switch. The link and speed keywords allow you to hardcode half duplex/full duplex (HD/FD) and the speed when needed. The auto-negotiation options enable (the default) or disable auto-negotiation.

Juniper's Auto-Negotiation Stance

Over the years, there has been a fair bit of discussion about whether it's best to manually hardcode various Ethernet parameters to ensure compatibility, or to simply enable auto-negotiation and trust that the two devices will find their highest common ground.

Given the maturity of current Ethernet technology, JUNOS enables auto-negotiation by default, and in the vast majority of cases, the process works as expected. Although taking the magic out of the recipe may seem like a good idea, studies have shown that it is common to make mistakes when manually configuring a compatible mode, and in the case of a duplex mismatch, these mistakes can lead to a poorly performing link that works just well enough to go unnoticed.

If there are known problems with a device's auto-negotiation capabilities, or if auto-negotiation is simply not supported, manual configuration makes sense.

Note that using auto-negotiation at one end and manual settings at the other end is not recommended, as this is known to lead to problems. For example, pairing a switch that is hardcoded to 100 Mbps FD with a Fast Ethernet network interface card (NIC) set to auto-negotiate results in the NIC falling back to HD mode. This is due to lack of auto-negotiation, and it causes the switch to fall back to the lowest ability of HD, which in turn results in a duplex mismatch.

Logical properties

All network interfaces require a logical unit to send or receive traffic. As noted previously, the logical unit abstract carves a single physical interface into multiple logical interfaces. For instance, an Ethernet interface can be subdivided into multiple VLANs, each requiring its own logical unit.

Recall that unlike other vendors' software, JUNOS requires a unit number and a logical interface definition even when a single logical entity is desired; this is because logical properties *must* be defined after the unit number hierarchy, and one such property is the choice of protocol family. An interface with no protocol family cannot send or receive anything. The most common types of logical properties include:

Protocol family
> Indicates whether the interface operates in Layer 2 or Layer 3 mode, and in the latter case, specifies what Layer 3 protocols can operate on the logical interface. You place an interface into Layer 2 mode by adding the `ethernet-switching` family. This mode permits a single logical unit, which must be 0, and does not permit VLAN encapsulation.

> When you omit the `ethernet-switching` family, the result is a Layer 3 interface. Such an interface can have one or more logical units, and each unit can have one or more protocol families. The most common of these is `family inet`, which enables the sending and receiving of IPv4 packets in the Transmission Control Protocol/Internet Protocol (TCP/IP) suite (e.g., TCP, User Datagram Protocol [UDP], Internet Control Message Protocol [ICMP], and IP). Other common families are `inet6` (IPv6) and `iso` (Intermediate System to Intermediate System [IS-IS] packets).

Protocol address
> Is the Layer 3 family address, such as a `family inet` IP address and related network mask.

Virtual circuit ID
> Is used to multiplex traffic from multiple logical units over a single interface device. For an Ethernet-based device, the virtual circuit ID takes the form of a VLAN ID.

The current best practice, however, is to keep the circuit address the same as the unit number for easier troubleshooting. So, if you have a VLAN ID of 40 configured on your interface, the logical interface should also be a unit of 40, although it's not required. Again, note that when you configure a non-VLAN tagged Ethernet, the logical unit number *must* be 0. Think of this unit as a placeholder for all the logical properties that will need to be configured on that interface.

EX Interface Configuration Examples

This section provides examples of typical EX interface configuration for both Layer 2 and Layer 3 modes of operation.

Layer 2 interface

We begin with a sample Layer 2 interface that is used for bridging and switching:

```
{master}
lab@Vodkila> configure
Entering configuration mode

{master}[edit]
lab@Vodkila# edit interfaces ge-0/0/0

{master}[edit interfaces ge-0/0/0]
```

You begin by entering configuration mode and parking yourself at the [edit interfaces <interface-name>] hierarchy. This example configures the ge-0/0/0 interface.

Some physical properties are set and then displayed:

```
{master}[edit interfaces ge-0/0/0]
lab@Vodkila# set ether-options speed 100m
{master}[edit interfaces ge-0/0/0]
lab@Vodkila# set ether-options link-mode full-duplex
{master}[edit interfaces ge-0/0/0]
lab@Vodkila# set mac 00:1f:12:3d:b4:dd
{master}[edit interfaces ge-0/0/0]
lab@Vodkila# show
mac 00:1f:12:3d:b4:dd;
ether-options {
    link-mode full-duplex;
    speed {
        100m;
    }
}
```

 By default, all JUNOS network interfaces are administratively enabled. If an interface needs to be administratively disabled, issue a configuration mode set interfaces <interface name> disable command. To reenable the interface, issue a delete interfaces <interface name> disable command, or perform a rollback 1 (and commit). Although having to *delete* a *disable* seems a bit like a double negative, if you think about it it's much the same as a *"no shutdown"* in IOS.

And now the definition of the logical unit and its binding to a (Layer 2) protocol family:

```
{master}[edit interfaces ge-0/0/0]
lab@Vodkila# set unit 0 family ethernet-switching
```

The complete Layer 2 interface definition is displayed, and committed:

```
{master}[edit interfaces ge-0/0/0]
lab@Vodkila# show
mac 00:1f:12:3d:b4:dd;
ether-options {
    link-mode full-duplex;
    speed {
        100m;
    }
}
unit 0 {
    family ethernet-switching;
}
```

The show interfaces command is used for operational status verification:

```
{master}[edit interfaces ge-0/0/0]
lab@Vodkila# run show interfaces ge-0/0/0 terse
Interface          Admin Link Proto   Local              Remote
ge-0/0/0           up    up
ge-0/0/0.0         up    up   eth-switch
```

The output confirms that the interface device and its logical unit are both administratively and operationally up. In the case of plain Ethernet, "being up" at the Link level simply means there are no obvious local interface faults. The lack of Ethernet keepalive, which is performed by Bidirectional Forwarding Detection (BFD) or Operational Administration and Maintenance (OAM), means it's easy to have an interface that has a status of physically up because the local end is attached to a switch—even though the remote end of the switch link may be unplugged, thereby preventing communications. For more detail, omit the terse switch:

```
{master}[edit interfaces ge-0/0/0]
lab@Vodkila# run show interfaces ge-0/0/0
Physical interface: ge-0/0/0, Enabled, Physical link is Up
  Interface index: 129, SNMP ifIndex: 110
  Link-level type: Ethernet, MTU: 1514, Speed: 100mbps, MAC-REWRITE Error: None,
  Loopback: Disabled, Source filtering: Disabled, Flow control: Enabled,
  Auto-negotiation: Enabled, Remote fault: Online
  Device flags   : Present Running
  Interface flags: SNMP-Traps Internal: 0x0
  Link flags     : None
  CoS queues     : 8 supported, 8 maximum usable queues
  Current address: 00:1f:12:3d:b4:dd, Hardware address: 00:1f:12:3d:b4:c1
  Last flapped   : 2005-01-30 07:59:57 UTC (00:00:13 ago)
  Input rate     : 512 bps (1 pps)
  Output rate    : 0 bps (0 pps)
  Active alarms  : None
  Active defects : None
```

```
Logical interface ge-0/0/0.0 (Index 65) (SNMP ifIndex 111)
  Flags: SNMP-Traps Encapsulation: ENET2
  Input packets : 6
  Output packets: 1
  Protocol eth-switch, MTU: 0
    Flags: None
```

The highlighted sections point out various details, such as confirmation of speed, auto-negotiation, SMAC filtering, and so on. The altered MAC address is also seen, as is a single logical interface that supports Layer 2 switching. Use the extensive switch, or specify the media switch, to see details on the current link properties:

```
lab@Vodkila# run show interfaces ge-0/0/0 media
Physical interface: ge-0/0/0, Enabled, Physical link is Up
  Interface index: 129, SNMP ifIndex: 110
  Link-level type: Ethernet, MTU: 1514, Speed: 100mbps, MAC-REWRITE Error: None,
  Loopback: Disabled, Source filtering: Disabled, Flow control: Enabled,
  Auto-negotiation: Enabled, Remote fault: Online
  Device flags   : Present Running
  Interface flags: SNMP-Traps Internal: 0x0
  Link flags     : None
  CoS queues     : 8 supported, 8 maximum usable queues
  Current address: 00:1f:12:3d:b4:dd, Hardware address: 00:1f:12:3d:b4:c1
  Last flapped   : 2005-01-30 07:59:57 UTC (00:11:06 ago)
  Input rate     : 0 bps (0 pps)
  Output rate    : 0 bps (0 pps)
  Active alarms  : None
  Active defects : None
  MAC statistics:
    Input bytes: 4226028, Input packets: 51477, Output bytes: 3074021,
    Output packets: 27429
  Filter statistics:
    Filtered packets: 0, Padded packets: 0, Output packet errors: 0
  Autonegotiation information:
    Negotiation status: Complete
    Link partner:
        Link mode: Full-duplex, Flow control: None, Remote fault: OK,
        Link partner Speed: 100 Mbps
    Local resolution:
        Flow control: None, Remote fault: Link OK
```

Layer 3 interface

This example demonstrates a Layer 3 interface that uses VLAN tagging to support two logical interfaces, each running the IPv4 protocol, and one also running IPv6. Start by deleting the existing Layer 2 configuration:

```
{master}[edit interfaces ge-0/0/0]
lab@Vodkila# delete
Delete everything under this level? [yes,no] (no) yes
```

Then move on to define the two logical units and related protocol family properties. This example leaves the physical interface properties at their default values, with the exception that specifying VLAN tagging is in effect:

```
{master}[edit interfaces ge-0/0/0]
lab@Vodkila# set vlan-tagging
```

And now the logical interface properties:

```
{master}[edit interfaces ge-0/0/0]
lab@Vodkila# set unit 1 family inet address 200.0.0.1/24

{master}[edit interfaces ge-0/0/0]
lab@Vodkila# set unit 1 vlan-id 1

{master}[edit interfaces ge-0/0/0]
lab@Vodkila# set unit 2 family inet address 10.0.0.1/16

{master}[edit interfaces ge-0/0/0]
lab@Vodkila# set unit 2 family inet address 10.0.0.1/16 arp 10.0.0.2 mac
    00:1f:12:3d:b4:ff

{master}[edit interfaces ge-0/0/0]
lab@Vodkila# set unit 2 family inet6 address 200::1/64

{master}[edit interfaces ge-0/0/0]
lab@Vodkila# set unit 2 vlan-id 2
```

The results are displayed:

```
{master}[edit interfaces ge-0/0/0]
lab@Vodkila# show
vlan-tagging;
unit 1 {
    vlan-id 1;
    family inet {
        address 200.0.0.1/24;
    }
}
unit 2 {
    vlan-id 2;
    family inet {
        address 10.0.0.1/16 {
            arp 10.0.0.2 mac 00:1f:12:3d:b4:ff;
        }
    }
    family inet6 {
        address 200::1/64;
    }
}
```

The output confirms a VLAN tagged interface device supporting two units, each with IPv4, and one with both IPv4 and IPv6. This example also includes a static ARP definition for the IP address 10.0.0.2 on unit 1. Note that best practice is followed here by matching unit numbers to the Layer 2 virtual circuit ID (the VLAN tag in this case), but such matching is not strictly required. Also note that VLAN 1 is used for the first unit. This VLAN ID has no special significance in JUNOS, but does in IOS, as explained in Chapter 5.

Note that JUNOS takes network masks in "/" or Classless Inter-Domain Routing (CIDR) notation only, and that failing to specify a mask results in an assumed /32 host address. Such an address may not commit on an Ethernet interface with some versions of JUNOS unless that version happens to support unnumbered /32 Ethernet addressing.

The Layer 3 interface's operational status is quickly confirmed:

```
{master}[edit interfaces ge-0/0/0]
lab@Vodkila# run show interfaces ge-0/0/0 terse
Interface          Admin Link Proto    Local                  Remote
ge-0/0/0           up    up
ge-0/0/0.1         up    up   inet     200.0.0.1/24
ge-0/0/0.2         up    up   inet     10.0.0.1/16
                                inet6    200::1/64
                                         fe80::21f:1200:23d:b4c1/64
ge-0/0/0.32767     up    up
```

The display confirms a single interface with two logical units, and reflects the various IP and IPv6 properties that were configured. The ge-0/0/0.32767 logical interface is automatically created to support OAM when VLAN tagging is in effect. In this JUNOS version, Ethernet OAM is not supported, but when it is, this unit will handle that traffic separately from the Layer 3 user traffic on the other units. At this stage, you could test Layer 3 connectivity using a ping, but this presumes the remote end of the link is also configured, which is not yet the case here. For now, display the route table to confirm the presence of IPv4 and IPv6 directly connected routes for the ge-0/0/0 interface:

```
{master}[edit interfaces ge-0/0/0]
lab@Vodkila# run show route protocol direct

inet.0: 27 destinations, 28 routes (27 active, 0 holddown, 0 hidden)
Restart Complete
+ = Active Route, - = Last Active, * = Both

10.0.0.0/16        *[Direct/0] 00:05:26
                    > via ge-0/0/0.2
. . .

200.0.0.0/24       *[Direct/0] 00:05:26
                    > via ge-0/0/0.1
. . .

inet6.0: 4 destinations, 4 routes (4 active, 0 holddown, 0 hidden)
Restart Complete
+ = Active Route, - = Last Active, * = Both

200::/64           *[Direct/0] 00:05:26
                    > via ge-0/0/0.2
fe80::/64          *[Direct/0] 00:05:26
                    > via ge-0/0/0.2
```

As expected, a direct route is present for the 10.0/16 IPv4 route pointing to the ge-0/0/0.1 interface. Note how the 200.0.0/24 IPv4 and the 200::/64 IPv6 routes point to unit 2 on the same interface device.

In most cases, serious interface configuration mistakes will generate an error and fail to commit. For example, here is a combination of Layer 2 and Layer 3 protocol families that are currently not supported on the same interface device:

```
{master}[edit interfaces ge-0/0/0]
lab@Vodkila# set unit 10 family ethernet-switching
{master}[edit interfaces ge-0/0/0]
lab@Vodkila# show
vlan-tagging;
unit 1 {
    vlan-id 1;
    family inet {
        address 200.0.0.1/24;
    }
}
unit 2 {
    vlan-id 2;
    family inet {
        address 10.0.0.1/16 {
            arp 10.0.0.2 mac 00:1f:12:3d:b4:ff;
        }
    }
    family inet6 {
        address 200::1/64;
    }
}
unit 10 {
    ##
    ## Warning: An interface cannot have both family ethernet-switching and vlan-
tagging configured
    ##
    family ethernet-switching;
}
```

Note that the CLI generates a warning for an objectionable statement when viewing the configuration. At commit time, such problems result in an error, and therefore, no changes are activated until the issue is corrected:

```
{master}[edit interfaces ge-0/0/0]
lab@Vodkila# commit check
[edit interfaces ge-0/0/0 unit 10 family]
  'ethernet-switching'
    An interface cannot have both family ethernet-switching and vlan-tagging
configured
error: configuration check-out failed: (statements constraint check failed)
```

Interface Troubleshooting

Interfaces and links can have a variety of issues that can stem from simple misconfigurations, broken hardware, or the more difficult area of vendor interoperability problems. Listing all the possibilities would require a separate book! In fact, much information is already available in our companion book in this series, *JUNOS Enterprise Routing*. Generally speaking, Layer 3 problems are easier to isolate because of the ability to use pings and traceroutes to locate the problem area. With Layer 2 forwarding, either things work end to end, or they simply don't, and gaining insight as to where the forwarding problem is can be very difficult. In some cases, you must resort to monitoring packet counts until you find the node that receives more than it sends. Such a technique is hard enough to perform in a test network, but on live networks where user traffic cannot be controlled, packet-count-based fault isolation is all but impossible.

You will be exposed to a multitude of real-world interface (and other) troubleshooting techniques throughout this book, in the form of various case studies or deployment scenarios that invariably go astray. Really, it's almost as though they were planned to break just to afford you the benefit of the resulting exercise in fault isolation. To keep things moving, this section highlights the primary tools available for troubleshooting in a JUNOS environment, and leaves demonstration of those tools to later chapters.

JUNOS troubleshooting tools

JUNOS is a powerful operating system, and its Unix underpinnings provide a great many troubleshooting tools, most of which are available through the CLI. Although the EX product is somewhat new, JUNOS is now nearing its first decade of deployed use. Much information is already available on general JUNOS troubleshooting, and the single-image model means this is directly portable to EX troubleshooting in general.

 Direct use of shell commands should be performed only under the supervision of Juniper support personnel. There is a lot of rope in a root shell, after all; if used incorrectly, the switch could be rendered inoperable.

Here's a list of key JUNOS fault isolation tools, in no particular order.

Syslog. The system logfiles often contain explicit information about hardware or software problems, and may also display errors for operational problems such as detection of duplicate IP addressing. Display the *messages* log and filter the output based on the interface name, or keywords such as error and fail. Monitor the syslog in real time as you commit the changes to spot any issues early. This example searches the *messages* file for instances of either error or fail:

```
{master}
lab@Vodkila> show log messages | match "(fail|error)"
Jan 30 01:10:03  Vodkila /kernel: RT_PFE: RT msg op 1 (PREFIX ADD) failed, err 5
```

```
(Invalid)
. . .
```

Monitor interface. This `iftop`-like utility quickly displays interface packet and byte counts, and also displays any overt alarms or warnings. It's useful when tracking data loss or testing traffic rates/interface loads. The command can also be performed on a logical interface to limit the statistics to just that unit:

```
lab@Rum> monitor interface ge-0/0/4

Rum                             Seconds: 6              Time: 12:00:51
                                                        Delay: 11/0/11
Interface: ge-0/0/4, Enabled, Link is Up
Encapsulation: Ethernet, Speed: 1000mbps
Traffic statistics:                                     Current delta
  Input bytes:              5839292 (0 bps)                      [0]
  Output bytes:             9324028 (0 bps)                      [0]
  Input packets:              91076 (0 pps)                      [0]
  Output packets:            100314 (0 pps)                      [0]
Error statistics:
  Input errors:                   0                             [0]
  Input drops:                    0                             [0]
  Input framing errors:           0                             [0]
  Policed discards:               0                             [0]
  L3 incompletes:                 0                             [0]
  L2 channel errors:              0                             [0]
  L2 mismatch timeouts:           0  Carrier transitions        [0]


Next='n', Quit='q' or ESC, Freeze='f', Thaw='t', Clear='c', Interface='i'
```

Monitor traffic. This `tcpdump`-like command is akin to `debug IP` on IOS-based routers. It works only for traffic sent to or from the local RE, however. To monitor transient traffic, you need to configure a monitor port and direct the traffic to an external analyzer. This example uses the `detail` switch for added info and shows a Spanning Tree Protocol (STP) message being transmitted:

```
lab@Rum> monitor traffic interface ge-0/0/4 detail
Address resolution is ON. Use <no-resolve> to avoid any reverse lookup delay.
Address resolution timeout is 4s.
Listening on ge-0/0/4, capture size 1514 bytes

12:06:41.406162 Out STP 802.1d, Config, Flags [none], bridge-id
8000.00:19:e2:56:ee:80.8205, length 35
        message-age 0.00s, max-age 20.00s, hello-time 2.00s, forwarding-delay
15.00s
        root-id 8000.00:19:e2:56:ee:80, root-pathcost 0
```

Operational mode show commands. There are numerous operational mode commands, generally beginning with `show`, that provide insight into interface or general networking/platform problems. Commands such as `show route`, `show interface`, `show chassis alarms`, `show log messages`, and `show ethernet-switching` are particularly useful.

Ethernet OAM. Beginning with Release 9.4, EX platforms support Ethernet OAM (on Layer 2 interfaces only) to provide fault detection for the forwarding plane between two peers (point-to-point [P-to-P] Ethernet). Because many link or interface hardware issues have no obvious signs, other than a lack of data/connectivity, the ability to eliminate the data plane from a Layer 2 protocol-agnostic point of view is a great help in isolating faults. An in-depth discussion of OAM is outside the scope of this section. For details, see the IEEE 802.3ah standard. For now, suffice it to say that OAM provides an in-band heartbeat, or keepalive function, in addition to basic management functions such as setting or clearing remote loopbacks.

Because Ethernet OAM operates *in-band*, which means over the same data link as the upper-layer user traffic, if OAM shows no issues there is little point in double-checking Physical level interface parameters, given that communication is successfully occurring across the link. Instead, focus should be placed on mistakes in the Layer 2 configuration or a downstream node. Currently, EX switches support interface loopbacks only through the use of OAM, and then only a remote loopback is supported. JUNOS routing devices have a (local) `loopback` configuration option for Ethernet interfaces. It's best to perform local loopbacks with an external loopback anyway, given the fact that an internal local loopback does not really validate all interface hardware, but such a loopback requires physical proximity to the port being tested. In this example, a Layer 2 interface is not running Spanning Tree, and has no user traffic flowing; as a result, the interface traffic counters cannot be used to determine link status, given that they are 0:

```
[edit protocols oam]
regress# run clear interfaces statistics all

[edit protocols oam]
regress# run show interfaces ge-0/0/4 extensive
Physical interface: ge-0/0/4, Enabled, Physical link is Up
  Interface index: 133, SNMP ifIndex: 118, Generation: 136
  Link-level type: Ethernet, MTU: 1514, Speed: Auto, Duplex: Auto, BPDU Error:
None, MAC-REWRITE Error: None, Loopback: Disabled,
  Source filtering: Disabled, Flow control: Enabled, Auto-negotiation: Enabled,
Remote fault: Online
  Device flags   : Present Running
  Interface flags: SNMP-Traps Internal: 0x0
  Link flags     : None
  CoS queues     : 8 supported, 8 maximum usable queues
  Hold-times     : Up 0 ms, Down 0 ms
  Current address: 00:1f:12:35:31:c4, Hardware address: 00:1f:12:35:31:c4
  Last flapped   : 2009-02-17 16:42:46 UTC (1d 01:21 ago)
  Statistics last cleared: 2009-02-18 18:03:47 UTC (00:00:02 ago)
  Traffic statistics:
   Input  bytes  :                0                    0 bps
   Output bytes  :                0                    0 bps
   Input  packets:                0                    0 pps
   Output packets:                0                    0 pps
   IPv6 transit statistics:
    Input  bytes  :               0
    Output bytes  :               0
    Input  packets:               0
```

```
  Output packets:                 0
 Input errors:
   Errors: 0, Drops: 0, Framing errors: 0, Runts: 0, Policed discards: 0, L3
incompletes: 0, L2 channel errors: 0,
   L2 mismatch timeouts: 0, FIFO errors: 0, Resource errors: 0
 Output errors:
   Carrier transitions: 0, Errors: 0, Drops: 0, Collisions: 0, Aged packets: 0,
FIFO errors: 0, HS link CRC errors: 0,
   MTU errors: 0, Resource errors: 0
 . . .
```

The display indicates that all is well with the interface, but it could just be that the local link is up while the remote end is down. OAM is added to the link, and in this case is combined with an action profile, which, as per its name, defines the set of actions to take in the event that OAM thresholds are crossed:

```
[edit protocols oam]
regress# show
ethernet {
    link-fault-management {
        action-profile down_link {
            event {
                link-adjacency-loss;
            }
            action {
                link-down;
            }
        }
        interface ge-0/0/4.0 {
            apply-action-profile down_link;
            link-discovery active;
            negotiation-options {
                allow-remote-loopback;
            }
        }
    }
}
```

The results are quickly confirmed:

```
[edit protocols oam]
regress# run show oam ethernet link-fault-management
  Interface: ge-0/0/4.0
    Status: Running, Discovery state: Send Any
    Peer address: 00:21:59:c0:ba:c4
    Flags:Remote-Stable Remote-State-Valid Local-Stable 0x50
    Remote entity information:
      Remote MUX action: forwarding, Remote parser action: forwarding
      Discovery mode: active, Unidirectional mode: unsupported
      Remote loopback mode: supported, Link events: supported
      Variable requests: unsupported
  Application profile statistics:
    Profile Name            Invoked      Executed
    down_link                  0            0
```

The output of the show oam ethernet link-fault-management command confirms that the OAM neighbor relationship is operational, and that the associated down_link action profile has not yet been triggered. The in-band operation of OAM results in traffic activity on the interface:

```
regress# run show interfaces ge-0/0/4 extensive
Physical interface: ge-0/0/4, Enabled, Physical link is Up
 . . .
  Statistics last cleared: 2009-02-18 18:03:47 UTC (00:01:12 ago)
  Traffic statistics:
   Input  bytes  :              3012                512 bps
   Output bytes  :              3524                512 bps
   Input  packets:                47                  1 pps
   Output packets:                55                  1 pps
   IPv6 transit statistics:
    Input  bytes  :               0
    Output bytes  :               0
 . . .
```

Not wanting to let well enough alone, we test the action profile by disabling OAM on the remote end of the link. The OAM status is again displayed:

```
[edit protocols oam]
regress# run show oam ethernet link-fault-management
  Interface: ge-0/0/4.0
    Status: Running, Discovery state: Active Send Local
    Peer address: 00:00:00:00:00:00
    Flags:0x8
Application profile statistics:
  Profile Name                  Invoked    Executed
  down_link                           1           1
```

As expected, the OAM neighbor is lost, as evidenced by a null MAC address and the lack of remote capability and status indication. The down_link profile has been executed once, and as a result the ge-0/0/4.0 logical interface is declared down:

```
[edit protocols oam]
regress# run show interfaces ge-0/0/4.0
  Logical interface ge-0/0/4.0 (Index 64) (SNMP ifIndex 119)
    Flags: Device-Down 0x0 Encapsulation: ENET2
    Input packets : 1571
    Output packets: 1769
    Protocol eth-switch
      Flags: Is-Primary
```

Diagnostic commands. The EX platform has a built-in fault isolation wizard that you can run from the J-Web or CLI interface. This utility includes a basic Time Domain Reflectometer (TDR) function that's useful in determining the general nature and location of cable faults. The following example is executed on a known working link, and the results are quite informative, with details provided on overall status as well as on a per-MDI-wire-pair basis:

```
lab@Rum> request diagnostics tdr start interface ge-0/0/0
Test successfully executed
```

. . .

```
lab@Rum> show diagnostics tdr interface ge-0/0/0

Interface TDR detail:
Interface name              : ge-0/0/0
Test status                 : Started
```

The output of the show diagnostics tdr command shows that the test is still running, so you patiently bide your time. Some things cannot be rushed, after all, and it's good to stop and smell life's TDRs every now and then:

```
lab@Rum> show diagnostics tdr interface ge-0/0/0

Interface TDR detail:
Interface name              : ge-0/0/0
Test status                 : Passed
Link status                 : UP
MDI pair                    : 1-2
  Cable status              : Normal
  Distance fault            : 0 Meters
  Polartiy swap             : Normal
  Skew time                 : 8 ns
MDI pair                    : 3-6
  Cable status              : Normal
  Distance fault            : 0 Meters
  Polartiy swap             : Normal
  Skew time                 : 0 ns
MDI pair                    : 4-5
  Cable status              : Normal
  Distance fault            : 0 Meters
  Polartiy swap             : Normal
  Skew time                 : 8 ns
MDI pair                    : 7-8
  Cable status              : Normal
  Distance fault            : 0 Meters
  Polartiy swap             : Normal
  Skew time                 : 8 ns
Channel pair                : 1
  Pair swap                 : MDI
Channel pair                : 2
  Pair swap                 : MDIX
Downshift                   : No Downshift
```

The completed test results confirm that all is fine and dandy with the interface and its link.

Loopbacks. Performing local and remote loopbacks is an invaluable technique when you need to isolate between interface hardware/configuration and cable/link faults. You can perform a hardwired loopback on any FD interface with an external loop plug. You may need to attenuate the signal on optical interfaces.

Unlike M, T, MX, and J Series devices, EX switches don't support a command-initiated local loopback for Ethernet interfaces. On the upside, you can perform a remote loopback on Layer 2 interfaces by enabling OAM and issuing the appropriate command. Once looped, you can use link statistics or pings (as applicable) to determine whether traffic can be sent and received. When both ends pass a local test, the link is bad or the endpoints are misconfigured. If either end fails the local test, that interface (or configuration) is bad.

Here's an OAM-based Layer 2 loopback example. The OAM configuration specifies that OAM should run on the ge-0/0/4.0 interface, and includes the allow-remote-loopback negotiation option to allow remote initiated loopbacks. The remote-loop back keyword is used to place the remote end into the loopback:

```
[edit protocols oam]
regress# show
ethernet {
    link-fault-management {
        interface ge-0/0/4.0 {
            link-discovery active;
            remote-loopback;
            negotiation-options {
                allow-remote-loopback;
            }
        }
    }
}
```

 EX support for Ethernet OAM started in JUNOS Release 9.4. The 9.2 release used in this test bed does not support OAM, so another pair of switches are used.

The output of a show oam ethernet link-fault-management command confirms that the loopback request is successful:

```
[edit protocols oam]
regress# run show oam ethernet link-fault-management
  Interface: ge-0/0/4.0
    Status: Running, Discovery state: Send Any
    Peer address: 00:21:59:c0:ba:c4
    Flags:Remote-Stable Remote-State-Valid Local-Stable 0x50
    Remote loopback status: Disabled on local port, Enabled on peer port
    Remote entity information:
      Remote MUX action: discarding, Remote parser action: loopback
      Discovery mode: active, Unidirectional mode: unsupported
      Remote loopback mode: supported, Link events: supported
      Variable requests: unsupported
```

After interface statistics are cleared, the matched send and receive count, along with a lack of any error indications, shows that all is well with the local interface, the link, and remote loopback circuitry at the far end of the link:

```
regress# run clear interfaces statistics all
. . .

[edit protocols oam]
regress# run show interfaces ge-0/0/4 extensive
Physical interface: ge-0/0/4, Enabled, Physical link is Up
    . . .
   Statistics last cleared: 2009-02-18 18:43:54 UTC (00:00:41 ago)
   Traffic statistics:
    Input  bytes  :              2624             512 bps
    Output bytes  :              2624             512 bps
    Input  packets:                41               1 pps
    Output packets:                41               1 pps
    . . .
```

Hard loops. The best way to confirm that an EX port is truly operational is to confirm that it can actually send and receive traffic outside the port itself. Although an internal loopback (not supported on EX ports) is convenient, it does not really confirm the port's external transmit and receive functions, which can result in a defective port passing such a loopback test. This section describes how you can attach an external loopback cable/plug to an EX switch port to truly test its ability to send and receive. Ideally, such a loop is performed at the end of a cable that is half as long as the real cable, as the round trip accurately simulates the loss through the real link; in most cases, a bad port will fail even with a short length of cable, which means you can carry a small RJ-45 jack with looped pins as part of your troubleshooting toolkit.

Here are some things to watch for when performing this type of test:

Does not work for GE on copper
> GE over twisted-pair copper requires extensive Near End Cross Talk (NEXT) circuitry. This function typically prevents a successful external loopback because the circuitry is designed to filter the local end's transmit from the much weaker signal that's received from the remote end. This means you will need to configure copper GE ports to operate at 100 Mbps, as the NEXT functionality cannot currently be disabled on EX switches.

Requires full duplex
> Any loopback requires FD. This may require a change in configuration or restart of auto-negotiations if an interface is operating in HD mode.

Best when combined with Layer 3
> By adding Layer 3 (IP) to the interface being tested, you can use utilities such as ping to generate and confirm test traffic. This takes a bit of finesse, as we will demonstrate shortly, but the results are true indicators of the port's external and PFE-facing functionality.

Figure 3-8 shows the printout of a typical Ethernet loopback plug.

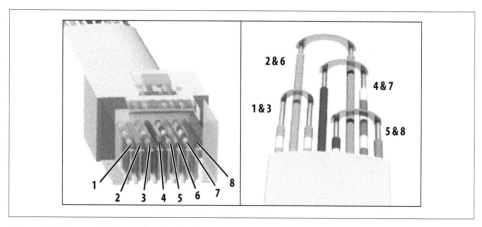

Figure 3-8. Copper Ethernet loopback plug

All eight wires are used because the wires are cheap, and it allows usage on all flavors of copper-based Ethernet (with the GE and NEXT caveat noted earlier).

To perform the loopback, you first configure the test interface for 100 Mbps operation. FD is the default, so an explicit duplex setting is not needed. This example adds an IP configuration to the interface to permit testing with IP packets:

```
[edit]
lab@Rum# show interfaces ge-0/0/20
ether-options {
    speed {
        100m;
    }
}
unit 0 {
    family inet {
        address 200.0.0.1/24 {
            arp 200.0.0.2 mac 00:19:e2:56:ee:95;
        }
    }
}
```

The static ARP entry is needed because you cannot send an IP packet out an Ethernet interface without a next hop MAC address. Because you expect the sending interface to also be the receiving interface, this MAC address *must* match the current hardware address used by the test interface. Using the local interface's own MAC address ensures that the frame will make it past the selective listening of the MAC layer so that it can be processed by the IP layer:

```
[edit]
lab@Rum# run show interfaces ge-0/0/20 | match hardware
    Interface flags: Hardware-Down SNMP-Traps Internal: 0x0
    Current address: 00:19:e2:56:ee:95, Hardware address: 00:19:e2:56:ee:95
```

With this configuration and loopback plug applied, the interface is confirmed to be in an operational state. So far, so good:

```
[edit interfaces ge-0/0/20]
lab@Rum# run show interfaces ge-0/0/20
Physical interface: ge-0/0/20, Enabled, Physical link is Up
  Interface index: 149, SNMP ifIndex: 135
  Link-level type: Ethernet, MTU: 1514, Speed: 100mbps, MAC-REWRITE Error: None,
  Loopback: Disabled, Source filtering: Disabled, Flow control: Enabled,
  Auto-negotiation: Enabled, Remote fault: Online
  Device flags   : Present Running
  Interface flags: SNMP-Traps Internal: 0x0
  Link flags     : None
  CoS queues     : 8 supported, 8 maximum usable queues
  Current address: 00:19:e2:56:ee:95, Hardware address: 00:19:e2:56:ee:95
  Last flapped   : 2008-08-29 13:11:51 UTC (00:06:05 ago)
  Input rate     : 0 bps (0 pps)
  Output rate    : 0 bps (0 pps)
  Active alarms  : None
  Active defects : None

  Logical interface ge-0/0/20.0 (Index 67) (SNMP ifIndex 149)
    Flags: SNMP-Traps 0x0 Encapsulation: ENET2
    Input packets : 516
    Output packets: 516
    Protocol inet, MTU: 1500
      Flags: None
      Addresses, Flags: Is-Preferred Is-Primary
        Destination: 200.0.0/24, Local: 200.0.0.1, Broadcast: 200.0.0.255
```

The highlights confirm that no error conditions are reported; the 100 Mbps mode; the presence of an IP address; and that no traffic is currently being sent or received. FD mode is also verified with the media switch:

```
[edit interfaces ge-0/0/20]
lab@Rum# run show interfaces ge-0/0/20 media
Physical interface: ge-0/0/20, Enabled, Physical link is Up
  Interface index: 149, SNMP ifIndex: 135
. . .
  Autonegotiation information:
    Negotiation status: Complete
    Link partner:
        Link mode: Full-duplex, Flow control: None, Remote fault: OK,
        Link partner Speed: 100 Mbps
    Local resolution:
        Flow control: None, Remote fault: Link OK
. . .
```

The key to the upcoming Layer 3 confirmation is that you will generate a ping request to a *remote* address that reflects some host on the directly connected network. Because of the loopback plug, there is *no remote* host, which is part of the method to this madness. When the packet is transmitted to the fictitious remote IP address, the loopback returns the frame to the receive circuitry of that same port. Because Layer 3 forwarding does not have issues with loops and redundant paths, the receipt of such a packet is not considered an error, and the traffic is handled normally. This means the frame is stripped, and a longest match is performed against the destination address of the IP packet. As a result, the router once again transmits the packet out the same interface, except this time with a TTL that is one less than when the packet was originally transmitted.

This process continues until the packet's TTL is decremented to 0, at which point an ICMP error message is generated to report the TTL expiration. Oddly enough, it's the *receipt* of this very *error* message that indicates a successful loopback test! If anything should go wrong with packet processing in any of the loopback iterations, the result is a silent discard, given the datagram nature of IP. Such a discard results in a timeout error, which in this case indicates packet loss. By altering the TTL value, you can control how likely a loss is to occur on a marginal circuit/interface. For example, a TTL of 254 means the packet must be sent and received 254 times before the expected TTL expired message is generated, whereas a TTL of 2 requires far fewer such iterations, and is therefore that much more likely to succeed.

Before generating the test traffic, the route to the target remote IP address is confirmed active and pointing out the looped interface:

```
[edit interfaces ge-0/0/20]
lab@Rum# run show route 200.0.0.2

inet.0: 5 destinations, 5 routes (5 active, 0 holddown, 0 hidden)
+ = Active Route, - = Last Active, * = Both

200.0.0.0/24        *[Direct/0] 00:05:06
                    > via ge-0/0/20.0
```

In this example, any host address in the range of 2 to 254 could be used. To work, it must not match the local IP address and it must have a static ARP assignment, however. Two test pings are generated with an explicit TTL setting of 254, to create a more strenuous test by overriding the default value of 64:

```
lab@Rum# run ping 200.0.0.2 count 2 ttl 254
PING 200.0.0.2 (200.0.0.2): 56 data bytes
36 bytes from 200.0.0.1: Time to live exceeded
Vr HL TOS  Len   ID Flg  off TTL Pro  cks      Src        Dst
 4  5  00 0054 3e1f   0 0000  01  01 eb86 200.0.0.1  200.0.0.2

36 bytes from 200.0.0.1: Time to live exceeded
Vr HL TOS  Len   ID Flg  off TTL Pro  cks      Src        Dst
 4  5  00 0054 3e21   0 0000  01  01 eb84 200.0.0.1  200.0.0.2


--- 200.0.0.2 ping statistics ---
2 packets transmitted, 0 packets received, 100% packet loss
```

Despite the 100% loss in the resulting output, the receipt of the TTL expired messages confirms proper port operation. Traffic statistics on the test interface confirm that each test packet was looped TTL times, at wire rate, which again is a very good indication that the interface and its PFE links are operational:

```
[edit]
lab@Rum# run show interfaces ge-0/0/20
Physical interface: ge-0/0/20, Enabled, Physical link is Up
  Interface index: 149, SNMP ifIndex: 135
  Link-level type: Ethernet, MTU: 1514, Speed: 100mbps, MAC-REWRITE Error: None,
  Loopback: Disabled, Source filtering: Disabled, Flow control: Enabled,
. . .

  Logical interface ge-0/0/20.0 (Index 67) (SNMP ifIndex 149)
    Flags: SNMP-Traps 0x0 Encapsulation: ENET2
    Input packets : 510
    Output packets: 510
    Protocol inet, MTU: 1500
      Flags: None
      Addresses, Flags: Is-Preferred Is-Primary
        Destination: 200.0.0/24, Local: 200.0.0.1, Broadcast: 200.0.0.255
```

In this case, you may have expected to see a packet count of 254 x 2 (508), but instead you observe 510. The extra two packets reflect the ICMP error message that was sent, and received, twice, resulting in a total of four extra packets in total.

EX Interface Summary

The EX platform supports only Ethernet interface types, which greatly reduces the scope of interface configuration and fault isolation. In JUNOS, an interface configuration involves both a device and at least one logical unit. Device-level settings are placed under the interface itself, whereas protocol families and their related properties are specified at the logical unit level.

JUNOS software provides numerous operational and diagnostics tools to assist in fault isolation and correction. The tools and techniques we discussed in this section are demonstrated in later real-world networking scenarios when things do not go as planned.

Basic Switch Maintenance

Once the switch is configured and ready for deployment, some basic checks should be performed, and logging and recovery configuration should be considered. We begin this section by issuing a few chassis health-check commands. We then discuss some additional syslog, SNMP, and Network Time Protocol (NTP) configurations, which are strongly recommended because they enhance troubleshooting and security-related activities. Lastly, we look at some additional (and cool) JUNOS features that can save the day, and occasionally your bacon, in cases of switch configuration deletion, mistakes, or just strange cosmic-ray-based corruption, as has been known to spread through IOS networks in the past.

Chassis Health Check

When the chassis powers up for the first time, issue a few basic commands to verify that all the components are working properly. You can do this before or after the initial configuration. First, check to see whether all the proper hardware pieces are working and recognized by the system; you can do this by issuing the show chassis hardware command. This is also a great command to record into your inventory sheet, as it will list the serial and part numbers for each piece of hardware. As displayed, the new switch is an EX4200-24T with a Gigabit Ethernet uplink module with two Small Form-factor Pluggable (SFP) optics and a 320-watt AC power supply:

```
lab@Tequila > show chassis hardware
Hardware inventory:
Item            Version  Part number  Serial number   Description
Chassis                               BM0208269767    EX4200-24T
FPC 0           REV 11   750-021256   BM0208269767    EX4200-24T, 8 POE
  CPU                    BUILTIN      BUILTIN         FPC CPU
  PIC 0                  BUILTIN      BUILTIN         24x 10/100/1000 Base-T
  PIC 1         REV 03B  711-021270   AR0208163417    4x GE SFP
    Xcvr 0      REV 01   740-011613   AM0813S8Z6W     SFP-SX
    Xcvr 1      REV 01   740-011613   AM0813S8Z26     SFP-SX
Power Supply 0  REV 03   740-020957   AT0508248670    PS 320W AC
Fan Tray                                              Fan Tray
```

 SFPs are sold separately from the uplink modules.

Verify that there are no system or chassis alarms. If there were any fan failures, power supply failures, and so forth, they should show up in the chassis command:

```
lab@Tequila> show system alarms
No alarms currently active

lab@Tequila> show chassis alarms
No alarms currently active
```

The show system storage command is a quick way to gauge flash memory fill levels. The flash is partitioned in various directories with the total storage size of 1 GB:

```
lab@Tequila > show system storage
fpc0:
--------------------------------------------------------------
Filesystem    Size  Used  Avail  Capacity  Mounted on
/dev/da0s2a   184M  83M   87M       49%    /
devfs         1.0K  1.0K  0B       100%    /dev
devfs         1.0K  1.0K  0B       100%    /dev/
/dev/md0      31M   31M   0B       100%    /packages/mnt/jbase
/dev/md1      13M   13M   0B       100%    /packages/mnt/jcrypto-ex-9.2R1.9
/dev/md2      4.3M  4.3M  0B       100%    /packages/mnt/jdocs-ex-9.2R1.9
/dev/md3      88M   88M   0B       100%    /packages/mnt/jkernel-ex-9.2R1.9
```

```
/dev/md4          13M     13M     0B    100%  /packages/mnt/jpfe-ex42x-9.2R1.9
/dev/md5          33M     33M     0B    100%  /packages/mnt/jroute-ex-9.2R1.9
/dev/md6          12M     12M     0B    100%  /packages/mnt/jswitch-ex-9.2R1.9
/dev/md7          15M     15M     0B    100%  /packages/mnt/jweb-ex-9.2R1.9
/dev/md8          63M     8.0K    58M     0%  /tmp
/dev/da0s2f       123M    1.2M    112M    1%  /var
/dev/da0s3e       55M     58K     51M     0%  /config
/dev/da0s3d       314M    4.0K    289M    0%  /var/tmp
/dev/md9          118M    8.1M    100M    7%  /var/rundb
procfs            4.0K    4.0K    0B    100%  /proc
/var/jail/etc     123M    1.2M    112M    1%  /packages/mnt/jweb-ex-9.2R1.9/\
    jail/var/etc
/var/jail/run     123M    1.2M    112M    1%  /packages/mnt/jweb-ex-9.2R1.9/\
    jail/var/run
/var/jail/tmp     123M    1.2M    112M    1%  /packages/mnt/jweb-ex-9.2R1.9/\
    jail/var/tmp
/var/tmp          314M    4.0K    289M    0%  /packages/mnt/jweb-ex-9.2R1.9/\
    jail/var/tmp/uploads
devfs             1.0K    1.0K    0B    100%  /packages/mnt/jweb-ex-9.2R1.9/jail/
```

Here, the various filesystems are based on slices (partitions) of the flash memory, and we can see that, for example, the */var* filesystem has used only 1% of its 123 MB slice. When you install or upgrade JUNOS software, the disk partitions may be resized, and information in user home directories (*/var/home/*) may be lost. Where possible, JUNOS attempts to keep SSH keys. When needed, use the request system storage cleanup command to delete unneeded files and free up space, such as might be needed to make room for a future software upgrade:

```
lab@Tequila > request system storage cleanup

List of files to delete:

    Size Date         Name
    112B Sep 21 08:03 /var/log/default-log-messages.0.gz
   8639B Sep 21 08:03 /var/log/interactive-commands.0.gz
   8429B Aug 22 21:00 /var/log/interactive-commands.1.gz
   8238B Aug 18 11:00 /var/log/interactive-commands.2.gz
   8205B Aug 13 09:00 /var/log/interactive-commands.3.gz
   8155B Aug  7 22:00 /var/log/interactive-commands.4.gz
   8211B Aug  4 06:00 /var/log/interactive-commands.5.gz
   8260B Jul 31 14:00 /var/log/interactive-commands.6.gz
   8185B Jul 27 09:00 /var/log/interactive-commands.7.gz
   8569B Jul 22 10:00 /var/log/interactive-commands.8.gz
   9008B Jul 17 12:00 /var/log/interactive-commands.9.gz
   8335B Sep 21 08:03 /var/log/messages.0.gz
   14.9K Aug 24 08:00 /var/log/messages.1.gz
   12.7K Aug  1 22:00 /var/log/messages.2.gz
    27B Sep 15 04:00 /var/log/wtmp.0.gz
   4232B Aug 25 06:38 /var/log/wtmp.1.gz
   15.4K Aug 16 04:01 /var/log/wtmp.2.gz
    137B Jul 17 08:42 /var/log/wtmp.3.gz
Delete these files ? [yes,no] (no) yes
```

Verify the software version running on the switch:

```
lab@Tequila > show version
fpc0:
--------------------------------------------------------------------------
Model: ex4200-24t
JUNOS Base OS boot [9.2R1.9]
JUNOS Base OS Software Suite [9.2R1.9]
JUNOS Kernel Software Suite [9.2R1.9]
JUNOS Crypto Software Suite [9.2R1.9]
JUNOS Online Documentation [9.2R1.9]
JUNOS Enterprise Software Suite [9.2R1.9]
JUNOS Packet Forwarding Engine Enterprise Software Suite [9.2R1.9]
JUNOS Routing Software Suite [9.2R1.9]
JUNOS Web Management [9.2R1.9]
```

Lastly, it's a good idea to verify that the fans are working properly and temperatures are well below alarm thresholds. Notice that only a single power supply was purchased:

```
lab@Tequila > show chassis environment
Class Item                     Status      Measurement
Power FPC 0 Power Supply 0     OK
      FPC 0 Power Supply 1     Absent
Temp  FPC 0 CPU                OK          37 degrees C / 98 degrees F
      FPC 0 EX-PFE1            OK          46 degrees C / 114 degrees F
      FPC 0 EX-PFE2            OK          47 degrees C / 116 degrees F
      FPC 0 GEPHY Front Left   OK          30 degrees C / 86 degrees F
      FPC 0 GEPHY Front Right  OK          31 degrees C / 87 degrees F
      FPC 0 Uplink Conn        OK          32 degrees C / 89 degrees F
Fans  FPC 0 Fan 1              OK          Spinning at normal speed
      FPC 0 Fan 2              OK          Spinning at normal speed
      FPC 0 Fan 3              OK          Spinning at normal speed
```

Also, for those who prefer the J-Web GUI, there is a nice dashboard that shows many of these commands in a more graphical format. Figure 3-9 shows an example.

Syslog

Syslog was originally developed as a method to send information for the sendmail application in BSD, but it was so useful that it was extended to other applications and operating systems. Essentially, syslog is a standard way to send log messages across an IP network. Syslog describes the actual transport mechanism used to send these messages, and is often used to describe the actual application that is sending these messages. Originally, it was an "industry" standard, and was not attached to an informational RFC until 2001, with RFC 3164, "The BSD Syslog Protocol."

Syslog messages are sent over UDP with a destination port of 514. The IP transport mechanism is defined and not the actual syslog content. It is left to the discretion of the application or system coder to create an informative message for the receiver. The message always contains a message severity level and a facility. The *facility* can be defined as the type of message that is being sent, and the *severity level* indicates the message's importance. Table 3-2 defines the severity levels.

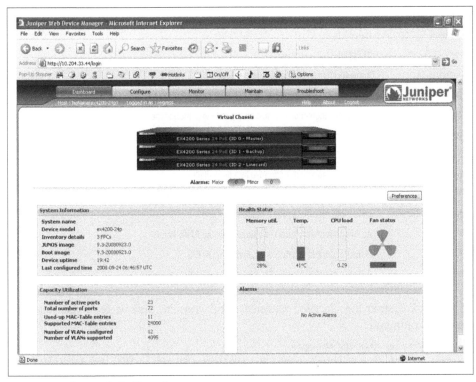

Figure 3-9. The J-Web dashboard

Table 3-2. Syslog severity levels

Numerical code	Severity
0	Emergency: system is unusable
1	Alert: action must be taken immediately
2	Critical: critical conditions
3	Error: error conditions
4	Warning: warning conditions
5	Notice: normal but significant condition
6	Informational: informational messages
7	Debug: debug-level messages

Table 3-3 lists the facilities that are available in JUNOS.

Table 3-3. Syslog facilities

Facility	Description
Any	All facilities (all messages)
Authorization	Authentication and authorization attempts
Change-Log	Changes to the configuration
Conflict-Log	Specified configuration is invalid on the routing platform type
Daemon	Actions performed or errors encountered by system processes
DFC	Events related to dynamic flow capture
Firewall	Packet filtering actions performed by a firewall filter
FTP	Actions performed or errors encountered by the FTP process
Interactive commands	Commands executed by the user interface
Kernel	Actions performed or errors encountered by the JUNOS kernel
PFE	Actions performed or errors encountered by the PFE
User	Actions performed or errors encountered by user-space processes

The default system log is called *messages*, and you can view it with the show log messages command:

```
root@Tequila> show log messages
Aug  5 16:14:28  Tequila chassisd[636]: CHASSISD_SNMP_TRAP7: SNMP trap generated:
FRU removal (jnxFruContentsIndex 7, jnxFruL1Index 1, jnxFruL2Index 0,jnxFruL3Index
0, jnxFruName FPC: EX4200-24T, 8 POE @ 0/*/*, jnxFruType 3,
jnxFruSlot 1)
Aug  5 16:14:29  Tequila chassisd[636]: CHASSISD_TIMER_VAL_ERR: Null timer ID
Aug  5 16:14:29  Tequila chassisd[636]: CHASSISD_SNMP_TRAP7: SNMP trap generated:
FRU insertion (jnxFruContentsIndex 7, jnxFruL1Index 1, jnxFruL2Index 0,
jnxFruL3Index 0, jnxFruName FPC: EX4200-24T, 8 POE @ 0/*/*, jnxFruType 3,
jnxFruSlot 1)
Aug  5 16:14:30  Tequila chassisd[636]: CHASSISD_SNMP_TRAP10: SNMP trap generated:
FRU power on (jnxFruContentsIndex 8, jnxFruL1Index 1, jnxFruL2Index 1,
jnxFruL3Index 0, jnxFruName PIC:  @ 0/0/*, jnxFruType 11, jnxFruSlot 1,
jnxFruOfflineReason 2, jnxFruLastPowerOff 243976, jnxFruLastPowerOn 250250)
Aug  5 16:14:30  Tequila chassisd[636]: CHASSISD_SNMP_TRAP10: SNMP trap generated:
FRU power on (jnxFruContentsIndex 8, jnxFruL1Index 1, jnxFruL2Index 2,
jnxFruL3Index 0, jnxFruName PIC: 4x GE SFP @ 0/1/*, jnxFruType 11, jnxFruSlot 1,
jnxFruOfflineReason 2, jnxFruLastPowerOff 243976, jnxFruLastPowerOn 250250)
Aug  5 16:14:30  Tequila chas[527]: PS 0: Transitioning from empty to online
Aug  5 16:14:30  Tequila chassisd[636]: CHASSISD_IFDEV_CREATE_NOTICE: create_pics:
created interface device for ge-0/0/0
Aug  5 16:14:30  Tequila chassisd[636]: CHASSISD_IFDEV_CREATE_NOTICE: create_pics:
created interface device for ge-0/0/1
Aug  5 16:14:30  Tequila chassisd[636]: CHASSISD_IFDEV_CREATE_NOTICE: create_pics:
created interface device for ge-0/0/2
Aug  5 16:14:30  Tequila chassisd[636]: CHASSISD_IFDEV_CREATE_NOTICE: create_pics:
created interface device for ge-0/0/3
Aug  5 16:14:30  Tequila chassisd[636]: CHASSISD_IFDEV_CREATE_NOTICE: create_pics:
created interface device for ge-0/0/4
```

```
Aug  5 16:14:30  Tequila chassisd[636]: CHASSISD_IFDEV_CREATE_NOTICE: create_pics:
created interface device for ge-0/0/5
Aug  5 16:14:30  Tequila chassisd[636]: CHASSISD_IFDEV_CREATE_NOTICE: create_pics:
created interface device for ge-0/0/6
Aug  5 16:14:30  Tequila chassisd[636]: CHASSISD_IFDEV_CREATE_NOTICE: create_pics:
created interface device for ge-0/0/7
Aug  5 16:14:30  Tequila chassisd[636]: CHASSISD_IFDEV_CREATE_NOTICE: create_pics:
created interface device for ge-0/0/8
Aug  5 16:14:30  Tequila chassisd[636]: CHASSISD_IFDEV_CREATE_NOTICE: create_pics:
created interface device for ge-0/0/9
Aug  5 16:14:30  Tequila chassisd[636]: CHASSISD_IFDEV_CREATE_NOTICE: create_pics:
created interface device for ge-0/0/10
Aug  5 16:14:30  Tequila chassisd[636]: CHASSISD_IFDEV_CREATE_NOTICE: create_pics:
created interface device for ge-0/0/11
Aug  5 16:14:30  Tequila chassisd[636]: CHASSISD_IFDEV_CREATE_NOTICE: create_pics:
created interface device for ge-0/0/12
Aug  5 16:14:30  Tequila chassisd[636]: CHASSISD_IFDEV_CREATE_NOTICE: create_pics:
created interface device for ge-0/0/13
Aug  5 16:14:30  Tequila chassisd[636]: CHASSISD_IFDEV_CREATE_NOTICE: create_pics:
created interface device for ge-0/0/14
Aug  5 16:14:30  Tequila chassisd[636]: CHASSISD_IFDEV_CREATE_NOTICE: create_pics:
created interface device for ge-0/0/15
Aug  5 16:14:30  Tequila chassisd[636]: CHASSISD_IFDEV_CREATE_NOTICE: create_pics:
created interface device for ge-0/0/16
Aug  5 16:14:30  Tequila chassisd[636]: CHASSISD_IFDEV_CREATE_NOTICE: create_pics:
created interface device for ge-0/0/17
Aug  5 16:14:30  Tequila chassisd[636]: CHASSISD_IFDEV_CREATE_NOTICE: create_pics:
created interface device for ge-0/0/18
Aug  5 16:14:30  Tequila chassisd[636]: CHASSISD_IFDEV_CREATE_NOTICE: create_pics:
created interface device for ge-0/0/19
Aug  5 16:14:30  Tequila chassisd[636]: CHASSISD_IFDEV_CREATE_NOTICE: create_pics:
created interface device for ge-0/0/20
Aug  5 16:14:30  Tequila chassisd[636]: CHASSISD_IFDEV_CREATE_NOTICE: create_pics:
created interface device for ge-0/0/21
Aug  5 16:14:30  Tequila chassisd[636]: CHASSISD_IFDEV_CREATE_NOTICE: create_pics:
created interface device for ge-0/0/22
Aug  5 16:14:30  Tequila chassisd[636]: CHASSISD_IFDEV_CREATE_NOTICE: create_pics:
created interface device for ge-0/0/23
Aug  5 16:14:30  Tequila chassisd[636]: CHASSISD_IFDEV_CREATE_NOTICE: create_pics:
created interface device for ge-0/1/0
Aug  5 16:14:30  Tequila chassisd[636]: CHASSISD_IFDEV_CREATE_NOTICE: create_pics:
created interface device for ge-0/1/1
Aug  5 16:14:31  Tequila mib2d[639]: SNMP_TRAP_LINK_DOWN: ifIndex 116,
ifAdminStatus up(1), ifOperStatus down(2), ifName ge-0/0/3
Aug  5 16:14:31  Tequila mib2d[639]: SNMP_TRAP_LINK_DOWN: ifIndex 118,
ifAdminStatus up(1), ifOperStatus down(2), ifName ge-0/0/4
Aug  5 16:15:05  Tequila mgd[678]: UI_DBASE_LOGIN_EVENT: User 'root' entering
configuration mode
Aug  5 16:15:17  Tequila mgd[678]: UI_LOAD_EVENT: User 'root' is performing a
'load override'
Aug  5 16:15:30  Tequila mgd[678]: UI_COMMIT: User 'root' requested 'commit'
operation (comment: none)
Aug  5 16:15:30  Tequila chassisd[895]: CHASSISD_IFDEV_DETACH_FPC: ifdev_detach(1)
Aug  5 16:15:30  Tequila chassisd[895]: CHASSISD_IFDEV_DETACH_FPC: ifdev_detach(2)
Aug  5 16:15:30  Tequila chassisd[895]: CHASSISD_IFDEV_DETACH_FPC: ifdev_detach(3)
```

```
Aug  5 16:15:30  Tequila chassisd[895]: CHASSISD_IFDEV_DETACH_FPC: ifdev_detach(4)
Aug  5 16:15:30  Tequila chassisd[895]: CHASSISD_IFDEV_DETACH_FPC: ifdev_detach(5)
Aug  5 16:15:30  Tequila chassisd[895]: CHASSISD_IFDEV_DETACH_FPC: ifdev_detach(6)
Aug  5 16:15:30  Tequila chassisd[895]: CHASSISD_IFDEV_DETACH_FPC: ifdev_detach(7)
Aug  5 16:15:30  Tequila chassisd[895]: CHASSISD_IFDEV_DETACH_FPC: ifdev_detach(8)
Aug  5 16:15:30  Tequila chassisd[895]: CHASSISD_IFDEV_DETACH_FPC: ifdev_detach(9)
Aug  5 16:15:33  Tequila vccp[879]: ISIS initialization complete
Aug  5 16:15:38  vccp[879]: ISIS initialization complete
Aug  5 16:15:39  vccp[879]: TASK_SCHED_SLIP: 4 sec scheduler slip, user: 0 sec 0
usec, system: 0 sec, 5694 usec
Aug  5 16:15:39  mgd[678]: UI_DBASE_LOGOUT_EVENT: User 'root' exiting
configuration mode
Aug  5 16:16:03  login: LOGIN_INFORMATION: User lab logged in from host [unknown]
on device ttyu0
Aug  5 16:16:49  LIBJNX_EXEC_WEXIT: Command exited: PID 1128, status 1, command
'/usr/bin/scp'
Aug  5 17:09:08  mgd[1125]: UI_DBASE_LOGIN_EVENT: User 'lab' entering
configuration mode
Aug  5 17:11:09  mgd[1125]: UI_DBASE_LOGOUT_EVENT: User 'lab' exiting
configuration mode
```

Many of the syslog messages will have headers specified in uppercase letters that you can input into the command, specifying on which facility the message was logged, the severity level, a description, and a recommended action. Looking at the log entry for August 5, one such header is noted as CHASSISD_IFDEV_CREATE_NOTICE:

```
Aug  5 16:14:30  Tequila chassisd[636]: CHASSISD_IFDEV_CREATE_NOTICE: create_pics:
created interface device for ge-0/1/1
```

You can examine this message using the help syslog command, which indicates that the chassisd software process created a new interface:

```
lab@Tequila> help syslog CHASSISD_IFDEV_CREATE_NOTICE
Name:         CHASSISD_IFDEV_CREATE_NOTICE
Message:      <function-name>: created <device-name> for <interface-name>
Help:         chassisd created interface device
Description:  The chassis process (chassisd) created the initial interface
              device for the indicated newly installed Physical Interface Card
              (PIC) or pseudodevice.
Type:         Event: This message reports an event, not an error
Severity:     notice
```

You can create custom logs by specifying a filename, facility, message facility, and location to send the message. The message can be stored in a local file, sent to a syslog server, sent to the console, or sent to a user or group of users when logged in to the switch.

The factory-default configuration enables three system logs: two logs that are sent to a file, and one log that is sent to any user that is logged in. Although the default system log receives all information as specified with the any keyword, you can create other files for easier log parsing:

```
syslog {
    user * {
        any emergency;
```

```
    }
    file messages {
        any any;
        authorization info;
    }
    file interactive-commands {
        interactive-commands any;
    }
}
```

Syslog case study

To avoid having to specify every syslog option available, let's examine a realistic example with specific goals. The goals are as follows:

- Increase the default size of the *messages* file to 1 MB and the number of archives to 15.

- Send all messages to a syslog server with a domain name of *syslog.underdogssf.com*.

- Ensure that all messages sent to the syslog server are in the same format as the Cisco switches in the network.

- Create a syslog file to log all firewall filter log information.

Each syslog file that is created on a Juniper Networks switch is stored in the file directory *var/log* and is given a size of 128 KB on a J Series switch and 1 MB on an M Series switch. When the file is full, the file is cleared, an archive is created of the old data, and the file is written to again. For example, once 128 KB of data is written into the *messages* file, that file will be cleared and the information will be moved into a *messages.0* file. When the *messages* file is filled up again, the old data is archived into *messages.0* and the old *messages.0* now becomes *messages.1*. This will continue for 10 archives until the data is written. In the case study, we will increase the default number of archives to 15 and the file size to 1 MB. We can do this with the following archive configuration:

```
[edit system syslog]
lab@TEQUILA# set file messages archive files 15 size 1M
[edit system syslog]
lab@TEQUILA# show file messages
any notice;
authorization info;
archive size 1m files 15;
```

Next, syslog messages need to be sent to a syslog server:

```
[edit system syslog]
lab@TEQUILA# set host syslog.underdogssf.com any any
```

The default JUNOS message does not send the priority (facility value and severity) of the syslog message, which could cause issues when trying to parse the output at the receiver. Cisco switches by default do send this priority message; to ensure that both vendors send the same message format, configure the `explicit-priority` keyword:

```
[edit system syslog]
lab@TEQUILA# set host syslog.underdogssf.com explicit-priority
```

Lastly, a new syslog file is created to log firewall entries:

```
[edit system syslog]
lab@TEQUILA# set file fw-log firewall info
```

Here is the complete stanza:

```
[edit system syslog]
lab@TEQUILA# show
user * {
    any emergency;
}
host syslog.underdogssf.com {
    any any;
    explicit-priority;
}
file messages {
    any notice;
    authorization info;
    archive size 1m files 15;
}
file interactive-commands {
    interactive-commands any;
}
file fw-log {
    firewall info;
}
```

SNMP

SNMP is a standard protocol used for a network management station to receive information for the switch (agent), as shown in Figure 3-10. The manager can poll the switch for switch health information such as memory utilization, link status, or firewall filter statistics in the form of a GET command. The switch can also send event information to the network manager without polling, in a process called a *trap*.

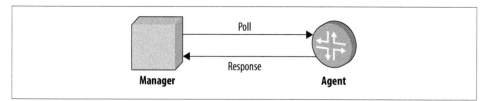

Figure 3-10. SNMP concept

The data structure that is used to carry information is called a Management Information Base (MIB). A MIB has a structure in the format of a tree that defines groups of objects into related sets. These MIBs are identified by an Object Identifier (OID), which names the object. The leaf of the OID contains the actual managed objects. MIBs are defined into two categories: standard MIBs and enterprise-specific MIBs. Standard MIBs are defined by the IETF in various RFCs, whereas enterprise-specific MIBs are defined by the vendor and must be compiled into the management station. Here is an example of MIB data taken from a network manager:

```
SNMPv2-MIB::sysDescr.0 = STRING: Mx480 - Okemos, MI
SNMPv2-MIB::sysObjectID.0 = OID: JUNIPER-MIB::jnxProductNameM480
SNMPv2-MIB::sysUpTime.0 = Timeticks: (80461526) 9 days, 7:30:15.26
SNMPv2-MIB::sysContact.0 = STRING: Doug Marschke - x8675309
SNMPv2-MIB::sysName.0 = STRING: TEQUILA-3
SNMPv2-MIB::sysLocation.0 = STRING: Okemis, MI USA - Rack 4
SNMPv2-MIB::sysServices.0 = INTEGER: 4
```

To configure SNMP on a Juniper switch, you must specify a community string on the switch. This acts as a password to verify incoming SNMP information on the management station:

```
[edit snmp]
root@Tequila# set community sample

[edit snmp]
root@Tequila# show
community sample;
```

 JUNOS software supports SNMPv1, SNMPv2, and SNMPv3.

With this basic configuration, SNMP GETs can be received on any interface from any management statement. It is recommended that access be restricted to particular interfaces and clients, such as the management network:

```
root@Tequila# show
interface me0.0;
community sample {
    clients {
        172.16.69.0/24;
        0.0.0.0/0 restrict;
    }
}
```

Also, the switch may want to initiate some information in the form of traps. Traps are sent to a specified list of targets and are defined by categories. Possible categories include:

Authentication
   User login authentication failures

Chassis
   Chassis and environmental notifications

Configuration

Notification of configuration changes

Link

Link status changes

Remote operations

Remote operation notifications

Rmon-alarm

Events for remote monitoring (RMON) alarms

Routing

Routing protocol information such as neighbor status changes

Services

Events for additional JUNOS services such as Network Address Translation (NAT) and SFW

Sonet-alarm

A variety of Synchronous Optical Network (SONET) alarms such as loss of light, bit error rate (BER) defects, and so on

Start-up

Warm and cold boots

VRRP events

Virtual Router Redundancy Protocol (VRRP) events such as mastership changes

In the following example, a trap group called `health` is added to the SNMP configuration, which sends chassis and link traps to station 10.10.12.4:

```
root@Tequila# show
interface fe-0/0/0.1141;
community sample {
    clients {
        10.10.12.4/32;
        0.0.0.0/0 restrict;
    }
}
trap-group health {
    categories {
        chassis;
        link;
    }
    targets {
        10.10.12.4;
    }
}
```

 By default, both SNMP v1 and v2 traps are sent. Since v3 has certain security parameters that must be configured, these are not sent automatically. You can overwrite this by specifying a version under the trap group.

It may also be useful to walk down the MIB tree to verify information in the MIB and for troubleshooting purposes. To perform an SNMP walk on the switch, issue the show snmp mib <object> command. In this case, the system MIB is examined on the switch:

```
root@Tequila> show snmp mib walk system
sysDescr.0     = Juniper Networks, Inc. ex4200-24t internet router, kernel JUNOS
9.2R1.9 #0: 2008-08-05 07:25:22 UTC
builder@amalath.juniper.net:/volume/build/junos/9.2/release/9.2R1.9/obj-
powerpc/sys/compile/JUNIPER-EX Build date: 2008-08-05 07:40:05 UTC Copyright (
sysObjectID.0 = jnxProductNameEX4200
sysUpTime.0    = 569337
sysContact.0
sysName.0      = Tequila
sysLocation.0
sysServices.0 = 6
```

NTP

When examining logs, it is essential to ensure that the proper date and time are recorded for each event; otherwise, an event on one system component will be very hard to compare on another! You can set the time and date manually on each switch using the set date command:

```
root@Tequila> set date ?
Possible completions:
  <time>  New date and time (YYYYMMDDhhmm.ss)
  ntp     Set system date and time using Network Time Protocol
servers
```

However, since many devices are likely to be managed at once, each with slightly different clock speeds and drift, it is virtually impossible to keep all the clocks on every device synchronized. NTP was developed for the purpose of clock synchronization. NTP works in one of three modes:

Client
    A client has a one-way synchronization with a server.

Symmetric active
    There is equal peer synchronization with each other's local clock.

Broadcast
    The server sends periodic broadcast messages on shared media, and clients listen to these messages for synchronization.

NTP uses a concept of *clock strata* to define the distance from the clock reference and the accuracy. A stratum 0 clock is the reference clock (such as an atomic clock) and each level of peering relationship decreases in accuracy and stratum level (see Figure 3-11).

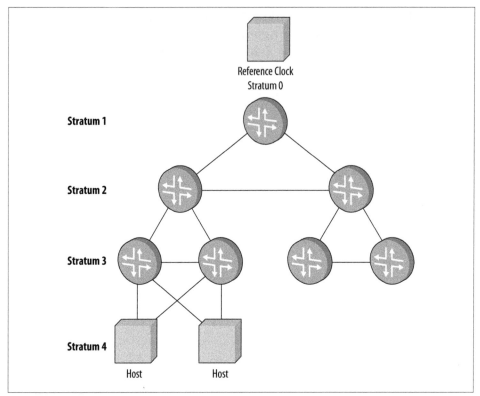

Figure 3-11. NTP stratum levels

All NTP configurations are set under [edit system ntp]. In the following configuration, Tequila is configured in client mode with a server of 172.16.69.254. Also, a boot server is configured to allow the initial clock setting to be set at boot time:

```
root@Tequila> show configuration system ntp
boot-server 172.16.69.254;
server 172.16.69.254;
```

If a switch is configured for NTP and the clocks are more than 128 seconds apart, the synchronization process will fail. In the past, to recover from that scenario the operator either rebooted the device with the boot server configuration or set the date manually within 128 seconds. JUNOS software now allows you to synchronize the device by simply issuing the set date ntp command and avoiding a reboot:

```
root@Tequila> set date ntp 172.16.69.254
10 Nov 22:50:21 ntpdate[794]: step time server 172.16.69.254 offset 0.000163 sec
```

To verify that NTP has worked correctly, issue the show ntp associations command and look for the * next to the remote IP:

```
lab@tequila> show ntp associations
     remote     refid  st t when poll reach  delay    offset  jitter
```

```
======================================================================
*172.16.69.254  LOCAL(0)  11 u   10   64   17   0.491  12.991 10.140
```

Check the correct time:

```
root@Tequila> show system uptime
Current time: 2008-11-22 03:53:35 UTC
System booted: 2008-11-20 04:58:58 UTC (1d 22:54 ago)
Protocols started: 2008-11-20 04:59:24 UTC (1d 22:54 ago)
Last configured: 2008-11-22 03:40:02 UTC (00:13:33 ago) by lab
 3:53AM  up 1 day, 22:55, 1 user, load averages: 0.19, 0.10, 0.03
```

You also can change the time zone in the switch by issuing a set system time-zone command:

```
root@Tequila# set system time-zone ?
Possible completions:
  <time-zone>           Time zone name or POSIX-compliant time zone string
  Africa/Abidjan
  Africa/Accra
  Africa/Addis_Ababa
  Africa/Algiers
  Africa/Asmera
---(more 5%)---[abort]
```

Is NTP really working?

The show ntp associations command is often a source of mass confusion and terror for operators, as there is no distinct "broken field." The synchronization process will be indicated by interpreting the delay and offset fields, as well as by noting the presence or absence of a * character.

Here is an example of an association that has failed. Notice the space in front of the 172.16.69.254, as well as the zeros in the delay and offset fields. This is an indication that no messages have been sent at all!

```
root@Tequila> show ntp associations
     remote       refid  st t when poll reach   delay   offset  jitter
======================================================================
 172.16.69.254 0.0.0.0      0 u  12   64    0   0.000   0.000 4000.00
```

In comparison, here is another association that failed; however, notice that there are values in the delay and offset fields. These indicate that NTP messages have been exchanged but synchronization has not been achieved, as no * has been displayed next to the remote peer. The large offset is usually an indication that the clocks are too far apart (above the 128-second threshold):

```
root@Tequila> show ntp associations
     remote     refid  st t when poll reach   delay   offset  jitter
======================================================================
 172.16.69.254 LOCAL(0)    11 u  25   64   37  0.492  2542804 4000.00
```

After issuing a set date ntp command, the clocks synchronize without having to reboot the switch. Note the more sane offset value and the presence of the illustrious star next to the remote peer address:

```
lab@r1> show ntp associations
     remote      refid  st t when poll reach  delay   offset  jitter
==================================================================
*172.16.69.254 LOCAL(0)    11 u   10   64   17   0.491   12.991  10.140
3:53AM  up 1 day, 22:55, 1 user, load averages: 0.19, 0.10, 0.03
```

> Since NTP uses a step process to synchronize the clocks after issuing the set date ntp command, the association could still appear to be broken. This is normal for NTP, so just sit back, enjoy a drink, and after three to five minutes, everything should be working as normal.

Rescue Configuration

The last item that should be set is the rescue configuration, which should be considered a "known good" configuration. This configuration is usually enough to get basic connectivity to the switch if the original configuration was deleted or lost. For security, the rescue configuration must contain a root password.

By default, there is no rescue configuration, so it must be set after the initial configuration. To save a configuration as the rescue configuration, use either *jweb config-management*→Rescue, or the CLI command:

```
--- JUNOS 9.2R1.9 built 2008-08-05 07:25:22 UTC
lab@Tequila> request system configuration rescue save
```

The rescue configuration is now stored in */config*:

```
lab@Tequila> file list /config/

/config/:
.snap/
db /
juniper.conf.1.gz
juniper.conf.2.gz
juniper.conf.gz
rescue.conf.gz
ssh_host_dsa_key
ssh_host_dsa_key.pub
ssh_host_key
ssh_host_key.pub
ssh_host_rsa_key
ssh_host_rsa_key.pub
usage.db
vchassis/
```

If the rescue configuration needs to be removed, issue a request system configuration rescue delete command.

Password Recovery

If you're locked out of your switch due to lack of password knowledge, and you don't want to return the switch to a factory or rescue configuration from the LCD panel, password "recovery" is the ticket. *Recovery* is really a misused word here, as you can never really recover a lost password, but you can *change* the passwords of user root or local users to a new value that hopefully they will pay attention to this time!

 Due to a bug in JUNOS v9.2, which was fixed in JUNOS v9.3, the password recovery output is displayed on a system running v9.3. This system is not used in the rest of this chapter's topology.

To recover a lost password:

1. Connect to the router via the console port through a direct connection from your PC or a term server.

2. Power-cycle the switch and wait for the following prompt:

   ```
   Hit [Enter] to boot immediately, or space bar for command prompt.
   Booting [kernel] in 1 second...
   ```

3. After pressing the space bar, enter single-user mode by typing **boot -s**:

   ```
   U-Boot 1.1.6 (Feb  6 2008 - 11:27:42)

   Board: EX4200-24T 2.11
   EPLD:  Version 6.0 (0x85)
   DRAM:  Initializing (1024 MB)
   FLASH: 8 MB
   USB:   scanning bus for devices... 2 USB Device(s) found
          scanning bus for storage devices... 1 Storage Device(s) found

   Consoles: U-Boot console
   Found compatible API, ver. 7

   FreeBSD/PowerPC U-Boot bootstrap loader, Revision 2.1
   (marcelm@apg-bbuild01.juniper.net, Wed Feb  6 11:23:55 PST 2008)
   Memory: 1024MB
   Loading /boot/defaults/loader.conf
   /kernel data=0x9ec818+0x6eb6c syms=[0x4+0x888e0+0x4+0x8f04d]

   Hit [Enter] to boot immediately, or space bar for command prompt.
   Booting [/kernel] in 1 second...

   Type '?' for a list of commands, 'help' for more detailed help.
   loader> boot -s
   ```

4. After the system boots up, start the recovery script:

   ```
   Kernel entry at 0xa0000100 ...
   GDB: no debug ports present
   KDB: debugger backends: ddb
   KDB: current backend: ddb
   ```

JUNOS 9.3-20080719.0 #0: 2008-07-19 06:10:30 UTC

builder@elliath.juniper.net:/volume/build/junos/9.3/production/20080719.
0/obj-powerpc/sys/compile/JUNIPER-EX
WARNING: debug.mpsafenet forced to 0 as ipsec requires Giant Timecounter
"decrementer" frequency 50000000 Hz quality 0
cpu0: Freescale e500v2 core revision 2.2
cpu0: HID0 80004080<EMCP,TBEN,EN_MAS7_UPDATE>
real memory = 512753664 (489 MB)
avail memory = 501309440 (478 MB)
ETHERNET SOCKET BRIDGE initialising
nexus0: <PPC e500 Nexus device>
ocpbus0: <on-chip peripheral bus> on nexus0
openpic0: <OpenPIC in on-chip peripheral bus> iomem
0xfef40000-0xfef600b3 on ocpbus0
uart0: <16550 or compatible> iomem 0xfef04500-0xfef0450f irq 58 on ocpbus0
uart0: console (115207,n,8,1)
uart1: <16550 or compatible> iomem 0xfef04600-0xfef0460f irq 58 on ocpbus0
lbc0: <Freescale 8533 Local Bus Controller> iomem 0xfef05000-
0xfef05fff,0xff000000-0xffffffff irq 22 on ocpbus0
cfi0: <AMD/Fujitsu - 8MB> iomem 0xff800000-0xffffffff on lbc0 syspld0 iomem
0xff000000-0xff00ffff on lbc0
tsec0: <eTSEC ethernet controller> iomem 0xfef24000-0xfef24fff irq 45,46,50 on
ocpbus0
tsec0: hardware MAC address 00:19:e2:50:71:5f
miibus0: <MII bus> on tsec0
e1000phy0: <Marvell 88E1112 Gigabit PHY> on miibus0
e1000phy0: 10baseT, 10baseT-FDX, 100baseTX, 100baseTX-FDX, 1000baseTX-FDX, auto
pcib0: <Freescale MPC8544 PCI host controller> iomem 0xfef08000-
0xfef08fff,0xf0000000-0xf3ffffff on ocpbus0
pci0: <PCI bus> on pcib0
pci0: <serial bus, USB> at device 18.0 (no driver attached)
ehci0: <Philips ISP156x USB 2.0 controller> mem 0xf0001000-0xf00010ff irq 22 at
device 18.2 on pci0
usb0: EHCI version 0.95
usb0: <Philips ISP156x USB 2.0 controller> on ehci0
usb0: USB revision 2.0
uhub0: Philips EHCI root hub, class 9/0, rev 2.00/1.00, addr 1
uhub0: 2 ports with 2 removable, self powered
umass0: STMicroelectronics ST72682 High Speed Mode, rev 2.00/2.10, addr
2
pcib1: <Freescale MPC8544 PCI Express host controller> iomem 0xfef0a000-
0xfef0afff,0xe0000000-0xe3ffffff,0xec000000-0xec0fffff on ocpbus0
pci1: <PCI bus> on pcib1
pcib2: <PCI-PCI bridge> at device 0.0 on pci1
pci2: <PCI bus> on pcib2
mpfe0: <Juniper EX-series Packet Forwarding Engine> mem 0xa4000000-
0xa40fffff,0xa0000000-0xa3ffffff irq 20 at device 0.0 on pci2
pcib3: <Freescale MPC8544 PCI Express host controller> iomem 0xfef09000-
0xfef09fff,0xe4000000-0xe7ffffff,0xec100000-0xec1fffff on ocpbus0

```
pci3: <PCI bus> on pcib3
pcib4: <PCI-PCI bridge> at device 0.0 on pci3
pci4: <PCI bus> on pcib4
mpfe1: <Juniper EX-series Packet Forwarding Engine> mem 0xac000000-
0xac0fffff,0xa8000000-0xabffffff irq 18 at device 0.0 on pci4
pcib5: <Freescale MPC8544 PCI Express host controller> iomem 0xfef0b000-
0xfef0bfff,0xe8000000-0xebffffff,0xec200000-0xec2fffff on ocpbus0
pci5: <PCI bus> on pcib5
pcib6: <PCI-PCI bridge> at device 0.0 on pci5
pci6: <PCI bus> on pcib6
mpfe2: <Juniper EX-series Packet Forwarding Engine> mem 0xb4000000-
0xb40fffff,0xb0000000-0xb3ffffff irq 19 at device 0.0 on pci6
i2c0: <MPC85XX OnChip i2c controller> iomem 0xfef03000-0xfef03014 irq 59 on
ocpbus0
i2c1: <MPC85XX OnChip i2c controller> iomem 0xfef03100-0xfef03114 irq 59 on
ocpbus0
idma0: <mp85xxx DMA Controller> iomem 0xfef21000-0xfef21300 irq 36 on ocpbus0
bme0:Virtual BME driver initializing Timecounters tick every 1.000 msec Loading
common multilink module.
IPsec: Initialized Security Association Processing.
if_pfe_open: listener socket opened, listening on ...
da0 at umass-sim0 bus 0 target 0 lun 0
da0: <ST ST72682 2.10> Removable Direct Access SCSI-2 device
da0: 40.000MB/s transfers
da0: 500MB (1024000 512 byte sectors: 64H 32S/T 500C) Trying to mount root from
ufs:/dev/da0s2a Attaching /packages/jbase via /dev/mdctl...
Mounted jbase package on /dev/md0...
System watchdog timer disabled
Enter full pathname of shell or 'recovery' for root password recovery or RETURN
for /bin/sh: recovery
```

5. The system will do some file-checking and then dump you into the CLI:

```
Performing filesystem consistency checks ...
/dev/da0s2a: FILE SYSTEM CLEAN; SKIPPING CHECKS
/dev/da0s2a: clean, 1005 free (13 frags, 124 blocks, 0.0% fragmentation)
/dev/da0s3e: FILE SYSTEM CLEAN; SKIPPING CHECKS
/dev/da0s3e: clean, 14031 free (15 frags, 1752 blocks, 0.1%
fragmentation)
/dev/da0s2f: FILE SYSTEM CLEAN; SKIPPING CHECKS
/dev/da0s2f: clean, 29276 free (124 frags, 3644 blocks, 0.4%
fragmentation)
/dev/da0s3d: FILE SYSTEM CLEAN; SKIPPING CHECKS
/dev/da0s3d: clean, 80452 free (20 frags, 10054 blocks, 0.0%
fragmentation)

Performing mount of main filesystems ...
Mounted jkernel-ex package on /dev/md3...
Mounted jpfe-ex42x package on /dev/md4...
Mounted jroute-ex package on /dev/md5...
Mounted jcrypto-ex package on /dev/md6...
Mounted jswitch-ex package on /dev/md7...
machdep.bootsuccess: 0 -> 0

Performing initialization of management services ...
mgd: error: could not open database: /var/run/db/schema.db: No such file or
```

```
                directory
                mgd: error: Database open failed for file '/var/run/db/schema.db': No such file
                or directory
                mgd: error: could not open database schema: /var/run/db/schema.db
                mgd: error: could not open database schema
                mgd: error: database schema is out of date, rebuilding it
                mgd: error: could not open database: /var/run/db/juniper.data: No such file
                or directory
                mgd: error: Database open failed for file '/var/run/db/juniper.data': No such
                file or directory
                mgd: error: Cannot read configuration: Could not open configuration database
                mgd: error: Couldn't open lib /usr/lib/dd//libjdocs-dd.so
                mgd: error: Couldn't open lib /usr/lib/dd//libjdocs-dd.so

                Performing checkout of management services ...

                NOTE: Once in the CLI, you will need to enter configuration mode using
                NOTE: the 'configure' command to make any required changes. For example,
                NOTE: to reset the root password, type:
                NOTE:    configure
                NOTE:    set system root-authentication plain-text-password
                NOTE:    (enter the new password when asked)
                NOTE:    commit
                NOTE:    exit
                NOTE:    exit
                NOTE: When you exit the CLI, you will be asked if you want to reboot
                NOTE: the system
```

6. Enter configuration mode and set a root password:

```
                Starting CLI ...
                root> configure
                Entering configuration mode

                [edit]
                root# set system root-authentication plain-text-password
                New password:
                Retype new password:
```

7. Don't forget to commit the configuration:

```
                [edit]
                root# commit and-quit
                error: could not open database: /var/run/db/juniper.data: No such file or
                directory
                error: Database open failed for file '/var/run/db/juniper.data': No such file or
                directory

                commit complete
                Exiting configuration mode

                root@148p2-sys>
```

8. Then reboot the switch to get it out of single-user mode:

```
                root@148p2-sys> request system reboot
                Reboot the system ? [yes,no] (no) yes
```

```
Shutdown NOW!
[pid 358]
Sep 25 07:35:42 shutdown: reboot by root:"libthr.so.2" not found, required by
"rpdc"
--- JUNOS 9.2R1.9 built 2008-08-05 07:25:22 UTC
lab@tequila> request system halt
Halt the system ? [yes,no] (no)

*** FINAL System shutdown message from lab@teqila ***

System going down IMMEDIATELY


Shutdown NOW!
[pid 663]
```

Now that the new switch is up and running with an initial configuration, take note that it's always best to perform a graceful shutdown to allow the multitasking JUNOS to close the file and gracefully terminate the various daemon processes that run in the background. Although rare, filesystem damage can occur with an abrupt power off, which may cause problems on the next boot. Use the `request system halt` or `request system reboot` command to gracefully shut down or reboot the OS. Once the OS is halted, it is safe to remove power.

Switch Maintenance Summary

This section pointed out the command used to determine overall hardware status, as well as additional configuration for syslogging, network management, and time synchronization, which can aid in later diagnostic activities. The section ended with a discussion of rescue configuration, which should always be saved, should you need it later, and password recovery for those aging hippies among us.

Conclusion

This chapter demonstrated the addition of a Juniper Networks EX switch into an existing all-Cisco network. First we examined two ways to get the initial configuration into the router, either with EZSetup or by manually entering the related CLI command, which in the end was not bad at all. With the initial configuration in place, additional users, remote access, DHCP, and OoB management were also configured. The install continued with verification of the switch's health, as well as adding a syslog, SNMP, and NTP configuration for enhanced switch health monitoring—for example, to allow proactive notice when the switch's health or operational status changes. We ended with a look at ways to recover the switch to a rescue or factory-default configuration in the event that the current configuration is corrupted, changed, or deleted.

Later chapters build upon this fine base by adding various Layer 2 and Layer 3 protocols and services.

Chapter Review Questions

1. What is the default password on the switch?

    a. Juniper

    b. Cisco

    c. There is no password

    d. Enable

2. Which predefined login class allows the user to have access rights to any login command?

    a. privileged

    b. super-user

    c. privileged exec

    d. power-user

3. Which interface on a EX Series switch is set aside for OoB management?

    a. fxp0

    b. fxp1

    c. bcm0

    d. me0

4. In which DHCP modes can an EX Series switch operate? (Choose two.)

    a. client

    b. server

    c. relay

    d. bootstrap

5. Which command is issued to view the DHCP configuration on an EX Series switch?

    a. show dhcp

    b. show system services dhcp

    c. show system services dhcp global

    d. show dhcp global

6. True or false: the switch is preloaded with a default rescue configuration.

    a. True

    b. False

    c. This question is totally unfair; I refuse to answer it!

7. In what ways can you configure the initial configuration settings on the switch? (Choose two.)

    a. CLI

    b. LCD panel

c. EZSetup

d. QuickSetup

e. BSD shell

8. Choose two items that are included in the factory-default configuration.

a. LLDP

b. OoB management

c. RSTP

d. Telnet

9. Which syslog facility logs all CLI commands?

a. `cli-commands`

b. `accounting`

c. `change-log`

d. `interactive-commands`

10. In which directory are all logfiles stored?

a. */var/home/user*

b. */log*

c. */var/home/log*

d. */var/log*

e. */syslog*

11. Which feature of SNMPv2 acts as a password to authenticate SNMP messages?

a. MIBs

b. Communities

c. OID

d. Traps

12. Which command allows NTP synchronization without a switch reboot?

a. `set system ntp`

b. `request system time update`

c. `set date ntp`

d. `set ntp boot-server`

Chapter Review Answers

1. Answer: C. There is no default password on a Juniper switch in the factory-default configuration. A single user, root, will be configured with no password.

2. Answer: B. The class of super-user allows users to issue any command that they desire on the switch. The other options listed are not supported classes.

3. Answer: D. The other answers are valid interfaces for JUNOS, but none of them are used for OoB management on an EX Series switch. For instance, fxp0 is used for OoB management on M/T Series routers.

4. Answer: B and C. The switch either can be set as a DHCP server or can act as a relay agent. It cannot be a server and a relay agent at the same time, however.

5. Answer: C. The other commands are not valid commands.

6. Answer: B. The switch does not come with a default rescue configuration, so it is very important that you set this after your initial installation.

7. Answer: A and C. The initial configuration can be done via the CLI or EZSetup. Remember that EZSetup can be invoked from the shell or the LCD panel.

8. Answer: A and C. The factory-default configuration contains many default parameters, including RSTP and LLDP. Remote access such as Telnet and an OoB management interface would have to be configured via the CLI or using EZSetup.

9. Answer: D. The facility `interactive-commands` will log any commands that were typed via any user interface method, including the CLI.

10. Answer: D. This is the directory for all syslog and `traceoptions` files.

11. Answer: B. A community will act as a password for SNMP messages. This community value is sent in clear text on the wire, which could easily be captured. The next version of SNMP corrects this issue.

12. Answer: C. If the NTP server is reachable, `set date ntp` will restart the NTP update process without having to reboot the switch, thus eliminating the need for a `boot-server` configuration statement.

## CHAPTER 4

EX Virtual Chassis

EX4200 switches support clustering of up to ten 4200 chassis into a single Virtual Chassis (VC), which provides significant High Availability (HA) benefits in addition to simplified network management. VC capabilities are included in the base 4200 model. A VC can be built incrementally, which means you can grow the VC by adding one or more switch chassis at any time, until the 10-chassis limit is reached, and you can mix and match any EX4200 model as part of the same VC. This capability means an enterprise can start at a modest scale with a single 4200 chassis, and then expand into a full-blown VC offering a local switching capacity of 1.36 Tbps/1.01 Bpps, with 128 Gbps of throughput for switching within the VC.

The topics covered in this chapter include:

- EX VC operation and deployment designs
- Configuration, operation, and maintenance
- VC deployment case study

The EX Virtual Chassis

The EX4200 VC is an exciting concept. By simply attaching a few rear-panel cables, you can turn any mix of 10 standalone 4200s into a single logical entity that both simplifies management and increases resiliency to hardware- and software-related faults.

Virtual Chassis Overview

The individual member switches comprising a VC can be any type of EX4200, with any mix of supported power supply units (PSUs), Power over Ethernet (PoE) options, and uplink modules. A fully blown VC can offer 480 × 1 GE and 20 × 10 GE ports with 128 Gbps of full duplex (FD) switching capacity between any pair of adjacent nodes. When desired, you can use ports on either the 2 × 10 GE or 4 × 1 GE uplink modules to form a VC Extension (VCE), which, as its name implies, supports extended distances (up to 400 meters) between VC members.

Here are some key VC capabilities and functional highlights:

- You can interconnect from 2 to 10 EX4200s to operate as though they are a single chassis.

- Management is simplified via a single management interface, a common JUNOS software version, a single configuration file, and an intuitive chassis-like slot and module/port interface numbering scheme.

- The design is simplified through a single control plane and the ability to aggregate interfaces across VC members.

- Increased availability and reliability is available through N:1 redundant routing engines. Also, JUNOS supports Graceful Restart, Graceful Routing Engine Switchover, and Non-Stop Routing (GR, GRES, and NSR).

- Performance and flexibility accommodate a grow-as-you-go design with no upfront investment in costly chassis hardware.

At this time, Link Aggregation Control Protocol (LACP)/aggregated Ethernet (AE) across multiple VCs is not supported. Member links of a bundled interface can be housed in different switch members *within* a VC, however.

Each EX4200 ships with a .5-meter Virtual Chassis Port (VCP) cable; VCP cables are also available in 1- and 3-meter lengths. This 3-meter length limit is the only restriction on physical member placement, and with some of the creative cabling schemes discussed later, a VCP ring can be built spanning some 13.5 meters (~44 feet), which is quite a respectable distance and more than suitable for a typical top-of-rack deployment scenario. A VC design based on a chain extends this distance to some 27 meters (88.5 feet), but such a design comes at the cost of a 50% reduction in VC trunk bandwidth and reduced reliability, as there is no tolerance for single VCP cable faults in such an arrangement.

 The VCP cables use a 68-pin connector and are considered proprietary, and therefore are available only through Juniper Networks and its authorized resellers. The user manual provides pin outs for the VCP cables, however.

Interchassis distances greater than 6 meters require use of a VCE, which has the disadvantages of requiring uplink module hardware, and the resulting reduction in trunk capacity, as determined by the speed of the uplink module used (e.g., 20 Gbps with 2 × 10 GE or 2 Gbps with 2 × 1 GE ports). Currently, the maximum supported VCE distance is 500 meters. Figure 4-1 illustrates these key VC capabilities.

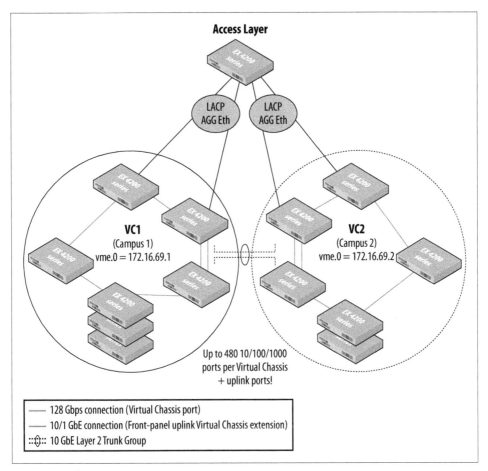

Figure 4-1. The EX VC

Figure 4-1 shows two EX4200 VCs. Within each VC, some switches are not collocated, hence the use of both VCP ring cabling and VCEs to tie in the remote switches to the main VC location. The VCEs could be 10 GE or 1 GE front-panel uplinks, and uplink speed can be mixed and matched within a single VC. In this example, the two VCs are interconnected using a Layer 2 Redundant Trunk Group (RTG), which is a Juniper proprietary link redundancy scheme that provides rapid failover convergence without the need for Spanning Tree Protocol (STP) given its primary/forwarding/secondary blocking operation. An RTG is similar in functionality to Cisco's *flexlink* feature. If desired, you can define an AE link to add additional inter-VC bandwidth, but depending on interconnection specifics, STP may be required to prevent loops.

A key aspect of Figure 4-1 is that each VC is associated with a single virtual management IP address that represents the entire VC cluster, thereby greatly simplifying network management. Lastly, note that an access layer switch is shown being dual-homed to each VC using an AE bundle. Such an AE link can contain from two to eight members, yielding as much as 80 Gbps, with added redundancy, as the Ethernet bundle can survive the loss of individual member links until the minimum link threshold is crossed; the minimum number of member links can be set to the range of 1 to 8 inclusive.

 Getting the 80 Gbps aggregated link mentioned previously on an EX4200 requires use of a four-node VC, with each member having a 2 × 10 GE uplink module. Recall that an AE bundle can span members within a chassis, allowing you to define an eight-member bundle that uses both 10 GE uplink ports on all four members. For large-scale 10 GE aggregation scenarios, consider an MX platform.

Also, note that each VC member can contain an uplink module, yielding a maximum of forty 1 GE or twenty 10 GE uplink ports per VC, in addition to the 480 GE ports supported in a single VC (each of the 10 member chassis can support 48 front-panel GE ports).

Figure 4-2 provides another VC deployment example.

Figure 4-2 shows a dual-VC design based on a top-of-rack deployment scenario. In this example, the access layer VCs are in turn dual-homed into a redundant aggregation/ core layer to eliminate single points of failure (assumes redundant power feeds) for maximum reliability and uptime. The single IP address used to manage and configure each VC greatly simplifies network management and support activities.

"Virtual Chassis Design and Deployment Options" on page 198 provides tips on how to get maximum bang from the limited length of VCP cables, and details design options that combine VCP and VCE links to optimize both performance and VC coverage area.

Virtual Chassis Control Protocol

The heart of Juniper VC technology is the Juniper proprietary link state (LS) Virtual Chassis Control Protocol (VCCP). VCCP functions to automatically discover and maintain VC neighbors, and to flood VC topology information that permits shortest-path switching between member switches using either internal or external (VC trunk) switch paths.

VCCP is not user-configurable, and operates automatically on the rear-panel VC ports. VCCP also operates over uplink ports when they are configured as VCE ports.

As with any LS protocol, the net effect is that VCCP rapidly detects and reacts to changes in the VC topology to ensure maximum connectivity over optimal paths in the face of VC moves and additions, because of switch or VC backbone failures. The loop-free switching topology that results from the Shortest Path First (SPF) calculations allows VCCP to "do the right thing" in almost any VC topology cabling scheme imaginable.

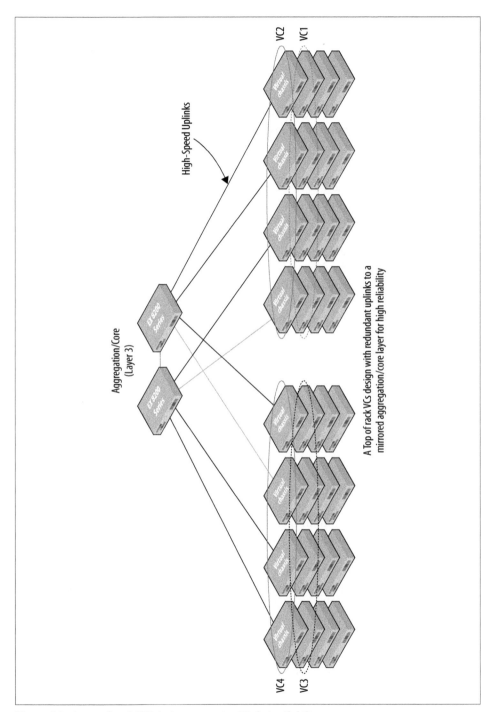

Figure 4-2. A top-of-rack VC design promoting High Availability

VCCP uses a link metric that's scaled to interface speed when calculating its SPF tree. Load balancing is currently not supported; a single best path is installed for each known destination, even though multiple equal-costs paths may exist.

Member roles within a VC

As mentioned previously, a VC consists of any 2 to 10 EX4200 switches. Within each VC, there are three distinct roles: master routing engine (RE), backup RE, and Line Card Chassis (LCC):

Master RE

The master RE runs the VC show, so to speak, by actively managing the VC switch members as far as VC operations go, and as importantly, by maintaining the master copy of the switching/routing table (RT). Because this table is in turn copied to each remaining VC member, the master RE controls overall packet and frame forwarding based on the operation of its switching and routing protocols. Within a VC, hardware-related commands are normally executed on the master RE, which then conveys the instructions and results over an internal communications path. Virtual console and Out of Band (OoB) management capability is also available, again via the master RE and its internal communications channels to all VC members. When a new switch member attempts to join a VC, the master RE is responsible for determining its compatibility with the VC and the resulting assignment of a member ID and VC role.

A functional VC must contain a master RE. You can configure mastership parameters that ensure a deterministic election behavior, or rely on the built-in tiebreaking algorithms, which ultimately favor the first switch powered up. We provide details on mastership election in a later section.

Backup RE

The backup RE, as its name implies, is the second most-preferred switch member that stands ready to take over chassis operations if the current master RE should meet an untimely demise. With default parameters, the backup RE is the second switch that is powered up in the VC. At a minimum, the backup RE maintains a copy of the active configuration (through use of `commit synchronize` on the master RE); if GR is enabled, the backup RE also maintains copies of the forwarding table (FT) to enable Non-Stop Forwarding (NSF) through a GRES event. Alternatively, with NSR enabled both the FT and control plane state—for example, Open Shortest Path First (OSPF) adjacency status or STP and learned Media Access Control (MAC) address state—are mirrored to provide a truly hitless GRES experience.

LCC

An LCC is any switch member that is not currently acting as the master or backup RE. This may simply be because it was the third through tenth switch member powered up, meaning it could one day become a master or backup RE, or because a configuration constraint bars it from any such ascendancy. An LCC accepts (and stores) its member ID from the current master, and then proceeds to perform as instructed with regard to hardware operations and FT entries. The LCC runs only a subset of JUNOS. For example, it does not run the chassis control daemon (chassid).

The receipt of exception traffic—for example, a newly learned MAC address or local hardware error condition—results in intrachassis communications between that switch member and the master RE, which may then mirror the change to the backup RE when GR or NSR is enabled. After processing the update, any related actions, such as updating the FT or taking a failed piece of hardware offline, are then communicated back to affected member switches, thereby keeping everything tidy and in sync within the VC.

Member ID

When an EX4200 is powered on and attached to a VC, it determines whether it should be master, and if not it's assigned the next available member ID. The assigned member ID is displayed on the front-panel LCD. When powered up as a standalone switch, the member ID is always 0. A VC master assigns member IDs based on various factors, such as the order in which the switch was added to the VC. Generally speaking, as each switch is added and powered on, it receives the next available (unused) member ID unless the VC configuration specifically maps that switch's serial number to a specific value.

Promoted stability ID assignments are *sticky*, meaning that the ID is *not* automatically reused if the corresponding switch is removed from the VC. A later section describes how you can clear or recycle VC switch member IDs when desired.

The member ID distinguishes the member switches from one another, and is used to:

* Assign a mastership priority value to a member switch
* Configure interfaces for a member switch (the function is similar to a slot number on Juniper Networks routers)
* Apply some operational commands to a member switch
* Display the status or characteristics of a member switch

The switch member ID is a logical function and is independent of any particular VC member role, or physical location along a VC ring or chain. Although it is a best practice to have the master RE assigned ID 0, this is not mandatory.

Mastership priority

You can designate the role (master, backup, or LCC) that a member switch performs within a VC by explicitly configuring its mastership priority. The priority ranges from 1 to 255, and larger values are preferred. The mastership priority value has the greatest influence over VC mastership election, and so is a powerful knob. The default value for mastership priority is 128. The current best practice is to assign explicit priority values for the master and backup roles, which should be the same, and should be set to the highest value (255) to avoid preemption after a GRES and to then ensure that any new LCC members have an explicit priority configured before they are attached to the VC. These procedures are described in detail in a later section and are intended to prevent undesired master RE transitions.

Default election algorithm

The default parameters ensure that even with no explicit configuration the VCP protocol will correctly detect and assign chassis member roles, such that there will be a master and a backup RE, and one or more LCCs if at least three member switches are present. Variations among switch models, such as whether the switch has 24 or 48 ports, have no impact on the master election process. The steps of the master election algorithm are:

1. Choose the member with the highest user-configured mastership priority (255 is the highest possible value).
2. Choose the member that was master the last time the VC configuration booted (retained in each switch's private configuration).
3. Choose the member that has been included in the VC for the longest period of time, assuming there was at least a one-minute uptime difference; the power-up sequence must be staggered by a minute or more for uptime to be factored.
4. Choose the member with the lowest MAC address, always a guaranteed tiebreaker.

Virtual Chassis Identifier

All members in a VC share a common Virtual Chassis Identifier (VCID) that is derived from internal parameters and is not directly configurable by the user. Various VC monitoring commands display the VCID as part of the command output.

Although it's clear that you can form a functional VC by simply slapping together 10 EX4200 switches (via their VC ports) with default configurations, such an arrangement is generally less than ideal. For example, you need to configure a virtual management interface to avail yourself of the benefits of a single IP management entity per VC, N:1 RE redundancy may not be desired, and you may wish to exert control over which members provide what VC functions, perhaps to maximize reliability in a given design. Later sections provide VC design guidelines that promote high levels of reliability while also alleviating potential confusion through explicit configuration of VC member roles.

Virtual Chassis Design and Deployment Options

This section details various VC design options and alternatives that you should carefully factor before deploying a VC in your network. Although the physical topology of the VC is a significant component of a VC's design, several aspects of a VC's operation can be controlled through configuration. This section explores VC topology and configuration options, with a focus on current best practices relating to overall VC design and maintenance.

In many cases, the basic choice of a VC topology is determined by the degree of separation between VC switch members. The approach used for a single closet will likely differ for a top-of-rack design in a data center, and both differ from a VC design that extends over a campus area (multiple wiring closets). In addition, some designs promote optimal survivability in the event of VC backplane faults, a design factor that is often overlooked.

It is worth pointing out that the Juniper VC architecture is such that local switching between the chassis front-panel ports, to include the uplinks, does not involve use of the VC trunks. As a result, when EX4200 switches are interconnected as part of a VC, each individual switch is still capable of local switching at the maximum standalone switching capacities detailed in Chapter 2. Therefore, in a 10-member VC where all traffic is locally switched, the maximum total switching capacity is 1.36 Tbps (10 × 136 Gbps), with an aggregate throughput of 1.01 *billion* packets per second!

VCP topologies

As described in Chapter 2, each EX4200 chassis has two rear-panel VCPs. Each VCP operates at 32 Gbps FD, which translates to 64 Gbps of throughput when you consider that a single VCP can be simultaneously sending *and* receiving 32 Gbps of traffic. The *combined* throughput of both VCPs is therefore 128 Gbps, or 64 Gbps FD. Note that the maximum FD throughput for any single flow between a sender and receiver that are housed on different VC switch members is 32 Gbps.

When deployed in the recommended ring topology, each VC member *could* be simultaneously switching among its front-panel and uplink ports, while also sending *and* receiving 128 Gbps of traffic (64 Gbps per VCP) to *and* from other VC members. When interconnected as a chain, or as a result of a VCP ring break, trunk capacity is reduced to 64 Gbps at the ends, while the switches in the middle still enjoy the ability to use both VCP ports.

Note that the actual VCP throughput is a function of the ingress and egress points associated with the traffic being switched (or routed), both locally and by other VC members. This is in part because currently EX switches do not support load balancing or congestion avoidance across VC trunks; for each destination, a given switch installs a *single* best path as determined by the lowest VCCP path metric to reach the switch associated with that destination. In the event of a metric tie, the first path learned is installed and used exclusively, until the VCCP signals a topology change, resulting in a new SPF calculation and subsequent update to the shortest path tree.

A VCP ring offers 128 Gbps of throughput capacity between any two member switches, but usage patterns and SPF switching between ring members may prevent any two members from being able to use the full 128 Gbps VC trunk capacity. Similarly, aggregate VC throughput can be *higher* than 128 Gbps when traffic patterns are carefully crafted to prevent congestion by having each pair of adjacent switches sink each other's traffic, which effectively yields an aggregate throughput potential of $n \times 64$ Gbps, where n is the number of member switch adjacencies in the VCP ring topology.

To help demonstrate this traffic dependency, consider the VC topology shown in Figure 4-3.

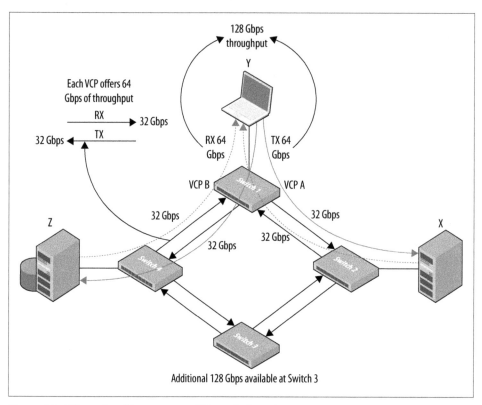

Figure 4-3. Effects of traffic patterns on VC trunk throughput

Given the VC ring topology, a total of 128 Gbps of VC switching capacity is available between member switches. However, because two VCP ports are used to instantiate the VC ring, only 64 Gbps of capacity is available in each direction.

Figure 4-3 shows a somewhat simplified scenario involving 64 Gbps of traffic arriving at Switch 1, of which 32 Gbps is destined to Address X on Switch 2, while the remaining 32 Gbps is for Address Z on Switch 4. Given the topology shown in Figure 4-2, and knowledge that the VCP installs routes to other destinations based on path metric, it's safe to assume that Switch 1 directs 32 Gbps of traffic out its VCP A port on to Switch 2 while simultaneously sending the 32 Gbps of traffic to Switch 4 via its VCP B port. The next result is a total of 64 Gbps of traffic leaving Switch 1, which is split over its two VCP ports and effectively consumes all available transit trunk bandwidth at Switch 1, given that each VCP port operates at 32 Gbps FD.

Because the VCP ports are FD, at this same time both Switch 2 and Switch 4 *can* be sending 32 Gbps of traffic to Destination Y on Switch 1. In this state, the trunk bandwidth between Switch 1, Switch 2, and Switch 4 is consumed, and there is a total of 128 Gbps of traffic being switched, which is the stated VC trunk capacity.

However, because the traffic that is sent by Switch 1 is pulled off the ring at both Switch 2 and Switch 4, the traffic sent by Switch 1 to Address X and Address Z does not consume VC trunk bandwidth between Switch 2, Switch 3, and Switch 4. It's therefore possible to switch an *additional* 128 Gbps of traffic, as long as this traffic is confined to sources and destinations on Switch 2, Switch 3, and Switch 4. In this somewhat contrived scenario, a total of 256 Gbps of traffic is switched over the VC ring. This is truly a case of the "under-promise and over-deliver" philosophy users have come to expect from Juniper Networks gear.

Although the previous example shows how you may get more than 128 Gbps of throughput over a VC ring, there are situations where the VC topology and traffic patterns are such that congestion occurs on some VC trunk segments, while other segments are otherwise idle. For example, if a source on Switch 3 were to send 32 Gbps to Destination Y on Switch 1, rather than some destination on Switch 4 or 2, we have the case of the VC link between Switch 3 and Switch 4 remaining idle (assumes that the VCCP at Switch 3 has installed the path through Switch 2 to reach Switch 1), while a total of 64 Gbps of traffic, 32 Gbps sourced by Source X at Switch 2 and another 32 Gbps from the remote Switch 3, begins queuing up for the 32 Gbps VCP port linking Switch 2 to Switch 1.

Generally speaking, there are two ways to form VCP connections: via a ring, and via a braid. The differences primarily deal with how many switches are spanned by the longest VCP cable used. We examine these options in the next section.

VCP single rack rings

The most direct way to form a VCP ring is to simply connect each switch to the next, with the last switch in the VC tied back to the first, as shown in Figure 4-4.

This type of linear ring cabling is best suited to a single-rack deployment, given that the maximum separation between VC members is limited to 3 meters/9.8 feet (the maximum supported VCP cable length). All three VCP cabling arrangements in Figure 4-4 are functionally identical. The approaches are termed a *linear ring* because the flow of traffic is sequential, passing through each switch until it—or, more correctly, some other traffic inserted by another switch—loops back at the end of the ring to start its journey anew at the first switch. Figure 4-4 also provides a top-of-rack equivalent that is spaced horizontally.

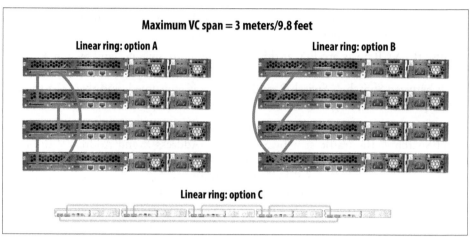

Figure 4-4. A linear ring; the longest cable spans the VC

VC cable can be arranged in many ways; the takeaway here is that VCP ports just work: as long as each switch is connected to the next and the last switch is tied to the first, a functional VCP ring will form.

VCP multiple rack rings

As stated previously, the longest supported VCP cable is only 3 meters, or some 9.8 feet. However, with some creative cabling, it's possible to create a ring that spans as much as 15 meters/49.2 feet. This respectable distance brings a VC ring topology into the realm of many multirack/data center deployment scenarios. Figure 4-5 depicts such braided ring cabling at work.

The key to the braided VC ring is that the longest VCP cable now spans only three switch members. Where desired, you can form a braided ring using a mix of short and long cables to save money, given that a short VCP cable ships with each EX4200 switch. As before, Figure 4-5 also provides a horizontally oriented top-of-rack equivalent. Trying to trace the packet flow through a braided ring can be a bit confusing, but overall performance and VCCP operation remain unchanged, and things simply work. In this example, the top switch sees the third switch as its VCP B neighbor, and packets take a somewhat convoluted path as they wind their way around the ring, sometimes being sent to a physically adjacent neighbor (at the ring's ends) and other times being sent past the *physically* adjacent member switch and onto the VCCP neighbor at the other end of the VCP cable.

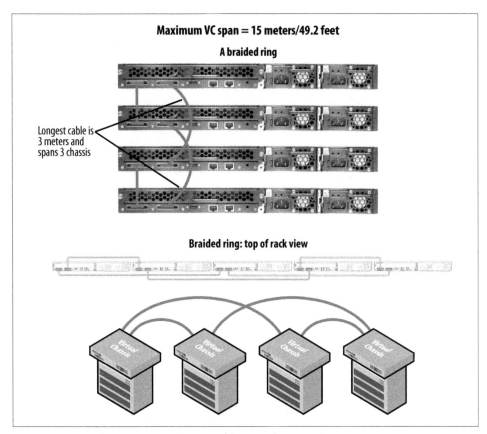

Maximum VC span = 15 meters/49.2 feet

A braided ring

Longest cable is 3 meters and spans 3 chassis

Braided ring: top of rack view

Figure 4-5. A braided VC ring; the longest cable spans three switches

VCP serial chain

The final VCP cabling option is a serial chain. Figure 4-6 shows this arrangement.

The obvious advantage to the serial chain is the ability to achieve the largest possible VC diameter, which is now some 27 meters/88.5 feet. The equally obvious downside is the use of a single VCP port at each switch, which halves the total VC trunk throughput to 64 Gbps. Another significant drawback is the lack of tolerance to any VCP cable fault, which results in a bifurcated VC, and the potential for unpredictable operation, given that a split VC is currently not supported.

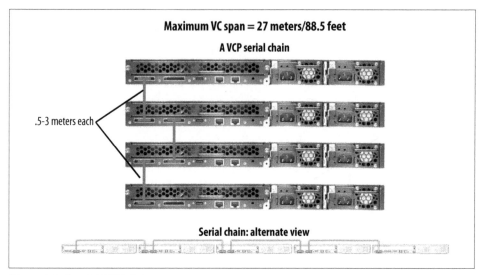

Figure 4-6. A serial VCP chain

VCE topologies

VCE topologies are formed by setting one or more ports on the (optional) uplink mod-ule to function in VCCP mode. Oddly, this is done with an operational rather than a configuration mode command, in the form of a `request virtual-chassis vc-port` statement, as we describe in a later section. Once placed into VCCP mode the related port cannot be used for normal uplink purposes until the VCE mode is reset. You can configure all uplink module ports as VCEs, and you can also mix and match normal and VCE modes on a per-port basis.

VCE functionality is supported on both the 2 × 10 GE and 4 × 1 GE uplink modules, and a mix of 1 GE and 10 GE VCE links can be used as part of an extended VCE design. It should be noted that even the rear-panel VCP ports, which offer an aggregate 128 Gbps of throughput, can become congested in some cases. Therefore, careful thought should be given before attempting to deploy a VC design incorporating 1× GE uplink ports because of the obvious potential for serious performance bottlenecks should there be a significant amount of interswitch member traffic over these links.

You can use any type of supported optics modules (SFP/XFP) and fiber for VCE links, but the maximum supported distance is currently limited to 500 meters. Although this may not reflect the maximum distance the related optics can drive, at 500 meters per VCE link, and with as many as 9 such links in a 10-member serial VC chain configu-ration, you are talking 4,500 meters (14,763 feet), which is some serious VC-spannage indeed!

A VCE-based topology can be either a ring (recommended), or a serial bus, as we described for VCP-based topologies earlier. Because even a 10× GE port represents reduced bandwidth, and due to the increased probability of a fault occurring on an extended length of fiber optic cable, you should always deploy a VCE ring topology, and should never rely on a single VCE to tie together to remote member clusters. A pure VCE ring configuration requires two VCE ports, and therefore consumes all uplink ports when using the 2 × 10 GE uplink module. However, a hybrid VCP/VCE design can allow the benefits of a ring topology while using only a single VCE port in some cases.

Extending the VC

A VC design based solely on VCE links is considered rare; the far more common case is a single- or multiple-rack deployment using only VCP ports, as this eliminates the need for uplink ports and their related SFPs, and also yields maximum performance with plug-and-run simplicity. Recall that a VCP ring can span 15 meters, and a VCP chain can extend up to 27 meters, distances that bring most data center and top-of-rack designs well within reach.

By combining VCP and VCE ports as part of your VC design, you can achieve the best of both worlds: high-speed VCP-based and local switching within a wiring closet and lower-speed VCE-based trunks linking the closets together. Figure 4-7 shows such a design.

In this example, the VCE links are shown as 10 GE, but 1 GE links could be substituted with no change in core functionality. Note how some of the VC members use only VCP ports, whereas other switch members use both VCP and VCE ports. An advantage of this design is that it preserves one of the two supported 10 GE uplinks for use in, well, uplinking non-local traffic to the distribution layer, where it's then sent to the core or to other distribution layer nodes as needed. The use of two VCE links between different VC members provides added capacity and redundancy in the event of VCE link failure. You can use as many parallel VCE links as desired to maximize these characteristics.

When deploying an extended VC, give careful thought to typical and projected traffic patterns. Where possible, you should follow the 80/20 rule, which is to say that ideally 80% of traffic is locally switched, which in this case refers to not having to be trunked over relatively low-bandwidth VCE links. Recall that local switching capacity is unaffected, and therefore remains wire-speed for intraswitch traffic. If the majority of traffic can be kept local to each member switch, bandwidth on the VC trunks is not a significant factor. The design shown in Figure 4-7 is optimized when most of the traffic is switched *within* that closet/VCP wiring domain, as this takes advantage of local switching and the high-speed VCP links.

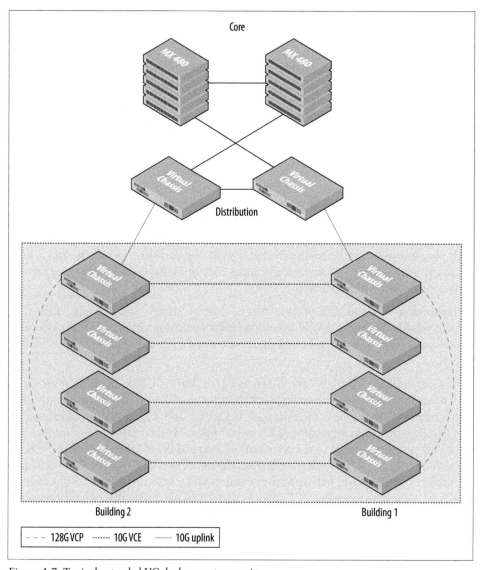

Figure 4-7. Typical extended VC deployment scenario

Packet Flow in a Virtual Chassis

This section focuses on packet flow through a VC, during both normal and VCP link failure conditions. Note that Chapter 3 details Layer 2 and Layer 3 packet flows *within* a standalone EX switch. Here the focus is on VC topology discovery and communications *between* switch members making up a VC.

Virtual chassis topology discovery

When a VC is brought up, each member switch floods VCCP packets over its VC trunk ports. Although proprietary, it can be said that VCCP is based on the well-known Intermediate System to Intermediate System (IS-IS) routing protocol. VCCP automatically discovers VC neighbors, builds adjacencies with these neighbors, and then floods link state packets (LSPs) to facilitate automatic discovery of the VC's topology, as well as rapid detection and reaction to changes in the VC topology due to a switch or VC trunk failure. Figure 4-8 illustrates this process in the context of a VC comprising three switch members arranged in a VCP-based ring topology.

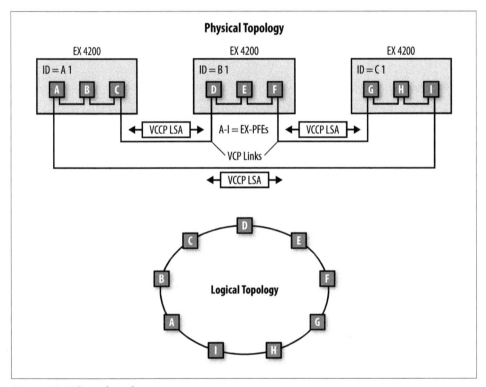

Figure 4-8. VC topology discovery

In this example, each member switch is a 48-port model, and therefore contains three EX-PFEs that work together to drive the 48 front-panel 1× GE ports, the optional uplink ports, as well as the internal and external VC trunks. Each PFE is identified as A–I, and each switch member is identified by a member ID, here shown as A1, B1, and C1. Note that in Figure 4-8, the B, E, and H entities are EX-PFE application-specific integrated circuits (ASICs) that have only internal or front-panel links.

Figure 4-8 also shows the resulting logical topology that is formed through the VC ring cabling. Note that a break in the VCP ring creates a serial chain, and a resulting halving of VC trunk bandwidth for destinations near the break, as their maximum VC bandwidth drops to 64 Gbps given the single communications path at the ends of the chain. Because each VC trunk segment operates point to point, those switches that are still connected by two functional VCP segments could still send and receive as much as 128 Gbps of traffic—for example, 32 Gbps (FD) to and from each adjacent neighbor. But if this traffic is destined for the ends of the chain, all VC trunk bandwidth is consumed, and the switches at either end of the break are each limited to 64 Gbps of switching throughput.

Because each switch member is using two VCP ports, there are two communications paths to choose from for any given destination; for example, member A1 can send out the VCP port connected to PFE B, or it can send out its other port to I. The destination-to-port mapping at any given time is a function of each switch's SPF calculation, as described next.

The SPF calculation

Figure 4-9 shows how the example's logical topology is in turn viewed as a source-rooted SPF tree at each EX-PFE to all other PFE destinations in the VC.

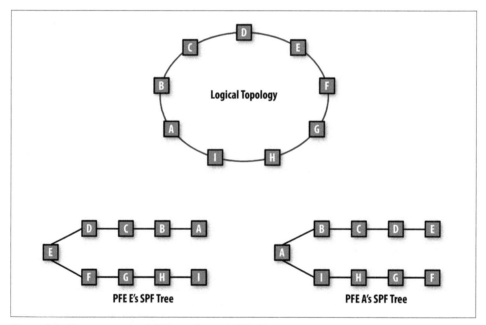

Figure 4-9. The source-rooted SPF tree for inter-VC destinations

Once the VC topology has stabilized, and all PFEs have sent and received each other's VCCP link state advertisements (LSAs), the result is a replicated link state database (LSDB) at each PFE. Changes to the VC topology are rapidly communicated by flooding updated VCCP LSPs, which in turn trigger new SPF runs at each PFE as an updated SPF tree is calculated.

Currently, load balancing is not supported. In the event of a hop-count tie, the winner is selected with preference to the PFE associated with the lowest member ID. Figure 4-9 shows the result for PFEs E and A, and depicts how their respective SPF trees reach the other PFEs that comprise the VC. The lack of load balancing results in a single forwarding next hop for each PFE destination. However, although there is no load balancing to a specific PFE, load balancing is possible to *different* PFE destinations. This is because VCPs are always in a forwarding state for some subset of the VC's PFE destinations, such that one port is used as the forwarding next hop for one-half of the VC's destinations and vice versa. Based on Figure 4-9, we see that packets in PFE A that need to be switched toward a destination housed on PFE D are sent out the upper VCP link toward PFE B over the *internal* VC trunk within switch member A1.

A topology change in the form of a ring break triggers a new SPF recalculation by all PFEs. For those stations adjacent to the break, the result is that the surviving VCP port is used to forward to *all* remaining PFE destinations. Figure 4-10 shows this state.

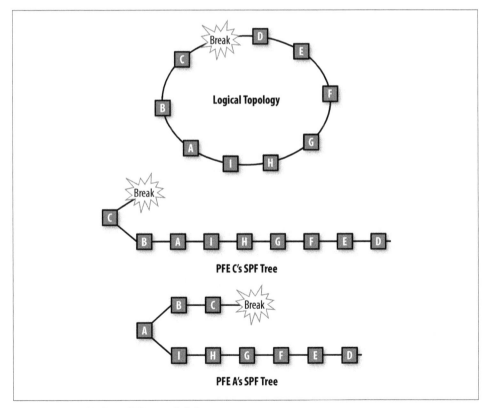

Figure 4-10. SPF after a VCP trunk failure

Figure 4-10 shows an updated SPF tree for PFE C, which is adjacent to the break, and A, which is shielded from the break through its adjacent PFE neighbors B and C. Because this is a ring rather than a switch break, all PFEs and all destinations remain reachable, albeit at reduced VC trunk capacity, as described previously. A similar situation results in the event of a VC switch member failure, except that the PFEs associated with the failed switch are no longer reachable and are removed from the VC's topology. Connectivity to remaining PFEs continues, and based on design redundancy, connectivity to endpoints can reconverge through an alternative path.

It should be noted that VC switch member roles are not affected by changes in VC topology or costs to reach a given PFE. A switch member's VC role is linked to its member ID and associated priority. Thus, a member switch need only remain reachable for it to retain its current VC role. As a result, only the failure of the current master, or two VCP trunk failures that serve to isolate the master from the rest of the VC, can result in the need to promote a backup RE to the role of master RE.

A bifurcated VC: It's a bad thing

When two VC trunk failures occur between two pairs of adjacent member switches, the result can be a bifurcated VC. In such a state, PFEs that were formerly part of the same VC lose contact. This can result in multiple master REs becoming active, each feeding its VC members with a copy of the VC configuration file, which can result in unpredictable and even network-disruptive behavior. In this example, in addition to both VCs potentially forwarding the same traffic, this condition also results in a duplicated IP management address; recall that the single vme.0 address is shared by all members in a VC, but we now have two VCs running the same configuration, which includes the virtual management address.

It's rare to actually encounter this condition, because it requires the simultaneous failure of two VC switch members, or two VC trunk cable segments between two different pairs of adjacent VC members. Given the relative infrequency of this failure mode, and the complexity of making things work in such a state, a bifurcated VC is currently not supported. With that said, the best practice in a vertical VC design has the master RE at either the top or the bottom, and the desired backup RE in the middle so that it's equidistant from either end. The rationale is that in the event of a bifurcation there is maximum probability that each of the split VCs will continue to operate using one of the two RE-capable switch members; this is especially important when deploying a redundancy scheme that permits only two of the VC members to function in the RE role.

Virtual chassis packet walk-through

This section details packet processing in a number of different scenarios. Figure 4-11 provides a macro view of VC packet forwarding.

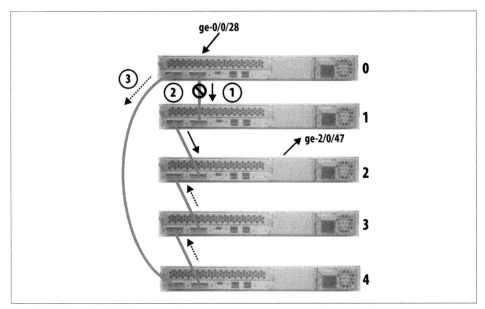

Figure 4-11. VC packet forwarding

As described previously, VC member discovery results in an SPF tree rooted at each EX-PFE that is optimized on the path metric and, in the case of a VCP ring, points to one of two VC port interfaces for forwarding between member switches within the VC. In this example, a source on Switch 0's ge-0/0/28 interface is sending to a destination on Switch 2's ge-2/0/47; Sequence Number 1 and the related solid arrows show the prefailure forwarding state, which—being based on path metric optimization—has Switch 0 forwarding toward Switch 2 through its VCP link to Switch 1 with a hop count of 2.

 When all interfaces have the same bandwidth, as is the case here, the SPF result is effectively based on hop count.

At step 2, the VCP cable between Switch 0 and Switch 1 suffers a fault, resulting in flooding of updated VCCP LSAs. After the new VC topology has stabilized, the updated calculation at Switch 0 causes it to begin forwarding toward Switch 2 via its VCP link to Switch 4, shown at step 3. The dashed arrows show that the remaining switches have also converged on the new topology, with the result being a sane forwarding path that, albeit no longer at three hops, permits ongoing communications between the source and destination VC interfaces.

Intersystem packet flows

This section builds on the previous, high-level example of interswitch packet flow by detailing how unicast and multicast packet flows are processed within each switch and its respective EX-PFEs. Figure 4-12 starts the discussion with a unicast flow for a known source and destination address.

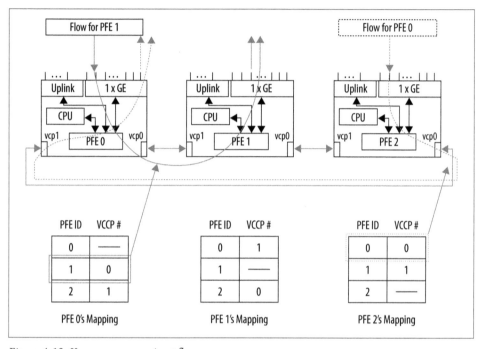

Figure 4-12. Known source: unicast flow

Figure 4-12 shows a somewhat simplified VC consisting of three EX switches, with each switch member having a single PFE. In this example, the switch member ID and PFE ID are the same, and range from 0 to 2. The three switches are connected in a ring, and each switch's PFE to VCP port mapping is shown. This is a function of a hop-count-optimized SFP run at each member switch given that all interface metrics are the same.

Figure 4-12 shows two simultaneous unicast flows. The solid flow ingresses at Switch 0 and egresses at Switch 1, and the dashed flow begins at Switch 2 and terminates at Switch 0. The MAC address to switch member/PFE ID mapping is assumed to be in place, meaning that both the Source MAC (SMAC) and Destination MAC (DMAC) addresses for both flows have been learned and the appropriate TCAM entries are in place and are used to map a MAC address to a destination PFE. In a normal case, many MAC addresses can map to the same PFE ID, and therefore to the same forwarding VCP next hop.

The highlighted entry in Switch 0's mapping table shows its mapping of Switch ID 1 to its VCP port 0, and the solid line shows the resulting unicast flow. In similar fashion, the dashed line at Switch 2 highlights the mapping table entry that causes it to use its VCP 0 to reach PFE 0.

Multicast, broadcast, or unknown DMAC address flows require special handling, as they must be flooded within the VC while taking safeguards to ensure that endless packet loops/broadcast storms do not occur. This is because such flows need to be flooded to all ports associated with the ingress port's virtual LAN (VLAN). This causes such traffic to be flooded to all remote PFEs, which may or may not have local VLAN members resulting in the decision to either replicate and forward, or discard, respectively. The lack of a Time to Live (TTL) field in Ethernet frames makes a forwarding loop particularly nasty and can easily bring a Layer 2 network to a grinding halt. Figure 4-13 details how EX-PFEs solve this problem.

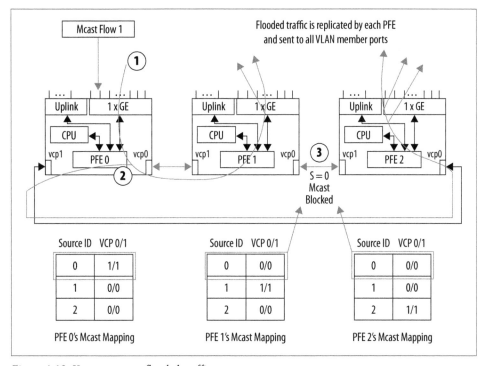

Figure 4-13. Known source: flooded traffic

The key to preventing broadcast storms is the use of a source ID mapping table in each PFE that ensures that a *single* copy of each flooded frame is received by every PFE within the VC. Step 1 in Figure 4-13 shows a single multicast stream sourced at Switch 0. The highlighted source ID mapping tables show how each switch forwards the associated traffic. Switch 0's locally originated traffic is flooded out through both its VCP 0 and

VCP 1 ports at step 2. This traffic, when received by PFEs 1 and 2, results in blocking at Switch 1's VCP 0 port, and also at Switch 2's VCP 1 port, which is shown in step 3.

The result is that each switch/PFE in the VC receives a single copy of each flooded frame, where specifics determine whether local replication or discard is appropriate for its front-panel ports. Forwarding of this traffic out the remaining VCP port to another PFE is constrained by a source ID mapping table built within each PFE that is based on VCCP exchanges and the resulting topology database. Changes in the VC topology trigger VCCP updates and a resulting modification to the source ID mapping tables to ensure continued connectivity.

Figure 4-14 expands on the preceding discussion with details of the flow of exception traffic within a VC.

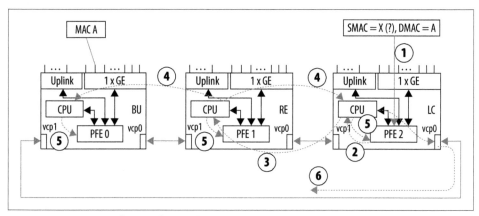

Figure 4-14. Exception flows within a VC

Exception traffic refers to packets that need to be shunted out of the ASIC fast path and up to the RE for processing that is outside the realm of pure packet forwarding. For example, when in Layer 2 bridging mode, each SMAC address must be learned so that forwarding decisions based on the DMAC can be intelligently made. When in Layer 3 mode, a similar exception flow occurs for certain IP packets—for example, those with a source route or Router Alert (RA) option, or in the case of traffic that is addressed to the switch itself.

 Exception flows are rate-limited and policed within both the PFE and the RE to prevent the lockout of critical control plane processes during periods of abnormally high levels of exception traffic, such as might occur during a denial of service (DoS) attack. You can view exception traffic statistics, including policed drops, using the output of the show pfe statistics and show system statistics commands.

Figure 4-14 details how an unrecognized SMAC address results in the need to redirect traffic to the control plane for additional processing. Things start at step 1, when a frame is received with a known DMAC address of A but an unknown SMAC address of X. The ingress PFE performs both SMAC and DMAC lookups for each frame as part of its Layer 2 forwarding and learning functions. Upon seeing the unknown SMAC, the frame is shunted out of the ASIC forwarding path and into the control plane, where it is sent to the local switch's RE at step 2 in Figure 4-14. At step 3, the LC switch constructs a notification message from the buffered frame's particulars. The notification is then sent to the VC's master RE using the SPF between the ingress switch's CPU and the master RE. The path chosen can contain both internal PFE-PFE and external VCP links, as shown in this example.

Step 4 has the VC's RE perform the needed accounting and admission control functions. For example, a MAC limit parameter may be set that prevents this SMAC from being learned, or perhaps a Layer 2 firewall filter is in place that indicates this SMAC should be blocked. Assuming no forwarding restrictions exist, the master RE sends an update to all PFEs in the VC, which is shown at step 5. This update instructs all PFEs to update their TCAM with the newly learned SMAC. Things conclude at step 6 when the ingress switch reinjects the previously buffered frame back into the local PFE complex, where it's now switched toward the egress PFE, and then toward the egress port, based on the frame's DMAC, using the SPF between the two PFEs. When the frame is received by the egress switch, which functions as a backup RE in this example, no additional learning/intra-VC communications are needed because the SMAC was programmed into all PFEs back at step 5, so the egress PFE simply forwards the frame out the port associated with Destination A.

Virtual Chassis Summary

This section detailed the architectural and design aspects of the EX4200 VC. With an understanding of VC member roles, member ID, and mastership priority, in addition to VCP topologies and cabling schemes, you are no doubt ready to move on to the act of configuring and maintaining a VC. You are in luck, because the next section provides this very information.

Configuration, Operation, and Maintenance

You configure and manage almost all aspects of a VC via the VC's master RE. However, you can also configure VC parameters when an EX4200 is a standalone switch and not actually attached to other switches. This is because each EX4200 switch has some innate characteristics of a VC by default. A standalone EX4200 switch is assigned member ID 0 and is the master of its own (and therefore single-member) VC, which allows you to configure VC parameters on a standalone switch.

When the previously standalone switch is interconnected with an existing VC, the VC configuration statements and any VCE uplink settings that you previously specified on the standalone switch remain part of its configuration, where they can do such things as influence mastership reelections. Once a switch becomes part of a VC, the current VC master synchronizes its configuration copy to all member switches, overwriting any local changes that conflict with the current master's settings.

As with configuration, VC operation and diagnostic troubleshooting is generally performed through the master switch. In a typical configuration, a single virtual IP address is shared among all VC switch members, and any incoming traffic received by a VC member for that shared address is automatically redirected to the current master RE. In a similar manner, a virtual console service redirects console input on any VC switch member to the master RE's console using an internal communications path.

There is a configuration option to disable the internal management VLAN, which allows you to access each VC switch member through its (now) individually IP-addressed OoB management Ethernet port. This option is generally used only in advanced VC troubleshooting situations, and is described later.

The next section details the parameters that are normally configured when deploying a VC with a preprovisioned configuration file.

Virtual Chassis Configuration Modes

Broadly speaking, there are two primary ways to deploy a VC: via plug-and-play with a default or *non-provisioned* configuration, or via *preprovisioned* mode. The latter must include explicit configuration that can force VC member ID and roles in a deterministic manner. Although the anything-goes nature of a default configuration VC is guaranteed to work (when using all VCP cabling), such a design is generally not considered to be a best practice. Note that use of VCE ports requires explicit actions because, by default, these ports operate in uplink mode.

In non-provisioned mode with all switches running default configurations, all will have the same mastership priority. As such, the master switch will be the first one powered up, and the backup RE is the second switch powered on. Should the current master later fail, the backup RE becomes the master and any of the remaining switches could become the new backup, assuming they were all powered up at nearly the same time; recall that uptime can influence mastership tiebreaking, as described previously. Although some might argue that having 1:10 RE redundancy (meaning that any of the 10 VC members can become the active RE) is a good thing, others argue that the law of diminishing returns kicks in when you actually factor the probability of a failure mode that manages to take down two switches while somehow managing to spare all other VC members. In most cases, whatever causes you to lose both your current and backup master REs at the same time—for example, the loss of AC power distribution—is going to take down the entire VC anyway.

In contrast, a deterministic design uses a preprovisioned VC member-to-role mapping that, in the best practice case, assigns a master and backup RE with all remaining VC members forced into a Line Card (LC) role. In most cases, you will also want to statically assign switch member IDs. Such an approach removes the significance of the switch power-up/VC attachment sequence, which, by default, assigns member IDs in the sequence in which each switch is powered up/attached to the VC. This is a significant point, because in addition to its use in various operational and maintenance commands, the member ID also impacts the interface names for that switch's front-panel ports. For example, the first GE port on the VC member with ID 3 is ge-3/0/0, and you may want this ID to be deterministically assigned to the third EX switch in a vertical stack, and to ensure that it can never change by nailing the value down with explicit configuration.

Hot or cold insertion: when does a VC addition become a VC merge?

Regardless of whether you use a default or prestaged configuration, when it comes to expanding a VC you will need to physically attach a new switch member to an existing VC. When performing a VC addition, you must ensure that the new switch is powered off before attaching it to the existing VC.

You *can* attach a powered-on switch to an existing VC, but this operation is disruptive because it constitutes the merging of two VCs into one. Recall that every standalone EX4200 is, by definition, the master of its own domain. Therefore, when attached to a VC *after* being powered on, there are two active master REs, which forces a contention situation, with the result being the reboot of the losing RE and its VC members, along with the loss of its configuration. Because the old master's configuration is lost, the final VC configuration will be that of the winning master, and it's likely that the member ID will also change. Because of the potential for resulting network disruption, cold insertion is always recommended when adding a new member to a VC. In a cold insertion, the newly attached and powered-on switch does not assume a master role, and instead listens to VCCP messages to learn its role in the VC. Assuming the new switch has a lower priority, there will be no master contention process, as would occur with a host insertion.

Virtual Chassis Configuration

Several parameters can be configured to control and manage the operation of a VC. At a minimum, you should assign a virtual management address that represents the VC itself and is serviced by whatever member switch is the VC's current master RE.

Most VC designs, even when using non-provisioned mode, typically add additional configuration to help promote some degree of deterministic operation. This section details VC configuration options, along with some best practice design tips to keep things interesting. Note that with a few exceptions, all of these parameters are configured through the master RE, which then pushes the changes to the affected members when the configuration is synchronized.

Virtual management address

You configure a virtual management IP address as you would on any real interface, except the vme interface does not accept any encapsulation options, and forces the use of unit 0, given that VLAN tagging is currently not supported. Once in place, any matching traffic received on the physical me0 interface at any switch member is automatically redirected to the vme.0 interface on the active VC master, regardless of the member switch on which the traffic ingresses.

You can configure both an IPv4 and IPv6 address if needed; an advanced license is not required for IPv6 on the OoB management interface. The example shown here performs IPv4 address configuration for the vme.0 interface:

```
[edit]

lab@Vodka# edit interfaces

[edit interfaces]
lab@Vodka# set vme unit 0 family inet address 172.16.69.34/24

[edit interfaces]
lab@Vodka# show vme
unit 0 {
    family inet {
        address 172.16.69.34/24;
    }
}
```

The result is a vme.0 interface instantiation on the current master RE. Note that non-master member switches do not display a vme.0 interface, and show their respective me0 interfaces as having no configuration. A later section details use of the no-management-vlan option to permit explicit IP addresses of a member switch's me0 interface while part of a VC.

Don't Use Both the vme and me0 Interfaces

The 9.2R1.9 used in this book permitted explicit configuration of both the vme.0 and me0 interfaces. However, even when a different subnet was used, the presence of an IP address on the me0 interface prevents proper operation of the vme.0 interface. The me0 interface worked as expected. Be sure to remove any explicit me0 configuration when planning a VC with a virtual vme0 interface. This behavior may change as the code evolves:

```
[edit]
lab@Vodka# run show interfaces vme.0
  Logical interface vme.0 (Index 8) (SNMP ifIndex 36)
    Flags: Link-Layer-Down SNMP-Traps Encapsulation:
Unspecified
    Input packets : 82895
    Output packets: 51005
    Protocol inet, MTU: 1500
      Flags: Is-Primary
      Addresses, Flags: Dest-route-down Is-Preferred Is-
```

```
Primary
        Destination: 172.16.69/24, Local: 172.16.69.34,
Broadcast: 172.16.69.255

[edit]
lab@Vodka# delete interfaces me0

[edit]
lab@Vodka# commit
fpc0:
configuration check succeeds
fpc1:
commit complete
fpc0:
commit complete

[edit]
lab@Vodka# run show interfaces vme.0 | match flags
    Flags: SNMP-Traps Encapsulation: Unspecified
      Flags: Is-Primary
      Addresses, Flags: Is-Preferred Is-Primary
```

Virtual chassis member parameters

Most of the VC configuration parameters are found at the [edit virtual-chassis] hierarchy, as you might expect. The options are shown here with the command-line interface (CLI) help function:

```
[edit virtual-chassis]
lab@Vodka# set ?
Possible completions:
+ apply-groups          Groups from which to inherit configuration data
+ apply-groups-except   Don't inherit configuration data from these groups
> mac-persistence-timer How long to retain MAC address when member leaves virtual
chassis
> member                Member of virtual chassis configuration
  preprovisioned        Only accept preprovisioned members
> traceoptions          Global tracing options for virtual chassis
```

The mac-persistence-timer knob determines how long a new master continues to use the MAC address that was owned by the previous master. The default is 10 minutes, and can be set to 0 for an immediate switch or to some really long value so that the manual claims are unlimited. The goal is to avoid a MAC address change and the resulting reflooding/relearning, in the event that the old master's departure is short-lived and it can return to operation without any external device being the wiser.

The preprovisioned option establishes how deterministic you wish to be. When this is enabled, you must map each member switch's serial number to a member ID and a specific role of either RE or LC. You cannot specify a member priority in a preprovisioned configuration. The priority is assigned based on a default mapping-to-member role (i.e., the RE role is assigned 129 while the LC role is assigned 0). In preprovisioned mode, a matching serial number must be found or the new member cannot become part of the active VC topology. Note that a VCCP adjacency is still formed in this case, and the unwelcome switch will quickly be relegated to an LC role where it can do little damage. Note also that the virtual console/management functionality will not work from the non-provisioned switch, and its LSAs are ignored during SPF calculation, which effectively eliminates it from the active VC's topology.

By default, the VC configuration mode is considered non-provisioned, which is synonymous with the factory default in that no explicit VC configuration is required. The default VC mode permits the mapping of specific member IDs to a mastership priority value, but this mapping is not required. Member IDs with no mapping receive the default value of 128; you can prevent a member from being able to become the master or backup RE by assigning a value of 0. The highest priority of 255 should be assigned to the members that you wish to function as the master and backup REs. You cannot manually assign a VC member role, or specify a serial number when in non-provisioned mode.

The current best practice is to have both a master and a backup RE in each VC. When there are more than three VC members, the remaining members are forced into an LC role. Both of the candidate RE members are assigned the same priority, which is typically the highest possible value of 255. This is done to prevent RE mastership preemption in the case that a primary RE suffers a transient failure, such as a reboot, and is known as *non-revertive behavior*. When the old master comes back online, its equal priority setting prevents it from forcing another mastership change, which promotes general stability while allowing human intervention (e.g., logfile analysis) before the decision is made to revert to the original mastership roles, which can then be performed in a maintenance window.

In a non-provisioned mode, the master RE assigns the next sequential member ID (and any associated priority that is set in the configuration) to each member switch as it is attached to the VC and powered on. The new switch's priority value is then used to determine its VC member role based on the mastership election algorithm. You control member ID assignment in this mode by powering up the member switches in the desired order so that each switch receives the next sequential (and desired) member ID. After the fact, you can still use the CLI to alter the membership values.

The member subhierarchy houses the remaining VC configuration options:

```
[edit virtual-chassis]
lab@Vodka# set member 0 ?
Possible completions:
+ apply-groups         Groups from which to inherit configuration data
+ apply-groups-except  Don't inherit configuration data from these groups
  mastership-priority  Member's mastership priority (1..255)
  no-management-vlan    Disable magagement VLAN
  role                 Member's role
  serial-number        Member's serial number
```

Here, you set parameters such as member ID to mastership priority when in non-provisioned mode, or in preprovisioned mode by binding a serial number to either an RE or an LC role. A typical non-provisioned mode configuration is as follows:

```
[edit virtual-chassis]
lab@Vodka# show
member 0 {
    mastership-priority 255;
}
member 1 {
    mastership-priority 255;
}
member 2 {
    mastership-priority 1;
}
```

Note again that to make the priority values matter you must control member ID assignment via the power-on sequence or via CLI commands after things have settled. In this example, members 0 and 1 are the primary REs and member 1 could become an RE only in the event of both members 0 and 1 failing. You cannot assign a 0 priority in non-provisioned mode, which means all VC members are candidates to become the master RE. Here is a sample preprovisioned configuration:

```
[edit virtual-chassis]
lab@Vodka# show
preprovisioned;
member 1 {
    role routing-engine;
    serial-number BM0208269767;
    no-management-vlan;
}
member 0 {
    role routing-engine;
    serial-number BM0208269834;
}
member 2 {
    role line-card;
    serial-number BM0208269888;
}
```

The preprovisioned keyword places the VC into non-default mode, where each member must be explicitly listed by serial number and allowed role. Priority assignments are now based on default values for the RE and LC roles. The example forces member 2 to function as an LC with a derived priority of 0, while members 0 and 1 can be REs. All candidate REs are assigned the same priority, leading to a non-revertive mode of operation.

VCEs

The use of a front-panel uplink as a VCE is an optional aspect of a VC design. As noted previously, the rear-panel VCP ports simply work, and accept no user configuration. Anytime you attach cables to these ports, a VC will attempt to form.

By default, an uplink port does not run the VCCP; hence it is called an "uplink" and not a VCE port. You use the request virtual-chassis vc-port operational mode command to activate or deactivate VCCP/uplink mode on these ports:

```
lab@Vodka# run request virtual-chassis vc-port ?
Possible completions:
  delete              Delete a member's virtual chassis port
  set                 Set a member's virtual chassis port
```

When performed in operational mode, these commands write the VCE port information into the affected switches' private configuration, where it persists until removed.

While in uplink mode the various ports are indexed as ge-*id*/1/*n*, where *id* is the switches' member ID and *n* is the port number, which can range from 0 to 3, depending on uplink module type. Once placed into VCE mode, these same ports are renamed and become vcp-255/1/*n*, and the corresponding uplink port device is removed. This process is shown in the following code for a switch with member ID 0:

```
lab@Vodka> show interfaces ge-0/1/0 terse
Interface              Admin Link Proto    Local                Remote
ge-0/1/0               up    up
ge-0/1/0.0             up    up   eth-switch
lab@Vodka> request virtual-chassis vc-port set pic-slot 1 port 0
```

Once placed into VCP mode, the previous uplink port is no longer available and a new VCP device is created:

```
lab@Vodka> show interfaces ge-0/1/0 terse
error: device ge-0/1/0 not found

lab@Vodka> show interfaces vcp-255/1/0 terse
Interface              Admin Link Proto    Local                Remote
vcp-255/1/0            up    up
vcp-255/1/0.32768      up    up
```

When you plan to use a VCE to attach a new switch to an existing VC, you should power up the new switch and configure both ends of the VCE (one end on the new switch, the other on an existing VC member) *before* you physically attach the new switch to the VC using the VCE link. This ensures that both ends agree to use VCCP over the shared link, and that proper VC topology discovery and member role determination is performed. Failing to perform these steps may cause master reelection and result in disruption to traffic forwarding until the VC topology stabilizes.

Once a switch is part of a VC, additional VCE ports can be provisioned or removed through the master RE by adding the desired member ID to the request virtual-chassis commands. In most cases the master RE will be member ID 0, but this is not mandatory. When a member ID is not specified, the local switch is assumed and the command is performed on the local master RE:

```
lab@Vodka> request virtual-chassis vc-port set pic-slot 1 port 1 ?
Possible completions:
  <[Enter]>    Execute this command
  all-members  Set virtual chassis port on all virtual chassis members
```

```
local       Set virtual chassis port on local virtual chassis member
member      Set virtual chassis port on specific virtual chassis member (0..9)
. . .
```

Using Synchronized Commits

With the 9.2 version used in this book, it's important that you perform a commit
synchronize operation when you have made any changes to the virtual-chassis stanza
on the master RE. Failing to do this can result in a member switch retaining the previous
VC configuration, which causes the changes to not take effect. Later JUNOS versions
may perform these operations without needing the synchronize option. Using
synchronize is always recommended in a dual-RE environment, but in a VC it's also
needed when making VC-related changes, even when there is only one RE. Consider
adding set system commit synchronize to your configuration to make every commit
behave as though you added synchronize.

Virtual chassis configuration summary

Well, that's pretty much it; there really isn't much to configuring a VC. In fact, if determinism is not your bag, you can just plug a bunch of EX4200s together using just
about any VCP cabling scheme, and things will just work. With that said, you really
should configure a virtual management address, and at a minimum map member IDs
to priorities as part of a non-provisioned deployment, to ensure consistency across
various reboot scenarios. Once a member is assigned it becomes sticky, meaning that
the associated priority remains fixed across reboots and power-downs, which tends to
ensure the same master, backup, and LC roles when all members are up and running.

The next section details operational analysis and VC maintenance procedures.

Virtual Chassis Operation and Maintenance

This section details commands and techniques used to perform operational analysis,
troubleshooting, or VC moves and changes. The next section demonstrates most of
these commands and techniques as part of a VC deployment case study.

To help keep you interested, the material is presented in the context of a VC discovery
task, in which your goal is to reverse-engineer a VC you know nothing about using only
CLI operational mode show commands. We suggest you use a sheet of paper to diagram
your discoveries as they are made. You start by displaying the VC configuration:

```
[edit]
regress@EX4200_VC_demo# show virtual-chassis

[edit]
```

From the preceding code, note that you are dealing with a default virtual-chassis
stanza, and therefore a non-provisioned deployment scenario. The lack of configuration details means there is not much more to be learned here, so let's move along. Your

VC discovery diagram remains largely blank, except that the name of the VC and its mode of deployment are known.

Operational mode commands with member context

Several CLI commands support member-specific context when executed in a VC. Generally speaking, when the member argument is omitted, the command acts on the local master RE. Adding the all keyword ensures that a command runs on all members, and in similar fashion, specifying a specific member ID constrains the command to that member.

For a complete list of VC member-aware commands, consult the JUNOS documentation matching the EX software release on your switch. For Release 9.2, this information is available at *http://www.juniper.net/techpubs/en_US/junos9.2/topics/reference/general/virtual-chassis-command-forwarding.html*. Most of the CLI's request system and show system commands support the member keyword. The reboot command is demonstrated:

```
regress@EX4200_VC_demo > request system reboot ?
Possible completions:
  <[Enter]>              Execute this command
  all-members            Reboot all virtual chassis members
  at                     Time at which to perform the operation
  in                     Number of minutes to delay before operation
  local                  Reboot local virtual chassis member
  media                  Boot media for next boot
  member                 Reboot specific virtual chassis member (0..9)
  message                Message to display to all users
  other-routing-engine   Reboot the other Routing Engine
  partition              Partition on boot media to boot from
  |                      Pipe through a command
regress@ EX4200_VC_demo > request system reboot member 1
Reboot the system ? [yes,no] (no) yes

Rebooting fpc1

regress@ EX4200_VC_demo >
```

A show version command runs on all members by default, and is a good way to confirm software compatibility, as all VC members should run the same code and members with mismatched software versions are typically not allowed to actively join the VC:

```
regress@EX4200_VC_demo> show version
fpc0:
--------------------------------------------------------------------
Hostname: EX4200_VC_demo
Model: ex4200-48t
JUNOS Base OS boot [9.2R1.9]
. . .

fpc1:
--------------------------------------------------------------------
Hostname: -fpc1-BK
Model: ex4200-48t
```

```
JUNOS Base OS boot [9.2R1.9]
. .
```

Notice that fpc1 has automatically been assigned a hostname that reflects its member ID (1) and its current VC role of BacKup (BK) RE. From the vacant configuration and its current backup status, you can deduce that fpc0 has a higher mastership priority, was powered on first, has been booted the longest, or simply has the lowest MAC address, as it has won the master election process. Further investigation shall reveal the truth.

Your discovery diagram now shows a single VC, with two members, both as candidate masters. You can also add that both VC members are 48-port models, but there's not much else to be learned here.

The show chassis lcd command is a fine way to view the information displayed on the EX's LCD panel, assuming you are not around to gaze upon its cheery countenance directly:

```
regress@EX4200_VC_demo> show chassis lcd fpc-slot 0
Front panel contents for slot: 0
---------------------------------
LCD screen:
    00:RE EX4200_VC_
    LED:SPD ALARM 00
LEDs status:
    Alarms LED: Off
    Status LED: Green
    Master LED: Green
Interface       LED(ADM/SPD/DPX/POE)
------------------------------------
ge-0/0/0        On:3 blinks per sec
ge-0/0/1        On:3 blinks per sec
. . .
```

The information displayed confirms that slot 0 is the active RE, the model number, and that no alarm-level events have been detected. The port status shows that the port is up, but shows no activity (the first LED is on), and that the port is in 1,000 Mbps mode (three blinks per second).

VC monitoring commands

Use the show virtual-chassis command to obtain VC-specific information. Given your current task, this seems a fine place to explore:

```
regress@EX4200_VC_demo> show virtual-chassis ?
Possible completions:
  <[Enter]>         Execute this command
  active-topology   Virtual chassis active topology
  member-config     Show virtual chassis member configuration from specified member
  protocol          Show virtual chassis protocol information
  status            Virtual chassis information
  vc-port           Virtual chassis port information
  |                 Pipe through a command
```

Note that the full range of command options is available only at the master RE. Non-RE switches run only a subset of the JUNOS processes, and therefore support a subset of the commands shown here for a master RE. Your discovery proceeds with a display of VC ports:

```
regress@EX4200_VC_demo> show virtual-chassis vc-port
fpc0:
--------------------------------------------------------------------
Interface      Type           Status
or
PIC / Port
vcp-0          Dedicated      Up
vcp-1          Dedicated      Down

fpc1:
--------------------------------------------------------------------
Interface      Type           Status
or
PIC / Port
vcp-0          Dedicated      Up
vcp-1          Dedicated      Down
```

The display tells you a few things. It seems that neither member has any VCE ports defined, given that only VCP port-related information is displayed. Also, each member switch is showing a single VCP port in the Up state, indicating that the two switches are attached in a serial VCP chain, which in this case is attached to the VCP 0 port on both members. Additional information is displayed to confirm that VCCP packets are being sent and received over member 0's VCP 0 interface:

```
regress@EX4200_VC_demo> show virtual-chassis vc-port statistics vcp-0 member 0


Member ID: 0    Port: vcp-0
                          RX                     TX
Total octets:             20807032               7654180
Total packets:            77187                  67757
```

Note that show virtual-chassis and show virtual-chassis status commands are synonymous in this release:

```
regress@EX4200_VC_demo> show virtual-chassis

Virtual Chassis ID: 3f04.fea3.76b3
                                    Mastership              Neighbor List
Member ID Status Serial No    Model    priority  Role    ID  Interface
0 (FPC 0) Prsnt  BP0208207200 ex4200-48t  128    Master*  1  vcp-0
1 (FPC 1) Prsnt  BP0208207201 ex4200-48t  128    Backup   0  vcp-0

Member ID for next new member: 2 (FPC 2)
```

There's some real gold here. You see confirmation of member ID assignment, member operational status, and serial number (handy, as this information is needed for a pre-provisioned deployment); the mastership priority (set to default for non-provisioned mode); the member's role within the VC; and the member's adjacency status. In this example, you see that member 0 connects to member 1 using VCP 0, which meshes nicely with your discoveries to date. Also note that the master switch plans to assign member ID 2 to the next switch that successfully joins this VC.

The active-topology switch displays the results of the VCCP that operates between EX-PFEs to ensure optimal paths and loop-free switching among all member switches. The active-topology option is available only on the master RE because LC members do not run any switching/routing daemons:

```
regress@EX4200_VC_demo> show virtual-chassis active-topology
   Destination ID        Next-hop

1(vcp-0)
```

Wow, this is somewhat of a letdown, especially if you have ever had to wade through the route table of an Internet router sporting several full Border Gateway Protocol (BGP) routing feeds. Still, this is a simple two-node VC with a single interconnection link, and therefore the active topology consists of Switch 0 connected to Switch 1, with a single link, as displayed.

Monitor the VC control protocol

The show virtual-chassis protocol command has several arguments that help get to the meat of a VC's operation:

```
regress@EX4200_VC_demo> show virtual-chassis protocol ?
Possible completions:
  adjacency     Show virtual chassis adjacency database
  database      Show virtual chassis link-state database
  interface     Show virtual chassis protocol interface information
  route         Show virtual chassis routing table
  statistics    Show virtual chassis performance statistics
```

As with any LS protocol, displaying adjacency status is always a good place to start when performing operational analysis.

A Word on VCCP and IS-IS

VCCP is a proprietary adaptation of the IS-IS routing protocol. Therefore, knowing IS-IS operation definitely helps in your understanding of VCCP. A complete description of IS-IS operation is beyond the scope of this book. Suffice it to say that IS-IS has traditionally served as an Interior Gateway Protocol (IGP) for the IPv4 and IPv6 (and even ISO/CNLS) protocols in large service provider or government networks. IS-IS uses hellos to discover and maintain adjacencies and reliably floods LSPs that are stored in a replicated LSDB among all routers in the same level. Each IS-IS node then computes a shortest path from itself to all other IS-IS nodes to form an RT. Unlike IP, IS-IS routers do not assign an address to each interface; instead, a single system-level ID, called a Network Entity Title (NET), is used to represent each node in the level.

IS-IS uses a Designated Intermediate System (DIS) concept similar to OSPF's Designated Router (DR); however, all stations on a shared LAN form an adjacency with each

---

other, which differs from OSPF, where DR-other routers see each other as neighbors but form an adjacency only with the LAN's DS and BDRs (Designated and Backup Designated Routers). IS-IS supports levels, which are somewhat analogous to OSPF's areas, provide hierarchy routing.

The nature of IS-IS is to use an easily extensible type length value (TLV) format that often results in it being used as a mule, in that it transports information that is opaque to the IS-IS protocol itself. The inherent transport flexibility, combined with its address-per-node rather than per-interface model, allowed Juniper to leverage its prior IS-IS development work for VCCP. This is good, as Juniper's IS-IS implementation is proven to offer high performance while also being robust and reliable.

This display shows several VCCP adjacencies, and the good news is they are all up:

```
regress@EX4200_VC_demo> show virtual-chassis protocol adjacency
fpc0:
-------------------------------------------------------------------
Interface          System          State     Hold (secs)
internal-0/24      001f.1234.8842  Up             65535
internal-1/25      001f.1234.8842  Up             65535
internal-2/24      001f.1234.8840  Up             65535
internal-2/27      001f.1234.8841  Up             65535
vcp-0              001f.1234.7dc0  Up                58
]
fpc1:
-------------------------------------------------------------------
Interface          System          State     Hold (secs)
internal-0/24      001f.1234.7dc2  Up             65535
internal-1/25      001f.1234.7dc2  Up             65535
internal-2/24      001f.1234.7dc0  Up             65535
internal-2/27      001f.1234.7dc1  Up             65535
vcp-0              001f.1234.8840  Up                58
```

At first glance, this output may seem odd. Where did all these devices and interfaces come from? You can find the answer, in part, in Chapter 3.

Recall that an EX4200-48 comprises three EX-PFEs, with each PFE dividing the work associated with front-panel, uplink, and VC ports. Each PFE has an internal device ID and an associated MAC address that serves as its VCCP system ID. By default, the EX itself is represented by the lowest-numbered PFE, which effectively functions as the DIS for the internal LAN connections.

Each PFE has both internal links and external links to other PFEs, with the latter in the form of VCE or VCP ports. Looking in detail at fpc0, we see a total of five adjacencies, only one of which is to an external PFE, in this case reachable over the VCP-0 port. The remaining adjacencies must therefore represent internal PFE connections, and in fact this is found to be the case given the internal interface designations.

Although you could make some discovery diagram updates based on the adjacency information, you opt to take the next analysis step by displaying the VCCP LSDB. In this example, the LSDB is displayed for switch member 1 only; given that all members have an identical LSDB when things are working, it should not matter which copy you view:

```
regress@EX4200_VC_demo> show virtual-chassis protocol database member 0
fpc0:
-----------------------------------------------------------------------
LSP ID                   Sequence Checksum Lifetime
001f.1234.7dc0.00-00       0xa7c   0xe1e9      116
001f.1234.7dc1.00-00       0x31b   0xb393      114
001f.1234.7dc2.00-00       0x315   0xaceb      115
001f.1234.8840.00-00       0xa6e   0x701c      117
001f.1234.8841.00-00       0x31b   0x869c      111
001f.1234.8842.00-00       0x318   0x4d99      112
    6 LSPs
```

The results show there are six LSPs. This makes sense, given that there are six EX-PFEs in this VC, and each runs a VCCP instance, and therefore each forms adjacencies with its neighbor PFEs and floods LSPs as a result. Based on this information, your VC discovery diagram begins to take the shape of a network consisting of two LAN segments (one per member switch), each with three VCCP entities (one per EX-PFE), which are in turn connected by a serial link (VCP-0).

The extensive switch is used to display as much VCCP LSDB information as Juniper sees fit to make available. The display is taken from member 1, in part to show it has the same database entries as did member 0.

Note that some TLVs are not completely decoded.... How mysterious!

```
regress@EX4200_VC_demo> show virtual-chassis protocol database extensive member 1
| no-more
fpc0:
-----------------------------------------------------------------------

001f.1234.7dc0.00-00 Sequence: 0xa3a, Checksum: 0x49ed, Lifetime: 117 secs
    Neighbor: 001f.1234.7dc2.00  Interface: internal-0/24  Metric:       10
    Neighbor: 001f.1234.8840.00  Interface: vcp-0          Metric:       10

  Header: LSP ID: 001f.1234.7dc0.00-00, Length: 356 bytes
    Allocated length: 1492 bytes,
    Remaining lifetime: 117 secs, Interface: 0
    Estimated free bytes: 1075, Actual free bytes: 1136
    Aging timer expires in: 117 secs

  Packet: LSP ID: 001f.1234.7dc0.00-00, Length: 356 bytes, Lifetime : 118 secs
    Checksum: 0x49ed, Sequence: 0xa3a,
    Fixed length: 27 bytes, Version: 1, Sysid length: 0 bytes
    Packet type: 18, SW version: 9.2

  TLVs:
    Node Info: Member ID: 1, VC ID: 3f04.fea3.76b3, Flags: 1, Priority: 128
        System ID: 001f.1234.7dc0, Device ID: 3
        System ID: 001f.1234.7dc1, Device ID: 4
        System ID: 001f.1234.7dc2, Device ID: 5
    Neighbor Info: 001f.1234.7dc2.00, Interface: internal-0/24, Metric: 10
```

```
        Neighbor Info: 001f.1234.8840.00, Interface:          vcp-0, Metric: 10
        Unknown TLV, Type: 24, Length: 1
        Unknown TLV, Type: 28, Length: 112
    No queued transmissions

001f.1234.7dc1.00-00 Sequence: 0x307, Checksum: 0x1deb, Lifetime: 111 secs
    Neighbor: 001f.1234.7dc2.00  Interface: internal-1/25  Metric:    10

    Header: LSP ID: 001f.1234.7dc1.00-00, Length: 195 bytes
        Allocated length: 1492 bytes,
        Remaining lifetime: 111 secs, Interface: 0
        Estimated free bytes: 1233, Actual free bytes: 1297
        Aging timer expires in: 111 secs

    Packet: LSP ID: 001f.1234.7dc1.00-00, Length: 195 bytes, Lifetime : 118 secs
        Checksum: 0x1deb, Sequence: 0x307,
        Fixed length: 27 bytes, Version: 1, Sysid length: 0 bytes
        Packet type: 18, SW version: 9.2

    TLVs:
        Node Info: Member ID: 1, VC ID: 3f04.fea3.76b3, Flags: 1, Priority: 128
            System ID: 001f.1234.7dc0, Device ID: 3
            System ID: 001f.1234.7dc1, Device ID: 4
            System ID: 001f.1234.7dc2, Device ID: 5
        Neighbor Info: 001f.1234.7dc2.00, Interface: internal-1/25, Metric: 10
    No queued transmissions

001f.1234.7dc2.00-00 Sequence: 0x301, Checksum: 0x1644, Lifetime: 111 secs
    Neighbor: 001f.1234.7dc0.00  Interface: internal-2/24  Metric:    10
    Neighbor: 001f.1234.7dc1.00  Interface: internal-2/27  Metric:    10

    Header: LSP ID: 001f.1234.7dc2.00-00, Length: 239 bytes
        Allocated length: 1492 bytes,
        Remaining lifetime: 111 secs, Interface: 0
        Estimated free bytes: 1188, Actual free bytes: 1253
        Aging timer expires in: 111 secs

    Packet: LSP ID: 001f.1234.7dc2.00-00, Length: 239 bytes, Lifetime : 118 secs
        Checksum: 0x1644, Sequence: 0x301,
        Fixed length: 27 bytes, Version: 1, Sysid length: 0 bytes
        Packet type: 18, SW version: 9.2

    TLVs:
        Node Info: Member ID: 1, VC ID: 3f04.fea3.76b3, Flags: 1, Priority: 128
            System ID: 001f.1234.7dc0, Device ID: 3
            System ID: 001f.1234.7dc1, Device ID: 4
            System ID: 001f.1234.7dc2, Device ID: 5
        Neighbor Info: 001f.1234.7dc0.00, Interface: internal-2/24, Metric: 10
        Neighbor Info: 001f.1234.7dc1.00, Interface: internal-2/27, Metric: 10
    No queued transmissions

001f.1234.8840.00-00 Sequence: 0xa2e, Checksum: 0xaa09, Lifetime: 116 secs
    Neighbor: 001f.1234.7dc0.00  Interface: vcp-0          Metric:    10
    Neighbor: 001f.1234.8842.00  Interface: internal-0/24  Metric:    10
```

```
Header: LSP ID: 001f.1234.8840.00-00, Length: 542 bytes
  Allocated length: 542 bytes,
  Remaining lifetime: 116 secs, Interface: 26
  Estimated free bytes: 42, Actual free bytes: 0
  Aging timer expires in: 116 secs

Packet: LSP ID: 001f.1234.8840.00-00, Length: 542 bytes, Lifetime : 118 secs
  Checksum: 0xaa09, Sequence: 0xa2e,
  Fixed length: 27 bytes, Version: 1, Sysid length: 0 bytes
  Packet type: 18, SW version: 9.2

TLVs:
  Node Info: Member ID: 0, VC ID: 3f04.fea3.76b3, Flags: 3, Priority: 128
     System ID: 001f.1234.8840, Device ID: 0
     System ID: 001f.1234.8841, Device ID: 1
     System ID: 001f.1234.8842, Device ID: 2
  Neighbor Info: 001f.1234.8842.00, Interface: internal-0/24, Metric: 10
  Neighbor Info: 001f.1234.7dc0.00, Interface:      vcp-0, Metric: 10
  Master Info: System ID: 001f.1234.8840
  Backup Info: System ID: 001f.1234.7dc0
  Member Info: System ID: 001f.1234.7dc0, Member ID: 1 Member role: Backup
     System ID: 001f.1234.7dc0, Device ID: 3
     System ID: 001f.1234.7dc1, Device ID: 4
     System ID: 001f.1234.7dc2, Device ID: 5
  Member Info: System ID: 001f.1234.8840, Member ID: 0 Member role: Master
     System ID: 001f.1234.8840, Device ID: 0
     System ID: 001f.1234.8841, Device ID: 1
     System ID: 001f.1234.8842, Device ID: 2
  Unknown TLV, Type: 24, Length: 1
  Unknown TLV, Type: 28, Length: 112
No queued transmissions

001f.1234.8841.00-00 Sequence: 0x308, Checksum: 0x11cd, Lifetime: 111 secs
  Neighbor: 001f.1234.8842.00  Interface: internal-1/25  Metric:     10

Header: LSP ID: 001f.1234.8841.00-00, Length: 195 bytes
  Allocated length: 284 bytes,
  Remaining lifetime: 111 secs, Interface: 26
  Estimated free bytes: 89, Actual free bytes: 89
  Aging timer expires in: 111 secs

Packet: LSP ID: 001f.1234.8841.00-00, Length: 195 bytes, Lifetime : 118 secs
  Checksum: 0x11cd, Sequence: 0x308,
  Fixed length: 27 bytes, Version: 1, Sysid length: 0 bytes
  Packet type: 18, SW version: 9.2

TLVs:
  Node Info: Member ID: 0, VC ID: 3f04.fea3.76b3, Flags: 3, Priority: 128
     System ID: 001f.1234.8840, Device ID: 0
     System ID: 001f.1234.8841, Device ID: 1
     System ID: 001f.1234.8842, Device ID: 2
  Neighbor Info: 001f.1234.8842.00, Interface: internal-1/25, Metric: 10
No queued transmissions

001f.1234.8842.00-00 Sequence: 0x305, Checksum: 0xfaa2, Lifetime: 115 secs
```

```
Neighbor: 001f.1234.8840.00  Interface: internal-2/24  Metric:      10
Neighbor: 001f.1234.8841.00  Interface: internal-2/27  Metric:      10

Header: LSP ID: 001f.1234.8842.00-00, Length: 239 bytes
  Allocated length: 284 bytes,
  Remaining lifetime: 115 secs, Interface: 26
  Estimated free bytes: 45, Actual free bytes: 45
  Aging timer expires in: 115 secs

Packet: LSP ID: 001f.1234.8842.00-00, Length: 239 bytes, Lifetime : 118 secs
  Checksum: 0xfaa2, Sequence: 0x305,
  Fixed length: 27 bytes, Version: 1, Sysid length: 0 bytes
  Packet type: 18, SW version: 9.2

TLVs:
  Node Info: Member ID: 0, VC ID: 3f04.fea3.76b3, Flags: 3, Priority: 128
    System ID: 001f.1234.8840, Device ID: 0
    System ID: 001f.1234.8841, Device ID: 1
    System ID: 001f.1234.8842, Device ID: 2
  Neighbor Info: 001f.1234.8840.00, Interface: internal-2/24, Metric: 10
  Neighbor Info: 001f.1234.8841.00, Interface: internal-2/27, Metric: 10
No queued transmissions
```

The display is rather long, but also largely repetitive. We highlighted the key points. For example, within each switch, the PFE with the lowest MAC address serves as the primary system ID. For switch member 0, this is 001f.1234.8840, which is lower than both 001f.1234.8841 and 001f.1234.8842. This PFE acts as the DIS for the internal network by advertising the list of internal device (PFE) IDs along with their adjacent neighbors. The master RE places additional information into its LSPs that details the role of each VC member. This information represents the results of mastership election, and is used to confirm how each member switch should operate. Each LSP also reports the software version that it's running. This information is used to detect software incompatibilities among VC members.

Figure 4-15 represents the result of your VC discovery exercise.

Considering that no details were originally provided, it's clear you can learn a lot about a VC's operation (or lack thereof when it's broken), using the provided CLI commands and tracing tools. Figure 4-15 shows that switch member 0 is the master, and that it contains three EX-PFEs. The lowest-numbered PFE MAC address is used as the system identifier, and is shown in the truncated form of 8840. Each internal link, which is numbered based on PFE ID, has a pair of adjacent PFEs, which accounts for four of the five adjacencies shown within each switch. The fifth adjacency is via the external VCP0 connection to member 1.

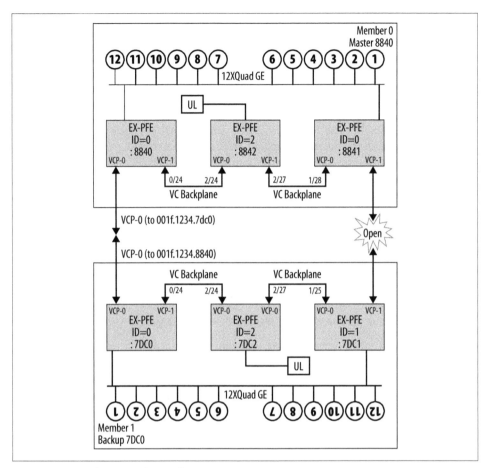

Figure 4-15. VC discovery: the result

The LSDB information indicates that member 0 is the current master and member 1 is a backup master. Similar information is deduced for switch member 1, using the same combination of CLI commands and LSDB analyses. Based on the number of LSPs generated by member 1, you can conclude that it is a three-PFE system, for example. This information allows the complete VC internal topology to be documented in Figure 4-15.

VC tracing

Tracing the operation of a VC is an excellent way to learn more about its operation and to troubleshoot problems when things don't go according to plan. The discovery VC is updated with the following tracing configuration:

```
[edit]
regress@EX4200_VC_demo# show virtual-chassis
```

```
traceoptions {
    file vc_trace;
    flag hello;
    flag lsp;
    flag me;
    flag error;
    flag psn;
    flag csn;
    }
```

In the trace config, the me flag traces master election. The error flag should call out any errors, and the hello, lsp, and sequence number (psn/csn) flags result in tracing hello packet exchanges, LSP flooding, and sequence number exchanges, respectively. Note that in IS-IS, DIS and non-DIS nodes periodically send CSN and PSN packets (respectively) to perform LSP acknowledgment.

After committing, the vc_trace logfile is monitored as the virtual-chassis process is restarted, which is a good way to shake VC things up:

```
regress@EX4200_VC_demo> monitor start vc_trace

regress@EX4200_VC_demo> restart virtual-chassis-control
*** vc_trace ***
Oct  5 14:35:28.207928 TASK_SIGNAL_TERMINATE: first termination signal received
Oct  5 14:35:28.208897
Oct  5 14:35:28.208897 VCCPD_PROTOCOL_ADJDOWN: Lost adjacency to 001f.1234.8842 on
internal-0/24,
. . .

Oct  5 14:35:28.514872 vccpd_provision_cfg_apply: pre_provisioned_cfg 0
. . .

Oct  5 14:35:28.517918 VCCPD_PROTOCOL_ADJUP: New adjacency to 001f.1234.8842 on
internal-0/24
Oct  5 14:35:28.517918
Oct  5 14:35:28.518292
Oct  5 14:35:28.518292 VCCPD_PROTOCOL_ADJUP: New adjacency to 001f.1234.8841 on
internal-2/27
Oct  5 14:35:28.518292
Oct  5 14:35:28.518700
Oct  5 14:35:28.518700 VCCPD_PROTOCOL_ADJUP: New adjacency to 001f.1234.8840 on
internal-2/24
. . .
Oct  5 14:35:28.525734 Sending PTP IIH on vcp-0.32768
Oct  5 14:35:28.525787     packet length 1492
Oct  5 14:35:28.527487 Received PTP IIH, source id 001f.1234.7dc0 on vcp-0.32768
Oct  5 14:35:28.527620 Sending PTP IIH on vcp-0.32768
Oct  5 14:35:28.527663     packet length 1492
Oct  5 14:35:28.528845 Received PTP IIH, source id 001f.1234.7dc0 on vcp-0.32768
Oct  5 14:35:28.529222
. . .

Oct  5 14:35:28.547225 isis_install_lsp: starting me link election
Oct  5 14:35:28.547282 vccpd_run_me_link_election: found 1 members
Oct  5 14:35:28.547304   member: id 0, role 5, link 0, me_ifl 0, bme_ifl 0
```

```
Oct  5 14:35:28.547376 vccpd_run_me_link_election: me_link_owner is 65535
Oct  5 14:35:28.547399 vccpd_run_me_link_election: my member role 5
. . .
```

The truncated output shows the detection of a non-provisioned installation, the formation of new adjacencies, and the exchange of VCCP hello packets, which are displayed as IIH (which stands for Intermediate System to Intermediate System Hello, by the way). The capture also shows the beginnings of the chassis mastership election process. The mastership election process is traced with all other flags removed to reduce clutter. Recall that at the time the VC process was restarted, FPC0 was acting as the backup RE:

```
. . .
Oct  5 14:49:44.261942 vccpd_run_me_link_election: found 1 members
Oct  5 14:49:44.262000  member: id 0, role 5, link 0, me_ifl 0, bme_ifl 0
Oct  5 14:49:44.262019 vccpd_run_me_link_election: me_link_owner is 65535
Oct  5 14:49:44.262040 vccpd_run_me_link_election: my member role 5
Oct  5 14:49:44.293543 isis_run_me
Oct  5 14:49:44.293604 isis_run_me: starting exit_vc_mode_startup timer

Oct  5 14:49:44.293636 exit_vc_mode_startup timer running, exit ME
Oct  5 14:49:44.348031 vccpd_run_me_link_election: found 1 members
Oct  5 14:49:44.348089  member: id 0, role 5, link 0, me_ifl 0, bme_ifl 0
Oct  5 14:49:44.348109 vccpd_run_me_link_election: me_link_owner is 65535
Oct  5 14:49:44.348129 vccpd_run_me_link_election: my member role 5
Oct  5 14:49:44.348363 vccpd_run_me_link_election: found 1 members
Oct  5 14:49:44.348390  member: id 0, role 5, link 0, me_ifl 0, bme_ifl 0
Oct  5 14:49:44.348409 vccpd_run_me_link_election: me_link_owner is 65535
Oct  5 14:49:44.348428 vccpd_run_me_link_election: my member role 5
Oct  5 14:49:44.348660 vccpd_run_me_link_election: found 1 members
Oct  5 14:49:44.348687  member: id 0, role 5, link 0, me_ifl 0, bme_ifl 0
Oct  5 14:49:44.348706 vccpd_run_me_link_election: me_link_owner is 65535
Oct  5 14:49:44.348725 vccpd_run_me_link_election: my member role 5
Oct  5 14:49:44.404440 isis_run_me
Oct  5 14:49:44.404499 exit_vc_mode_startup timer running, exit ME
Oct  5 14:50:24.293693 isis_exit_vc_mode_startup: mastership mode to
vc_mode_line_card
```

At this time, FPC0 has returned to the default LC role, where it waits to determine whether any master REs are present:

```
regress@EX4200_VC_demo> show virtual-chassis

Virtual Chassis ID: 3f04.fea3.76b3
                                            Mastership        Neighbor List
Member ID Status  Serial No    Model        priority   Role   ID Interface
0 (FPC 0)  Prsnt  BP0208207200 ex4200-48t       128   Linecard*
```

The tracing continues:

```
Oct  5 14:50:24.294832 isis_run_me
Oct  5 14:50:24.294889  001f.1234.7dc0.00 selected as master
Oct  5 14:50:24.295008  001f.1234.8840.00 selected as backup
Oct  5 14:50:24.295034  001f.1234.7dc0.00 over 001f.1234.8840.00 vc_mode_master
1>5
```

```
Oct  5 14:50:24.295058   001f.1234.7dc0.00 over 001f.1234.8840.00 vc_mode_master
1>5
Oct  5 14:50:24.295162 Assignment: role-2, mid-0, devs-3, devid1-0, devid2-1
Oct  5 14:50:24.295193 mastership mode changed from line_card to backup

Oct  5 14:50:24.295291 isis_run_me: starting me link election
Oct  5 14:50:24.295321 vccpd_run_me_link_election: found 2 members
Oct  5 14:50:24.295342  member: id 1, role 1, link 1, me_ifl 0, bme_ifl 0
Oct  5 14:50:24.295362  member: id 0, role 5, link 0, me_ifl 0, bme_ifl 0
Oct  5 14:50:24.295381 vccpd_run_me_link_election: me_link_owner is 1
Oct  5 14:50:24.295401 vccpd_run_me_link_election: my member role 5
Oct  5 14:50:24.396111 isis_run_me
Oct  5 14:50:24.396204   001f.1234.7dc0.00 selected as master
Oct  5 14:50:24.396231   001f.1234.8840.00 selected as backup
. . .
```

At the end, FPC0 transitions from LC to backup RE role; given that the current master RE never restarted, its uptime was higher, and therefore it was not preempted. Had both FPCs restarted, the result would be the same, as the tiebreaker is the lowest MAC address, which again favors FPC0 in this example.

VC maintenance

Maintaining a VC is much like maintaining a standalone router or switch. As noted previously, many operational commands have a member context to allow execution on one or all VC members.

When performing a reboot or shutdown operation, you should factor the potential effects of mastership transitions in the event of only some VC members being rebooted. As a general rule, it's always best to reboot or shut down all members of a VC at the same time, especially when not using a preprovisioned mode:

```
lab@Vodkila> request system halt all-members
warning: This command will halt all the members.
If planning to halt only one member use the member option
Halt the system ? [yes,no] (no) yes

Halting fpc0

*** FINAL System shutdown message from lab@Vodkila ***
System going down IMMEDIATELY

*** FINAL System shutdown message from lab@Vodkila ***
System going down IMMEDIATELY

Shutdown NOW!
[pid 677]
```

You can perform software upgrades or downgrades on selected members by adding the appropriate member switch to the request system software command. Future releases may offer automatic software download to members with a mismatched version. In the current release, the user is responsible for ensuring version compatibility:

```
regress@EX4200_VC_demo> request system software add jinstall-ex-9.2R1.9-domestic-
signed.tgz member ?
Possible completions:
  <member>              Install package on VC Member (0..9)
regress@EX4200_VC_demo> ...install-ex-9.2R1.9-domestic-signed.tgz member 1
Pushing bundle to fpc1
WARNING: A reboot is required to install the software
WARNING:    Use the 'request system reboot' command immediately
```

VC adds, moves, and changes

There are a few scenarios where a VC member switch may be removed from the VC. In some cases, the switch is taken away, never to return, resulting in a smaller VC. In other cases, a switch may be removed temporarily—for example, to have repair actions performed after a hardware failure. In yet other cases, one member switch may be replaced with a different switch—for example, swapping out a 24-port model with a 48-port version.

In all of these cases, you should understand the default member ID assignment function, and how CLI commands are used to alter and control member IDs. By default, a switch member writes its ID into a private area of its configuration, and this value is retained through removal and reinsertion events. To prevent conflicts, the master RE retains a list of all assigned values and assigns the next highest available value when a new switch joins the VC.

Use the `request virtual-chassis recycle` command when a switch is removed from a VC to free up a previously assigned member ID, thus making it available for use by another switch. Use the `request virtual-chassis renumber` command to change a member's current assignment to a new value; an error is returned if the new value is currently in use.

It's a best practice to assign member ID 0 to the primary master RE, and then number all switches sequentially based on their vertical or horizontal placement relative to member 0. If you assume a three-member VC, you would thus have switch IDs 0, 1, and 2 assigned.

If switch member 1 is to be removed from the VC for good, you perform these actions, which are not disruptive to the VC's overall operation:

1. Power down and remove Switch 1 (the second switch). Adjust the cabling to account for its loss.

2. Use the `request virtual-chassis renumber member-id 2 new-member-id 1` command to maintain best-practice sequential numbering.

Note that as a result of a member ID change, you will need to update the master's interface configuration to reflect the new member ID, which recalls functions such as the FPC component in a traditional JUNOS interface name. The CLI conveniently warns the user of this necessity, and even offers sample syntax to make the job easier:

```
lab@switch> request virtual-chassis renumber member-id 1 new-member-id 2
To move configuration specific to member ID 1 to member ID 2, please
use the replace command. e.g. replace pattern ge-1/ with ge-2/

Do you want to continue ? [yes,no] (no)no
```

 In typical JUNOS software fashion, you can configure interfaces that relate to a VC switch member that has not yet been attached. Such a configuration is simply ignored until the related hardware is present, at which point the related configuration is activated.

To replace switch member 1 with a new switch that is intended to take over its role in the VC, perform the following steps:

1. Power down and remove Switch 1 (the second switch).

2. Load the default config on the replacement switch, power it down, and attach it to the VC.

3. Apply power to the new switch. The new switch receives the next available member ID, which will be 3 in this example. Use the `request virtual-chassis renumber member-id 3 new-member-id 1` command to reassign the member ID to 1, the value that is associated with the switch being replaced.

 When using preprovisioned mode, you will need to update the configuration to reflect the serial number of the replacement switch before it can join the VC. At this point, the previous commands are used to reassign the member ID as desired.

Connecting to non-master members

You can use the `request session member <id>` command to connect to a non-master switch member using an internal communications path. This command operates regardless of whether you have configured the `no-management-vlan` option for that member switch. This capability can be useful when you are troubleshooting an issue with a specific VC member and wish to view its private configuration or view its local operational status.

Using the no-management-vlan option

In most cases, you will assign a virtual management address that redirects your session from any management port to the VC's current master. You can alter this behavior and assign explicit me0 OoB management addresses to the master and backup REs so that you can access each simultaneously. Note that configuring the me0 interface disables the vme0 functionality.

In either case, you may want to access the OoB port of an LC member without having to connect to the master and then request a session. This capability is also useful during VC communications failures that may prevent you from connecting through the master switch. In addition to specifying the no-management-vlan option for the desired member, you must also access a root shell on the desired member switch so that you can assign an IP address to the me0 interface using the ifconfig command. Note that in the current release you cannot use the CLI to assign an me0 address to an LC member in a VC. Although this procedure also works for a backup RE, it's not needed because you can use the CLI on a master or backup master to assign an me0 address.

The process is demonstrated here, but is performed on a backup RE as the demonstration VC has only two members, and neither is in the LC role:

```
[edit virtual-chassis]
regress@EX4200_VC_demo# set member 1 no-management-vlan

[edit virtual-chassis]
regress@EX4200_VC_demo# show
member 1 {
    no-management-vlan;
}

[edit virtual-chassis]
regress@EX4200_VC_demo# commit synchronize
fpc0:
configuration check succeeds
fpc1:
commit complete
fpc0:
commit complete
```

You start by placing the no-management-vlan option into effect for member 1. Note the use of commit synchronize to push the change to other members, in this case member 1. Next, request a session to member 1. Note that you cannot use the console to perform this operation because of virtual console redirection, which places you right back at the master, unless a communications problem prevents the redirection, in which case you are presented with a local login prompt:

```
[edit virtual-chassis]
regress@EX4200_VC_demo# run request session member 1
€
--- JUNOS 9.2R1.9 built 2008-08-05 07:25:22 UTC
regress@EX4200_VC_demo-fpc1-BK> show interfaces terse me0
Interface               Admin Link Proto    Local               Remote
me0                     up    up
me0.0                   up    up   eth-switch

regress@EX4200_VC_demo-fpc1-BK> show interfaces terse vme0
error: device vme0 not found
```

Once you are connected, the CLI confirms that no me0 or vme0 interface address is currently assigned. To run ifconfig, you must access a root shell:

```
regress@EX4200_VC_demo-fpc1-BK> start shell

% su
Password:
```

As root, the ifconfig command is used to assign and then display an IPv4 address to the me0 device:

```
root@EX4200_VC_demo-fpc1-BK% ifconfig me0.0 add inet 172.69.16.6 netmask
255.255.255.0
root@EX4200_VC_demo-fpc1-BK% ifconfig me0
me0:    encaps: ether; framing: ether
        flags=0x3/0x8000 <PRESENT|RUNNING>
        curr media: i802 0:1f:12:34:7e:3e
me0.0:  flags=0x8000 <UP|MULTICAST>
        inet primary mtu 1500 local=172.69.16.6 dest=172.69.16.0/24
bcast=172.69.16.255
        <unknown> primary mtu 0
root@EX4200_VC_demo-fpc1-BK% exit
%exit
```

The root shell is exited and the CLI at member 1 confirms that the address is in effect:

```
regress@EX4200_VC_demo-fpc1-BK> show interfaces me0
Physical interface: me0, Enabled, Physical link is Up
  Interface index: 1, SNMP ifIndex: 33
  Type: Ethernet, Link-level type: Ethernet, MTU: 1514, Speed: 100mbps
  . . .
      Addresses, Flags: Is-Preferred Is-Primary
        Destination: 172.69.16/24, Local: 172.69.16.6, Broadcast: 172.69.16.255
      Protocol eth-switch, MTU: 0
        Flags: Is-Primary
```

The result of all this is the ability to telnet or SSH directly to VC member 1 using the assigned 172.16.69.6 address. Once you are connected, the lack of management VLAN prevents redirection to the current master RE, thereby allowing you to work on the member switch directly. Generally speaking, this capability is used for specific troubleshooting or fault isolation procedures. For example, if a new JUNOS software bundle cannot be pushed from the master RE to some switch member using the internal communications paths, perhaps due to a software version issue, you may need to FTP a matching software bundle directly to a switch member. Without the no-management-vlan option, you can FTP only to the active or backup REs, and this is possible only when you are *not* using a virtual management interface. When you are using a virtual management interface, you can FTP only to the active RE.

Configuration, Operation, and Maintenance Summary

The EX4200 VC does not require any explicit configuration to just work, out of the box, so to speak. The use of uplink ports as a VC extension does require explicit configuration, however. There are two main options when deploying a VC: the non-provisioned and preprovisioned modes. The former, which can use a factory default, can also include a member ID to priority mapping, which, given the sticky nature of member IDs, helps to promote stability of the original VC configuration through reboots and power-down events.

The current best practice is to use preprovisioned mode, which explicitly maps member serial numbers to a desired VC role. This mode provides maximum determinism, and also security, in that a switch must be preconfigured as a member before it is allowed to actively join the VC.

The operation and maintenance of a VC is much like that of any standalone EX, which is the beauty of the design. Various operational mode commands exist to specifically monitor and control VC operation, and many support a member context to allow global versus targeted actions.

The next section combines the information we've covered to this point as part of a VC deployment case study involving the book's topology.

Virtual Chassis Case Study

This section combines the information we've covered thus far in the form of a practical VC deployment case study based on the book's topology. Here are the design requirements:

1. Rename Switch 3, Vodka, to Vodkila, to reflect the impending merger of the two previously standalone switch names.

    There is an old saying that mixing beer and whiskey is risky; this is sage advice and we can only hope that no ill effects result from the vodka-tequila combination that is soon to be born!

2. Assign 172.16.69.34/24 to the OoB interface; ensure that incoming Telnet and console connections to any switch members are redirected to the current VC master.

3. Switch 3 and Switch 4 share a closet; the third member is located in a remote closet.

4. Ensure that Switch 3 and Switch 4 can function as the master RE in a non-revertive manner.

5. Ensure that the third VC member functions as an LC whenever either or both Switch 3 and Switch 4 are operational.

6. Assign Switch 3 and Switch 4, and the new VC member IDs 0, 1, and 2, respectively.

7. Ensure that no additional switches can join the VC without management intervention.

8. The design must have no single point of VC failure, and the loss of any VC trunk should not prevent ongoing communications.

You may need the following information to complete your task:

1. All three VC member switches have a 4 × 1 GE uplink module installed.
2. The EX4200 serial numbers are:
   - Switch 3 = BM0208269780
   - Switch 4 = BM0208269767
   - Third VC member = BM0208269780

Figure 4-16 shows the before and after topologies.

In this example, the VC deployment occurs in two phases. Phase 1 combines the two existing switches, Vodka and Tequila, into the mythical Vodkila, and phase 2 brings in a third VC member that is housed in a remote closet. Note how the interface names change on switch member 1 to reflect its shift from a standalone VC with member ID 0 to member ID 1. Except for LLDP updates that report the connected peer's interface name change, there is minimal disturbance to the surrounding switches. The same level of connectivity is maintained, and for the most part things just get simpler, given that there is one less bridge/router in the network to generate protocol updates and to contend for the role of designated bridge.

Combining the two switches may result in changes to the spanning tree forwarding topology, especially if either Switch 3 or Switch 4 is the current designated bridge. Regardless, the nature of Layer 2 forwarding and its propensity to form loops means there will be fewer forwarding or root bridge ports in the "after" topology. For example, assuming that Whiskey is the current root bridge, before the merge Switch 3 and Switch 4 have a root port that can forward toward the root. After the merge the combined VC has a single root port; assuming default parameters, this will be either the ge-0/0/3 *or* the ge-0/0/4 interface. Despite the potential for some reduction in the aggregate number of forwarding ports, the VC provides many administrative gains, and if desired the physical topology could be altered to provide added capacity, likely in the form of an aggregated bundle. In fact, with the addition of AE bundles, more capacity can be easily attached.

Note that in Layer 3 mode the redundant links can still be used and the ability to load-balance means there is little net effect to forwarding capacity. This is yet another reason router geeks are prone to saying, "Bridge when you must and route when you can."

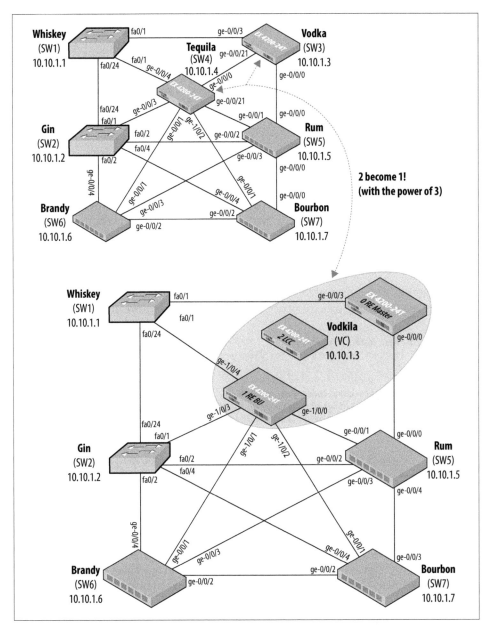

Figure 4-16. The results of VC deployment

Prepare for the Merge

Before rolling out the VC bandwagon, there are a few things you should do to ensure a smooth migration:

1. Back up both standalone switch configurations to a remote location.
2. Return one of the switches to a factory-default configuration, and power off.
3. Attach VCP cables.

The first step is just a good practice and is pretty much common sense. After the merge, you will need to modify the interface assignments in one of the switches to match its new member ID. The master switch overwrites the public configuration portion of all VC members, resulting in loss of the original configuration. Lastly, the master could (in some future release) automatically push a new JUNOS install image into any members with a mismatched software version. The install process may only attempt to save SSH keys from the original configuration, so be sure to move anything you plan to keep to safe storage off the EX device itself.

In this example, the OoB is used to quickly copy a saved configuration from Vodka to an FTP server. The same process is performed at Tequila, but is not shown:

```
[edit]
lab@Vodka# save Vodka_pre_vc_config
Wrote 172 lines of configuration to 'Vodka_pre_vc_config'

[edit]
lab@Vodka# run file copy Vodka_pre_vc_config
ftp://instructor:training1@ftp_server/juniper/vodka_13_vc
ftp://instructor:training1@ftp_server/juniper/100% of 3346  B  917 kBps
```

 In case you're curious, a static host mapping is configured to allow use of the hostname:

```
[edit]
lab@Vodka# show system static-host-mapping
ftp_server inet 172.16.69.254;
```

Next, you load and commit a factory default on Tequila. Given that **root** is the only user in the new config, you log out as user **lab**, so you can return as **root** to perform a graceful shutdown (the non-existent **lab** user is no longer permitted to perform this operation):

```
lab@Tequila> configure
Entering configuration mode

[edit]
lab@Tequila# load factory-default
warning: activating factory configuration

[edit]
```

```
lab@Tequila# set system root-authentication plain-text-password
New password:
Retype new password:

[edit]
lab@Tequila# commit and-quit
commit complete
Exiting configuration mode

lab@Tequila> quit

login:
Amnesiac (ttyu0)
root@%
root@% root
root@% OS 9.0R2.10 built 2008-03-06 10:31:45 UTC
root@% cli
root> request system halt
Halt the system ? [yes,no] (no) yes

*** FINAL System shutdown message from root@ ***
System going down IMMEDIATELY

Shutdown NOW!
[pid 1198]
. . .
```

Though not easy to show in this book, the final preparation step is the attachment of the VCP cables between Switch 3 and Switch 4. Given the redundancy aspects of the design, and their close proximity, you opt for a VCP ring. In this case, there is no merit to any particular VCP cabling scheme, as distance is not an issue and all VCP ports operate the same. As a result, you end up attaching the VCP 0 port of both switches with one VCP cable, and their VCP 1 ports with the other.

Preparation is complete; you are now ready to actually deploy the VC!

Configure VC Parameters

Now the fun can officially start. Your first configuration choice is to perform a non-provisioned versus a preprovisioned installation. Thinking back, the choice is clear given the requirement that no switch be allowed to join the VC without management action. The non-provisioned mode does not offer you the ability to list allowed members by serial number, and will therefore allow any switch with compatible software into the VC. So, preprovisioned it is.

The preprovisioned mode allows you to explicitly bind each member's serial number with a VC member ID and VC role, which is just the ticket here. This helps to ensure that you also meet the stated member ID requirements, which will no longer be based on VC attachment sequence. This mode also handles priority based on role, taking care of the requirement that member 3 must be an LC when either member 1 or member 2 is acting as the master RE.

The first step is to assign the new hostname:

```
lab@Vodka> configure
Entering configuration mode

[edit]
lab@Vodka# set system host-name Vodkila
```

The next configuration task removes the standalone me0 OoB management interface configuration to replace it with a virtual one that is shared among REs. You configure the vme interface as though it were any other interface, but currently there is no VLAN encapsulation and you must use unit 0:

```
[edit interfaces]
lab@Vodka# show me0
unit 0 {
    family inet {
        address 172.16.69.3/24;
    }
}

[edit interfaces]
lab@Vodka# delete me0
```

With the physical me0 interface out of the way, you are clear to add the vme configuration. Note that committing with both prevents the virtual interface from being created:

```
[edit interfaces]
lab@Vodka# set vme unit 0 family inet address 172.16.69.34/24

[edit interfaces]
lab@Vodka# show vme
unit 0 {
    family inet {
        address 172.16.69.34/24;
    }
}
```

 Rather than deleting the me0 and adding a new vme, you might consider using the CLI's rename function. After renaming me0 to vme you will likely have to reassign the IP address, an operation that can also be performed using rename.

Next, position yourself at the [edit virtual-chassis] hierarchy to begin configuration of the VC-specific parameters:

```
[edit interfaces]
lab@Vodka# top edit virtual-chassis

[edit virtual-chassis]
lab@Vodka# set preprovisioned
```

The **preprovisioned** statement does what it says, and once in this mode you must list each member that is allowed to join the VC by its serial number, which is then mapped to an ID and VC role. You start with the current switch, Vodka, because it's supposed to be assigned ID 0, and because it's the only VC member currently powered on. As previously explained, this switch is already the master of its one-VC domain using ID 0. Because you will assign this switch an RE role, and because it's been powered up first, expect this switch to *remain* as master once construction of the new Vodkila is complete.

The definition for switch member 0 is added:

```
[edit virtual-chassis]
lab@Vodka# set member 0 role routing-engine

[edit virtual-chassis]
lab@Vodka# set member 0 serial-number BM0208269834

[edit virtual-chassis]
lab@Vodka# show
preprovisioned;
member 0 {
    role routing-engine;
    serial-number BM0208269834;
}
```

That was pretty straightforward, so you move on to complete the definition for member ID 1. The updated configuration is displayed:

```
[edit virtual-chassis]
lab@Vodka# set member 1 role routing-engine serial-number BM0208269766

[edit virtual-chassis]
lab@Vodka# show
preprovisioned;
member 0 {
    role routing-engine;
    serial-number BM0208269834;
}
member 1 {
    role routing-engine;
    serial-number BM0208269766;
}
```

Happy with your work, you decide to commit the changes. Before doing so, add the **commit synchronize** option so that all commit events are treated as though the **synchronize** switch was included. Recall that in any dual-RE configuration it's recommended that you always synchronize the configuration at every commit, for obvious reasons. As a rule, if you do not have a specific need to maintain distinctly different configs on the master and backup REs, don't. Note that NSR/Non-Stop Bridging support mandates this option, but general GRES and/or protocol GR does not:

```
[edit virtual-chassis]
lab@Vodka# top
```

```
[edit]
lab@Vodka# set system commit synchronize

[edit]
lab@Vodka# commit and-quit
error: Could not connect to fpc-1 : Can't assign requested address
warning: Cannot connect to other RE, ignoring it
commit complete
```

The commit warning is expected; given that only member 0 is powered up, the attempt
to synchronize the configuration to non-existent VC members is in vain, and fails accordingly. Before powering up Switch 4, you quickly access VC status at member 0:

```
lab@Vodkila> show virtual-chassis

Preprovisioned Virtual Chassis
Virtual Chassis ID: 001f.123d.b4c0
                                        Mastership         Neighbor List
Member ID  Status  Serial No    Model   priority  Role  ID  Interface
0 (FPC 0)  Prsnt   BM0208269834 ex4200-24t        129   Master*
```

The VC status is as expected. One switch is up; it's the current master RE, and it has
an ID of 0:

```
lab@Vodkila> show virtual-chassis vc-port
fpc0:
--------------------------------------------------------------------
Interface       Type            Status
or
PIC / Port
vcp-0           Dedicated       Down
vcp-1           Dedicated       Down
```

There is nothing to be alarmed about here; VCP ports can be up only when attached
to another switch that has power applied:

```
lab@Vodkila> show virtual-chassis protocol adjacency
fpc0:
--------------------------------------------------------------------
Interface        System          State    Hold (secs)
internal-0/27    001f.123d.b4c1 Up           65535
internal-1/24    001f.123d.b4c0 Up           65535
```

The VCCP adjacency status confirms that the switch has two PFEs and that both are
internally adjacent. Again, all is as expected. Note that the presence of only two PFEs
indicates this must be a 24-port model, which in fact is the case:

```
lab@Vodkila> show chassis hardware
Hardware inventory:
Item            Version  Part number  Serial number   Description
Chassis                               BM0208269834    EX4200-24T
```

Confirm initial VC operation

With the single-node VC doing all it can, the time has come to power up the old Tequila. Before flipping its power switch, attach to its console to monitor its boot progress and look for any error messages:

```
. . .
FLASH: 8 MB
USB:    scanning bus for devices... 2 USB Device(s) found
        scanning bus for storage devices... 1 Storage Device(s) found
. . .
starting local daemons:.
Sat Jan  1 01:41:09 UTC 2005

Amnesiac (ttyu0)
```

Nothing noteworthy shows in the boot-up messages, except that things do not end well given the display of an Amnesiac prompt. The expectation was a virtual console redirection to the current master RE, which clearly did not happen. This does not bode well for the nascent VC. You log in to the now factory-default configuration as root:

```
root@% root
root@% OS 9.0R2.10 built 2008-03-06 10:31:45 UTC
root@% cli
root> show virtual-chassis
                              ^
syntax error, expecting <command>.
```

After logging in as root, you find it odd that the show virtual-chassis command does not complete as it did on Switch 3. You decide to create a lab/super-user account so that you can log in via internal communications paths via Switch 3/Vodkila; root logins are not currently permitted on the intra-VC communications path:

```
root> configure
Entering configuration mode

root# ...user lab class super-user authentication plain-text-password
New password:
Retype new password:

[edit]
root# commit
commit complete
```

Back on member 0/Switch 3, you decide to do some troubleshooting:

```
lab@Vodkila> show virtual-chassis

Preprovisioned Virtual Chassis
Virtual Chassis ID: 001f.123d.b4c0
                                     Mastership              Neighbor List
Member ID Status Serial No    Model     priority  Role      ID  Interface
0 (FPC 0) Prsnt  BM0208269834 ex4200-24t    129   Master*   1   vcp-0
                                                            1   vcp-1
```

```
  1 (FPC 1) Prsnt  BM0208269767 ex4200-24t    129  Linecard   0  vcp-0
                                                              0  vcp-1
```

The output of the show virtual-chassis command works again, and interestingly it shows two members, each connected by two VC ports, which is pretty much as expected. The only detail that seems off is that member 1 is shown as an LC, when the preprovisioned configuration stated it should function in an RE role; the expected status for member 1 is therefore backup. Something is not right. You next display VC adjacency status:

```
lab@Vodkila> show virtual-chassis protocol adjacency
fpc0:
--------------------------------------------------------------------
Interface        System        State       Hold (secs)
internal-0/27    001f.123d.b4c1 Up              65535
internal-1/24    001f.123d.b4c0 Up              65535
vcp-0            001f.123d.d281 Up                 58
vcp-1            001f.123d.d280 Up                 57


fpc1:
--------------------------------------------------------------------
Interface        System        State       Hold (secs)
internal-0/27    001f.123d.d281 Up              65535
internal-1/24    001f.123d.d280 Up              65535
vcp-0.32768      001f.123d.b4c1 Up                 58
vcp-1.32768      001f.123d.b4c0 Up                 59
```

The output confirms that all expected adjacencies for two EX4200-24s connected via two VCP ports are present. Each switch has two internal and two external adjacencies, which is expected given this configuration and cabling scheme. VCCP LSP flooding also seems to be working, given that the expected number of LSPs (one from each PFE in the VC) is present in both members:

```
lab@Vodkila> show virtual-chassis protocol database
fpc0:
--------------------------------------------------------------------
LSP ID                    Sequence Checksum Lifetime
001f.123d.b4c0.00-00        0x2cd   0xea48     117
001f.123d.b4c1.00-00        0x2c9   0xef34     117
001f.123d.d280.00-00        0x1e1   0xe412     115
001f.123d.d281.00-00        0x1e3   0xce41     116
  4 LSPs

fpc1:
--------------------------------------------------------------------
LSP ID                    Sequence Checksum Lifetime
001f.123d.b4c0.00-00        0x2cd   0xea48     115
001f.123d.b4c1.00-00        0x2c9   0xef34     115
001f.123d.d280.00-00        0x1e1   0xe412     117
001f.123d.d281.00-00        0x1e3   0xce41     118
  4 LSPs
```

The first clue as to what is wrong comes from a detailed analysis of the LSP generated by the DIS/pseudonode at member 0/Switch 3. The output is truncated to save space:

```
lab@Vodkila> show virtual-chassis protocol database extensive member 0
fpc0:
------------------------------------------------------------------------
. . .

  Packet: LSP ID: 001f.123d.b4c0.00-00, Length: 584 bytes, Lifetime : 118 secs
    Checksum: 0x84cc, Sequence: 0x2d6,
    Fixed length: 27 bytes, Version: 1, Sysid length: 0 bytes
    Packet type: 18, SW version: 9.2

  TLVs:
    Node Info: Member ID: 0, VC ID: 001f.123d.b4c0, Flags: 3, Priority: 129
       System ID: 001f.123d.b4c0, Device ID: 0
       System ID: 001f.123d.b4c1, Device ID: 1
    Neighbor Info: 001f.123d.b4c1.00, Interface: internal-0/27, Metric: 10
    Neighbor Info: 001f.123d.d281.00, Interface:        vcp-0, Metric: 10
    Master Info: System ID: 001f.123d.b4c0
    Backup Info: System ID: 0000.0000.0000
    Member Info: System ID: 001f.123d.d280, Member ID: 1 Member role: Not Part of
Virtual Chassis
       System ID: 001f.123d.d280, Device ID: 3
       System ID: 001f.123d.d281, Device ID: 4
    Member Info: System ID: 001f.123d.b4c0, Member ID: 0 Member role: Master
       System ID: 001f.123d.b4c0, Device ID: 0
       System ID: 001f.123d.b4c1, Device ID: 1
    Unknown TLV, Type: 25, Length: 40
    Unknown TLV, Type: 24, Length: 1
    Unknown TLV, Type: 28, Length: 112
  No queued transmissions
  . . .
```

Given the highlighted output, it appears the VC's master feels that member 1 is not a functional part of the VC. This condition should help drive home how the *operation* of VCCP is independent of the *results*; this is to say that VCCP appears to be working fine here, but the result is not the expected two-member VC.

You decide to monitor the *messages* log to see whether anything is up, and the following messages are found to be repeating:

```
lab@Vodkila> monitor start messages

lab@Vodkila>
*** messages ***
Jan  1 02:13:48  Vodkila fpc1 (mrvl_cos_egr_buf_init) Egress buffer limits on 0
error 0
Jan  1 02:13:48  Vodkila /kernel: invalid fpc 1, closing connection
Jan  1 02:13:48  Vodkila /kernel: pfe_listener_disconnect: conn dropped: listener
idx=3, tnpaddr=0x11, reason: none
Jan  1 02:13:48  Vodkila fpc1 (mrvl_cos_egr_buf_init) Egress buffer limits on 1
error 0
Jan  1 02:13:48  Vodkila fpc1 PFEMAN: Master socket closed
Jan  1 02:13:48  Vodkila fpc1 PFEMAN disconnected; PFEMAN socket closed abruptly
Jan  1 02:13:48  Vodkila fpc1 pfeman_get_server_addr: selecting master RE
Jan  1 02:13:48  Vodkila fpc1 Routing engine PFEMAN reconnection succeeded after 1
```

```
tries
Jan  1 02:13:48 Vodkila fpc1 PFEMAN master RE reconnection made
```

Clearly something is not right with member 1. You add tracing to the VC, and see what it has to say:

```
[edit]
lab@Vodkila# show virtual-chassis
preprovisioned;
traceoptions {
    file vc_trace;
    flag error;
}
member 0 {
    role routing-engine;
    serial-number BM0208269834;
}
member 1 {
    role routing-engine;
    serial-number BM0208269767;
}

[edit]
lab@Vodkila# run monitor start vc_trace

[edit]
lab@Vodkila#
*** vc_trace ***
Jan  1 02:38:56.329244 WARNING: LSP from 001f.123d.d281 on interface vcp-0 with
bad packet version 9.0
Jan  1 02:38:56.329451 WARNING: LSP from 001f.123d.d280 on interface vcp-1 with
bad packet version 9.0
. . .
```

The VC trace spews out errors that serve to make the issue somewhat clear. The bad version messages jog your memory; what was that about a VC needing compatible software versions again? You internally connect to member 1 and have a real "d'oh!" moment as its version is displayed:

```
ab@Vodkila> request session member 1
€
--- JUNOS 9.0R2.10 built 2008-03-06 10:31:45 UTC
lab>
```

Breaking the connection and back on the master RE, you are relieved to note a compatible Jinstall package on its local storage:

```
lab> exit
rlogin: connection closed

lab@Vodkila> file list

/var/home/lab/:
.ssh/
Vodka_pre_vc_config
jinstall-ex-9.2R1.9-domestic-signed.tgz
```

```
l2_base
l2_base_vc
. . .
```

And now the beauty of VCCP strikes you. It provides an opaque transport of VC-related information, and therefore can operate even when VC parameters do not match. Although a VC may not form, the continued operation of the VCCP protocols permits ongoing diagnostics and facilitates maintenance activities, such as use of internal links to allow login into member switches or, luckily for you, the pushing and subsequent installation of software upgrades:

```
lab@Vodkila> request system software add jinstall-ex-9.2R1.9-domestic-signed.tgz
member 1 reboot
Pushing bundle to fpc1
WARNING: A reboot is required to install the software
WARNING:    Use the 'request system reboot' command immediately
Rebooting ...
shutdown: [pid 720]
Shutdown NOW!

lab@Vodkila>
```

The `software add` command makes use of the `member` switch, and correctly identifies member 1, preventing the software from being installed on member 0, which is already on that version and would serve only to force a needless reboot. Meanwhile, you switch attention back to Switch 4's console port to again monitor its boot progress:

```
. . .
Verified SHA1 checksum of jswitch-ex-9.2R1.9.tgz
Verified SHA1 checksum of jweb-ex-9.2R1.9.tgz
Running requirements check first for jbundle-ex-9.2R1.9-domestic...
Running pre-install for jbundle-ex-9.2R1.9-domestic...
Installing jbundle-ex-9.2R1.9-domestic in /tmp/pa2350.21/jbundle-ex-9.2R1.9-
domestic.x2350...
Running post-install for jbundle-ex-9.2R1.9-domestic...
Adding jkernel-ex...
. . .
```

 Updated JUNOS versions may automate the pushing of compatible software to new switch members. In the 9.2 release, this process must be performed manually.

So far, so good; all indications are that the matched software version is being installed. It makes watching water boil seem...*enticing*, doesn't it? Sometime later when the upgrade is complete, the initial console login is again `Amnesiac`, causing a pang of fear to run up your spine:

```
Tue Aug  5 08:59:53 UTC 2008

Amnesiac (ttyu0)
```

Your concerns melt away in true "Calgon bath" fashion, when you see that the console login is now correctly redirected to the VC's master. Eureka! A Vodkila is born and you can't help declaring, "It's alive!"

```
login: lab

Logging to master
€ƒPassword:

--- JUNOS 9.2R1.9 built 2008-08-05 07:25:22 UTC
lab@Vodkila>
```

With things apparently working, the VC status is displayed to confirm that member 1 is now correctly listed as a backup RE:

```
lab@Vodkila> show virtual-chassis

Preprovisioned Virtual Chassis
Virtual Chassis ID: 001f.123d.b4c0
                                            Mastership              Neighbor List
Member ID Status Serial No    Model         priority Role       ID  Interface
0 (FPC 0) Prsnt  BM0208269834 ex4200-24t         129 Master*    1   vcp-0
                                                                1   vcp-1
1 (FPC 1) Prsnt  BM0208269767 ex4200-24t         129 Backup     0   vcp-0
                                                                0   vcp-1
```

The assignment of the same priority (based on the VC member role in preprovisioned mode) results in a non-revertive mastership behavior, which is a required aspect of the VC deployment. Should the current master fail and then come back online, it will not overthrow the new master, given that both have the same priority. The show chassis mac-addresses command provides final proof that you now have a functional two-member VC:

```
lab@Vodkila> show chassis mac-addresses
    FPC 0   MAC address information:
      Public base address     00:1f:12:3d:b4:c0
      Public count            64
    FPC 1   MAC address information:
      Public base address     00:1f:12:3d:d2:80
      Public count            64
```

Recall that in the event of an RE switch, the new master begins using its own MAC address after the configured age-out time. During the age-out it continues to use the old master's MAC address.

To complete the VC migration, the original interface assignments in Tequila's original (and luckily enough, saved) configuration are modified to replace slot/FPC number 0 with 1, resulting in the list shown. While you're at it, you decide to add some descriptions to help keep things straight in the new world order that is Vodkila:

```
ge-1/0/0 {
    description "To Rum also";
    unit 0 {
        family inet {
```

```
                address 10.4.5.4/24;
            }
        }
    }
    ge-1/0/1 {
        description "To Brandy";
        unit 0 {
            family inet {
                address 10.4.6.4/24;
            }
        }
    }
    ge-1/0/2 {
        description "To Bourbon";
        unit 0 {
            family inet {
                address 10.4.7.4/24;
            }
        }
    }
    ge-1/0/3 {
        description "To Gin";
        unit 0 {
            family inet {
                address 10.2.4.4/24;
            }
        }
    }
    ge-1/0/8 {
        description "To Whiskey too";
        unit 0 {
            family inet {
                address 10.1.4.4/24;
            }
        }
    }
```

A load merge terminal relative is entered at the [edit interfaces] hierarchy to paste the modified interface name into the configuration, and the change is committed:

```
[edit interfaces]
lab@Vodkila# load merge terminal relative
[Type ^D at a new line to end input]
    ge-1/0/0 {
        description "To Rum also";
        unit 0 {
            family inet {
                address 10.4.5.4/24;
            }
        }
    }
    ge-1/0/1 {
        description "To Brandy";
        unit 0 {
            family inet {
                address 10.4.6.4/24;
```

```
            }
        }
    }
    ge-1/0/2 {
        description "To Bourbon";
        unit 0 {
            family inet {
                address 10.4.7.4/24;
            }
        }
    }
    ge-1/0/3 {
        description "To Gin";
        unit 0 {
            family inet {
                address 10.2.4.4/24;
            }
        }
    }
    ge-1/0/8 {
        description "To Whiskey too";
        unit 0 {
            family inet {
                address 10.1.4.4/24;
            }
        }
    }
load complete

[edit interfaces]
lab@Vodkila#

[edit interfaces]
lab@Vodkila# commit and-quit
fpc0:
configuration check succeeds
fpc1:
commit complete
fpc0:
commit complete
Exiting configuration mode

lab@Vodkila>
```

Note how the commit synchronize setting results in the configuration being mirrored
to both REs. You will be happy you did this should the mastership later change. Not
being one to leave well enough alone, you decide to test a mastership change now using
a request chassis routing-engine command:

```
lab@Vodkila> request chassis routing-engine master release no-confirm
lab@Vodkila>
Vodkila (ttyu0)
```

The virtual management session is broken during the mastership change. Upon recon-
nection, you are again attached to the newly active master:

```
login: lab
Password:

--- JUNOS 9.2R1.9 built 2008-08-05 07:25:22 UTC
lab@Vodkila>

lab@Vodkila> show virtual-chassis

Preprovisioned Virtual Chassis
Virtual Chassis ID: 001f.123d.b4c0
                                      Mastership           Neighbor List
Member ID  Status Serial No   Model   priority  Role    ID  Interface
0 (FPC 0)  Prsnt  BM0208269834 ex4200-24t   129  Backup   1  vcp-0
                                                          1  vcp-1
1 (FPC 1)  Prsnt  BM0208269767 ex4200-24t   129  Master*  0  vcp-0
                                                          0  vcp-1
```

Given the change in mastership that was just triggered, it is no surprise to see that
member 1 is now acting as the VC's master RE while member 0 stands by waiting to
return to its former glory. Having confirmed proper mastership switchover (GRES) and
the desired non-revertive behavior, you return to the original member roles, but using
a slightly different approach to keep it real, yo:

```
lab@Vodkila> request session member 0
€
--- JUNOS 9.2R1.9 built 2008-08-05 07:25:22 UTC
lab@Vodkila-fpc0-BK> request chassis routing-engine master acquire no-confirm

lab@Vodkila-fpc0-BK>
Vodkila-fpc1-BK (ttyu0)

login: lab

Logging to master
€Password:

--- JUNOS 9.2R1.9 built 2008-08-05 07:25:22 UTC
lab@Vodkila> show virtual-chassis

Preprovisioned Virtual Chassis
Virtual Chassis ID: 001f.123d.b4c0
                                      Mastership           Neighbor List
Member ID Status Serial No    Model   priority  Role    ID  Interface
0 (FPC 0) Prsnt  BM0208269834 ex4200-24t   129  Master*  1  vcp-0
                                                          1  vcp-1
1 (FPC 1) Prsnt  BM0208269767 ex4200-24t   129  Backup   0  vcp-0
```

The highlights call out the redirection of the console port on Switch 4 to the current
VC master, which is always called Vodkila. The current master is again member 0. In
the next section the two-member VC is expanded to pick up a remote member.

Newer versions of JUNOS helps to disambiguate which member is the active RE by adding the member ID to the hostname prompt.

Expand the VC with VCE Links

In this section, you will complete the VC deployment by adding a third member switch, located some distance away from the current VC members. Given the distance needed, you will need to use VCE ports, which involves redefining one or more uplink module ports to operate in VCCP mode.

Figure 4-17 shows the VCE design specifics needed to meet the stated redundancy requirements.

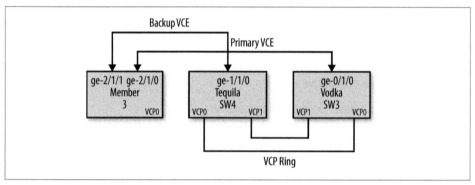

Figure 4-17. Expanding the VC with member 3

The key point is the use of *two* VCE ports at member 3, with each port attached to a *different* VC member, thereby meeting the stated redundancy objectives given that this approach ensures no single point of failure for the VC, and that the VC can continue to communicate with the loss of any one VC trunk.

You can define as many VCEs as you have uplink ports, but note that VCCP routing will use only one such link for any given destination, based on the results of its shortest-path calculation. VCCP factors link speed so that the high-speed VCP backplane ports are generally preferred over any VCE link when an all-VCP path exists.

Prepare the new switch

As before, you should return the new switch to a factory default, after archiving any configuration it may have. In this case, you also have the foresight to ensure that it's already running a matched software version.

You can configure the new switch's uplinks now, as long as you do not attach its uplink cables to the other VC members *before* their uplinks are set to VCCP mode. Failing to do this may prevent proper VCCP auto-sense, which may force a `restart virtual-chassis-control` to get things working properly.

Configure the VCE ports

In our lab, the cables are already attached, and the lab is too far away to warrant a trip just for this. To help ensure that things work to plan, the approach taken is to configure the VC's VCE ports first. In effect, this places them in auto-discovery mode. The new switch does not have any uplink configuration, so currently only light traffic is sent over its uplink ports, which should avoid any potential confusion to the VCCP auto-sense function. When ready, you plan to activate the VCE ports on the new switch and things *should* work.

You start at the master RE, where you configure member 0's VCE link to Switch 3. Note that given that this is member 0, you are configuring PIC 0/1, which is the uplink module:

```
lab@Vodkila> request virtual-chassis vc-port set ?
Possible completions:
  interface          Member's virtual chassis interface
  pic-slot           Member's PIC slot
lab@Vodkila> request virtual-chassis vc-port set pic-slot 1 port 0
```

The `pic-slot` argument correctly identifies the uplink module on the local switch, and sets the first port to VCE mode. The change is confirmed for member 0:

```
lab@Vodkila> show virtual-chassis vc-port member 0
fpc0:
--------------------------------------------------------------------
Interface         Type              Status
or
PIC / Port
vcp-0             Dedicated         Up
vcp-1             Dedicated         Up
1/0               Configured        Up
```

The new VCE is correctly displayed and is in the Up state given that it's cabled to the new switch member and has light. This leaves the three remaining 1 GE ports available for normal uplink usage at member 0. The same command is used to create a VCE link at member 1. In this case, the `member` argument is used to push the change to the correct VC member:

```
lab@Vodkila> request virtual-chassis vc-port set pic-slot member 1 1 port 0
fpc1:
--------------------------------------------------------------------

lab@Vodkila>
```

And the results are confirmed as before:

```
lab@Vodkila> show virtual-chassis vc-port member 1
fpc1:
--------------------------------------------------------------------
```

```
Interface          Type            Status
or
PIC / Port
vcp-0              Dedicated       Up
vcp-1              Dedicated       Up
1/0                Configured      Up
```

As expected, no adjacencies have been formed over either VCE at this point:

```
lab@Vodkila> show virtual-chassis protocol adjacency member 0
fpc0:
------------------------------------------------------------------
Interface           System          State       Hold (secs)
internal-0/27       001f.123d.b4c1  Up              65535
internal-1/24       001f.123d.b4c0  Up              65535
vcp-0               001f.123d.d281  Up                 58
vcp-1               001f.123d.d280  Up                 59
vcp-255/1/0         001f.123d.d280  Down               11
```

With the VC end ready to go, it's time to power up the new switch and configure its uplinks for VCE operation. Note that until it's attached, the switch has member ID 0, which you must keep in mind when you configure its VCE ports. Once it becomes part of the VC it will receive its new member ID and will do the right thing to update its private configuration:

```
root> show virtual-chassis

Virtual Chassis ID: 001f.123d.e6c0
                                     Mastership           Neighbor List
Member ID Status Serial No    Model     priority  Role    ID Interface
0 (FPC 0) Prsnt  BM0208269780 ex4200-24t     128  Master*

Member ID for next new member: 1 (FPC 1)
```

Both VCE ports are activated. The 1/0 port goes to member 0 and the 1/1 port goes to member 1:

```
root> request virtual-chassis vc-port set pic-slot 1 port 0

root> request virtual-chassis vc-port set pic-slot 1 port 1
```

After a moment or two, the results are confirmed. You begin by logging out, at which time you note that the prompt changes, which indicates proper VC operation and is a good omen:

```
-fpc0-LC (ttyu0)


. . .
root@-fpc0-LC% .9 built 2008-08-05 07:25:22 UTC
root@-fpc0-LC% cli

root@-fpc0-LC> show virtual-chassis
Virtual Chassis ID: 001f.123d.b4c0
                                     Mastership           Neighbor List
Member ID Status Serial No    Model     priority  Role    ID Interface
0 (FPC 0) Prsnt  BM0208269780 ex4200-24t     128  Linecard*
```

But the show virtual-chassis display confirms that something is again wrong. The new switch member has the wrong ID, and fails to list the rest of the VC members that are known to exist. Back at the master we get some additional hints as to the nature of the problem:

```
lab@Vodkila> show virtual-chassis

Preprovisioned Virtual Chassis
Virtual Chassis ID: 001f.123d.b4c0
                                       Mastership          Neighbor List
Member ID Status Serial No    Model    priority  Role    ID  Interface
0 (FPC 0) Prsnt  BM0208269834 ex4200-24t  129    Master*  1  vcp-0
                                                           1  vcp-1
1 (FPC 1) Prsnt  BM0208269767 ex4200-24t  129    Backup   0  vcp-0
                                                           0  vcp-255/1/0
                                                           0  vcp-1

    -            Unprvsnd BM0208269780 ex4200-24t
```

Yep, the display makes the problem clear. We are in a preprovisioned mode but have failed to update the master configuration with the new member's serial number. The configuration mistake is rectified:

```
lab@Vodkila> configure
Entering configuration mode

[edit]
lab@Vodkila# set virtual-chassis member 2 role line-card serial-number BM0208269780

[edit]
lab@Vodkila# commit and-quit
fpc0:
configuration check succeeds
fpc1:
commit complete
fpc0:
commit complete
Exiting configuration mode
```

And the VC status is again displayed:

```
lab@Vodkila> show virtual-chassis
Preprovisioned Virtual Chassis
Virtual Chassis ID: 001f.123d.b4c0
                                       Mastership          Neighbor List
Member ID Status Serial No    Model    priority  Role      ID  Interface
0 (FPC 0) Prsnt  BM0208269834 ex4200-24t  129    Master*    1  vcp-0
                                                             2  vcp-255/1/0
                                                             1  vcp-1
1 (FPC 1) Prsnt  BM0208269767 ex4200-24t  129    Backup     0  vcp-0
                                                             2  vcp-255/1/0
                                                             0  vcp-1
2 (FPC 2) Prsnt  BM0208269780 ex4200-24t    0    Linecard   0  vcp-255/1/0
                                                             1  vcp-255/1/1
```

The results are as expected, and they confirm that a three-member VC is now operational. The neighbor ID list indicates that some neighbors are adjacent over a VCE, as indicated by the vcp-255/1/0 designation. Proper internal routing is confirmed by displaying the VC's active topology:

```
lab@Vodkila> show virtual-chassis active-topology
  Destination ID      Next-hop

  1                   1(vcp-0)  1(vcp-1)

  2                   (vcp-255/1/0)  1(vcp-1)
```

The display confirms that member 0 plans to reach member 1, which is collocated using the VCP cables. Its first choice is to use VCP Port 0, and it can fall back to Port 1 if needed. To reach the remote chassis, it prefers the direct VCE link vcp-255/1/0, but in the event of that next hop failing it is prepared to fall back to its vcp-1 link, which takes it through member 1 via the VCP connection, and then from member 1 to member 2 over the remaining VCE link.

A quick look at the VC RT for Destination 3 at member 0 confirms that VCE links have a higher cost, due to their reduced bandwidth:

```
lab@Vodkila> show virtual-chassis protocol route 2 member 0
fpc0:
--------------------------------------------------------------------

Dev 001f.123d.b4c0 ucast routing table         Current version: 79
----------------
System ID         Version   Metric Interface      Via
001f.123d.b4c1       79        10 internal-0/27 001f.123d.b4c1
001f.123d.d280       79        20 vcp-0         001f.123d.d281
                                  internal-0/27 001f.123d.b4c1
001f.123d.d281       79        10 vcp-0         001f.123d.d281
001f.123d.e6c0       79        63 vcp-255/1/0   001f.123d.e6c0
001f.123d.e6c1       79        73 vcp-255/1/0   001f.123d.e6c0
```

The display shows that internal and VCP links have a cost of 10. Thus, the route from PFE 0 in member 0 to PFE d280, which is in member 1, has a path cost of 20, given the use of two such links. The lower-speed 1 GE VCE is assigned a cost of 63 by the automatic scaling algorithm. As such, for member 0 to reach PFE 1 (e6c1) in member 2, the path cost is 73, reflecting the need to transit one VCE and one internal link.

These results confirm the operation of the new VC and complete the VC deployment case study.

Since you were the one to bring it up, who better to pull it all down? Note the use of the all-members switch to force the graceful shutdown of the entire VC:

```
lab@Vodkila> request system halt all-members
warning: This command will halt all the members.
If planning to halt only one member use the member option
Halt the system ? [yes,no] (no) yes
```

```
Halting fpc1

Halting fpc2
Shutdown NOW!
[pid 2413]

*** FINAL System shutdown message from lab@Vodkila ***
System going down IMMEDIATELY
```

Case Study Summary

The VC deployment case study demonstrated the simplicity and veritable plug-and-play nature of the EX4200 VC. Along with VC configuration, the commands and procedures used to verify a VC's operational status and maintenance procedures were also demonstrated.

Conclusion

The EX4200 VC is a powerful form of abstraction that permits you to start small and grow into a large-scale switching platform that offers the ease of use and management associated with a single entity.

The Juniper Networks VC is truly simple to deploy and maintain, given the CLI's support of a member context switch and the ability to connect to member switches over internal communications paths, even when there is a problem that may prevent that member from actively joining the VC. You have the choice of just plugging things together with no explicit VC configuration, or you can exercise a high degree of deterministic control over most aspects of a VC's operation. Besides being easier to manage, there are significant HA benefits to the VC, including redundant REs and the resulting ability to support GRES and NSR, but also for forwarding plane faults and the VC's ability to work around a failed PFE or VC trunk through LS VCCP.

The next chapter keeps things rolling by getting into the meat of STP, the heart and soul of any LAN switch. You will be glad that there is now one less switch to manage and troubleshoot when things invariably go astray, given that Vodkila lives strong, and lives large, and is far more than the sum of its individual parts.

Chapter Review Questions

1. How many EX4200s can be grouped into a VC?

    a. 2

    b. 6

    c. 8

    d. 10

2. Each VC port operates at:
   a. 16 Gbps, FD
   b. 32 Gbps, FD
   c. 64 Gbps, FD
   d. 128 Gbps, FD
3. True or false: the master RE must be assigned member ID 0.
4. Which is the correct order for master RE election when a VC is rebooted?
   a. Uptime, member ID, priority, lowest MAC address
   b. Priority, last role, uptime, lowest MAC address
   c. Last role, priority, uptime, highest MAC address
   d. Uptime, last role, lowest MAC address, priority
5. Which is true about VCE ports?
   a. Only the 2 × 10 GE uplink module can support VCE in Release 9.2
   b. All ports on an uplink module must operate in normal or VCE mode
   c. You must use a single VCE link when connecting wiring closets to avoid loops
   d. An operational mode configuration command is used to set or clear VCE mode
6. Which are true regarding VCCP? (Choose all that apply.)
   a. VCCP is defined in an Internet draft
   b. It's a Juniper proprietary protocol based on distance vector routing
   c. It's a Juniper proprietary protocol based on link state OSPF routing
   d. It's a Juniper proprietary protocol based on link state IS-IS routing
7. Which can be set in a non-provisioned deployment?
   a. A member ID to a priority
   b. A member ID to a VC role such as RE or LC
   c. A serial number to a priority and VC role
   d. You cannot configure any VC parameters in this mode
8. How is the member ID assigned in a non-provisioned deployment?
   a. Based on VC role
   b. Based on power-up sequence, with each new switch getting the next ID
   c. The serial number is mapped to the member ID
   d. You cannot control member ID in this mode
9. A new switch is added to a preprovisioned VC. Until the serial number is mapped:
   a. No VCCP adjacency can form
   b. The switch can enter the VC but as an LC only
   c. A VCCP adjacency will form but the switch will not be a member of the VC

d. The master configuration file is automatically updated with the new member's serial number, as learned through VCCP

10. Which command connects you to the current backup RE, from the current master?

    a. `request routing-engine login backup`

    b. `request routing-engine login other`

    c. `start shell pfe network member <id>`

    d. `request session member <id>`

11. You want to alter a serial number to VC role assignment, but after you commit the change, the related VC member does not take on the new role. What may be wrong?

    a. You need to use `commit synchronize` to push the change to the VC member's private configuration

    b. The VC backbone is malfunctioning, preventing intra-VC communications

    c. The EX4200 chassis type mapped to that member ID is not compatible with the new role; a hardware key is needed to function as the master RE

    d. You must use non-provisioned mode to map a member ID to a VC role

Chapter Review Answers

1. Answer: D. A VC consists of 2 to 10 EX4200 chassis.

2. Answer: B. Each VC port operates at 32 Gbps FD, yielding 64 Gbps of throughput per port. Two ports yield 128 Gbps of throughput between adjacent VC members.

3. Answer: False. Although it is a best practice to make the master RE member 0, this is not mandatory; the member ID must be unique, but has no direct relationship to the VC role.

4. Answer: B. Only B is correct, and the lowest MAC address is the tiebreaker.

5. Answer: D. The operational mode `request virtual chassis vc-port` command alters the private NVRAM configuration to set or clear VCE mode. Support for 1X GE VCE ports first appeared in Release 9.2. You can mix and match VCE and normal modes, and all ports can operate in VCE mode if desired.

6. Answer: D. VCCP is based on the IS-IS link state routing protocol, but is proprietary to Juniper.

7. Answer: A. In non-provisioned mode, you can map a member ID to a mastership priority.

8. Answer: B. In non-provisioned mode, the member ID is not set directly, but is instead controlled through the power-up sequence. You can later alter the assignment with the CLI `request chassis renumber` command.

9. Answer: C. VCCP can function even when mismatched software or configuration issues prevent a successful VC join.

10. Answer: A, B. From the current master you can request a login to the other RE, which will be a backup. You can also explicitly state that a connection should be formed to the backup RE. Option C is used to form internal VTY connections to PFE components, which may or may not be a master/backup RE. Option D allows you to access the CLI of a member switch using an internal communications path. As with option C, such a switch member may be acting in an LC role.

11. Answer: A. In the 9.2 release used for this book, member-specific VC changes committed on the master RE are pushed to the member's private configuration only when the `synchronize` switch is used; later releases may automate this function for LC-related changes, but synchronization is required between the master and backup REs as in some cases you may wish to have different configurations. Use of NSR requires that the system `commit synchronize` option be set, forcing all commits to synchronize.

## CHAPTER 5

Virtual LANs and Trunking

LAN switches filter and forward traffic based on learning Source MAC (SMAC) addresses, and then later filter traffic based on Destination MAC (DMAC) addresses. Although a regular switch does terminate a collision domain, it does not isolate broadcast and multicast traffic; such traffic must be flooded to all LAN members because these addresses are never used as an SMAC, and thus can never be learned. By definition, this Broadcast, Unknown, or Multicast (BUM) traffic by its nature has to be flooded to all stations anyway.

LAN virtualization, also known as a virtual LAN or VLAN, allows switches to logically group end stations to provide isolation of BUM traffic, enhancing both performance and security. This chapter covers VLAN tagging and the configuration and monitoring of VLAN switching on the EX platform in the context of integrating a JUNOS switch into a Cisco environment.

The topics covered in this chapter include:

- Virtual LANs and trunking
- EX to catalyst VLAN integration (and troubleshooting)

Virtual LANs and Trunking

Ethernet LAN and LAN switching concepts, also known as *bridging*, are described in Chapter 1. This chapter focuses on using VLAN tags to logically partition the LAN into multiple broadcast domains.

In a conventional Ethernet LAN, all nodes must be physically connected to the same MAC broadcast domain to communicate at Layer 2. With VLANs, the physical location of the nodes is not important, as you can group network devices in ways that make sense for your organization—for example, by department or business function—without regard for the physical station attachment point. A VLAN is identified by a specific tag value, as described shortly, and the stations that belong to a common VLAN will normally share a common IP network address, meaning they share a LAN from the IP perspective.

VLAN capability allows switching for multiple-LAN communities that share a physical infrastructure, while ensuring that no information is leaked between these communities. This *trunking* is achieved via VLAN tags, as defined in the IEEE 802.1Q standard; in fact, a trunk interface is often said to operate in *dot1Q* mode. The 802.1Q standard also characterizes the meaning of a VLAN with regard to Media Access Control (MAC) layer functions and protocols such as bridging or Spanning Tree Protocol (STP), and how stations on separate VLANs can communicate through the services of a Layer 3 device in the form of a router. In this case, the router uses a Routed VLAN Interface (RVI) to route between the otherwise ships-in-the-night separation of traffic riding over different VLANs. In IOS terminology, the Layer 3 interface that connects logically distinct Layer 2 domains with IP forwarding is referred to as a Switched Virtual Interface (SVI).

Port Modes

Switch ports, or interfaces, operate in either access or trunk mode. An interface in access mode connects to a network device, such as a desktop computer or an IP telephone. The interface itself belongs to a single VLAN, and the frames sent and received over that access interface are normal, untagged *native* Ethernet frames. In a default EX configuration, all interfaces are placed in access mode.

Trunk interfaces handle traffic for multiple VLANs, multiplexing the traffic for all those VLANs over the same physical connection. Trunk interfaces are generally used to interconnect with other switches, or routers, and use explicit tagging to segregate traffic on the shared trunks.

Tagging User Traffic

Traffic received on an access port, or from a particular MAC address when so configured, is mapped to an associated VLAN and the related VLAN tag is inserted before the frame is forwarded over trunk ports. Once tagged, the frame is forwarded only over those trunks that are defined to support that VLAN, and for which that VLAN is in a forwarding state when running Multiple Spanning Tree Protocol (MSTP), as described in Chapter 6. When a switch has local access ports that belong to the associated VLAN, and the frame needs to be sent to those ports (based on DMAC and learning), the tag is removed and the frame is sent in its original (native) format out the access port.

ISL Versus 802.1Q Trunks

The Inter Switch Link (ISL) protocol predates the IEEE 802.1Q standard and is proprietary to Cisco Systems. ISL is considered a two-level encapsulation scheme and operates by adding a 26-byte header and 4-byte CRC trailer to each frame. The header contains a 10-bit VLAN identifier to support 1,024 VLANs, as well as a user priority indication.

JUNOS software does not support ISL. Therefore, you must configure .1Q trunking on IOS devices to interoperate with JUNOS switching gear by issuing a `switchport trunk encapsulation dot1q` and `switchport mode trunk` configuration statement under the desired interface. Note that some IOS hardware is not capable of supporting the standards-based .1Q form of trunking, and that in some cases you can run both ISL and .1Q as part of a migration strategy.

> Given that the original IEEE 802.1D (Bridging) standard has been updated to support features such as MSTP and Rapid Spanning Tree Protocol (RSTP) in a .1Q trunk context, there is little need to continue deploying vendor-proprietary trunk protocols such as ISL.

Figure 5-1 shows the format of an IEEE 802.1Q tagged frame.

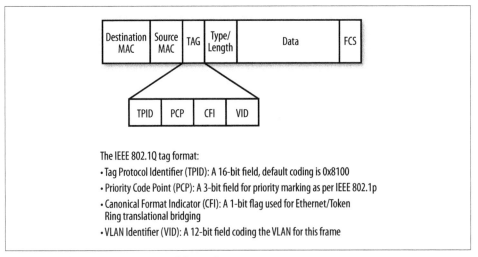

Figure 5-1. The IEEE 802.1Q tagged frame format

Figure 5-1 shows that 802.1Q tagging is not really an encapsulation scheme, but rather involves the insertion of a 4-byte field after the frame's SMAC address. The tag is inserted before the Type or Length field when sending Ethernet frames or IEEE 802.3 frames, respectively.

Although standards exist to define the transport of IPv4/IPv6 traffic over IEEE 802.3, typically using a combination of Logical Link Control (LLC/IEEE 802.2) and Subnetwork Access Protocol (SNAP) encapsulation, which ironically allows an escape back to EtherType codes, this method of IP transport is rarely used and is not supported by JUNOS devices. JUNOS devices always encapsulate IP in Ethernet v2 frames, which use an EtherType for explicit identification of the encapsulated protocol, thus eliminating the need for LLC/SNAP. Ethernet-based IP encapsulation is defined in RFCs 894 and 2464 for IPv4 and IPv6 using Type codes 0×0800 and 0×86DD, respectively.

The insertion of .1Q tags increases frame size, and necessitates recalculation and update of the frame check sequence (FCS). When received over a trunk port, the VLAN tag is removed and the FCS is restored before the (now) untagged frame is forwarded (as needed) to all access ports associated with that VLAN.

Adding four bytes to an Ethernet frame containing 1,500 bytes of user data creates a *jumbo frame*, which, according to the original Ethernet/802.3 standards, is a frame larger than 1,518 bytes (not counting the Start Frame Delimiter). Modern devices typically adjust the local definition of a jumbo frame when VLAN tagging is in effect to support so-called *baby giants* of 1,522 octets, but this is not always the case. Note that older LAN switching gear may not even support VLAN tagging/.1Q trunking. Although you can pass VLAN-tagged traffic through such a pre-VLAN switch, you must ensure that it's capable of being configured to support the larger frame sizes that result from tag insertion to ensure that you do not suffer jumbo-related frame discards.

The Tag Protocol Identifier (TPID) informs switches as to whether there is a .1Q tag field in the frame being processed; the default coding of 0×8100 is selected to not conflict with valid IEEE 802.3 Length indicators, which are always less than 0×0600, or with any preexisting EtherType codes. As a result, tag coding is unambiguous and ensures that proper actions are taken for tagged and untagged frames. Other TPID values exist to support stacked VLANs, which is also known as *QinQ* and is described in a later section.

The Priority Code Point (PCP) is defined in IEEE 802.1p, and provides eight levels (0–7) of priority indication to accommodate class of service (CoS) actions by Layer 2/3 devices. The use of .1p priority as part of a CoS solution is covered in Chapter 10. Note that a Layer 3 device can also use .1p marking when configured for VLAN tagging via 802.1Q, but more often, Layer 3 devices are set to look into the Layer 3 header, that is, at an IP packet's Precedence or DiffServ Code Points (DSCPs), for CoS classification.

The Canonical Format Identifier (CFI) is used to support translational bridging between Ethernet and Token Ring networks. Ethernet/802.3 LANs use a canonical address format, meaning that each octet of the MAC address is sent starting with the Least Significant Bit (LSB) first. In contrast, Token Ring networks (IEEE 802.5) send each address octet starting with the Most Significant Bit (MSB). This bit is sometimes referred to as the *Token Ring Encapsulation flag*, which denotes its historical use in this capacity. The term *historical* is used here because, despite its technical merits, the simple truth is that 802.5/Token Ring and the related Source Route Bridging (SRB) failed in the marketplace, and quite resoundingly, too. This fact makes the whole concept of *translational bridging* moot given the ubiquitous nature of Ethernet technologies.

The 12-bit VLAN Identifier field is the meat of the matter here; it codes the VLAN value in the range of 0 to 4,095 inclusive. Note that a coding of 0 indicates a frame that is not part of any VLAN. In this case, the .1Q tag can still indicate user priority and is referred to as a *priority tag*. Also, the all-1s value of 0xFFF (decimal 4,095) is reserved for implementation use. Together these restrictions yield a total of 4,094 usable VLANs. Note that in some implementations the VLAN number space is said to be global (chassis-based), and in some cases internal functions may consume additional VLAN IDs, further reducing the values that can be defined for user purposes; in the EX architecture, VLAN numbering is global, but the full range of VLAN IDs from 1 to 4,094 are available for use.

It's a common practice to see a VLAN, oftentimes VLAN 1, reserved for network management, especially on Cisco devices, which, unlike EX switches, often do not support a discrete Out of Band (OoB) management interface that is reserved for that purpose.

MX Platforms and Switching Domains

Although coverage of the MX platform is outside the scope of this book, it bears mentioning that the MX platform supports the notion of a virtual switching domain or instance. On an MX, *each* switching instance can support the full range of 1 to 4,094 VLAN IDs, *and* you can bridge/route between these switching domains using an Integrated Route Bridge (IRB) interface. EX platforms do not support virtual switches, or the IRB construct. On an EX you use an RVI to route between VLANs that are all part of its single switching instance.

QinQ, a.k.a. provider bridging

The term *QinQ*, also called Service VLANs (S-VLANs), defines a method of pushing multiple VLAN tags onto a single frame, typically at the edges of a service provider's network, to provide a VLAN "trunk of trunks" type of service that requires a minimum of cooperation between the service's users and the service provider. In some ways, this is similar to an ATM Virtual Path Indicator (VPI), which allowed a single virtual path to carry multiple Virtual Chassis (VCs) that were managed by the end user in a manner that was transparent to the provider of the ATM path service.

QinQ is defined in the IEEE 802.1ad-2005 specification and is gaining popularity in metropolitan and wide-area LAN emulation services, sometimes called a Transparent LAN Service (TLS) or a Virtual Private LAN Service (VPLS) because of the added flexibility to the service user and ease of management for the provider.

In the basic case, the service provider allocates a single outer tag, called a Service tag (S-VLAN), that's used to identify the customer access link; this leaves the customer free to add additional VLANs, called Customer VLANs (C-VLANs), which are then trunked over the provider's core in transparent fashion.

In most cases, an alternative TPID value is used for the outer tag to flag QinQ versus normal, single-tagged traffic. Although the 802.1ad specification defines 0×88a8 for service-provider outer tags, most QinQ-supporting equipment allows you to define the S-tag value, and 0×9100 or 0×9200 are commonly used in today's networks.

Figure 5-2 shows a typical QinQ scenario.

Figure 5-2 shows two provider customers: Customer 1 is using tagged frames and Customer 2 sends native Ethernet; within the service provider's network, an additional layer of tagging in the form of S-tags is added. This results in both the original C-tag as well as an S-tag for Customer 1, while Customer 2 has only the S-tag when in the provider's core. The additional layer of trunking allows the customers to manage their own VLAN trunking independent of the service provider, which greatly simplifies VLAN moves, adds, and changes. The provider gains the benefit of being able to offer an S-VLAN service, and can also take steps to normalize tagging for its customers and their varying service types, which in turn can simplify general operations and maintenance activities, creating the proverbial win-win situation for QinQ.

With all those great things to be said for S-VLANs, it has to be noted that in the 9.2 JUNOS release, EX platforms do not offer QinQ support. As always, this may change in a later release, so check back often. MX platforms feature full QinQ support and can be used to build the core/distribution layer in a large switching deployment involving both S-VLANs and large-scale routing services as well. By mixing and matching lower-cost EX platforms at the network while the more sophisticated (and expensive) MX platforms handle the core, you get the best of both worlds: many lower-cost ports at the edge, combined with fewer high-cost but also high-touch ports in the core that provide both value-added services and high-speed transport.

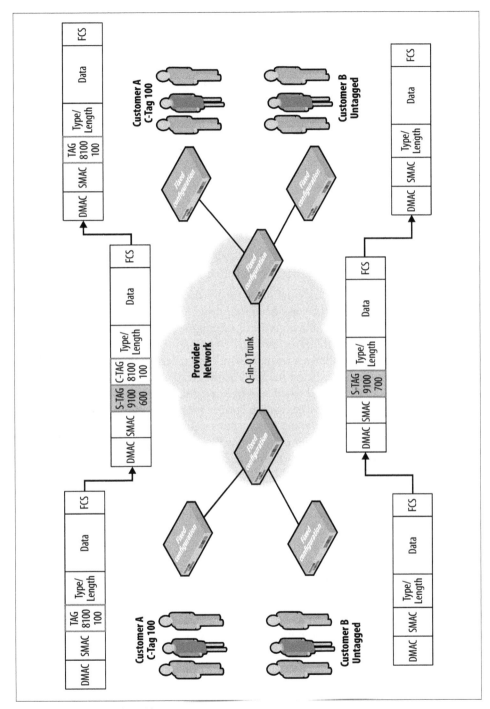

Figure 5-2. S-VLANs (QinQ) in action

The Native and Default VLANs

There is much confusion over the terms *default* VLAN and *native* VLAN, and whether this means tagged or untagged. In many cases, the confusion stems from the fact that the native and default VLANs *can* be different, but by default are the same in IOS devices, and because the native VLAN *can* use tagging, but typically is sent and received untagged.

The native VLAN

Clause 9 in the 802.1Q specification defines a native VLAN that is designed to support older switches that do not understand .1Q tagging. Although this is not a Cisco book, Cisco did invent much of what is now called bridging and routing, and many modern network concepts have roots in IOS. Cisco provides the following definition of the native VLAN:

> The native VLAN is defined as the VLAN to which a port returns when not trunking, and is the untagged VLAN on an 802.1Q trunk. By default, VLAN 1 is the native VLAN.

In other words, the native VLAN defines how a switch transmits and receives traffic associated with the native VLAN assigned to the switch port. At egress, such traffic is not tagged, and is therefore sent in native (i.e., untagged) format on that link. In reverse, at reception any untagged frame is assumed to belong to that port's native VLAN. Support for native VLAN allows a trunk to send and receive a mix of tagged and untagged traffic, with the latter being mapped to the trunk's native VLAN. Non-native traffic is always explicitly tagged when sent over a trunk interface, and upon reception the explicit tag value is used to associate the traffic with its destination VLAN. In contrast, the lack of an explicit tag *implicitly* associates traffic with that port's native VLAN. Although each trunk can have its own native VLAN, the current best practice is to define a single native VLAN within the bridge domain.

Native VLANs are significant when interoperating with Cisco devices because the switches have historically associated control plane traffic with a native VLAN in an effort to protect the (limited) control plane resources from the burden associated with having to process end-station multicast and broadcast traffic. Placing the management and Layer 2 control plane traffic in a specific VLAN, one that is not shared by normal user devices, alleviated this problem.

Historically, the native VLAN on a Cisco device is VLAN ID 1, and as noted, the associated traffic is typically sent and received untagged; note that you can alter this behavior on supported platforms with a `vlan dot1q tag native` configuration statement. You can also modify the native VLAN ID with the interface-level `switchport trunk native vlan <value>` command; but keep in mind that making changes to the native VLAN ID can lead to complexities with signaling protocols and should be approached with caution.

It should also be noted that in IOS, support for implicit tagging has resulted in an exploit that allows traffic to jump across VLAN boundaries without having to first cross a router. See *http://www.sans.org/resources/idfaq/vlan.php* for more information. Although it's unclear whether EX switches are vulnerable to this particular issue, the current best practice is to ensure that the native VLAN ID used for trunk ports is *never* assigned to user access ports, thus preventing a user from crafting traffic that seeks to avail itself of special treatment via the implicit tagging that was really intended for interswitch communications.

You need to watch out for the vendor's defaults here. In Cisco's case, the defaults have trunk ports accepting untagged traffic as part of the native VLAN, and all access ports belong to this native VLAN, a condition that can open you up to this exploit and put you in conflict with Cisco's stated best practice. In contrast, JUNOS trunk ports do not accept untagged traffic without explicit configuration to do so, and all interfaces are placed into a default and untagged VLAN.

The native VLAN concept is also used to support IP telephony devices that do not support Link Layer Discovery Protocol (LLDP). Here the switch port attached to the IP telephone is placed into trunk mode, and a native VLAN is defined to map untagged data traffic received via the phone's PC interface into a data VLAN, while the voice traffic is sent and received using the access port tag, which is in turn associated with voice traffic and specialized CoS treatment. We provide details on EX IP telephony and LLDP support in Chapter 10.

The default VLAN

A default VLAN describes a device's factory-default VLAN configuration and is not part of any LAN standard. Generally, this configuration places all Layer 2 interfaces into a common VLAN, one that is not tagged and which provides out-of-the-box connectivity among all ports.

On a Cisco switch, the default has all interfaces in access mode and associated with VLAN 1, the native VLAN. On EX switches, the factory default places all interfaces into an untagged VLAN called `default`. The JUNOS Layer 2 default configuration does not define a native VLAN, and explicit support for a native VLAN must be added to trunk interfaces with the `native-vlan-id <value>` statement.

Putting it all together

We've covered quite a few concepts getting to this point. It's time to step back and look at the big picture, and Figure 5-3 provides just such an overview of the key concepts we've covered thus far.

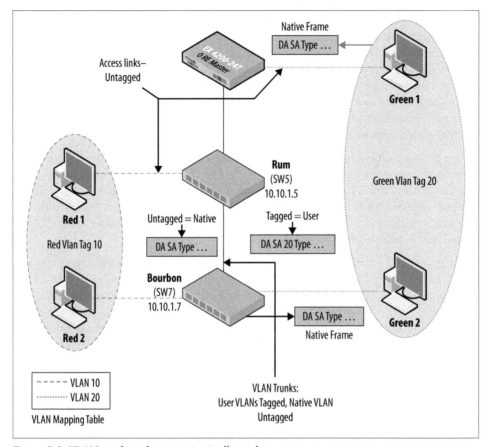

Figure 5-3. VLANs and trunking: putting it all together

Figure 5-3 shows two user communities named **red** and **green**. The two communities are associated with VLAN tags 10 and 20, respectively. The end-user devices are attached to access ports that send and receive untagged traffic, which in turn are associated with VLAN ID 10 or 20 based on user role. The trunk interface is configured with both the **green** and **red** VLANs, and user traffic is explicitly tagged based on the ingress access port's VLAN association.

In this example, a native VLAN has been defined (or is a default) on each trunk port and uses VLAN ID 1. Traffic that is intended for the other end of the link—for example, a Cisco Discovery Protocol (CDP) packet or an EX's LLDP message—is sent on the native VLAN, which means this traffic is not tagged.

The default/native VLAN handling behavior differs between an EX and an IOS device:

Cisco

- The default VLAN is untagged and *is* the native VLAN.
- Trunk ports default to accept all for VLANs in the range 1 to 4,095, which includes the native VLAN, which is also the default VLAN.
- Ports default to auto-negotiate and try to form trunks, or else operate as access ports.
- Access ports that are not reassigned to some specific VLAN default to the native VLAN.

JUNOS

- The default VLAN is named default and is untagged. There is *no* native VLAN by default.
- Ports default to access mode with no auto-negotiation; trunk ports do not support any VLANs, including the default VLAN.

The result of the Cisco default is that stations on "unused" ports can communicate with other "unused" ports using the native VLAN, both locally and remotely via trunk interfaces. Such communications can represent a security risk, especially from known exploits that allow VLAN hopping when trunked over the native VLAN, not to mention the potential for a denial of service (DoS) attack given that control plane protocols often ride over the native VLAN as well. To mitigate these risks, you could redefine the native VLAN, disable native VLAN trunking, or simply configure all unused access ports into a non-trunked VLAN that acts as a type of quarantine.

For JUNOS, explicit configuration is needed to create a native VLAN, and again to define a trunk port, and yet again to support the native VLAN on that trunk port. In addition, unused access ports default to the default VLAN, which, as previously stated, is not automatically supported by trunk interfaces, and as a result bars any remote connectivity for members of the default VLAN.

Generic Attribute Registration Protocol

Generic Attribute Registration Protocol (GARP) is defined in the IEEE 802.1D (Bridging) standard, and is intended to provide a generic framework for bridges to register and deregister attribute values such as VLAN tags or multicast group membership. GARP defines the architecture, rules of operation, state machines, and variables for the registration and deregistration of these attribute values, while the attributes themselves are opaque to GARP, much as a donkey has no opinion about what's in his saddle bags, as long as they're not too heavy.

One application for GARP is a VLAN management protocol called GARP VLAN Registration Protocol (GVRP), which can simplify the administration and management of VLAN membership information. As a network expands and the number of clients and

VLANs increases, VLAN administration becomes complex, and the task of efficiently configuring VLANs on multiple switches becomes increasingly difficult.

GVRP learns VLANs on a particular 802.1Q trunk port, and adds the corresponding VLAN to the trunk port if the learned VLAN is locally defined on that switch. For example, if two trunk-attached switches have a local VLAN 10 defined (the numeric tag must match, not necessarily the local symbolic name), and one of them associates its VLAN 10 with its trunk interface, the remote switch will automatically add its VLAN 10 to that trunk interface as well. The result is that one step is eliminated at an access switch. You still have to define the trunk port, enable GVRP, and define a local VLAN. But with GVRP, you are spared having to map the VLAN to the trunk interface.

Restating this a bit differently, in the current EX implementation, GVRP can automate trunk VLAN membership on one end of a trunk as a result of changes to the local end's VLAN definition (removing the VLAN indicates no interest, and that VLAN is pruned), or as a result of the remote end unbinding a VLAN from its trunk. In either case, that VLAN is automatically removed from the local switch's trunk interface. In addition to reducing administration overhead, GVRP helps to preserve trunk bandwidth by pruning VLANs that are of no interest to the downstream switch, thus eliminating any flooding of broadcast traffic for that VLAN over that trunk.

Cisco and GVRP

Although some IOS versions claim to support GVRP, most in-place boxes continue to use Dynamic Trunking Protocol (DTP) and VLAN Trunking Protocol (VTP) to automate trunk and VLAN administration. DTP first negotiates and ensures compatible trunk parameters (i.e., encapsulation type and native VLAN), and after trunk establishment VTP kicks in (when so configured) to automatically propagate VLAN membership information.

JUNOS devices do not support either DTP or VTP. You should disable both protocols to ensure proper operation in an EX network. This is especially true for DTP, which when run over a trunk attached to a non-IOS device will typically fail to activate the trunk. You should disable DTP on trunks that attach to an EX device using an interface-level `switchport nonegotiate` command.

It should be stressed that, unlike Cisco's VTP, GVRP does *not* support a client/server model, and the current implementation offers no support for authentication or encryption. Also, although VTP actually propagates VLAN definitions from the server to client switches, GVRP currently only handles *trunk membership* for VLANs that must still be locally defined on each switch.

VLAN and Trunking Summary

This section detailed the concept of virtual LANs, the related use of tagging to instantiate, and the use of trunk versus access interfaces in the role of a switch VLAN.

Although this is not a Cisco book, the reality is that many EX switches will be installed alongside some form of IOS-based switching device. This section detailed the Cisco and Juniper defaults for Layer 2 interfaces, and how these relate to interoperability and out-of-the-box operation.

In the next section, we put the theory to good use when we deploy VLANs and trunking in a Cisco/Juniper switching environment.

EX to Catalyst VLAN Integration

Having once again muddled our way through the boring background, it's time to jump on the test bed and put all this information to use. Figure 5-4 details the test topology used in the VLAN trunking integration task.

Before going any further, the following topology characteristics should be duly noted:

- There is a mix of IOS and Juniper EX devices; note that integration is fun.
- The switches and links are chosen to be loop-free and STP is disabled; STP is detailed in the next chapter.
- All switches are running a factory-default configuration with regard to Layer 2 interfaces, with the exception that STP is explicitly disabled.
- The four host machines share an IP subnet that permits full connectivity if allowed by the switched backbone.

Disabling STP in a Switched Network! Are You Crazy?

In a word, yes. That being said, STP serves a useful purpose in a switched Layer 2 network, and failing to run it when it's actually needed will quickly generate a deluge of flooded traffic that will crash your network. Loops are bad: very, very bad. But so too is running with a pair of scissors, usually.

In this case, we have specific pedagogical reasons for doing this. Chief among them is that STP is covered in the *next* chapter, it's not *technically* needed in a physically *loop-free* network, and it has little to do with the topics of VLAN tagging and trunking proper. Given that this is not a production network and the topology is known to be loop-free, and considering the flow of material coverage, it makes perfect sense to disable STP in this context.

In a similar fashion, a highly trained individual wearing appropriate safety attire in a controlled environment that's free of innocents may well have a good reason for wanting to run with a pair of scissors, and by golly, with these constraints, he ought to be free to do so!

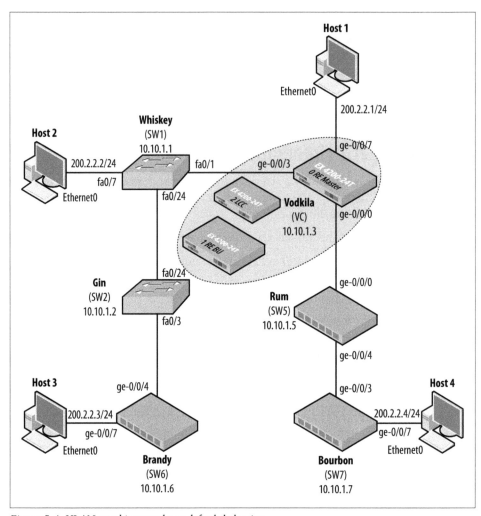

Figure 5-4. *VLAN trunking topology: default behavior*

Default VLAN/Trunking Behavior

Before making changes to add VLANs and trunking, it's a good idea to first get your head around each vendor's out-of-the-box Layer 2 behavior. Note again that aside from disabling STP, both the IOS and JUNOS EX boxes are in their factory-default mode of operation, which is based on Layer 2 switching.

The relevant portions of Whiskey, an IOS-based Catalyst 3550 switch, are shown here:

```
Whiskey#show running-config
Building configuration...
. . .
spanning-tree mode pvst
```

```
 spanning-tree extend system-id
 no spanning-tree vlan 1-4094
 !
 . . .
 interface FastEthernet0/1
  switchport mode dynamic desirable
 !
 . . .
 interface FastEthernet0/7
  switchport mode dynamic desirable
 !
 . . .
 interface FastEthernet0/24
  switchport mode dynamic desirable
 !
 . . .
 end
```

Based on the configuration, it can be said that STP has been disabled on all VLANs, that no specific VLAN definitions exist, and that all interfaces are in their default mode, which means they try to form a trunking relationship (via DTP), or the port falls back to access mode. Knowing that JUNOS does not speak DTP, it's safe to assume that Whiskey's Fa0/1 port is in access mode, which is confirmed:

```
Whiskey#show interfaces Fa0/1 switchport
Name: Fa0/1
Switchport: Enabled
Administrative Mode: dynamic desirable
Operational Mode: static access
Administrative Trunking Encapsulation: negotiate
Operational Trunking Encapsulation: native
Negotiation of Trunking: On
Access Mode VLAN: 1 (default)
Trunking Native Mode VLAN: 1 (default)
Administrative Native VLAN tagging: enabled
Voice VLAN: none
. . .
Trunking VLANs Enabled: ALL
Pruning VLANs Enabled: 2-1001
Capture Mode Disabled
Capture VLANs Allowed: ALL
```

The highlighted portions of the code confirm the port's dynamic/desirable setting, which makes it trunking-capable and always looking to establish a trunk when a compatible device is detected. The port is currently in access mode, which is expected given that the remote end does not speak DTP. As an access port, the current encapsulation is native, indicating that no tagging is in effect on this port.

Some other trunk defaults, such as default support for the native VLAN and allowed VLAN range, are also highlighted. The results are compared to the state of Whiskey's Fa0/24 port, which *is* attached to another IOS device, and therefore *was* able to form a trunk via DTP:

```
Whiskey#show interfaces Fa0/24 switchport
Name: Fa0/24
Switchport: Enabled
Administrative Mode: dynamic desirable
Operational Mode: trunk
Administrative Trunking Encapsulation: negotiate
Operational Trunking Encapsulation: isl
Negotiation of Trunking: On
Access Mode VLAN: 1 (default)
Trunking Native Mode VLAN: 1 (default)
. . .
```

The output confirms that DTP has found a soul mate in Gin, and as a result established an operational trunk. In this case, the default encapsulation is the proprietary ISL rather than the standards-based .1Q, but that's no great surprise, as this is a Cisco-to-Cisco link, and ISL is Cisco's own brand of dog food, so they may as well dine on it. Even though Fa0/24 is an operational trunk, given that only the native VLAN currently exists, the net result is that no VLAN tagging occurs on this port, either.

The default VLAN state for IOS is confirmed with a show vlan brief command; note that the Fa0/24 interface is omitted from the port listing due to its trunk status:

```
Whiskey#show vlan brief
VLAN Name                    Status    Ports
---- -------------------- --------- -------------------------------
1    default              active    Fa0/1, Fa0/2, Fa0/3, Fa0/4
                                    Fa0/5, Fa0/6, Fa0/7, Fa0/8
                                    Fa0/9, Fa0/10, Fa0/11, Fa0/12
                                    Fa0/13, Fa0/14, Fa0/15, Fa0/16
                                    Fa0/17, Fa0/18, Fa0/19, Fa0/20
                                    Fa0/21, Fa0/22, Fa0/23, Fa0/25
                                    Fa0/26, Fa0/27, Fa0/28, Fa0/29
                                    Fa0/30, Fa0/31, Fa0/32, Fa0/33
                                    Fa0/34, Fa0/35, Fa0/36, Fa0/37
                                    Fa0/38, Fa0/39, Fa0/40, Fa0/41
                                    Fa0/42, Fa0/43, Fa0/44, Fa0/45
                                    Fa0/46, Gi0/1, Gi0/2
. . .
```

The display confirms that only the default VLAN is defined, which here is also the native VLAN by virtue of its tag 1 assignment. Note that all non-trunk interfaces are listed as members of the default/native VLAN, to include the link to switch Vodkila given that trunking was not successfully negotiated there. Recall that the ISL trunk between Whiskey and Gin, the only trunk in the current network, has already been confirmed to support VLAN 1 as its native VLAN. The output of a show interfaces trunk command confirms its status:

```
Whiskey#show interfaces trunk
Port        Mode            Encapsulation   Status      Native vlan
Fa0/24      desirable       n-isl           trunking    1

Port        Vlans allowed on trunk
Fa0/24      1-4094
```

```
Port          Vlans allowed and active in management domain
Fa0/24        1

Port          Vlans in spanning tree forwarding state and not pruned
Fa0/24        1
```

As noted in a previous section, the net result here is that there will be *no* VLAN tagging because *only* the native VLAN is in effect, all access ports *belong* to the native VLAN, and all trunk ports consider VLAN 1 as their native VLAN and support trunking of this VLAN. Security issues aside, the net result is a plug-and-play switch that provides connectivity among all devices. In fact, to an external observer, the current operation of the IOS devices mirrors that of a non-VLAN-aware switch. All ports are allowed to communicate, both locally and via the "uplink" (ISL trunk) to remote ports; there is no tagging; and there is a single broadcast domain by virtue of there being only one virtual LAN, the native/default one.

Having seen the IOS default behavior, you do the same for the JUNOS switch Vodkila. First, the relevant bits of its configuration:

```
[edit]
lab@Vodkila# show interfaces
ge-0/0/0 {
    description "To Rum";
    unit 0 {
        family ethernet-switching;
    }
}
ge-0/0/3 {
    description "To Whiskey";
    unit 0 {
        family ethernet-switching;
    }
}
ge-0/0/7 {
    description "To Host 1";
    unit 0 {
        family ethernet-switching;
    }
}
. . .

[edit]
lab@Vodkila# show protocols
. . .
stp {
    disable;
}

[edit]
lab@Vodkila# show vlans

[edit]
lab@Vodkila#
```

Seems fair enough. STP is disabled globally, as was the case with Cisco; the pertinent interfaces are operating in their default mode; and no explicit VLAN configuration has been performed. In the Juniper case, all ports default to access mode. Although not the most direct characteristic to view, adding the extensive switch to a show vlans command does the trick:

```
[edit]
lab@Vodkila# run show vlans extensive
VLAN: default, Created at: Sat Jan  1 17:54:05 2005
Internal index: 6, Admin State: Enabled, Origin: Static
Protocol: Port Mode
Number of interfaces: Tagged 0 (Active = 0), Untagged  3 (Active = 3)
        ge-0/0/0.0*, untagged, access
        ge-0/0/3.0*, untagged, access
        ge-0/0/7.0*, untagged, access
. . .
```

The highlighted portions of this snippet confirm that no trunks are enabled in the default VLAN. The Protocol field identifies whether a VLAN is operating in the default port rather than MAC-based VLAN mode. Also note that all three interfaces associated with the default VLAN are in access mode and are operating in an untagged mode. Although the default VLAN is similar to IOS's native VLAN, a key difference here is that no VLAN ID is assigned to the default VLAN, and the default VLAN is not considered synonymous with a native VLAN, which in JUNOS has to be explicitly configured on each trunk. This all assumes, of course, that a native VLAN is even desired. It's IOS that seems to get wrapped around the whole native VLAN axle, after all.

Note that in the JUNOS case we have no trunk ports, even on the Juniper-to-Juniper link between Vodkila and Rum; this is expected given no support for a dynamic trunking protocol such as DTP in the 9.2 release being tested.

There are significant differences in the two vendors' defaults: for example, native VLAN 1 versus an untagged default VLAN, or IOS having ports that want to be trunks, and in fact actually having an active ISL trunk to its IOS brethren, while the EX switches have *only* access ports. The irony is that outside the respective chassis, no one is the wiser. Although the Cisco switch feels it has a trunk, the association with the native VLAN means there is no tagging, so in effect, the ISL trunk ends up operating as though it were an access port anyway!

In the factory-default integration scenario, you discover the same overall connectivity and forwarding behavior with both operating systems, which is to say that all ports are able to talk to all others, forming a single logical IP subnet (LIS) that permits open communications among all four hosts. The communications bit is confirmed from the viewpoint of Host1:

```
Host1#ping 200.2.2.2

Type escape sequence to abort.
Sending 5, 100-byte ICMP Echos to 200.2.2.2, timeout is 2 seconds:
!!!!!
```

```
Success rate is 100 percent (5/5), round-trip min/avg/max = 4/6/8 ms
Host1#ping 200.2.2.3

Type escape sequence to abort.
Sending 5, 100-byte ICMP Echos to 200.2.2.3, timeout is 2 seconds:
!!!!!
Success rate is 100 percent (5/5), round-trip min/avg/max = 4/14/48 ms
Host1#ping 200.2.2.4

Type escape sequence to abort.
Sending 5, 100-byte ICMP Echos to 200.2.2.4, timeout is 2 seconds:
!!!!!
Success rate is 100 percent (5/5), round-trip min/avg/max = 4/5/8 ms

Host1#show ip arp
Protocol  Address      Age (min)  Hardware Addr   Type   Interface
Internet  200.2.2.1        -      0010.7b3a.02ea  ARPA   Ethernet0
Internet  200.2.2.2        1      0060.7015.9576  ARPA   Ethernet0
Internet  200.2.2.3        1      0010.7b3a.0404  ARPA   Ethernet0
Internet  200.2.2.4        1      0010.7b3a.0399  ARPA   Ethernet0
```

The pings to all three remote hosts succeed, and the Address Resolution Protocol (ARP) cache at Host1 reflects direct reachability at the IP layer, as no gateway (router) is used. A traceroute confirms a single hop, belying the Layer 2 forwarding nature of the current network:

```
Host1#traceroute 200.2.2.2

Type escape sequence to abort.
Tracing the route to 200.2.2.2

  1 200.2.2.2 4 msec 4 msec *
```

Before moving on, clear the switching table of learned entries at Rum. Note that this command clears learned entries for *all* VLANs, which is currently only the default VLAN:

```
[edit]
lab@Rum# run clear ethernet-switching table

[edit]
```

The now-empty table is confirmed:

```
lab@Rum# run show ethernet-switching table
Ethernet-switching table: 1 entries, 0 learned
  VLAN         MAC address      Type        Age Interfaces
  default          *            Flood         - All-members
```

Well, it's empty save for the flood entry used for unknown/multicast MAC addresses, which results in the sending of that frame to all interfaces that are members of that VLAN. Without STP enabled, there is really no reason for an interface to be blocked. The forwarding state is confirmed on both of Rum's interfaces with a show ethernet-switching interfaces command:

```
lab@Rum# run show ethernet-switching interfaces
Interface  State  VLAN members        Blocking
ge-0/0/0.0  up     default             unblocked
ge-0/0/4.0  up     default             unblocked
```

To validate MAC address learning, you quickly generate another set of pings from Host1 to Host4:

```
Host1#ping 200.2.2.4

Type escape sequence to abort.
Sending 5, 100-byte ICMP Echos to 200.2.2.4, timeout is 2 seconds:
!!!!!
Success rate is 100 percent (5/5), round-trip min/avg/max = 4/29/128 ms
Host1#
```

and again display the switching table at Rum:

```
[edit]
lab@Rum# run show ethernet-switching table
Ethernet-switching table: 4 entries, 3 learned
  VLAN            MAC address        Type        Age Interfaces
  default         *                  Flood         - All-members
  default         00:0b:5f:c3:cc:81  Learn        31 ge-0/0/0.0
  default         00:10:7b:3a:02:ea  Learn        29 ge-0/0/0.0
  default         00:10:7b:3a:03:99  Learn        26 ge-0/0/4.0
```

Good; three MAC addresses have been learned, and two appeared as SMACs on Rum's ge-0/0/0 interface. Given the different OUI values, it's a safe bet that 00:0b:5f:c3:cc: 81 belongs to Host1 and 00:10:7b:3a:02:ea belongs to Vodkila. We can make this assumption because the third MAC address, which was learned on Rum's ge-0/0/4 interface, has the same 00:10:7b vendor (OUI) code, and from the topology it's safe to say that Rum does not connect to two Cisco switches. The theory is quickly confirmed with a show chassis mac-addresses command:

```
[edit]
lab@Vodkila# run show chassis mac-addresses

  FPC 0   MAC address information:
    Public base address      00:1f:12:3d:b4:c0
    Public count             64
  FPC 1   MAC address information:
    Public base address      00:1f:12:3d:d2:80
    Public count             64
```

The chassis MAC at Vodkila indeed matches one of the two addresses learned on Rum's ge-0/0/0 interface. Therefore, the other MAC must belong to Host1, which in this lab is actually a Cisco router, thus explaining the use of the 0010.7b OUI, a value that is registered to our good friends over on Tasman Drive (Cisco Systems, Inc.):

```
Host1#show interface ethernet 0 | include Hardware
  Hardware is Lance, address is 0010.7b3a.02ea (bia 0010.7b3a.02ea)
```

Having analyzed the default out-of-the-box Layer 2 behavior for both Cisco and Juniper, and finding it more or less the same despite some differences, you decide in all your pluckiness to proceed with the addition of VLAN and trunk interfaces. After all, what could go wrong?

Define VLANs

In this section, you'll modify the default configuration by adding VLAN definitions and associating access ports with the correct VLAN. Figure 5-5 details the logical partitioning needed in this example.

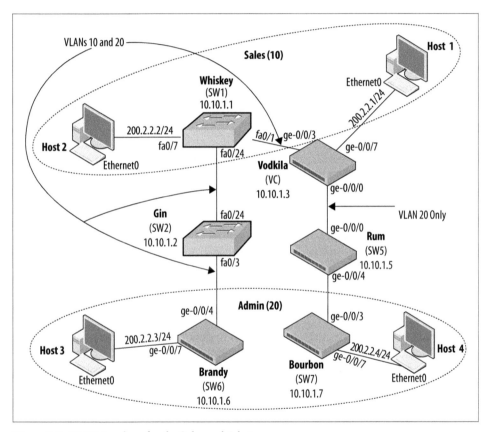

Figure 5-5. VLAN topology for the Sales and Admin groups

Figure 5-5 confirms that two VLANs are needed: one for the Sales group supporting Host1 and Host2, and another for the Admin group with Host3 and Host4. Two user communities, two VLANs: seems simple enough. In this example, IDs 10 and 20 are used for the Sales and Admin groups, respectively. Because of the specifics in this topology, some switches need to support both VLANs on their trunks whereas others need to support one trunked VLAN.

Perhaps noteworthy, if only by virtue of its omission, is the absence of trunking support for Cisco's native VLAN. Given that the native VLAN is primarily an issue for Cisco and its litany of proprietary signaling protocols, there is no real need for a native VLAN here, as the only two IOS devices are directly connected and most of the signaling that happens on the native VLAN is not supported by EX switches, and in this example there is no need to "trunk" CDP or the like over the EX core.

The plan in this example is to omit native VLAN definition at the EX switches, and disallow trunking of the native VLAN on the Cisco switches. In this manner, the native VLAN remains unchanged in the Cisco switches, but now functions as a type of quarantine used by unassigned access ports, which default to VLAN 1. By disabling trunk support, you ensure that "unused" access ports cannot communicate with core switch ports, or other remote "unused" access ports over the native VLAN.

Configure and confirm IOS VLANs and trunking

As usual, we start with the configuration needed for the Cisco switches, in this example Whiskey:

```
Whiskey#configure terminal
Enter configuration commands, one per line.  End with CNTL/Z.
Whiskey(config)#interface fastEthernet 0/7
Whiskey(config-if)# switchport mode access
Whiskey(config-if)#switchport access vlan 10
% Access VLAN does not exist. Creating vlan 10
```

The first set of configuration commands forces the Fa0/7 switch port into access mode, and associates the port with VLAN 10. The software notes that the related VLAN does not exist, and kindly creates one for you:

```
Whiskey(config-if)#exit
Whiskey(config)#vlan 10
Whiskey(config-vlan)#name "Sales_vlan"
```

The next set of commands adds a description to denote the VLAN's association with the sales team. So far, so good; you move on to configure the Admin VLAN. Although there are no local access ports for this VLAN, Whiskey still needs to trunk VLAN 20 to Vodkila to permit communications between Host3 and Host4, and therefore must know of the VLAN so that it can be made active:

```
Whiskey(config-vlan)#exit
Whiskey(config)#vlan 20
Whiskey(config-vlan)#name "Admin_vlan"
```

 Although the vlan commands are entered in configuration mode, subsequent display of the running configuration does *not* reflect the VLAN definitions. The output of a show vlan command confirms that they persist through reloads, however. To see locally defined VLANs in the running configuration you must be in VTP transparent mode.

With both VLANs defined, and the Host1 access port configured, confirm the access and trunk port settings at Whiskey:

```
Whiskey#show interfaces fastEthernet 0/7 switchport
Name: Fa0/7
Switchport: Enabled
Administrative Mode: static access
Operational Mode: static access
Administrative Trunking Encapsulation: negotiate
Operational Trunking Encapsulation: native
Negotiation of Trunking: Off
Access Mode VLAN: 10 (Sales_vlan)
Trunking Native Mode VLAN: 1 (default)
Administrative Native VLAN tagging: enabled
. . .
```

The output confirms that the access port is correctly configured. This leaves only the trunk definitions at Whiskey. Given that both Whiskey and Gin are compatible IOS devices, you opt to leave the defaults in place and initially focus on the shared trunk. The default settings are expected to yield a dynamically negotiated ISL trunk. The switch port settings are displayed with a show interfaces <name> switchport command:

```
Gin#show inter fastEthernet 0/24 switchport | include Trunking
Administrative Trunking Encapsulation: negotiate
Negotiation of Trunking: On
Trunking Native Mode VLAN: 1 (default)
Trunking VLANs Enabled: ALL
```

The results confirm that things pretty much "just worked" on the Cisco-to-Cisco trunk, with the exception that the trunk supports only the native VLAN, which is the default. This condition is quickly remedied with a switchport trunk allowed statement omitting VLAN 1:

```
. . .
Whiskey(config)#interface fa0/24
Whiskey(config-if)#switchport trunk allowed vlan 2-4094
Whiskey(config-if)#^Z
```

The modified trunk status is displayed:

```
Whiskey#show interfaces fastEthernet 0/24 switchport
Name: Fa0/24
Switchport: Enabled
Administrative Mode: dynamic desirable
Operational Mode: trunk
Administrative Trunking Encapsulation: negotiate
Operational Trunking Encapsulation: isl
Negotiation of Trunking: On
Access Mode VLAN: 1 (default)
Trunking Native Mode VLAN: 1 (default)
. . .
Trunking VLANs Enabled: 2-4094
Pruning VLANs Enabled: 2-1001
. . .
```

The `show interfaces` *<name>* trunk command is also useful when confirming IOS trunk status and VLAN settings. Here's some sample output:

```
Whiskey#show interfaces fastEthernet 0/24 trunk

Port        Mode            Encapsulation  Status      Native vlan
Fa0/24      desirable       n-isl          trunking    1

Port        Vlans allowed on trunk
Fa0/24      2-4094

Port        Vlans allowed and active in management domain
Fa0/24      10,20,666

Port        Vlans in spanning tree forwarding state and not pruned
Fa0/24      10,20,666
```

However, you notice that things are not so rosy on Whiskey's "trunk" connection to EX switch Vodkila:

```
Whiskey#show interfaces fastEthernet 0/1 switchport
Name: Fa0/1
Switchport: Enabled
Administrative Mode: dynamic desirable
Operational Mode: static access
Administrative Trunking Encapsulation: negotiate
Operational Trunking Encapsulation: native
Negotiation of Trunking: On
Access Mode VLAN: 1 (default)
. . .
```

The display confirms the expected: Cisco's proprietary DTP/ISL protocols failed to successfully negotiate a trunk link to a JUNOS device. There's no real surprise here, it would seem. To interoperate, you must configure compatible trunking parameters, which in this case require that you manually set trunk mode and use .1Q tagging.

You move on to make the changes needed for an interoperable VLAN trunk between an IOS- and JUNOS-based device:

```
Whiskey#configure terminal
Enter configuration commands, one per line.  End with CNTL/Z.
Whiskey(config)#interface fastEthernet 0/1
Whiskey(config-if)#no switchport
1d22h: %LINK-3-UPDOWN: Interface FastEthernet0/1, changed state to up
1d22h: %LINEPROTO-5-UPDOWN: Line protocol on Interface FastEthernet0/1, changed
state to up
Whiskey(config-if)#switchport
Whiskey(config-if)#switchport mode trunk
1d22h: %LINK-3-UPDOWN: Interface FastEthernet0/1, changed state to up
1d22h: %LINEPROTO-5-UPDOWN: Line protocol on Interface FastEthernet0/1, changed
state to down
1d22h: %LINEPROTO-5-UPDOWN: Line protocol on Interface FastEthernet0/1, changed
state to up
Whiskey(config-if)#switchport trunk encapsulation dot1q
Whiskey(config-if)#switchport nonegotiate
```

```
Whiskey(config-if)#switchport trunk allowed vlan 2-4094
Whiskey(config-if)#^Z
```

In this example, the no switchport command is issued first to clear previous switch port defaults, which in turn ensures no conflicts with the configuration commands that are then entered. Note that the trunk is set to switchport nonegotiate, which "nails up" the trunk's status while also ensuring that no resources are wasted on attempts to dynamically negotiate trunk mode and parameters; disabling negotiation can prevent unpredictable operation that may occur when both ends of a trunk don't support the same negotiation protocols. Note that native VLAN trunking is disabled via the switch port trunk allowed vlan 2-4094 statement, which is in keeping with the deployment plans described previously.

The new trunk settings at Whiskey are confirmed:

```
Whiskey#show interfaces fastEthernet 0/1 switchport
Name: Fa0/1
Switchport: Enabled
Administrative Mode: trunk
Operational Mode: trunk
Administrative Trunking Encapsulation: dot1q
Operational Trunking Encapsulation: dot1q
Negotiation of Trunking: Off
Access Mode VLAN: 1 (default)
Trunking Native Mode VLAN: 1 (default)
Administrative Native VLAN tagging: enabled
. . .
Trunking VLANs Enabled: 2-4094
Pruning VLANs Enabled: 2-1001
. . .
```

Similar changes are now made at switch Gin. Specifically, both the Sales and Admin VLANs are defined, and the Fa0/3 trunk to Brandy is configured for static .1Q operation. In this example, CDP is disabled on Gin's Fa0/3 link using a no cdp enable interface-level configuration statement, because CDP is not supported by the EX, and therefore the protocol would just waste bandwidth here. The VLAN settings are confirmed at Gin:

```
Gin#show vlan brief

VLAN Name                    Status    Ports
---- -------------------- --------- -------------------------------
1    default              active    Fa0/1, Fa0/2, Fa0/4, Fa0/5
                                    Fa0/6, Fa0/7, Fa0/8, Fa0/9
                                    Fa0/10, Fa0/11, Fa0/12, Fa0/13
                                    Fa0/14, Fa0/15, Fa0/16, Fa0/17
                                    Fa0/18, Fa0/19, Fa0/20, Fa0/21
                                    Fa0/22, Fa0/23, Fa0/25, Fa0/26
                                    Fa0/27, Fa0/28, Fa0/29, Fa0/30
                                    Fa0/31, Fa0/32, Fa0/33, Fa0/34
                                    Fa0/35, Fa0/36, Fa0/37, Fa0/38
                                    Fa0/39, Fa0/40, Fa0/41, Fa0/42
                                    Fa0/43, Fa0/44, Fa0/45, Fa0/46
                                    Gi0/1, Gi0/2
```

```
10   Sales_vlan            active
20   Admin_vlan            active
```

You can assume that Gin's ISL and .1Q trunks are configured as described for Whiskey, and are confirmed operational. With the IOS devices configured and confirmed, it's time to direct attention to getting JUNOS and the EX switches up and running in the trunking topology.

JUNOS VLAN and trunk configuration

Before moving on with the VLAN trunking configuration, first assess the state of host communications; recall that previously, with the default configurations, connectivity was confirmed among all four hosts:

```
Host1#ping 200.2.2.2

Type escape sequence to abort.
Sending 5, 100-byte ICMP Echos to 200.2.2.2, timeout is 2 seconds:
.....
Success rate is 0 percent (0/5)
Host1#ping 200.2.2.3

Type escape sequence to abort.
Sending 5, 100-byte ICMP Echos to 200.2.2.3, timeout is 2 seconds:
.....
Success rate is 0 percent (0/5)
Host1#ping 200.2.2.4

Type escape sequence to abort.
Sending 5, 100-byte ICMP Echos to 200.2.2.4, timeout is 2 seconds:
!!!!!
Success rate is 100 percent (5/5), round-trip min/avg/max = 4/4/8 ms
Host1#
```

The limited connectivity matches expectations, given the partial trunking configs now in place. Host1 and Host3 can still communicate because the EX switches they transit are still in the default any-to-any untagged mode. The other pings fail because this traffic leaves Vodkila untagged, and it is therefore assumed to be part of the native VLAN upon receipt at Whiskey. Because native VLAN trunking has been disabled, this traffic is dropped.

A similar fate befalls Host2 to Host3 pings:

```
Host2#ping 200.2.2.3

Type escape sequence to abort.
Sending 5, 100-byte ICMP Echos to 200.2.2.3, timeout is 2 seconds:
.....
Success rate is 0 percent (0/5)
Host2#
```

This traffic belongs to the Sales VLAN at Whiskey, and therefore arrives at Brandy with a VLAN tag of 10. However, Brandy has no such VLAN configured, and no trunk ports, for that matter, so the tagged traffic is dropped at ingress.

The JUNOS configuration begins at switch Vodkila, where the two VLANs are defined at the [edit vlans] hierarchy. The supported options to the vlan statement are displayed with the command-line interface's (CLI's) ? help function:

```
[edit]
lab@Vodkila# edit vlans

[edit vlans]
lab@Vodkila# set ?
Possible completions:
  <vlan-name>           VLAN name
+ apply-groups          Groups from which to inherit configuration data
+ apply-groups-except   Don't inherit configuration data from these groups
> traceoptions          VLAN trace options
[edit vlans]
```

The context-based help implies that the software is looking for a name, so one is provided:

```
lab@Vodkila# set Sales_vlan ?
Possible completions:
  <[Enter]>             Execute this command
+ apply-groups          Groups from which to inherit configuration data
+ apply-groups-except   Don't inherit configuration data from these groups
  description           Text description of the VLAN
> filter                Packet filtering
> interface             Name of interface that uses this VLAN
  l3-interface          Layer 3 interface for this VLAN
  mac-limit             Number of MAC addresses allowed on this VLAN (1..65535)
  mac-table-aging-time  MAC aging time (60..1000000 seconds)
  vlan-id               802.1Q tag (1..4094)
  vlan-range            VLAN range in the form '<vlan-id-low>-<vlan-id-high>'
  |                     Pipe through a command
```

The output confirms that a number of VLAN-specific options are available. In JUNOS you can map a VLAN name to a specific tag, or to a range of tags, using the vlan-id or vlan-range keyword, respectively. In addition, you can apply Layer 2/Layer 3 filters on a VLAN basis with the filter hierarchy. You can also set MAC address limits and aging parameters under the mac-limit and mac-table-aging-time hierarchies. We cover VLAN-based filters in Chapter 8.

The interface keyword is used to link the VLAN to an RVI (a.k.a. an SVI in IOS) that provides Layer 3 services (i.e., routing) for the VLAN. We cover the use of RVIs in Chapter 7.

For now, define the two VLANs needed in this example:

```
[edit vlans]
lab@Vodkila# set Sales_vlan vlan-id 10
```

```
[edit vlans]
lab@Vodkila# set Admin_vlan vlan-id 20
```

The result is confirmed:

```
[edit vlans]
lab@Vodkila# show
Admin_vlan {
    vlan-id 20;
}
Sales_vlan {
    vlan-id 10;
}
```

While in VLAN definition mode, add a single VLAN definition for Admin_vlan to Rum, Brandy, and Bourbon. Only Brandy is shown:

```
[edit]
lab@Brandy# edit vlans

[edit vlans]
lab@Brandy# load merge terminal relative
[Type ^D at a new line to end input]
Admin_vlan {
    vlan-id 20;
}
^D
load complete

[edit vlans]
lab@Brandy#
```

Only the Admin VLAN is defined on these three switches because all three service *only* Admin users. Adding full VLAN definitions to all switches is sometimes considered a good practice because it simplifies adds and changes, especially when using GVRP, as demonstrated in "Troubleshoot a VLAN problem" on page 301.

In this example, we are keeping things minimal, so only the needed VLAN definition is added.

> The symbolic names used to reference VLAN tag values are locally sig-
> nificant. For proper operation, the tag value must match between
> switches; the names used to reference these tags are not communicated
> between switches. For obvious reasons, it's considered a good practice
> to adopt a global naming strategy to avoid confusion as the network
> grows.

Meanwhile, back at Vodkila, the ge-0/0/7 interface is configured to belong to the Sales VLAN. This action removes the interface from the default VLAN, which is where the remaining access ports still reside:

```
[edit vlans]
lab@Vodkila# top
```

```
[edit]
lab@Vodkila# edit interfaces ge-0/0/7

[edit interfaces ge-0/0/7]

lab@Vodkila# set unit 0 family ethernet-switching vlan members Sales_vlan
```

Note that when configuring VLAN assignments to an interface, you can use either a symbolic name, as shown here, or the corresponding numeric tag value. The result is displayed:

```
[edit interfaces ge-0/0/7]
lab@Vodkila# show
description "To Host 1";
unit 0 {
    family ethernet-switching {
        vlan {
            members Sales_vlan;
        }
    }
}
```

With the VLANs and access link defined, all that remains is trunk definition. You start with the ge-0/0/3 link to IOS device Whiskey:

```
[edit interfaces ge-0/0/7]
lab@Vodkila# up

[edit interfaces]
lab@Vodkila# edit ge-0/0/3

[edit interfaces ge-0/0/3]
lab@Vodkila# set unit 0 family ethernet-switching port-mode trunk

[edit interfaces ge-0/0/3]
lab@Vodkila# set unit 0 family ethernet-switching vlan members Sales_vlan

[edit interfaces ge-0/0/3]
lab@Vodkila# set unit 0 family ethernet-switching vlan members Admin_vlan
```

The port-mode statement is used to place the interface's logical unit into trunk mode; recall that the JUNOS default is access mode. The trunk settings are confirmed for the ge-0/0/3 interface:

```
[edit interfaces ge-0/0/3]
lab@Vodkila# show
description "To Whiskey";
unit 0 {
    family ethernet-switching {
        port-mode trunk;
        vlan {
            members [ Sales_vlan Admin_vlan ];
        }
    }
}
```

Note that in this case, there is no native VLAN defined on the trunk interface. This is the JUNOS default, and is fine for our purposes; it's one less VLAN to have to later remove (i.e., *prune* in IOS-speak) from the trunks, after all. The CLI supports the notion of the native VLAN, as well as a wildcard match-all function, via the `native` and `all` keywords. A `set unit 0 family ethernet-switching vlan members all` statement quickly adds all defined VLANs to a trunk interface, which saves a significant degree of work when dealing with large numbers of VLANs.

This example shows the configuration of a single logical interface (`ifl`) that is assigned unit number 0. Note that in the 9.2 release used for this book, EX switches do *not* support the notion of *flexible Ethernet encapsulation*. The result is the mandate that any interface configured to support the `ethernet-switching` family must be assigned a single unit that is assigned number 0. This limitation is demonstrated at Rum:

```
[edit interfaces ge-0/0/0]
lab@Rum# show
unit 0 {
    family ethernet-switching {
        port-mode trunk;
    }
}
unit 1 {
    family inet {
        address 200.0.0.1/24;
    }
}

[edit interfaces ge-0/0/0]
lab@Rum# commit check
[edit interfaces ge-0/0/0]
  'unit 1'
    Only unit 0 is valid for this encapsulation
error: configuration check-out failed
```

To better illustrate this point, the following example is taken from a significantly more expensive MX platform that does offer support for flexible Ethernet services:

```
{master}[edit interfaces ge-5/0/3]
user@mx960# run show chassis hardware | match mx
Chassis                           JN10868FCAFA      MX960
. . .

{master}[edit interfaces ge-5/0/3]
regress@auror# show
vlan-tagging;
encapsulation flexible-ethernet-services;
unit 0 {
    encapsulation vlan-bridge;
    vlan-id-range 3-4094;
    family bridge;
}
unit 1 {
    vlan-id 2;
```

```
    family inet {
        address 200.0.0.1/24;
    }
}

{master}[edit interfaces ge-5/0/3]
user@mx960# commit check
configuration check succeeds
```

Getting back to the task at hand, the configuration of the ge-0/0/0 trunk is similar, except that it specifies only the Admin VLAN:

```
[edit interfaces ge-0/0/0]
lab@Vodkila# show
description "To Rum";
unit 0 {
    family ethernet-switching {
        port-mode trunk;
        vlan {
            members Admin_vlan;
        }
    }
}
```

The configuration is committed at Vodkila, which should provide a working VLAN for hosts in the Sales VLAN. Proper operation is verified at Host1:

```
Host1#ping 200.2.2.2

Type escape sequence to abort.
Sending 5, 100-byte ICMP Echos to 200.2.2.2, timeout is 2 seconds:
.!!!!
Success rate is 80 percent (4/5), round-trip min/avg/max = 4/5/8 ms
```

The pings succeed, thereby confirming proper Sales VLAN operation. This is a good sign. You confirm VLAN status back at Vodkila with a show vlans command:

```
[edit]
lab@Vodkila# run show vlans
Name          Tag    Interfaces
Admin_vlan    20
                     ge-0/0/0.0*, ge-0/0/3.0*
Sales_vlan    10
                     ge-0/0/3.0*, ge-0/0/7.0*
default
                     None
mgmt
                     bme0.32770, me0.0*
```

The output from the show vlans command confirms the presence of the default, Sales, and Admin VLANs. In this configuration, no unused interfaces are enabled for the ethernet-switching family, and as a result the default VLAN has no interface associations; the default VLAN still exists, however, and cannot be deleted. The asterisk (*) in the display denotes that unknown (BUM) traffic is flooded over the interface to facilitate learning and provide ongoing connectivity for traffic that can never be learned—for instance, Open Shortest Path First (OSPF) routing updates, which are multicast-based.

Ethernet switching information is displayed to complete validation of the Sales VLAN:

```
[edit]
lab@Vodkila# run show ethernet-switching interfaces
Interface    State    VLAN members         Blocking
bme0.32770   down     mgmt                 unblocked
ge-0/0/0.0   up       Admin_vlan           unblocked
ge-0/0/3.0   up       Admin_vlan           unblocked
                      Sales_vlan           unblocked
ge-0/0/7.0   up       Sales_vlan           unblocked
me0.0        up       mgmt                 unblocked
```

The display confirms that the ge-0/0/3 interface is operating as a trunk, albeit somewhat indirectly by virtue of the interface having more than *one* VLAN association. In this case, ge-0/0/0 is also a trunk, but due to the single VLAN assignment, you cannot confirm the port's role as access versus trunk interface in this display. For easy access to that information, use the extensive switch to the show vlans command:

```
[edit]
lab@Vodkila# run show vlans extensive Sales_vlan
VLAN: Sales_vlan, Created at: Wed Jan  5 00:09:42 2005
802.1Q Tag: 10, Internal index: 8, Admin State: Enabled, Origin: Static
Protocol: Port Mode
Number of interfaces: Tagged 1 (Active = 1), Untagged  1 (Active = 1)
        ge-0/0/3.0*, tagged, trunk
        ge-0/0/7.0*, untagged, access
```

This output clearly indicates that the Sales VLAN has a single untagged access link and one tagged trunk, which is in accordance with the test topology. The show ethernet-switching table detail command is executed to confirm forwarding and learning behavior for each active VLAN:

```
[edit]
lab@Vodkila# run show ethernet-switching table detail
Ethernet-switching table: 3 entries, 1 learned

  Admin_vlan, *
    Interface(s): ge-0/0/0.0
    Interface(s): ge-0/0/3.0
    Type: Flood
    Nexthop index: 1283

  Sales_vlan, *
    Interface(s): ge-0/0/3.0
    Interface(s): ge-0/0/7.0
    Type: Flood
    Nexthop index: 1293

  Sales_vlan, 00:10:7b:3a:02:ea
    Interface(s): ge-0/0/7.0
    Type: Learn, Age: 0, Learned: 39:08
    Nexthop index: 1285
```

The output confirms that `Admin_vlan` has two flood interfaces, as denoted by the *
character that denotes a wildcard (match-all) entry. As noted previously, the wildcard
entries use a *flood*-style next hop, which confirms that BUM traffic associated with this
VLAN will be flooded out of both the `ge-0/0/3.0` and `ge-0/0/7.0` interfaces; given this
topology, this behavior ensures proper ARP flooding and resulting IP operation. Note
that a flood next hop is always associated with an interface in Ethernet-switching mode,
even when that interface is not forwarding, for example. The lack of a learned MAC
entry on either of these ports implies there is no direct end-station attachment, which
in turn indicates both trunk ports (as was confirmed previously). `Sales_vlan` also shows
two flood next hops, but in addition displays a learned MAC address for its `ge-0/0/7`
access port.

With `Sales_vlan` now confirmed, it's time to shift your attention to the steps needed
to get `Admin_vlan` up and running. Recall that you have already added the `Admin_vlan`
definition to all switches, including `Rum`, `Brandy`, and `Bourbon`:

```
[edit]
lab@Rum# show vlans
Admin_vlan {
    vlan-id 20;
}
```

This means all that's left to get things up and running is trunk definition at all three
switches, and of course, the access port definitions for VLAN 20 at `Brandy` and
`Bourbon`. You begin at `Brandy` with both access and trunk port definition:

```
[edit interfaces]
lab@Brandy# set ge-0/0/4 unit 0 family ethernet-switching port-mode trunk

[edit interfaces]
lab@Brandy# set ge-0/0/7 unit 0 family ethernet-switching vlan members Admin_vlan
```

That wasn't too difficult, was it? You display the results and commit the changes:

```
[edit interfaces]
lab@Brandy# show
ge-0/0/4 {
    unit 0 {
        family ethernet-switching {
            port-mode trunk;
        }
    }
}
ge-0/0/7 {
    unit 0 {
        family ethernet-switching {
            vlan {
                members Admin_vlan;
            }
        }
    }
}
. . . .
```

```
[edit interfaces]
lab@Brandy# commit
```

Next, define both trunks at Rum:

```
[edit interfaces]
lab@Rum# set ge-0/0/0 unit 0 family ethernet-switching port-mode trunk

[edit interfaces]
lab@Rum# set ge-0/0/4 unit 0 family ethernet-switching port-mode trunk

[edit interfaces]
lab@Rum# show
ge-0/0/0 {
    unit 0 {
        family ethernet-switching {
            port-mode trunk;
        }
    }
}
ge-0/0/4 {
    unit 0 {
        family ethernet-switching {
            port-mode trunk;
        }
    }
}
. . .
```

The changes are committed, and you complete the configuration with the access and trunk port definitions at Bourbon:

```
[edit interfaces]
lab@Bourbon# set ge-0/0/3 unit 0 family ethernet-switching port-mode trunk

[edit interfaces]
lab@Bourbon# set ge-0/0/7 unit 0 family ethernet-switching vlan members 20

[edit interfaces]
lab@Bourbon# show
ge-0/0/3 {
    unit 0 {
        family ethernet-switching {
            port-mode trunk;
        }
    }
}
ge-0/0/7 {
    unit 0 {
        family ethernet-switching {
            vlan {
                members 20;
            }
        }
```

```
        }
    }
```

After committing the changes, test connectivity over the nascent Admin_vlan with a Host3 to Host4 ping:

```
Host3#ping 200.2.2.4

Type escape sequence to abort.
Sending 5, 100-byte ICMP Echos to 200.2.2.4, timeout is 2 seconds:
.....
Success rate is 0 percent (0/5)
Host3#
```

The results are less than satisfactory. It seems something is wrong, so you get to do some extra-credit troubleshooting.

Troubleshoot a VLAN problem

You begin fault analysis by displaying the VLAN status at Brandy:

```
[edit interfaces]
lab@Brandy# run show vlans
Name         Tag    Interfaces
Admin_vlan   20
                    ge-0/0/7.0*
default
                    None
```

The results seem pretty clear: only one interface, the ge-0/0/7 access port, is associated with Admin_vlan. But why; what could be wrong here?

The issue stems from the JUNOS default that effectively sets the allowed VLAN range to null on all trunk interfaces, which, note again, is the opposite of Cisco trunk defaults. So, although you have functional trunk interfaces on the EXs, they are currently not associated with any VLANs, so they do you little good. This condition is confirmed at Rum, where both of its VLANs are shown to have *no* interface associations:

```
[edit]
lab@Rum# run show vlans extensive
VLAN: Admin_vlan, Created at: Mon Aug 11 03:41:29 2008
802.1Q Tag: 20, Internal index: 7, Admin State: Enabled, Origin: Static
Protocol: Port Mode
Number of interfaces: Tagged 0 (Active = 0), Untagged  0 (Active = 0)

VLAN: default, Created at: Thu Aug  7 01:36:26 2008
Internal index: 6, Admin State: Enabled, Origin: Static
Protocol: Port Mode
Number of interfaces: Tagged 0 (Active = 0), Untagged  0 (Active = 0)
```

Given that Brandy is isolated from the remaining EX switches by a Cisco cloud, there is not much you can do there, except to add a static VLAN association to Brandy's trunk interface:

```
[edit interfaces ge-0/0/4]
lab@Brandy# set unit 0 family ethernet-switching vlan members 20

[edit interfaces ge-0/0/4]
lab@Brandy# commit
commit complete
```

After the change is committed, the result of a show vlans command confirms that tag 20 is now correctly associated with Brandy's trunk:

```
[edit interfaces ge-0/0/4]
lab@Brandy# run show vlans 20 detail
VLAN: Admin_vlan, 802.1Q Tag: 20, Admin State: Enabled
Number of interfaces: 2 (Active = 2)
  Untagged interfaces: ge-0/0/7.0*
  Tagged interfaces: ge-0/0/4.0*
```

While you are here, you decide to disable LLDP (i.e., 802.1ab) on the links that attach to Cisco switches, because their IOS version does not understand the protocol—so nothing is gained by running them on the Cisco attached links:

```
[edit]
lab@Brandy# show protocols lldp
interface all;
interface ge-0/0/7.0 {
    disable;
}
interface ge-0/0/4.0 {
    disable;
}
```

This completes the configuration of Brandy, and you begin to mull what's needed at the remaining EX switches. Thinking back to the previous discussion of GVMP, and the fact that Vodkila, Rum, and Bourbon share an all-EX path and share the Admin_vlan definition, this seems like a classic case of letting GVRP do some of the work by letting it automatically add interested VLANs to trunk interfaces. The current GVRP implementation is very basic, and pretty much only allows you to enable or disable the protocol and adjust some timers. Given there is little effort involved, you opt to deploy GVRP. Things begin at Vodkila, where you display the GVRP options:

```
[edit protocols gvrp]
lab@Vodkila# set ?
Possible completions:
+ apply-groups          Groups from which to inherit configuration data
+ apply-groups-except   Don't inherit configuration data from these groups
  disable               Disable GVRP
> interface             Configure interface options
  join-timer            Join timer interval (milliseconds)
  leave-timer           Leave timer interval (milliseconds)
  leaveall-timer        LeaveAll timer interval (milliseconds)
```

It seems that there's not much there, as previously noted. Seeing no need to alter the default timer, it seems that the interface hierarchy is where you want to go. You enable GVRP on Vodkila's trunk interface to Rum:

```
[edit protocols gvrp]
lab@Vodkila# set interface ge-0/0/0

[edit protocols gvrp]
lab@Vodkila# commit
fpc0:
. . .
```

After the commit, you confirm that GRVP is running with a few show gvrp commands:

```
[edit protocols gvrp]
lab@Vodkila# run show gvrp
Global GVRP configuration
  GVRP status  : Enabled
  GVRP Timers (ms)
    Join      : 200
    Leave     : 600
    LeaveAll  : 10000
Interface Name     Protocol Status
--------------     ---------------
ge-0/0/0.0         Enabled

[edit protocols gvrp]
lab@Vodkila# run show gvrp statistics
GVRP statistics
  Join Empty received      : 0
  Join In received         : 0
  Empty received           : 0
  Leave In received        : 0
  Leave Empty received     : 0
  LeaveAll received        : 0
  Join Empty transmitted   : 36
  Join In transmitted      : 72
  Empty transmitted        : 0
  Leave In transmitted     : 0
  Leave Empty transmitted  : 0
  LeaveAll transmitted     : 16
```

The display confirms that GVRP is running on the ge-0/0/0 interface. However, statistics indicate that *only* transmit activity is occurring on the trunk. Adding GVRP to Rum should remedy this:

```
[edit]
lab@Rum# set protocols gvrp interface all

[edit]
lab@Rum# commit
. . .
```

After the change, the statistics at Rum confirm that GVRP messages are now being exchanged in both directions on the trunk:

```
[edit]
lab@Rum# run show gvrp statistics
GVRP statistics
  Join Empty received      : 4
```

```
Join In received            : 4
Empty received              : 0
Leave In received           : 0
Leave Empty received        : 0
LeaveAll received           : 1
Join Empty transmitted      : 4
Join In transmitted         : 0
Empty transmitted           : 0
Leave In transmitted        : 0
Leave Empty transmitted     : 0
LeaveAll transmitted        : 2
```

The result is that Rum *automatically* adds the Admin VLAN to its ge-0/0/0 trunk interface:

```
[edit]
lab@Rum# run show vlans
Name            Tag     Interfaces
Admin_vlan      20
                        ge-0/0/0.0*
default
                        None
```

You're definitely getting close, but you're not quite there yet; the lack of GVRP operation on Rum's ge-0/0/4 trunk to access the Bourbon switch means there's no automatic tag association with this interface, given that GVRP cannot determine whether any interested switches are attached to the trunk. Adding GVRP to Bourbon is the final piece to this puzzle:

```
[edit]
lab@Bourbon# set protocols gvrp interface ge-0/0/3

[edit]
lab@Bourbon# commit
```

After the change, the VLAN-to-trunk associations are again displayed at Rum:

```
[edit]
lab@Rum# run show vlans
Name            Tag     Interfaces
Admin_vlan      20
                        ge-0/0/0.0*, ge-0/0/4.0*
default
                        None
```

Awesome! Both of Rum's trunks now have the correct tag association. Things should really turn around for the Admin folks. The ping is repeated at Host3 to confirm:

```
Host3#ping 200.2.2.4

Type escape sequence to abort.
Sending 5, 100-byte ICMP Echos to 200.2.2.4, timeout is 2 seconds:
.!!!!
Success rate is 80 percent (4/5), round-trip min/avg/max = 4/5/8 ms
```

As hoped for, the machines in the Admin VLAN are now able to communicate. For a final check, you confirm that no inter-VLAN communication is possible with a ping attempt from Host3 in the Admin group to Host2 in the Sales group:

```
Host3#ping 200.2.2.2
Type escape sequence to abort.
Sending 5, 100-byte ICMP Echos to 200.2.2.2, timeout is 2 seconds:
.....
Success rate is 0 percent (0/5)
Host3#
```

As expected, the pings fail. For these stations to communicate, they will need to share a VLAN, or a router will need to provide Layer 3 forwarding. We cover the use of an RVI to provide Layer 3 routing between VLANs in Chapter 7.

These results conclude the VLAN and trunking integration verification.

When All Looks Good but Things Don't Work

In some rare instances after making VLAN-related changes, you may find that switching is not working as expected. In many cases, you can issue a **restart ethernet-switching** operational mode command to restart the Layer 2 switching process to regain normal operation without having to reboot the switch:

```
lab@Rum> restart ethernet-switching
Ethernet Switching Process started, pid 665

lab@Rum>
```

This concept is similar to restarting the routing process when something goes awry at Layer 3. The modular nature of JUNOS software makes such incremental recovery possible, and in turn helps to minimize negative impact to the network.

Early adopters of any new product may find that they need to avail themselves of a restart function more often than they would like; as always, a Problem Report should be opened with the Juniper Technical Assistance Center (JTAC) to open a support case and get the defect repaired.

Add Native VLAN Support

Typically, folks are never happy with "good enough." In this case, after you successfully deploy the required VLAN connectivity, you are asked to support CDP between the two Cisco switches. Because CDP is sent and received on the native VLAN, this task requires that you add native VLAN support to the EX switches; you can leave native VLAN trunking disabled at both Cisco devices because the CDP traffic is locally originated and is not switched between trunk interfaces. Figure 5-6 provides details on the current task, along with the topology changes that need to be made.

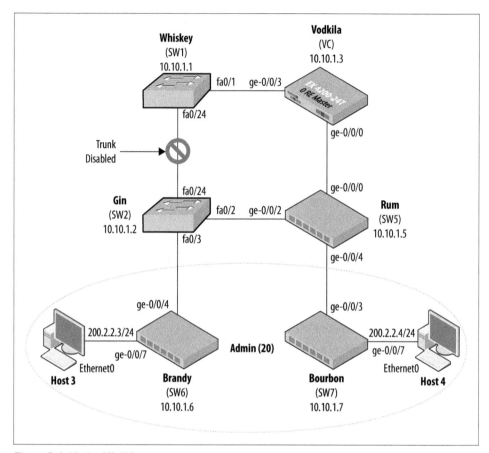

Figure 5-6. Native VLAN support

Figure 5-6 shows a simplified version of the VLAN topology. Note that the ISL trunk between the two Cisco devices is shut down, for reasons that will become clear shortly:

```
Whiskey#configure terminal
Enter configuration commands, one per line.  End with CNTL/Z.
Whiskey(config)#interface fastEthernet 0/24
Whiskey(config-if)#sh
2d04h: %LINK-5-CHANGED: Interface FastEthernet0/24, changed state to
administratively down
. . .
```

The reason for this change will become obvious a bit later. Also, a new link has been added between Gin and Rum. The trunk has been configured on both ends as per the previous EX examples. At the IOS end:

```
Gin(config-if)#switchport
Gin(config-if)# switchport trunk encapsulation dot1q
Gin(config-if)# switchport trunk allowed vlan 2-4094
Gin(config-if)# switchport mode trunk
```

```
Gin(config-if)# switchport nonegotiate
Gin(config-if)#no sh
2d04h: %LINK-3-UPDOWN: Interface FastEthernet0/2, changed state to up
. . .
```

And at the Juniper end:

```
[edit interfaces ge-0/0/2]
lab@Rum# show
unit 0 {
    family ethernet-switching {
        port-mode trunk;
        vlan {
            members 20;
        }
    }
}
```

The modified topology is confirmed with a host ping over the Admin VLAN:

```
Host3#ping 200.2.2.4

Type escape sequence to abort.
Sending 5, 100-byte ICMP Echos to 200.2.2.4, timeout is 2 seconds:
!!!!!
Success rate is 100 percent (5/5), round-trip min/avg/max = 4/4/8 ms
```

You now confirm baseline CDP operation at IOS switch Whiskey:

```
Whiskey#show cdp interface fa0/1
FastEthernet0/1 is up, line protocol is up
  Encapsulation ARPA
  Sending CDP packets every 60 seconds
  Holdtime is 180 seconds
```

The display confirms that CDP is enabled on its fa0/1 trunk, as is also the case with Gin's Fa0/2 trunk (not shown). Yet for some reason CDP has failed to detect the two as neighbors:

```
Whiskey#show cdp neighbors
Capability Codes: R - Router, T - Trans Bridge, B - Source Route Bridge
                  S - Switch, H - Host, I - IGMP, r - Repeater, P - Phone

Device ID       Local Intrfce   Holdtme   Capability  Platform  Port ID
INTERVLAN-SWITCH Fas 0/48        173          S I      WS-C3550- Fas 0/21
```

These results confirm that although CDP is enabled, it's not being trunked through the EX segment of the network. Thinking back on that native VLAN stuff, and the differences in default handling of all things native, a solution presents itself. You must enable native VLAN trunking on EX switches so that they can correctly process the untagged CDP traffic that's being sent by the Cisco devices. By adding VLAN 1 support as a native VLAN to the EX trunks, you enable them to switch this untagged traffic end to end.

You start by defining a native VLAN on the Vodkila and Rum switches. The VLAN does not technically have to be called *native*, but why not? This business is hard enough without additional complexity.

```
lab@Vodkila# show vlans
Admin_vlan {
    vlan-id 20;
}
Sales_vlan {
    vlan-id 10;
}
native {
    vlan-id 1;
}
```

Note again that, unlike IOS, JUNOS does not define a native VLAN in a default config, and by default the native VLAN is not trunked. To add trunking support, you must define the native VLAN on each trunk interface (or use config groups to catch them all):

```
[edit]
lab@Vodkila# show interfaces ge-0/0/0
description "To Rum";
unit 0 {
    family ethernet-switching {
        port-mode trunk;
        vlan {
            members Admin_vlan;
        }
        native-vlan-id 1;
    }
}

[edit]
lab@Vodkila# run show vlans native
Name           Tag     Interfaces
native         1
                       ge-0/0/0.0*, ge-0/0/3.0*
```

The results confirm that native VLAN support is added to both of Vodkila's interfaces. Similar changes are added to Rum, and a few moments later CDP status is again checked at the IOS switches:

```
Whiskey#show cdp neighbors
Capability Codes: R - Router, T - Trans Bridge, B - Source Route Bridge
                  S - Switch, H - Host, I - IGMP, r - Repeater, P - Phone

Device ID      Local Intrfce   Holdtme   Capability  Platform  Port ID
Gin.rtp.ietraining.net
               Fas 0/1         152        S I        WS-C3550- Fas 0/2
INTERVLAN-SWITCH Fas 0/48      155        S I        WS-C3550- Fas 0/21
```

The display confirms that CDP has formed a neighbor relationship between Whiskey and Gin, showing that the EX switches are now correctly handling native VLAN traffic on their trunk interfaces. Congratulations!

Trunking CDP Through a VLAN

In this section, we tried to get CDP running between the Host machines, as they are also IOS-based. This effort was foiled by the Catalyst 3550 switches used in the lab, which seemed bent on intercepting this traffic, even though it had been received on an EX access port and was not sporting a *non-native* VLAN tag. This traffic would pass end to end through an EX-only switch path, but upon ingress to the Cisco switch, the CDP messages were locally intercepted, ostensibly due to the use of a well-known Ethernet multicast address (0100.0ccc.cccc). This condition is shown here for Host4 and Whiskey, which have CDP enabled but are *not* directly connected:

```
Host4(config)#interface ethernet 0
Host4(config-if)#cdp enable
Host4(config-if)#
3d23h: CDP-PA: version 2 packet sent out on
Ethernet0^Z
Host4#
```

And back at Whiskey, we have a new CDP neighbor and an accompanying console warning that reports a (bogus) duplex mismatch, stemming from the belief that the two neighbors are directly connected when in fact they are not:

```
15:50:54: %CDP-4-DUPLEX_MISMATCH: duplex mismatch
discovered on FastEthernet0/1 (not half duplex), with Host4
Ethernet0 (half duplex).

Whiskey#show cdp neighbor
Capability Codes: R - Router, T - Trans Bridge, B -
Source Route Bridge
                S - Switch, H - Host, I - IGMP, r -
Repeater, P - Phone

Device ID      Local Intrfce    Holdtme
Capability  Platform  Port ID
Gin.rtp.ietraining.net
                Fas 0/24        162        S I
WS-C3550- Fas 0/24
INTERVLAN-SWITCH Fas 0/48       178        S I
WS-C3550- Fas 0/21
Host4          Fas 0/1          151           R
2520     Eth 0
```

IOS supports a feature called Layer 2 Protocol Tunneling (L2PT) that allows the tunneling of STP, CDP, and other control plane messages over a Layer 2 switched infrastructure. This feature is not needed on EX switches, as shown in this section. On the EX, you simply tag such traffic at ingress, where it can be trunked normally.

Getting Loopy with It

Before we part ways in this chapter, a word on Layer 2 switching and loops is in order. Recall that STP was disabled in this network, thereby eliminating any automatic protection from loops. Forwarding loops are particularly disastrous in a Layer 2 switched network because there is no Time to Live (TTL) mechanism to break the loop—so things tend to just keep getting worse until the weak link in the chain breaks, which temporarily breaks the loop, allowing a brief respite and return to normalcy. This relief is short-lived and lasts only until the next BUM packet, which, given the chatty nature of most LAN-based protocols, won't be long.

Currently, the test network is loop-free due to the disabling of the ISL trunk link between Whiskey and Gin. Before making any changes, get an idea of the normal baseline for host pings by starting pings from Host3 to Host4, while also monitoring traffic at Brandy's trunk interface:

```
Host3#ping
Protocol [ip]:
Target IP address: 200.2.2.4
Repeat count [5]: 500000
Datagram size [100]:
Timeout in seconds [2]:
Extended commands [n]:
Sweep range of sizes [n]:
Type escape sequence to abort.
Sending 500000, 100-byte ICMP Echos to 200.2.2.4, timeout is 2 seconds:
!!!!!!!!!!!!!!!!!!!!!!!!!!!!!!!!!!!!!!!!!!!!!!!!!!!!!!!!!!!!!!!!!!!!!!!!!
. . .
```

And back at Brandy, here are the effects:

```
                                                   Delay: 8/0/26
    Interface: ge-0/0/4, Enabled, Link is Up
    Encapsulation: Ethernet, Speed: 100mbps
    Traffic statistics:                            Current delta
      Input bytes:         324127 (144048 bps)        [131106]
      Output bytes:        139459 (143472 bps)        [128100]
      Input packets:           3694 (148 pps)           [1089]
      Output packets:          1101 (147 pps)           [1050]
    Error statistics:
      Input errors:             0                          [0]
      Input drops:              0                          [0]
      Input framing errors:     0                          [0]
      Policed discards:         0                          [0]
      L3 incompletes:           0                          [0]
      L2 channel errors:        0                          [0]
    . . .
```

The display confirms a rate of approximately 150 packets per second (pps). Big wow, I know. The pings at Host3 are stopped, and the ISL trunk is reenabled:

```
Whiskey(config)#inter fastEthernet 0/24
Whiskey(config-if)#no sh
Whiskey(config-if)#
2d05h: %LINK-3-UPDOWN: Interface FastEthernet0/24, changed state to up
```

Remember, folks, please don't try this at home. With redundant paths and no STP active, this is a recipe for disaster. That being said, clear the ARP cache at Host3, which should generate a single ARP broadcast at the next ping:

```
Host3#clear arp
```

The ping is started, and you notice that it fails:

```
Host3#ping 200.2.2.4

Type escape sequence to abort.
Sending 5, 100-byte ICMP Echos to 200.2.2.4, timeout is 2 seconds:
.....
Success rate is 0 percent (0/5)
```

The failure stems from loss related to the resulting broadcast storm; note that in this network we have 10 Mbps access links and a 1,000-fold multiplication factor when the packet starts looping through a GE-enabled core. Meanwhile, back at Brandy, things are looking mighty wet given that there is a lot of rain in this "storm":

```
Interface: ge-0/0/4, Enabled, Link is Up
Encapsulation: Ethernet, Speed: 100mbps
Traffic statistics:                                    Current delta
  Input bytes:         994191979 (16288856 bps)           [78485889]
  Output bytes:          9914725 (0 bps)                        [204]
  Input packets:        11759148 (26425 pps)                 [943254]
  Output packets:          81231 (0 pps)                          [3]
Error statistics:
  Input errors:                0                                 [0]
  Input drops:                 0                                 [0]
  Input framing errors:        0                                 [0]
  Policed discards:            0                                 [0]
  L3 incompletes:              0                                 [0]
  L2 channel errors:           0                                 [0]
  L2 mismatch timeouts:        0  Carrier transitions            [0]
  Input framing errors:        0                                 [0]
  Policed discards:            0                                 [0]
. . .
```

A total of 26,425 pps is a lot of traffic to stem from a few *wafer-thin* ARP packets! Note that each new BUM message simply piles on to increase the aggregate rate, until something gives. In this example, the following warnings are noted on the Cisco console:

```
2d05h: %SW_MATM-4-MACFLAP_NOTIF: Host 001d.b50e.9603 in vlan 666 is flapping
between port Fa0/24 and port Fa0/48
2d05h: %SW_MATM-4-MACFLAP_NOTIF: Host 0010.7b3a.0404 in vlan 20 is flapping between
port Fa0/24 and port Fa0/1
. . .
```

The messages are an indication that a MAC address is seen to be moving between ports too often, which is a sure sign of a loop. In this case, as the messages are logged, the forwarding loop is broken, which does save the network from the indefinite pain of a "death by a million floods." This IOS flood-protection feature works to break the loop, but kicks in only after the threshold is crossed and does nothing to prevent or limit the effect of the next loop until the threshold is triggered, again forcing a cessation of traffic on the affected VLAN to break the loop.

Note that during the storm, MAC learning errors are also reported by the EX switches in the output of a show ethernet-switching statistics mac-learning command:

```
[edit]
lab@Brandy# run show ethernet-switching statistics mac-learning
Learning stats: 9579 learn msg rcvd, 1204 error
  Interface        Local pkts      Transit pkts      Error
  ge-0/0/4.0       90              390               381
  ge-0/0/7.0       1               7                 0
```

Here, the errors are incrementing on the trunk interface because the attached host's SMAC keeps oscillating between the access and trunk ports thousands of times per second—a behavior that is deemed unnatural.

EX switches offer a storm-control feature that allows you to block all broadcast and all unknown unicast traffic, or to rate-limit both to some percentage of interface speed. You configure storm-control at the [edit ethernet-switching-options] hierarchy:

```
[edit]
lab@Rum# set ethernet-switching-options ?
Possible completions:
> analyzer               Analyzer options
+ apply-groups           Groups from which to inherit configuration data
+ apply-groups-except    Don't inherit configuration data from these groups
> bpdu-block             Block BPDU on interface (BPDU Protect)
> redundant-trunk-group  Redundant trunk group
> secure-access-port     Access port security options
> static                 Static forwarding entries
> storm-control          Storm control configuration
> traceoptions           Global tracing options for access security
> voip                   Voice-over-IP configuration
[edit]
lab@Rum#
```

Storm control is configured on a per-interface basic. The options are displayed:

```
[edit]
lab@Rum# edit ethernet-switching-options

[edit ethernet-switching-options]
lab@Rum# set storm-control interface ge-0/0/4 ?
Possible completions:
  <[Enter]>             Execute this command
+ apply-groups          Groups from which to inherit configuration data
+ apply-groups-except   Don't inherit configuration data from these groups
  level                 Percentage of link bandwidth (0..100)
  no-broadcast          Disable broadcast storm control
  no-unknown-unicast    Disable unknown unicast storm control
  |                     Pipe through a command
```

The CLI help function confirms options to block all broadcast or all unlearned unicast traffic:

```
edit ethernet-switching-options]
lab@Rum# set storm-control interface ge-0/0/4 ?
Possible completions:
```

```
    <[Enter]>              Execute this command
 +  apply-groups           Groups from which to inherit configuration data
 +  apply-groups-except    Don't inherit configuration data from these groups
    level                  Percentage of link bandwidth (0..100)
    no-broadcast           Disable broadcast storm control
    no-unknown-unicast     Disable unknown unicast storm control
    |                      Pipe through a command
```

In this example, we crank down BUM traffic to 1%, a low rate that still allows normal operation, assuming, of course, that for most networks BUM is the exception rather than the rule:

```
[edit ethernet-switching-options]
lab@Rum# show
storm-control {
    interface ge-0/0/0.0 {
        level 1;
    }
    interface ge-0/0/2.0 {
        level 1;
    }
    interface ge-0/0/4.0 {
        level 1;
    }
}
```

With the limits in place, the error messages cease to be logged on the Cisco devices. Although 1% of a 1,000 Mbps link is still a boatload of traffic, in this case the storm-control feature, applied on a single switch, greatly reduces the severity of the storm, almost tamping it down to the level of a refreshing spring shower. The interface monitor is started while pings are again attempted:

```
Brandy                     Seconds: 201            Time: 01:07:38
                                                   Delay: 0/0/18

Interface: ge-0/0/4, Enabled, Link is Up
Encapsulation: Ethernet, Speed: 100mbps
Traffic statistics:                              Current delta
  Input bytes:            996537526 (178976 bps)   [80831436]
  Output bytes:             9915473 (544 bps)           [952]
  Input packets:          11790031 (329 pps)        [974137]
  Output packets:            81242 (1 pps)              [14]
Error statistics:
  Input errors:                   0                       [0]
. . .
```

And as expected, the display confirms a dramatic decrease in packet count, which now hovers at a paltry 329 pps.

Even though the EX platform's storm-control feature has been demonstrated to work, every effort should be made to keep a Layer 2 network loop-free. Storm control is a safeguard that prevents total meltdown, but the presence of a Layer 2 loop is always disruptive and never beneficial. At the same time, you often need a rich interconnection to ensure no single points of failure. This is where STP comes in. It lets you have as many redundant paths as you can pay for, while ensuring that the end-to-end forwarding topology remains loop-free. When all goes well with the STP plan, storm control is an insurance policy you will never need.

Interested? I bet. And that's great because guess what's covered in the next chapter?

Enjoy the ride.

VLAN Integration Summary

This section examined out-of-the-box VLAN and trunking behavior for IOS and JUNOS, and went on to demonstrate a VLAN trunking scenario in a multivendor environment. Along with configuration, the various operational mode commands used to verify and troubleshoot a VLAN configuration on EX switches were also shown.

The section ended with the purposeful creation of a forwarding loop and demonstration of the mitigating EX storm-control feature, and hopefully motivated the reader for the ensuing STP discussion in the following chapter.

Conclusion

VLANs and trunking are the norm in today's Layer 2 switched networks. VLANs have standards-based support in multivendor networks, offering the ability to perform virtual adds, moves, and changes, and inherently limit BUM traffic to the interested user communities to help improve both performance and security.

This chapter demonstrated best-practice VLAN and trunking concepts for EX switches operating in a multivendor network.

Chapter Review Questions

1. In a default Layer 2 configuration, all EX interfaces:
   a. Belong to the native VLAN, which is untagged
   b. Belong to the native VLAN, which is VLAN 1
   c. Belong to the default VLAN, which is untagged
   d. Belong to the default VLAN, which is VLAN 1

2. Which of the following is true for the native VLAN?
   a. It is not always required
   b. It is defined by default on EX switches because switch-to-switch protocols use it
   c. User traffic can never use the native VLAN
   d. EX switches support native VLAN trunking by default, for interoperability with IOS devices

3. True or false: an access port is mapped to its VLAN through explicit tagging.

4. What is required for communications to occur between stations assigned to different VLANs?

    a. You must run multi-instance STP

    b. A standalone Layer 3 device is required

    c. This is never allowed, which is why the users are in different VLANs to start with

    d. A Layer 3 device, possibly housed within the switch itself, is required

5. What setting is required on an IOS switch to successfully trunk with an EX?

    a. Nothing; the defaults on the IOS device will automatically negotiate compatible settings

    b. You must manually configure ISL trunking because EX switches do not support VTP

    c. You must manually configure .1Q trunking because EX switches do not support DTP

    d. This is possible only through tentative VLAN, which requires no explicit configuration

6. Which are true regarding GVRP on EX switches?

    a. It can automate the addition and removal of a local VLAN to a trunk

    b. It can automate both the propagation of VLAN definitions and trunk binding

    c. It performs propagation of VLAN definitions only

    d. It operates in a client/server model with the ability to authenticate exchanges

    e. Both A and D

7. By default, when you pair two EX switches the link will be in:

    a. Trunk mode

    b. Access mode, untagged

    c. Down until you define either an access or a trunk mode role at both ends

    d. Access mode using native VLAN tagging

8. Based on the following capture, which is true?

```
[edit]
lab@Brandy# run show ethernet-switching interfaces
Interface    State    VLAN members        Blocking
ge-0/0/4.0   up       Admin_vlan          unblocked
                      foo                 unblocked
ge-0/0/7.0   up       Admin_vlan          unblocked
```

    a. Interface ge-0/0/4 is an access link

    b. Interface ge-0/0/7 is a trunk

    c. Interface ge-0/0/4 is a trunk

d. Not enough information is given to determine the interface's trunking status

9. Which is true with regard to MAC learning and VLANs on an EX?

    a. Learning is performed on a per-VLAN basis

    b. You can set MAC limits on a per-VLAN basis

    c. You can define storm-control parameters on a per-VLAN basis

    d. When configuring VLAN properties, you can specify either a symbolic name or its numeric tag

    e. Both A and B

10. Which of the following correctly assigns `Admin_vlan` to the `ge-0/0/0` interface?

```
lab@Rum# show vlans
Admin_vlan {
    vlan-id 20;
}

[edit interfaces ge-0/0/0 unit 0]
```

    a. `set family ethernet-switching vlan members Admin_vlan`

    b. `set family ethernet-switching vlan members 20`

        `[edit interfaces ge-0/0/0]`

    c. `set unit 0 vlan-id 20`

    d. `set unit 0 vlan-id Admin_vlan`

    e. Either A or B

Chapter Review Answers

1. Answer: C. In the default EX configuration, there is no native VLAN. The default VLAN is untagged, and in this manner it is similar but not equal to a native VLAN.

2. Answer: A. A native VLAN is not technically required, depending on what switch-to-switch protocols and what types of equipment you are working with. On the EX, you must specifically define and then add trunk support for a native VLAN when one is desired.

3. Answer: False. Traffic on access links is untagged. The access link is mapped to a VLAN at the `[edit interfaces <name> unit <number> family ethernet-switching]` hierarchy using the `vlan members` statement.

4. Answer: D. A Layer 3 device (a router) is needed, but this function does not have to be external and can be instantiated through the EX switch's RVI.

5. Answer: C. Only C is correct. EX switches do not have any dynamic trunking negotiation protocols, and do not support ISL encapsulation.

6. Answer: A. Currently, GVRP functions only to automate VLAN-to-trunk bindings. It does not operate in a client/server model, nor does it propagate VLAN definitions themselves, and it does not support authentication.

7. Answer: B. With no trunk negotiation protocols, the default is access mode, even on switch-to-switch links. As there is no native VLAN in a default configuration, the access port defaults to the default VLAN, and does not use tagging.

8. Answer: C. Only a trunk interface is allowed to have multiple VLAN associations. Use the output of a show vlans extensive command when in doubt:

```
[edit]
lab@Brandy# run show vlans extensive
VLAN: Admin_vlan, Created at: Sat Aug  9 22:05:10 2008
802.1Q Tag: 20, Internal index: 7, Admin State: Enabled, Origin: Static
Protocol: Port Mode
Number of interfaces: Tagged 1 (Active = 1), Untagged  1 (Active = 1)
      ge-0/0/4.0*, tagged, trunk
      ge-0/0/7.0*, untagged, access
```

9. Answer: E. Each VLAN provides the ability to specify MAC limits and aging. Storm control is done on an interface rather than VLAN basis. When you assign a VLAN to an interface you can use either a name or a tag, but when configuring the VLAN and its properties at the [edit vlans] hierarchy you must use the symbolic name.

10. Answer: E. When you assign VLAN membership to an interface you can use the symbolic name or numeric tag value, and this is done under the Ethernet-switching hierarchy. The VLAN ID at the interface's unit level is used to support a VLAN tagged interface, which is not the same thing as binding an access or trunk link to VLAN membership.

## CHAPTER 6

Spanning Tree Protocol

Switches have been used in networking for years and at times seem like a perfect solution to your problems. However, switches do have some limitations, and that is that they do exactly what they are told. (Don't you wish everybody did?) This can lead to some undesirable results and often can lead to network meltdowns. To avoid this, protocols such as Spanning Tree Protocol (STP) were developed to add some intelligence to the physical switching architecture—in other words, "Do what you are told, unless what I am telling you is incorrect."

The following topics are examined in this chapter:

- Switches and loop issues
- STP
- Rapid Spanning Tree Protocol (RSTP)
- Configuring and monitoring the spanning tree
- Multiple Spanning Tree Protocol (MSTP)
- Redundant Trunk Groups (RTGs)

Feeling a Little Loopy

Before diving into the problem, let's recap how a switch works:

- It uses Media Access Control (MAC) addresses to forward traffic toward its destinations.
- It "learns" station locations based on the Source MAC (SMAC) addresses.
- If a destination is not known, it will send the frame out all ports in that broadcast domain.

With these fundamental concepts in mind, consider the network in Figure 6-1. At first glance, it may seem like a very redundant, well-designed network. Let's examine what happens when unicast, multicast, and broadcast frames are sent from Station A to Station C.

First, the unicast packet is sent from A to C, both switches on the LAN. They will perform a Destination MAC (DMAC) lookup for Station C and determine that the outgoing port is Segment 2. In this case, it does not matter whether Station C had been learned yet, as there is only one possible forwarding direction onto Segment 2, since a frame will not be forwarded from the port that was received. The end result of this is that Station C receives two copies of the frame. This violates the duplication definition that a LAN should never receive multiple frames to a destination for a single frame transmission. Also, duplicate frames could place quite a burden on the upper layer protocols (perhaps breaking many), as well as increasing the network bandwidth on the LAN by a factor of two.

If things are not ominous enough yet, consider what happens if a multicast or broadcast frame is sent from Station A. Once again, both switches receive that frame and forward it on toward Segment 2. However, both switches will receive the broadcast frame from each other on Segment 2, and will forward it on to Segment 1. Then they will see the frame on Segment 1 and forward it back to Segment 2, and so on, and so forth, and a loop will be created. This will continue forever or until the switches' resources are consumed and they stop working properly.

Even worse, when the frame is circulating, the switches have to update their MAC table, and it appears that a station has moved. For example, Switch 1 will see Segment 1 as the source for Station A, but when receiving the broadcast frame on Segment 2, it will update its table to reflect the "move" even though the actual physical station has not moved. Then, moments later, Switch 1 will see a frame with Station A as a source on Segment 1, so it will update its MAC table. Then it will receive a frame on Segment 2 with the source of Station A—oh, the confusion! In other words, the switches see either Station A on two different segments at the same time, or two Station As, which causes enough uncertainty for the switches to want to stop working and head to the bar.

Stupid Is As Stupid Does

You may just say to yourself, "Well, I would *never* configure my LAN for such a harmful situation. I am the best network engineer in the world!" Before you start handing out blue ribbons to yourself (hey, you can't lose if nobody else is competing), consider these fundamental network rules:

- Two of anything is usually better than one (increased capacity).
- People will make mistakes, and they may connect the wrong physical port.
- Redundancy is good, and you need to design around the possibility of element failure.

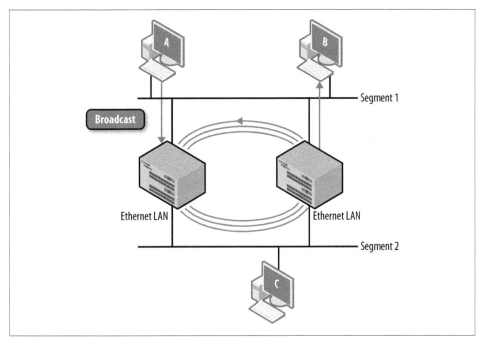

Figure 6-1. Sample LAN topology

Based on these rules, it is quite easy to create a loop in a network topology. So, how do you avoid this? Ideally, you will simply never configure the network to perform a loop, but with redundancy requirements this is often not possible. Manually disabling/enabling certain ports could be a solution, but not a very scalable one in large networks. The answer? A simple protocol: Spanning Tree Protocol.

Loop Issue Summary

Loops can be an issue in a LAN environment based on even simple topologies. Loops can cause:

- Unicast frame duplication
- Multicast and broadcast storms
- Flapping of switching tables

Loops are sometimes unavoidable due to a network's physical topology and requirements, so STP was created to intelligently avoid loops. We examine this protocol in detail in the next section.

Spanning Tree Protocol

The loop issue was detected early on and many switch vendors developed their own loop detection protocols. In 1985, the IEEE set out to develop switching standards that could be used on any vendor, and part of the standard (IEEE 802.1D) included STP.

STP Basics

So, how does STP prevent loops? In simple terms it creates a tree topology with only one available path between the root of a tree and a leaf (Figure 6-2). As the physical switched network evolves, STP reconverges and rebuilds the tree. It does this by electing a single "root" bridge and building paths with this bridge as the starting/ending point.

Although a single root bridge will be elected, the bridge can change over time due to topology or bridge alterations.

The root bridge itself does not prevent loops in our two-switch topology (as in Figure 6-1). The issue is that each bridge forwards a duplicate frame onto the same LAN segment. An easy fix is to simply elect a single switch to forward the frame onto the segment. In STP terminology, a single switch should be elected to forward from the root bridge to any single branch. This ensures that there is only a single path from one end station to another end station. The switch that is responsible for this is called the *designated bridge* (Figure 6-3).

In this chapter, we use the words *switch* and *bridge* interchangeably. Historically, a bridge has two ports and is used to interconnect LANs, usually in software-based devices. A switch is really a multiport bridge and is usually a hardware-based device. Because there is not much of a market for a two-port switch, bridges are no longer manufactured.

Since the designated bridge connects many paths between end stations, there can be many designated bridges. These are decided on a link-by-link basis. So, a switch can be a designated bridge for one link, many links, or no links at all. Also, the root bridge is always the designated bridge for all of its directly connected ports.

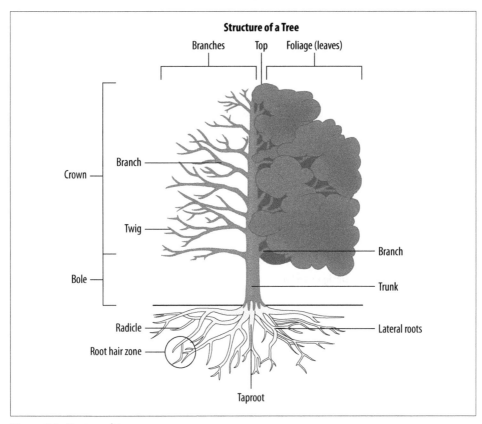

Figure 6-2. Roots and trees

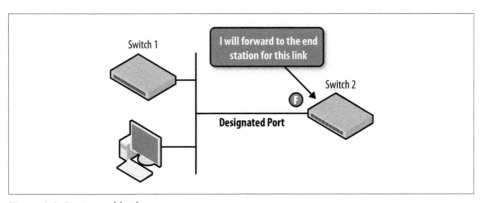

Figure 6-3. Designated bridge

The designated bridge can have three types of ports:

Designated port
> This is the port chosen for forwarding from the root bridge to a local segment (Figure 6-4).

Root port
> This is the port closest to the root bridge. Traffic is forwarded from the root port to the designated port and vice versa.

Inactive port
> If a port is not the designated or root port, it is not used for forwarding, so it is inactive.

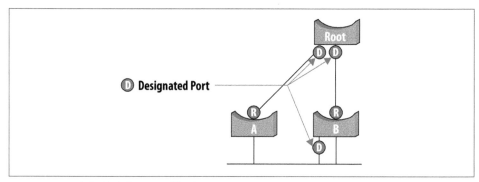

Figure 6-4. Link-by-link designated bridges

For the entire network to learn about the tree topology, messages called *Bridge Protocol Data Units* (BPDUs) are sent around the network, describing each bridge/switch. To identify each switch, a unique value called the *bridge identifier* is used. The original specification states that the bridge identifier is a 64-bit field consisting of a 16-bit bridge priority and a 48-bit MAC address. Since the switch will have many MAC addresses for each interface, switches usually choose the lowest numbered port in the broadcast domain as the bridge identifier. Since MAC addresses are simply identifiers, and do not carry any numerical significance, the 16-bit priority plays a very important role. The priority field is what gives the network administrator the control needed for spanning tree control. The bridge priority default value is in the center of the range, at 32,768.

Bridge Priority and STP

In 2004, the bridge priority was redefined to be four bits of priority and 12 bits of system ID. This system ID is used for 802.1Q MSTP, discussed later in this chapter. To be compatible with older implementations, the priority component is still considered, for management purposes, to be a 16-bit value. However, the values are restricted to the four most significant bits. So, in practice the bridge priority must be set in the range of 0 to 61,140 in increments of 4,096.

Although MAC address OUIs are not assigned in any particular order, the manufacturer-assigned field in the MAC address is generally assigned in numerical order as the switch is made. As a result, older switches usually have an advantage, due to default behaviors and STP tiebreaking rules. This is why the priority field plays such an important role, as an older, less-powerful switch may continue as the root bridge using default parameters.

Each port on the bridge is also assigned a 16-bit port identifier, which consists of an 8-bit port priority field and an 8-bit port number (see the sidebar "Port Numbers in STP" on page 326). The port priority is a value in the range of 0 to 255 with a default value of 128. The port number identifies the port on the switch, and can be any value depending on the vendor.

 Cisco switches begin counting at a value of 1, and Juniper begins at a value of 512.

Each link also receives a corresponding cost. In the original STP standard, the equation was very simple:

Link cost = 1,000 / Data rate (Mbps)

However, the IEEE 802.1 committee soon realized that LAN speeds were steadily increasing past 1 Gbps. Therefore, in 1998 they recommended a non-linear relationship between data rates and link costs while maintaining backward compatibility with the original specification. Table 6-1 lists STP link costs.

Table 6-1. STP link costs

Link speed (Mbps)	STP cost
4	250
10	100
16	62
45	39
100	19
155	14
200	12
622	6
1,000	4
2,000	3
10,000	2

These costs were updated yet again in 2004 when Rapid Spanning Tree Protocol was defined (we discuss RSTP later in this chapter).

Port Numbers in STP

In the original specification, the port number field was set to 8 bits. This seemed like an ample size, since most switches were two, four, or eight ports. The emergence of switched LANs, microsegmentation, and freaking big switches changed all of that, and the IEEE specification in 1998 set the port number field to 12 bits. However, to maintain backward compatibility with older implementations' management, the priority is still considered 8 bits but with restricted values; only the most significant 4 bits can be set.

In practical terms, this means priority values must be set in increments of 16 (0, 16, 32, 64, 128, 144, 160, 176, 192, 208, 224, 240).

Calculating and Maintaining the Spanning Tree

Now that the exciting definitions are out of the way, let's get down to the dirty business of how STP actually works. When STP is configured, each switch must perform three very simple functions:

1. Determine the root bridge.
2. Determine the designated bridges and ports for each link.
3. Maintain and react to topology changes.

All of the functions are done in parallel and independently on each switch. In step 1, determining the root bridge, an election occurs (Figure 6-5). The election is very simple: whichever bridge has the lowest bridge identifier becomes the root bridge. This root bridge maintains its monarchist status until another bridge with a lower identifier enters the fray, or the current root bridge is removed (or somebody unplugs it to use the outlet for a phone charger). You control the root bridge and root responsibilities by setting the bridge priority.

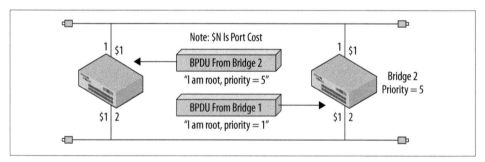

Figure 6-5. Root bridge election

After the election process, there is a root bridge for every link in the broadcast domain. By definition, the root bridge is the designated bridge for every attached link. For other switches, the designated bridge will be elected based on the cost back toward the root. The lowest cost (based on interface speed) will be the designated bridge. If two switches have the same cost back to the root bridge, the bridge with the lowest bridge identifier will become the bridge with the designated port, and thus will become the designated bridge (Figure 6-6).

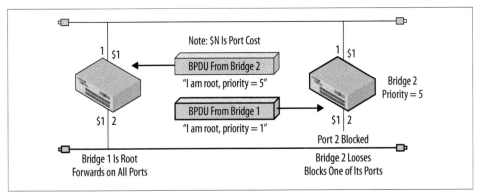

Figure 6-6. Election of designated ports

If a bridge has multiple ports on the same LAN segment (Figure 6-7), the port with the lowest port identifier becomes the designated port.

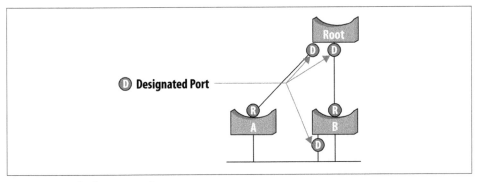

Figure 6-7. Multiple ports on the same segment

Once the tree has been created, it needs to be maintained to keep the topology loop-free. This is done by having the root bridge and all designated bridges advertise the current understanding of the spanning tree via configuration messages encoded as BPDUs. Every bridge listens for the configuration messages and compares the information it receives with its own internal information. If the bridge's internal information indicates that it is a better choice for the root bridge or designated bridge, it takes action to change the topology. In addition, these configuration messages act as keepalives for a segment. If no messages are received on a certain port, it could indicate a switch failure and cause that port to become a forwarding port and reconverge to a new topology.

In steady state operation, the root bridge is the source of the configuration messages. At every "Hello" instance (the default is two seconds), the root bridge will send out a configuration message indicating that it is the root (the root identifier will equal the bridge identifier) and the path cost is zero. All bridges sharing links with the root bridge receive the message, and the designated bridges create a new message based on the information from the root bridge.

 BPDUs are never forwarded through a bridge, but rather are re-created and updated based on received information.

The designated bridges, in turn, transmit this message out of their designated ports. Then the designated bridge for the next tier sends a message from its designated ports. This process continues until there are no more designated bridges and the tree has reached its final leaf.

During this process, each bridge compares the received information to its own internal information. In particular, each bridge will compare:

- The root identifier in the received message to its own root identifier. If the bridge has a better root identifier than that in the received configuration message, the bridge will attempt to become the root by issuing a topology change configuration message.

- The path cost received to the path cost available through any other ports. If the bridge believes it has a better path cost than the designated bridge, it will attempt to become the designated bridge by issuing a topology change configuration message.

Bridge Protocol Data Units

BPDUs are the messages that the switches send to learn about each other, create the spanning tree, and send configuration changes. There are two BPDUs: the configuration message and the topology change message. BPDUs have changed over time (see "Rapid Spanning Tree Protocol" on page 338), but the *original* fields are as follows (Figure 6-8).

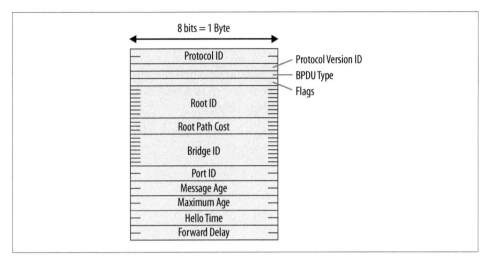

8 bits = 1 Byte

Protocol ID — Protocol Version ID
— BPDU Type
— Flags
Root ID
Root Path Cost
Bridge ID
Port ID
Message Age
Maximum Age
Hello Time
Forward Delay

Figure 6-8. Original STP BPDU

Protocol Identifier
> Identifies STP, and is always set to 00000000. Generic Attribute Registration Protocol (GARP) uses the same BPDU format with a different protocol identifier.

Protocol Version
> Gives the version of the protocols, and is set to 00000000.

BPDU Type
> Indicates the type of message (0 = configuration and 1 = topology change).

Flags
> Specify bit 1 for the topology change indicator and bit 8 for the topology change acknowledgment.

Root Identifier
> Specifies the current root bridge.

Root Path Cost
> Specifies the total cost back to the root bridge.

Bridge Identifier
> Provides the sender's bridge ID.

Port Identifier
> Provides the sending port's ID.

Message Age
> Indicates the amount of time that has elapsed since the root sent the configuration message on which the current configuration message is based.

Max Age
> Indicates when the current configuration message should be deleted.

Hello Time
> Indicates the time between root bridge configuration messages.

Forward Delay
> Indicates the length of time bridges should wait before transitioning to a new state after a topology change. This timer is also used to age-out MAC addresses more quickly than normal during a topology change.

The topology change BPDU shown in Figure 6-9 is much more basic and contains only the following fields:

* Protocol Identifier (set to 0)
* Protocol Version (set to 0)
* BPDU Type (set to 1)

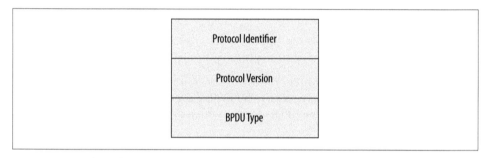

Figure 6-9. Topology change BPDU

BPDUs are encapsulated using Logical Link Control (LLC) Type 1 with a Destination Service Access Point (DSAP) and a Source Service Access Point (SSAP) of 0x42. The MAC source is the port in which the frame is transmitted, and the DMAC address is a well-known link local multicast address of 01-80-C2-00-00-00. This means a bridge does not need to know the unicast address of other bridges to run STP. Due to the link local scope, bridges running STP will not forward BPDUs beyond the significant link.

 Although STP uses a link local DMAC address, it is possible for devices not running STP to forward the BPDU as a normal multicast address.

BPDU Learning and Port States

When BPDUs are sent out from each switch, each port can be in one of five possible states: disabled, listening, learning, forwarding, or blocking (Figure 6-10).

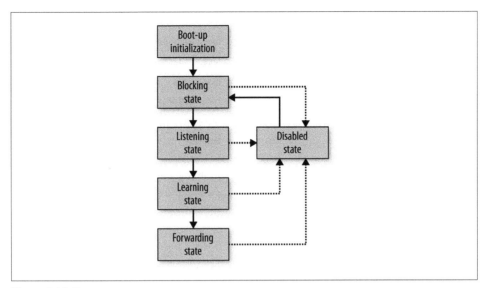

Figure 6-10. Port states

The disabled state is just as described—it is not involved in STP. A port is in this state due to a link failure, or if intentionally disabled by the administrator. A port in this state does not send or receive any BPDUs.

The first state in which the port will normally initiate is called the *blocking* state. This is a port that is neither the designated nor the root port. The blocking state will not forward any data plans or send any BPDUs. However, it will listen for BPDUs to determine whether it should be active in the spanning tree, as well as to make sure the neighbor switch is still "alive."

The first state a switch usually transitions to is the *listening* state. In this state, the port will not forward traffic but will continue to receive BPDUs. The switch may also send out BPDUs on this port to determine whether it should be the root or the designated bridge. This is a transient state, as the port will then transition to either the blocking state or the learning state.

The learning state indicates that the port is preparing to forward traffic. Since a new switch will have an empty address table, it waits a certain amount of time, called the *forwarding delay*, to build its address table. This is the same time a system will wait in the listening state.

After the switch's forwarding delay timer has expired, it is ready to forward data frames. This state is called the *forwarding* state and is the steady state for a port that is part of the active spanning tree. Table 6-2 shows the activity in each state.

Table 6-2. Behavior of switches in different states

State	Forward data	Receive BPDU	Send BPDU	Important timers
Disabled	No	No	No	None
Blocking	No	Yes	No	Hello
Listening	No	Yes	Yes	Forwarding delay
Learning	No	Yes	Yes	Forwarding delay
Forwarding	Yes	Yes	Yes	Max age

Once the spanning tree is established, topology changes should not regularly occur. The most common reasons for a topology change include the following:

- A new bridge or port is added to the network.
- An active link in the spanning tree fails.
- A root or designated bridge fails.
- A new link is added into the network.
- The administrator adjusts the bridge priorities, port priorities, or link costs that affect the current spanning tree.

Since a topology change can cause a temporary disruption in service, do not add any new switches or ports or change configurations until you reach a planned maintenance window. A failure can never be planned, though, so STP has timers in place to avoid transient loops as well as explicit topology change detection and notification. When a topology change is detected (due to better configuration data than what is available in the internal database), the switch will transmit a topology change message from its root port toward its designated bridge. It sends this message until an explicit acknowledgment from the designated bridge is received. This acknowledgment is determined by the `tc ac` bit being set in the topology change BPDU. The designated bridge then sends a topology change BPDU to its root port and waits for an acknowledgment. This process continues all the way up the tree until it reaches the root bridge. The root bridge then sets the topology change message in all of its configuration messages for some period of time so that all bridges are informed of the change (Figure 6-11). When this bit is set, the bridges will use a shorter address-aging time for their address tables.

Protocol Timers

A few important timers need to be adjusted for STP to operate efficiently. Some of these we have touched upon, but not fully examined. These timers affect convergence and are configurable as needed. The specification has recommended default values, however, which usually exhibit the needed behavior.

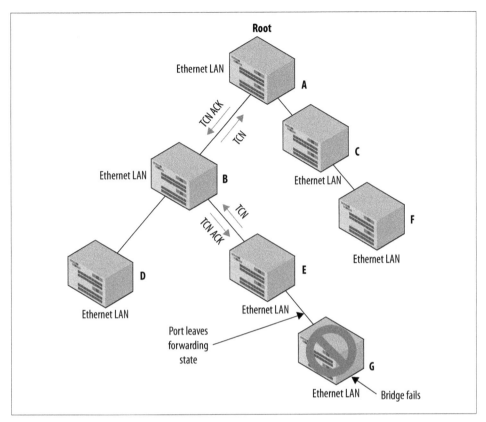

Figure 6-11. Topology change

Table age

Recall that switches are based on SMAC and DMAC addresses and receiving and sending port numbers. Since there could be thousands of stations on the segment and new ones arriving and leaving the network, it is important to keep the table as small and efficient as possible. Unlike IP routing, which can accomplish this efficiency via route summarization, MAC addresses cannot be summarized into blocks because a station can choose to be wired to just about any switch port. In order to increase table efficiency, switches use an aging rule, removing entries that have not been heard from for some time, called the *table aging* time. As in any relationship, a delicate balance of time and removal must be maintained. If the aging time is too short, the table will purge quickly and constant flooding will be required for learning (nag, nag). If the timer is too long, the table size may approach maximum limits due to stations that have not transmitted for hours (ignore, ignore). The most common aging time is 5 minutes, but this value is adjustable from 10 seconds up to several days.

Hello time

The hello time is the time between configuration messages sent by the root bridge, which in turn correlates to the time between configuration messages established by the designated bridges. Faster hello times will speed up topology changes, but will also increase messaging. The default hello time is 2 seconds, with a configurable range from 1 to 10 seconds.

Message age

Since there may be many switches in the network, propagation of new topology changes takes time. During this time, it is entirely possible that a new topology change could occur. This means before this new switch receives the information, it could be using stale information and making the incorrect spanning tree decision, forming transient loops. In order to prevent this eventuality, each configuration message carries a message age that starts from zero at the root bridge and increments at each designated bridge. When a bridge accepts the configuration data, the timer starts, and the message is aged out if no refresh message is received within the maximum age (the default is 20 seconds).

Forwarding delay

The forwarding delay is the time a switch waits to transition to a new state, as well as the MAC address aging time during a topology change. It's necessary because sometimes stations appear to move based on a spanning tree topology change.

In Figure 6-12, if Link 1 fails, Link 2 will be activated and added to the spanning tree. From the view of Switch 3, the stations appear to move from Port 1 to Port 2. In normal operations, the switch would have to wait five minutes to age-out the old entries or wait to receive data from each station and enable correct data flow. To remedy this, when switches receive a topology change they enter a "fast aging" time indicated by the forwarding delay, which is usually set to 15 seconds but can be in a range of 4 to 30 seconds. As previously mentioned, the forwarding delay takes on a dual role, as it is also the time that the switch is waiting to transition to each state (listening to learning, and learning to forwarding).

Putting the Theory Together

We have tossed a lot of concepts at you in this chapter, so let's look at an example and see how a spanning tree is built. Figure 6-13 shows a sample network. The bridge IDs, port IDs, and link costs are all specified. Please note that Figure 6-13 is for learning purposes only and is not necessarily what you will see in a real network. Also, notice that Segment A is attached via a hub that is not shown in the diagram.

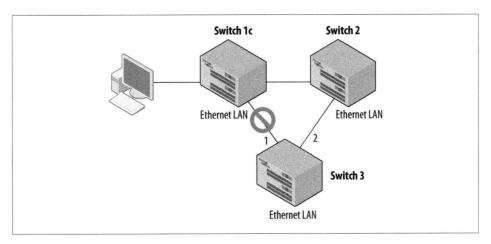

Figure 6-12. The incredible moving station

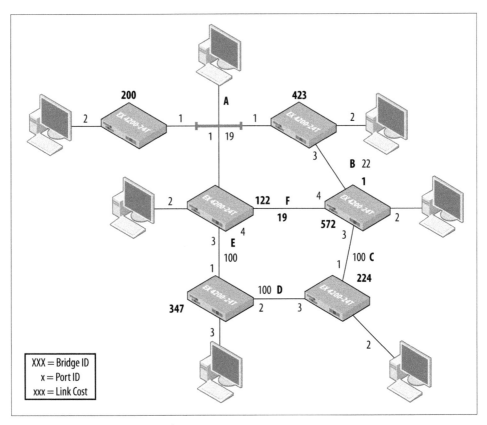

Figure 6-13. Spanning tree example

The first step is to determine the root bridge. In this example, the root bridge is Bridge 122, as it has the lowest bridge ID. By definition, all ports on that bridge will be designated ports, and the root ports of Bridge 200 and Bridge 423 will be on Port 1 Link A, as that is the lowest-cost path back to the root bridge.

Next, determine the root port—that is, the lowest cost back to the root bridge for each switch:

- Bridge 200 is Port 1 with a cost of 19.
- Bridge 423 is Port 1 with a cost of 19.
- Bridge 347 is Port 1 with a cost of 100.
- Bridge 572 is Port 4 with a cost of 19.
- Bridge 347 is Port 1 with a cost of 100.
- Bridge 224 is Port 1 with a cost of 119.

Lastly, calculate the designated bridge for Segment B, Segment C, and Segment D, as all other segments are connected to the root bridge.

 Recall that the root bridge is always the designated bridge for its local ports.

Segment B has two possible options, and both bridges have a cost of 19 back to the root bridge. In this case, the bridge with the lowest bridge ID, 423, wins. Bridge 572 is the designated bridge for Segment C because it has a lower cost. The same goes for Bridge 347, with a lower cost for Segment D. All other bridge ports that are not root or designated bridges will transition to the blocking state. Figure 6-14 shows the resulting topology. We highly suggest that you draw out Figure 6-13 and see whether you get the same spanning tree result.

STP Issues

STP seems like the perfect little protocol, right? It was created under the assumption that topology changes do not happen often, and as a result, it was designed for stability and loop-free reliability as opposed to speed.

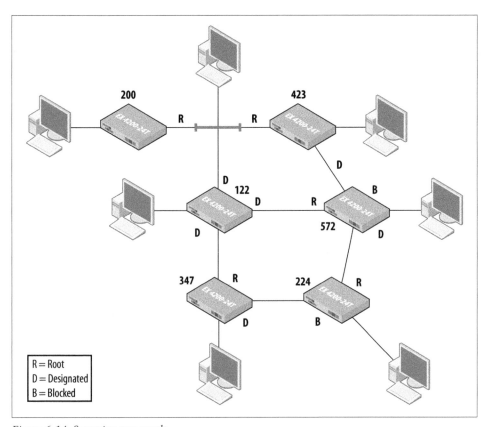

Figure 6-14. Spanning tree result

However, it does have limitations, some of which include the following:

- Sometimes suboptimal paths are taken. STP chooses all of its paths based on link speed, with no influence on the size or reliability of the link or switch.
- Convergence can be slow. Sometimes it can take up to 30 seconds for the spanning tree to reconverge after a change.
- There is no per-link protocol. As a result, "hellos" essentially originate from the root bridge and propagate to each designated bridge, which can slow down the detection of link failure.
- The transition to the forwarding state can be slow, based on the forwarding delay timer.

These issues are addressed in RSTP (IEEE 802.1w), and we discuss them shortly. Today most switches follow the 801.1D-2004 specification, which sets RSTP as the default mode for switches.

STP Summary

Spanning Tree Protocol was made to solve the "loop" issues in bridges by creating a single path between hosts with the reference point of the root bridge. It does some simple calculations based on user-controllable parameters, which creates a single path from one host to another to avoid any loops in the topology. However, STP was not written for fast convergence, a major drawback for modern-day networks. Rapid Spanning Tree Protocol, examined next, was created to fix those deficiencies.

Rapid Spanning Tree Protocol

In order to fix some of the deficiencies in the original STP, the IEEE defined 802.1w, Rapid Spanning Tree Protocol, later described in the 802.1D-2004 specification.

STP converges very slowly, after a topological change. When using default values for the max age (20 sec) and the forward delay (15 sec) parameters, it takes 50 seconds for the bridge to finally converge:

convergence delay = 2 × forward delay + max age

This setup was OK for early networks, but like the Commodore 64, it became unacceptable as technology advanced. RSTP vastly improves that time, often providing subsecond transitions.

Essentially, the major difference is a centralized versus distributed model. In STP, the root bridge has most of the responsibility for spanning tree maintenance and change notifications and is the "central" point. In RSTP, all bridges take an active role in the network's connectivity. This and other new protocol specifications lead to increased reconvergence times.

New BPDU Definition and Function

RSTP slightly modifies the BPDUs' format and function and sends them out at a quicker rate (two-second default). They are now used as "hellos," as opposed to in original STP, where BPDUs were relayed from the root bridge. Every bridge will send out the BPDU with current information every hello time, even if it does not receive a BPDU from the root bridge. This method is quicker and more distributed.

The first change is that the BPDU and type field are now version 2, which means a switch that does not support RSTP will drop the BPDU. The biggest change, however, is the flags field. STP defines only two bits, but in RSTP, all bits are defined. Three of the bits are used to indicate the port role, and two of the bits are for port states, as shown in Figure 6-15.

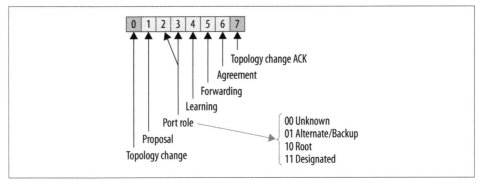

Figure 6-15. Option bits

The final two bits are used for proposal and agreement, which we discuss in the sections on topology change and transitions later in this chapter.

Interface Types and States

RSTP defines new interface types, as well as new port states. This allows for quicker direct and indirect link failure and recovery. The first new features define new interface types:

Point-to-point (P-to-P)
Direct connections between switches

Edge port
Port connection to end stations and not receiving BPDUs

Shared/non-edge
Connection to a LAN that is a shared medium and is receiving BPDUs, mostly likely a simple hub

Depending on the interface type, RSTP acts differently, based on the port roles. The port roles in RSTP are as follows:

Root port
Indicates the port with the lowest cost to the root bridge, as in original STP

Designated port
Indicates the designated port, as in original STP

Alternate port
Provides an alternate port to the root port in case of a failure

Backup port
Provides a redundant path to a segment, and acts as a backup in case of a designated port failure

Figure 6-16 shows the alternate and backup ports.

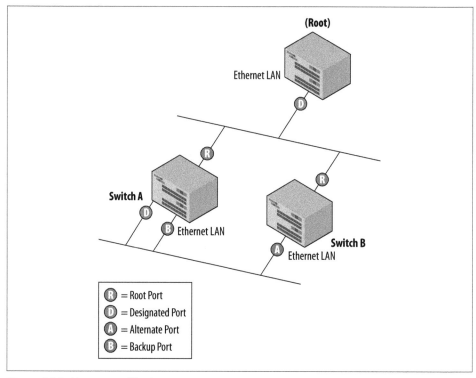

Figure 6-16. Alternate and backup ports

RSTP uses fewer port states than STP, defining only three states: *discarding*, *learning*, and *forwarding*. The discarding state is comparable to three of STP's states: disabled, blocking, and listening. Table 6-3 compares the states.

Table 6-3. RSTP versus STP port states

STP	RSTP
Disabled	Discarding
Blocking	Discarding
Listening	Discarding
Learning	Learning
Forwarding	Forwarding

RSTP Convergence

As previously mentioned, RSTP speeds up convergence by decreasing transition time and topology change notifications. Whereas STP relied heavily on timers, RSTP allows a feedback mechanism in order to transition a port. A new root port or designated port that is connected to a P-to-P link or configured as an edge port can be transitioned to the forwarding state without waiting for the timer to expire. The edge port connects directly to end stations and thus cannot create loops. Therefore, it can be placed in the forwarding state without any delay. Also, an alternate port and backup port can immediately transition to the root port or designated port in the event of a link failure.

A P-to-P link can transition to the forwarding state after it receives an acknowledgment from the neighboring bridge that it is attached to the link in the agreement/proposal mechanism. Basically, when a bridge needs to transition a port into the forwarding state, it sends out a configuration message with the proposal bit set. If a bridge receives this message and verifies that the information in the message is superior, it begins the *sync* operation. This operation places all non-edge ports in the blocked state and sends a configuration message with the agreement bit set to the original bridge. The agreement bit setting causes the interface to immediately transition to the forwarding state. If a bridge does not receive an acknowledgment of a proposal message it has sent, it returns to the original 802.1D convention and slowly transitions its port to the forwarding state through listening and then learning intermediate states.

 This method works only for P-to-P links; shared media follow the legacy STP method.

As an example, take a look at Figure 6-17. The root bridge and Bridge A exchange BPDUs. Initially, the link between the root bridge and Bridge A is in the blocked/learning state, as is normal for the initialization process. The BPDUs have the proposal bit set. Bridge A notices that the BPDU from the root bridge is superior, so it begins the sync process (Figure 6-18).

The sync process causes the bridge to put its links to Bridge B and Bridge C in the discarding state. It also sends a BPDU to the root bridge with the agreement bit set. This is a copy of the BPDU that was received from the root with only the agreement bit set. This causes the ports between the root bridge and Bridge A to go to the forwarding state.

Next, Bridge B and Bridge C receive the BPDU with the proposal bit set. Bridge B contains only edge ports, so it has no ports to block for the sync process. Bridge C has two edge ports and one non-edge port to Bridge D, which it blocks. Again, Bridge B and Bridge C send BPDUs back to Bridge A, which sets the ports to the forwarding state (Figure 6-19).

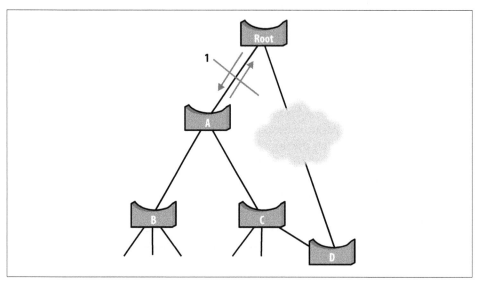

Figure 6-17. RSTP convergence, step 1

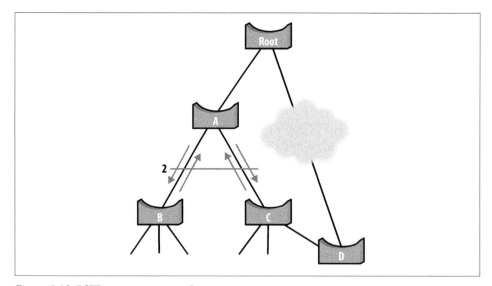

Figure 6-18. RSTP convergence, step 2

Finally, Bridge C sends a BPDU to Bridge D, which examines the BPDU and notices that it is inferior. It responds to Bridge C with its superior BPDU, which causes Bridge C to remain in the blocking state.

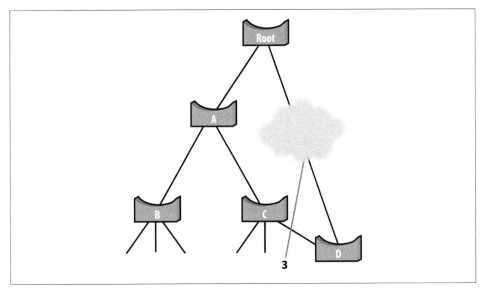

Figure 6-19. RSTP convergence, step 3

Topology changes

As mentioned previously, when an STP bridge detects that a topology change has occurred, it transmits a TCN message toward the root bridge and sets the TC flag in the configuration messages it transmits to the network. When network bridges receive a configuration message with the TC flag on, they also set this flag in the configuration messages they transmit on their designated ports toward the root bridge. This process continues until the root bridge is notified of the change. The root bridge then sets the TC flag in its BPDUs so that all bridges are notified of the event and the shortened aging time can be used.

RSTP bridges, on the contrary, use TCN messages only if an STP bridge needs to be notified of the topology change. Instead, each RSTP bridge that has detected a topology change starts a timer equal to twice the hello time for all its non-edge designated ports and its root port if needed, and then constantly transmits configuration messages on these ports with the TC flag on until the timer expires.

When a bridge receives a configuration message with the TC flag set, it first flushes all the entries in its filtering database except the entries containing MAC addresses that were learned via the port that received the configuration message with the TC flag on. Then that bridge starts a timer and transmits its own configuration messages with the TC flag set on all its designated ports and the root port until the timer expires.

 This is another example of RSTP's efficiency. STP sets a timer (forwarding delay) when the TC flag is received in order to age-out entries; in RSTP, on the other hand, the bridge ages out its database as soon as it receives the TC flag.

In this way, the news of the topology change quickly spreads over the entire network. This mechanism is faster and much more efficient than the one used by STP bridges. First, there is no need to wait until the TCN message reaches the root bridge, and then wait still longer until a configuration message with the TC flag on is received on the root port, and even longer until the shorter aging timer expires in order to delete old entries from the filtering database. Instead, the RSTP bridge immediately deletes old entries and notifies the other bridges to do the same. Second, a receipt of a configuration message with the TC flag set causes an STP bridge to age-out all the entries of its filtering database. The RSTP bridge, on the contrary, does not flush entries containing MAC addresses that were learned via the port that received the configuration message with the TC flag on. As a result, the number of flushed MAC addresses during a topology change is reduced.

Link failures

Let's look at two examples of direct link failures between switches and the root bridge. These scenarios are often known as "backbone fast" in the Cisco world. Figure 6-20 depicts the first scenario. The link between the root and Switch B fails. The process goes like this:

1. As soon as Switch B detects the link failure, the alternate port assumes the new role of root port. Depending on the failure type, this could be a subsecond transition.

2. Switch B sends out BPDUs with the TC flag on to Switch A, which sends the BPDU to the root. This causes the switches to age-out the MAC addresses and start the relearning process.

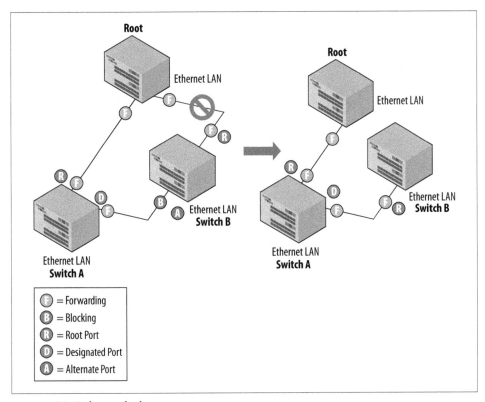

Figure 6-20. Failure with alternate port

A slightly different failure is observed if we look at the link between Switch A and the root bridge. The process, shown in Figure 6-21, is as follows:

1. Switch A's root port fails, assuming it's a new root, since it has no previous alternate port.
2. Switch B receives inferior BPDUs from Switch A on its alternate port, and sends a root link query to verify that the root bridge is still alive.
3. Upon root verification, Switch B immediately moves from the alternate port to the designated port role, and begins sending stored superior BPDUs to the downstream Switch A.
4. Switch A receives superior BPDUs, knows it is not the root, and places the port connecting to Switch B in the root port role.

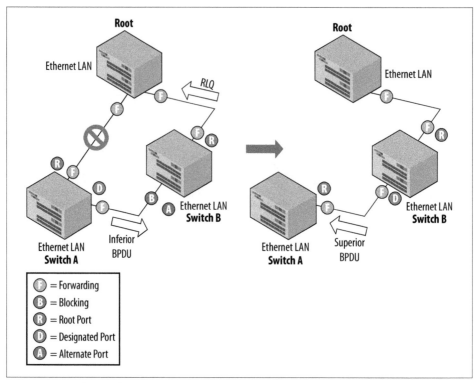

Figure 6-21. Failure without alternate port

Link Cost in RSTP

The original STP specification used a 16-bit unsigned integer for link speed, but RSTP uses the full 32 bits of range. This causes the default values to change according to the rules laid out in Table 6-4. If a bridge is encountered that uses the old 16-bit link speed field, it will use a 65,535 cost for any link speeds greater than 100 Mbps.

Table 6-4. RSTP link costs

Link speed	Recommended value
≤100 Kbps	200,000,000
1 Mbps	20,000,000
10 Mbps	2,000,000
100 Mbps	200,000
1 Gbps	20,000
10 Gbps	2,000
100 Gbps	200

Link speed	Recommended value
1 Tbps	20
10 Tbps	2

Compatibility with STP

As previously mentioned, RSTP is backward compatible with STP. If there are three bridges on the segment, A, B, and C, and Bridge C runs legacy STP, Bridge A and Bridge B will fall back to STP mode.

For instance, suppose Bridge A and Bridge B both run RSTP, with Switch A designated for the segment (Figure 6-22). Bridge C is brought online and runs legacy STP. Because 802.1D bridges ignore RSTP BPDUs and drop them, C believes there are no other bridges on the segment and starts to send its inferior 802.1D-format BPDUs. Switch A receives these BPDUs and, after twice the hello-time seconds maximum, changes its mode to 802.1D on that port only. As a result, C now understands the BPDUs of Switch A and accepts A as the designated bridge for that segment.

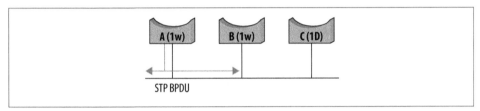

Figure 6-22. STP compatibility

Notice that in this particular case, if Bridge C is removed, Bridge A runs in STP mode on that port even though it is able to work more efficiently in RSTP mode with its unique neighbor B. This is because A does not know Bridge C is removed from the segment. In this situation, you'd have to manually restart the STP process in order for Bridge A and Bridge B to revert to RSTP.

Interoperability Between Juniper and Cisco

Before RSTP and MSTP (discussed shortly) were standardized, Cisco implemented its own proprietary protocols to try to solve these issues. One such protocol on Cisco is PVST+, which implements "multi-instance" spanning trees that allow multiple instances of STP on a single trunk. Essentially, virtual LAN (VLAN) 1 traffic uses the standard spanning tree destination multicast address 01:80:C2:00:00:00. The other VLAN(s) on the trunk advertise tagged PVST+ BPDUs to Cisco's reserved multicast address, 01:00:0C:CC:CC:CD. This allows Cisco to have a per-VLAN instance that can utilize multiple redundant trunk links by load-balancing the VLANs across them. The disadvantage of this protocol is that it has scaling issues, in addition to being proprietary.

Since this is the default configuration for many Cisco switches, it is important to note that any non-VLAN 1 traffic will simply be forwarded like any other multicast address and flooded on all forwarding ports within the VLAN in which the respective BPDUs are received (Figure 6-23). If the port is in the blocked state, no traffic will be forwarded, as usual.

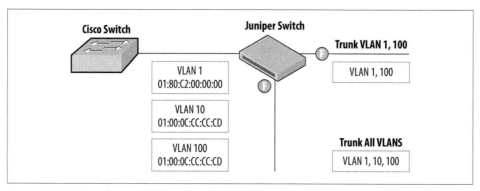

Figure 6-23. Juniper forwarding with PVST+

In Figure 6-24, Switch 1 and Switch 2 are Cisco switches running PVST+. Switch 1 is the STP root for VLANs 1, 5, and 77, with a bridge priority of 4096. It is also the STP backup root switch for VLANs 20 and 66, with a bridge priority of 8192. Switch 2 is the STP root switch for VLANs 20 and 66, with a bridge priority of 4096, and is the STP backup root switch for VLANs 1, 5, and 77, with a bridge priority of 8192.

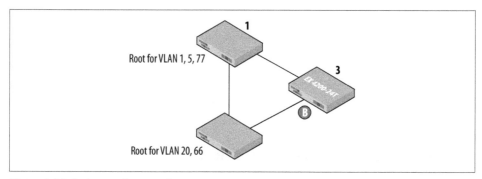

Figure 6-24. Juniper at the access layer

There is one additional caveat to PVST+. By default, the path cost assigned by PVST+ is based on Cisco's STP cost convention, not the IEEE STP cost. You must change the path cost so that it complies with the IEEE STP path cost and all similar links (such as FastE and GbE) have the same link-cost value. On the Cisco devices, the command `spanning-tree pathcost method long` must be entered to make it compliant with the current IEEE STP path cost. This command will be applied to all Cisco switch ports.

Switch 3, shown in Figure 6-25, is a Juniper Networks EX Series switch running RSTP. Since RSTP has only a single spanning tree instance for all VLANs across a trunk and there is no VLAN awareness, all traffic will be forwarded or blocked based on the VLAN 1 topology. Therefore, the link between Switch 1 and Switch 3 will be forwarded for all VLANs.

The net result is that Switch 3 is blocking one of the uplinks; there is no logical Layer 2 loop in the topology presented in Figure 6-25, regardless of which Cisco switch is the root for any of the VLANs (except VLAN 1) and regardless of the fact that these Cisco switches are running PVST+.

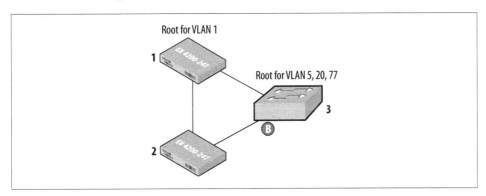

Figure 6-25. Cisco at the access layer

Again, RSTP is enabled on the EX Series switches (Switch 1 and Switch 2) and PVST+ is enabled on the Cisco switch (Switch 3). Switch 1 is the root of RSTP with a priority of 4096; Switch 2 is the backup root switch of the Common Spanning Tree (CST, discussed shortly) with a priority of 8192.

Switch 3 (the Cisco switch running PVST+) recognizes Switch 1 as the STP root for VLAN 1 and recognizes Switch 2 as having a lower bridge priority than itself on VLAN 1, therefore blocking its uplink trunk port to Switch 2 on VLAN 1.

For VLANs 5, 77, 20, and 66, Switch 3 sends out Cisco proprietary PVST+ BPDUs on the trunks. Because Switch 1 and Switch 2 do not understand PVST+ BPDUs, the switches will treat them as regular Layer 2 multicast packets and flood them out of the switch ports. Switch 3 is the PVST+ root for VLANs 5, 77, 20, and 66 and receives its own PVST+ BPDUs for the VLANs on both of its uplinks; so it ends up blocking one of these two uplink trunk ports.

Therefore, the result is a forwarding topology similar to that in Figure 6-25 with only one uplink trunk port forwarding on all VLANs on Switch 3 and the other uplink trunk port blocking on all VLANs.

RSTP Summary

Rapid Spanning Tree Protocol fixes the convergence issues found in the original Spanning Tree Protocol. It accomplishes this by simplifying states and eliminating the timers needed to transition from one state to another. Why not take a quick break and make sure you understand these concepts as we prepare to take a swan dive into a real live network.

Spanning Tree Configuration

To configure spanning in JUNOS software, first navigate to the [edit protocols stp] level. Enable each interface on which you would like to run STP, or alternatively, if you would like to run STP on every interface with Ethernet switching, simply enable the protocol itself:

```
[edit]
lab@Brandy# set protocols stp

[edit]
lab@Brandy# show protocols
lldp {
    interface all;
}
stp;
```

The all keyword is not configured, as every interface with Ethernet switching will have STP enabled.

This is also where timers, priorities, and other STP parameters can be set. Notice that some aspects of RSTP configuration are shown, since JUNOS implements 802.1D-2004:

```
lab@Brandy# set protocols stp ?
Possible completions:
  <[Enter]>           Execute this command
+ apply-groups        Groups from which to inherit configuration data
+ apply-groups-except Don't inherit configuration data from these groups
  bpdu-block-on-edge  Block BPDU on all interfaces configured as edge (BPDU
                      Protect)
  bridge-priority     Priority of the bridge (in increments of 4k - 0,4k,8k,..60k)
  disable             Disable STP
  forward-delay       Time spent in listening or learning state (4..30 seconds)
  hello-time          Time interval between configuration BPDUs (1..10 seconds)
> interface
  max-age             Maximum age of received protocol bpdu (6..40 seconds)
> traceoptions        Tracing options for debugging protocol operation
  |                   Pipe through a command
```

Let's look at how STP operates with default parameters. Recall the topology in Figure 6-25. All interfaces are enabled for STP using the default configuration.

Figure 6-26 shows the resulting STP topology, with Rum as the root bridge, since it has the lowest bridge ID (MAC address). Since all default parameters are used, the root bridge (Rum) is determined by the MAC address.

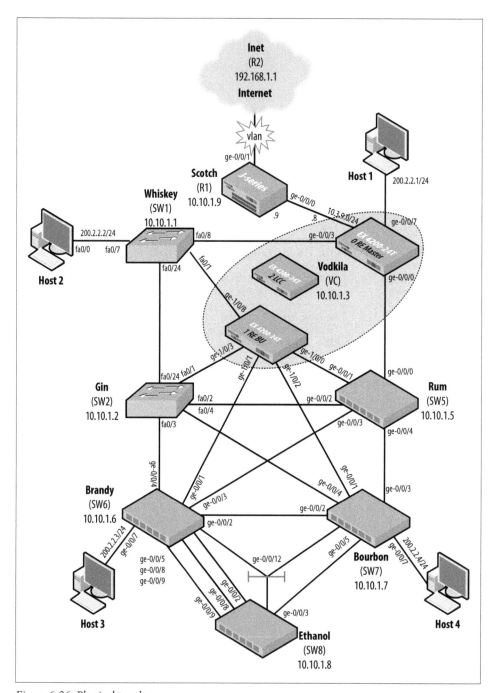

Figure 6-26. Physical topology

Figure 6-27 shows the default spanning tree.

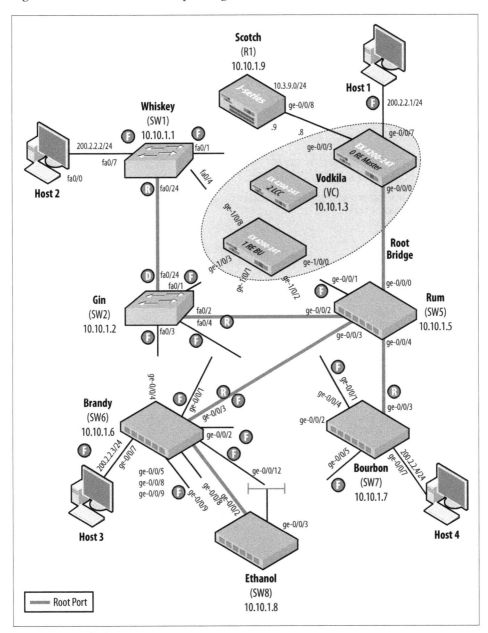

Figure 6-27. Default spanning tree

The loop-free paths between the hosts are as follows:

Host1→Vodkila, Rum, Bourbon→Host4
Host1→Vodkila, Rum, Gin, Whiskey→Host2
Host1→Vodkila, Rum, Brandy→Host3
Host2→Whiskey, Gin, Rum, Brandy→Host3
Host2→Whiskey, Gin, Rum, Bourbon→Host4
Host3→Brandy, Rum, Bourbon→Host4

Let's take a closer look at a few of the switches in the topology, namely, Rum, Brandy, and Gin, and see how some of the ports were assigned the states.

First, look at the default parameters for each switch. On Rum, notice that the bridge ID and the root ID are identical, indicating that it is indeed the root bridge:

```
lab@Rum> show spanning-tree bridge detail
STP bridge parameters
Context ID                        : 0
Enabled protocol                  : STP
  Root ID                         : 32768.00:19:e2:56:ee:80
  Hello time                      : 2 seconds
  Maximum age                     : 20 seconds
  Forward delay                   : 15 seconds
  Message age                     : 0
  Number of topology changes      : 1
  Time since last topology change : 1328 seconds
  Local parameters
    Bridge ID                     : 32768.00:19:e2:56:ee:80
    Extended system ID            : 0
    Internal instance ID          : 0
    Hello time                    : 2 seconds
    Maximum age                   : 20 seconds
    Forward delay                 : 15 seconds
    Path cost method              : 32 bit
```

In Juniper, the bridge ID is chosen as the first public MAC address assigned to the switch. A show chassis mac-addresses command will show you the first public address:

```
lab@Rum> show chassis mac-addresses
    FPC 0   MAC address information:
      Public base address     00:19:e2:56:ee:80
      Public count            64
```

Brandy also sees Rum as the root bridge with a MAC address of 32768.00:19:e2:56:ee:80:

```
lab@Brandy> show spanning-tree bridge
STP bridge parameters
Context ID                        : 0
Enabled protocol                  : STP
  Root ID                         : 32768.00:19:e2:56:ee:80
  Root cost                       : 20000
  Root port                       : ge-0/0/3.0
  Hello time                      : 2 seconds
  Maximum age                     : 20 seconds
  Forward delay                   : 15 seconds
```

```
Message age                    : 1
Number of topology changes     : 1
Time since last topology change : 1402 seconds
Local parameters
  Bridge ID                    : 32768.00:1f:12:30:96:80
  Extended system ID           : 0
  Internal instance ID         : 0
```

On Gin, the same timers are observed with this default configuration:

```
spanning-tree mode pvst
spanning-tree extend system-id
```

Notice there are two VLANs configured—the default VLAN 1 and management VLAN 666, as PVST+ is enabled by default:

```
Gin#show spanning-tree bridge de
VLAN0001
  Bridge ID  Priority    32769  (priority 32768 sys-id-ext 1)
             Address     000b.5fc6.9180
             Hello Time   2 sec  Max Age 20 sec  Forward Delay 15 sec
VLAN0666
  Bridge ID  Priority    33434  (priority 32768 sys-id-ext 666)
             Address     000b.5fc6.9180
             Hello Time   2 sec  Max Age 20 sec  Forward Delay 15 sec
```

It also sees the root bridge as Rum:

```
Gin#show spanning-tree root

                                 Root  Hello Max Fwd
Vlan                Root ID      Cost  Time  Age Dly  Root Port
-----------    --------------------  --------- ----- --- ---  -----------
VLAN0001       32768 0019.e256.ee80     19    2    20  15   Fa0/2
VLAN0666       33434 000b.5f01.f380     19    2    20  15   Fa0/48
```

Now examine the interface states on the switches. First, on the root bridge Rum, we expect to see all interfaces as designated ports and in the forwarding state, and this is the case except for ge-0/0/8, which is an administratively disabled port:

```
lab@Rum> show spanning-tree interface

Spanning tree interface parameters for instance 0

Interface   Port ID   Designated    Designated        Port    State Role
                      port ID       bridge ID         Cost
ge-0/0/0.0  128:513      128:513  32768.0019e256ee80  20000   FWD   DESG
ge-0/0/1.0  128:514      128:514  32768.0019e256ee80  20000   FWD   DESG
ge-0/0/2.0  128:515      128:515  32768.0019e256ee80  200000  FWD   DESG
ge-0/0/3.0  128:516      128:516  32768.0019e256ee80  20000   FWD   DESG
ge-0/0/4.0  128:517      128:517  32768.0019e256ee80  20000   FWD   DESG
ge-0/0/8.0  128:521      128:521  32768.0019e256ee80  20000   BLK   DIS
```

Notice the port cost is using the updated recommended values in 802.1D-2004. All 1 Gbps links have a port cost of 20,000, while the link to ge-0/0/2, which is a 100 Mbps link, has a port cost of 200,000.

Next, looking at Brandy, we see that it is the designated bridge for most of its ports, as it has a lower bridge ID (MAC address) than its neighbor switches. The priorities are the same, since they have not been configured, so this is the tiebreaker calculation. Also observe that ge-0/0/3 is the root port, as it is the lowest-cost path back to Rum. We also see a path cost of 2,000,000 for ge-0/0/7 and ge-0/0/12 because they connect to a 10 Mbps port:

```
lab@Brandy> show spanning-tree interface

Spanning tree interface parameters for instance 0

Interface   Port ID  Designated   Designated       Port     State Role
                     port ID      bridge ID        Cost
ge-0/0/1.0  128:514     128:514   32768.001f12309680   20000 FWD   DESG
ge-0/0/2.0  128:515     128:515   32768.001f12309680   20000 FWD   DESG
ge-0/0/3.0  128:516     128:516   32768.0019e256ee80   20000 FWD   ROOT
ge-0/0/4.0  128:517       128:3   32769.000b5fc69180  200000 BLK   ALT
ge-0/0/5.0  128:518     128:518   32768.001f12309680   20000 FWD   DESG
ge-0/0/7.0  128:520     128:520   32768.001f12309680 2000000 FWD   DESG
ge-0/0/8.0  128:521     128:521   32768.001f12309680   20000 FWD   DESG
ge-0/0/9.0  128:522     128:522   32768.001f12309680   20000 FWD   DESG
ge-0/0/12.0 128:525     128:525   32768.001f12309680 2000000 FWD   DESG
```

 In a Juniper device, port IDs begin counting at 513, whereas Cisco devices begin counting at 0.

Brandy's ge-0/0/4 interface is in the blocking state. Let's look at the neighbor switch to see why:

```
lab@Brandy> show spanning-tree interface ge-0/0/4 detail

Spanning tree interface parameters for instance 0

Interface name         : ge-0/0/4.0
Port identifier        : 128.517
Designated port ID     : 128.3
Port cost              : 200000
Port state             : Blocking
Designated bridge ID   : 32769.00:0b:5f:c6:91:80
Port role              : Alternate
Link type              : Pt-Pt/NONEDGE
Boundary port          : NA
```

Gin is the designated bridge on all its interfaces and has a root port of Fa0/2. One important difference from the Juniper switches is that the default cost is based on the STP cost and not the IEEE recommended cost. Notice the cost of 19 on the interfaces, which causes Gin to appear as the best root path cost for all ports:

```
Gin# show spanning-tree

VLAN0001
  Spanning tree enabled protocol ieee
  Root ID    Priority    32768
             Address     0019.e256.ee80
             Cost        19
             Port        2 (FastEthernet0/2)
             Hello Time   2 sec  Max Age 20 sec  Forward Delay 15 sec

  Bridge ID  Priority    32769  (priority 32768 sys-id-ext 1)
             Address     000b.5fc6.9180
             Hello Time   2 sec  Max Age 20 sec  Forward Delay 15 sec
             Aging Time 300

Interface          Role Sts Cost      Prio.Nbr Type
------------------ ---- --- --------- -------- --------------------
Fa0/1              Desg FWD 19        128.1    P2p
Fa0/2              Root FWD 19        128.2    P2p
Fa0/3              Desg FWD 19        128.3    P2p
Fa0/4              Desg FWD 19        128.4    P2p
Fa0/24             Desg FWD 19        128.24   P2p

VLAN0666
  Spanning tree enabled protocol ieee
  Root ID    Priority    33434
             Address     000b.5f01.f380
             Cost        19
             Port        48 (FastEthernet0/48)
             Hello Time   2 sec  Max Age 20 sec  Forward Delay 15 sec

  Bridge ID  Priority    33434  (priority 32768 sys-id-ext 666)
             Address     000b.5fc6.9180
             Hello Time   2 sec  Max Age 20 sec  Forward Delay 15 sec
             Aging Time 300

Interface          Role Sts Cost      Prio.Nbr Type
------------------ ---- --- --------- -------- --------------------
Fa0/48             Root FWD 19        128.48   P2p
```

In order to have a consistent view and to link costs in the topology, the Cisco switches should enable IEEE recommendations with the `spanning-tree pathcost method long` command. This small alteration causes the STP topology to change, as now slower 100 Mbps links have a higher link cost. Also notice that this enables the RSTP roles, as we saw on the Juniper switches (alternate, backup, etc.). However, RSTP is still not enabled on the switches:

```
Gin(config)#spanning-tree pathcost method long
Gin(config)#

Gin#show spanning-tree

VLAN0001
  Spanning tree enabled protocol ieee
  Root ID    Priority    32768
             Address     0019.e256.ee80
             Cost        200000
             Port        2 (FastEthernet0/2)
             Hello Time  2 sec  Max Age 20 sec  Forward Delay 15 sec

  Bridge ID  Priority    32769  (priority 32768 sys-id-ext 1)
             Address     000b.5fc6.9180
             Hello Time  2 sec  Max Age 20 sec  Forward Delay 15 sec
             Aging Time 300

Interface           Role Sts Cost      Prio.Nbr Type
------------------- ---- --- --------- -------- --------------------
Fa0/1               Altn BLK 200000    128.1    P2p
Fa0/2               Root FWD 200000    128.2    P2p
Fa0/3               Altn BLK 200000    128.3    P2p
Fa0/4               Altn BLK 200000    128.4    P2p
Fa0/24              Desg FWD 200000    128.24   P2p

VLAN0666
  Spanning tree enabled protocol ieee
  Root ID    Priority    33434
             Address     000b.5f01.f380
             Cost        200000
             Port        48 (FastEthernet0/48)
             Hello Time  2 sec  Max Age 20 sec  Forward Delay 15 sec

  Bridge ID  Priority    33434  (priority 32768 sys-id-ext 666)
             Address     000b.5fc6.9180
             Hello Time  2 sec  Max Age 20 sec  Forward Delay 15 sec
             Aging Time 300

Interface           Role Sts Cost      Prio.Nbr Type
------------------- ---- --- --------- -------- --------------------
Fa0/48              Root FWD 200000    128.48   P2p
```

This changes the paths over which each host communicates (see Figure 6-28). For example, Host2 now reaches Host3 via Whiskey, Vodkila, Rum, and then Brandy.

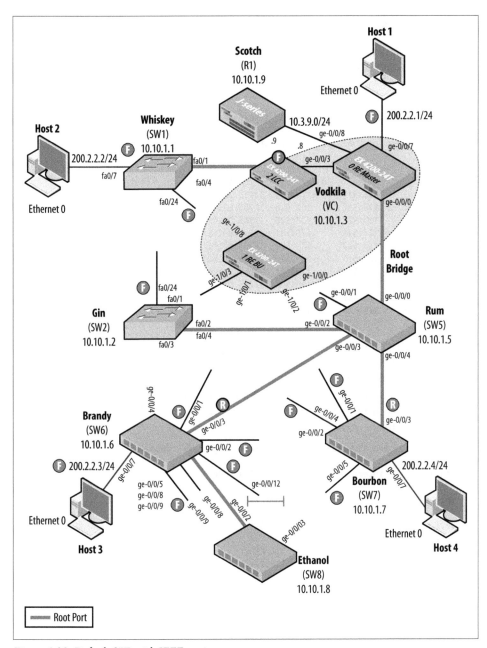

Figure 6-28. Default STP with IEEE settings

Failures with Default Parameters

Let's see how quickly the current topology can converge when there is a link failure between Host1 and Host4. Currently, Vodkila forwards the frames out its root port of ge-0/0/0 to get to Host4:

```
[edit]
lab@Vodkila# run show spanning-tree interface

Spanning tree interface parameters for instance 0

Interface    Port ID    Designated    Designated            Port      State  Role
                        port ID       bridge ID             Cost
ge-0/0/0.0   128:513    128:513    32768.0019e256ee80      20000    FWD    ROOT
ge-0/0/3.0   128:516    128:516    32768.001f123dd280      200000   FWD    DESG
ge-0/0/7.0   128:520    128:520    32768.001f123dd280      2000000  FWD    DESG
ge-0/0/8.0   128:521    128:521    32768.001f123dd280      20000    FWD    DESG
ge-0/0/9.0   128:522    128:522    32768.001f123dd280      200000   FWD    DESG
ge-0/1/1.0   128:610    128:610    32768.001f123dd280      20000    BLK    DIS
ge-1/0/0.0   128:625    128:514    32768.0019e256ee80      20000    BLK    ALT
ge-1/0/1.0   128:626    128:514    32768.001f12309680      20000    BLK    ALT
ge-1/0/2.0   128:627    128:514    32768.001f1230c480      20000    BLK    ALT
ge-1/0/3.0   128:628    128:628    32768.001f123dd280      200000   FWD    DESG
ge-1/0/8.0   128:633    128:633    32768.001f123dd280      20000    BLK    DIS
```

Host1 is going to ping Host4, while the ge-0/0/0 interface is disabled on Vodkila. This causes port ge-1/0/0 to transition to the new root port. However, this transition does not happen instantaneously. Recall that the port must transition from the blocking state to the forwarding state. Before this happens, it stays in the learning state for the length of the forwarding delay, which defaults to 15 seconds. The following sequence shows these states on Vodkila:

```
lab@Vodkila# run show spanning-tree interface ge-1/0/0 detail

Spanning tree interface parameters for instance 0

Interface name              : ge-1/0/0.0
Port identifier             : 128.625
Designated port ID          : 128.514
Port cost                   : 20000
Port state                  : Blocking
Designated bridge ID        : 32768.00:19:e2:56:ee:80
Port role                   : Root
Link type                   : Pt-Pt/NONEDGE
Boundary port               : NA
lab@Vodkila# run show spanning-tree interface ge-1/0/0 detail

Spanning tree interface parameters for instance 0

Interface name              : ge-1/0/0.0
Port identifier             : 128.625
Designated port ID          : 128.514
Port cost                   : 20000
Port state                  : Learning
```

```
Designated bridge ID      : 32768.00:19:e2:56:ee:80
Port role                 : Root
Link type                 : Pt-Pt/NONEDGE
Boundary port              : NA

[edit]
lab@Vodkila# run show spanning-tree interface ge-1/0/0 detail

Spanning tree interface parameters for instance 0

Interface name            : ge-1/0/0.0
Port identifier           : 128.625
Designated port ID        : 128.514
Port cost                 : 20000
Port state                : Forwarding
Designated bridge ID      : 32768.00:19:e2:56:ee:80
Port role                 : Root
Link type                 : Pt-Pt/NONEDGE
Boundary port              : NA
```

If you need more verification, notice that the ping from Host1 to Host4 times out and 15 pings are lost:

```
Reply to request 464 (4 ms)
Reply to request 465 (4 ms)
Request 466 timed out
Request 467 timed out
Request 468 timed out
Request 469 timed out
Request 470 timed out
Request 471 timed out
Request 472 timed out
Request 473 timed out
Request 474 timed out
Request 475 timed out
Request 476 timed out
Request 477 timed out
Request 478 timed out
Request 479 timed out
Request 480 timed out
Request 481 timed out
Reply to request 482 (172 ms)
```

This seems like a great time to deploy RSTP!

Configuring RSTP

In order to move from STP to RSTP on Juniper switches, simply delete the old STP configuration on each switch and move to RSTP. Here is an example taken from Rum:

```
[edit]
lab@Rum# delete protocols stp

[edit]
lab@Rum# set protocols rstp
```

```
[edit]

lab@Rum# commit
commit complete
```

The options are similar to what was shown in STP:

```
lab@Rum# set protocols rstp ?
Possible completions:
  <[Enter]>             Execute this command
+ apply-groups          Groups from which to inherit configuration data
+ apply-groups-except   Don't inherit configuration data from these groups
  bpdu-block-on-edge    Block BPDU on all interfaces configured as edge (BPDU
                        Protect)
  bridge-priority       Priority of the bridge (in increments of 4k - 0,4k,8k,..60k)
  disable               Disable STP
  forward-delay         Time spent in listening or learning state (4..30 seconds)
  hello-time            Time interval between configuration BPDUs (1..10 seconds)
> interface
  max-age               Maximum age of received protocol bpdu (6..40 seconds)
> traceoptions          Tracing options for debugging protocol operation
  |                     Pipe through a command
```

On the Cisco devices, enable rapid-pvst:

```
Whiskey(config)#spanning-tree mode rapid-pvst
```

Verify that RSTP is running in the Juniper switches:

```
[edit]
lab@Rum# run show spanning-tree bridge
STP bridge parameters
Context ID                         : 0
Enabled protocol                   : RSTP
  Root ID                          : 32768.00:19:e2:56:ee:80
  Hello time                       : 2 seconds
  Maximum age                      : 20 seconds
  Forward delay                    : 15 seconds
  Message age                      : 0
  Number of topology changes       : 4
  Time since last topology change  : 141 seconds
  Local parameters
    Bridge ID                      : 32768.00:19:e2:56:ee:80
    Extended system ID             : 0
    Internal instance ID           : 0
```

Also verify that RSTP is running for VLAN 1 on the Cisco switches:

```
Whiskey#show spanning-tree bridge
                                               Hello Max Fwd
Bridge Group             Bridge ID             Time  Age Dly  Protocol
---------------          --------------------  ----- --- ---  --------
Bridge group 1           128 000b.5fc3.cc80        1  15  30  dec
                                               Hello Max Fwd
Vlan                     Bridge ID             Time  Age Dly  Protocol
---------------          --------------------  ----- --- ---  --------
```

```
VLAN0001      32769 (32768,   1) 000b.5fc3.cc80   2   20   15   rstp
VLAN0666      33434 (32768, 666) 000b.5fc3.cc80   2   20   15   rstp
```

The topology has not changed from Figure 6-28. However, notice that on Vodkila, ge-1/0/0, ge-1/0/1, and ge-1/0/2 are in the alternate port roles. This means that upon a root port failure, the alternate port will automatically switch to the root port. Since Vodkila has multiple alternate ports, the default behavior is for the numerically lowest port to take precedence, which in this case is ge-1/0/0:

```
lab@Vodkila# run show spanning-tree interface

Spanning tree interface parameters for instance 0

Interface    Port ID   Designated   Designated        Port      State  Role
                       port ID      bridge ID         Cost
ge-0/0/0.0   128:513      128:513   32768.0019e256ee80   20000   FWD    ROOT
ge-0/0/3.0   128:516      128:516   32768.001f123dd280  200000   FWD    DESG
ge-0/0/7.0   128:520      128:520   32768.001f123dd280 2000000   FWD    DESG
ge-0/0/8.0   128:521      128:521   32768.001f123dd280   20000   FWD    DESG
ge-0/0/9.0   128:522      128:522   32768.001f123dd280  200000   FWD    DESG
ge-0/1/1.0   128:610      128:610   32768.001f123dd280   20000   BLK    DIS
ge-1/0/0.0   128:625      128:514   32768.0019e256ee80   20000   BLK    ALT
ge-1/0/1.0   128:626      128:514   32768.001f12309680   20000   BLK    ALT
ge-1/0/2.0   128:627      128:514   32768.001f1230c480   20000   BLK    ALT
ge-1/0/3.0   128:628      128:628   32768.001f123dd280  200000   FWD    DESG
ge-1/0/8.0   128:633      128:633   32768.001f123dd280   20000   BLK    DIS
```

Let's invoke the link failure again:

```
[edit]
lab@Vodkila# set interfaces ge-0/0/0 disable
[edit]
lab@Vodkila# commit
fpc1:
configuration check succeeds
fpc0:
commit complete
fpc2:
commit complete
fpc1:
commit complete
```

The interface switches over to the root port and forwarding mode without any timers expiring:

```
lab@Vodkila# run show spanning-tree interface ge-1/0/0 detail

Spanning tree interface parameters for instance 0

Interface name            : ge-1/0/0.0
Port identifier           : 128.625
Designated port ID        : 128.514
Port cost                 : 20000
Port state                : Forwarding
Designated bridge ID      : 32768.00:19:e2:56:ee:80
Port role                 : Root
```

```
Link type                    : Pt-Pt/NONEDGE
Boundary port                : NA
```

The Port Fast Feature

The port fast feature is enabled on non-edge P-to-P interfaces, which is the majority of connections in modern networks. Since the port was set for full duplex (FD) and BPDUs were received from neighbor switches, it is automatically set to P-to-P. In contrast, interface ge-0/0/7 does not receive BPDUs, as it contains only a host and is put in edge mode:

```
[edit]
lab@Vodkila# run show spanning-tree interface ge-
0/0/7 detail
Spanning tree interface parameters for instance 0

Interface name               : ge-0/0/7.0
Port identifier              : 128.520
Designated port ID           : 128.520
Port cost                    : 2000000
Port state                   : Forwarding
Designated bridge ID         : 32768.00:1f:12:3d:d2:80
Port role                    : Designated
Link type                    : SHARED/EDGE
Boundary port                : NA
```

On Host1, a single ping packet is dropped, which is a vast improvement over what happens in the STP case (15 drops). Also, remember that the switches will flush out their MAC tables in order to achieve proper forwarding. This is likely the cause for the single lost packet in the following code:

```
Reply to request 452 (8 ms)
Reply to request 453 (8 ms)
Reply to request 454 (8 ms)
Reply to request 455 (8 ms)
Reply to request 456 (4 ms)
Reply to request 457 (4 ms)
Request 458 timed out
Reply to request 459 (148 ms)
Reply to request 460 (4 ms)
Reply to request 461 (4
```

Notice that ge-0/0/0 is now disabled and is blocking, whereas ge-1/0/0 is now forwarding and is the root port. If that port failed, ge-1/0/1 would be next in line to succeed as the root port:

```
[edit]
lab@Vodkila# run show spanning-tree interface

Spanning tree interface parameters for instance 0

Interface    Port ID   Designated   Designated          Port     State  Role
                       port ID      bridge ID           Cost
ge-0/0/0.0   128:513       128:513  32768.0019e256ee80   20000    BLK    DIS
ge-0/0/3.0   128:516       128:516  32768.001f123dd280  200000    FWD    DESG
ge-0/0/7.0   128:520       128:520  32768.001f123dd280 2000000    FWD    DESG
ge-0/0/8.0   128:521       128:521  32768.001f123dd280   20000    FWD    DESG
ge-0/0/9.0   128:522       128:522  32768.001f123dd280  200000    FWD    DESG
ge-0/1/1.0   128:610       128:610  32768.001f123dd280   20000    BLK    DIS
ge-1/0/0.0   128:625       128:514  32768.0019e256ee80   20000    FWD    ROOT
ge-1/0/1.0   128:626       128:514  32768.001f12309680   20000    BLK    ALT
ge-1/0/2.0   128:627       128:514  32768.001f1230c480   20000    BLK    ALT
ge-1/0/3.0   128:628       128:628  32768.001f123dd280  200000    FWD    DESG
ge-1/0/8.0   128:633       128:633  32768.001f123dd280   20000    BLK    DIS
```

When RSTP isn't going to be rapid

Remember that this rapid transition happens only for ports that are in P-to-P mode. Ports in other modes, such as edge ports or shared LANs, do not enjoy this quick failover. For example, the shared LAN between Brandy, Bourbon, and Ethanol will simply use the standard STP timers due to the shared status of the interface:

```
[edit]
lab@Brandy# run show spanning-tree interface ge-0/0/12 detail

Spanning tree interface parameters for instance 0

Interface name        : ge-0/0/12.0
Port identifier       : 128.525
Designated port ID    : 128.525
Port cost             : 2000000
Port state            : Forwarding
Designated bridge ID  : 32768.00:1f:12:30:96:80
Port role             : Designated
Link type             : SHARED/NONEDGE
Boundary port         : NA
```

RSTP design consideration

So far in the example we have relied on the default parameters to calculate the spanning tree, but often the default parameters are not sufficient. This usually corresponds to the root bridge as well as interface costs. STP takes into account only link speeds, and not other important items such as monetary link cost or link reliability. In order for STP to use these factors in the calculation, you may have to manually configure the link cost:

```
lab@Rum# set protocols rstp interface ge-0/0/1 ?
Possible completions:
+ apply-groups          Groups from which to inherit configuration data
+ apply-groups-except   Don't inherit configuration data from these groups
> bpdu-timeout-action   Define action on BPDU expiry (Loop Protect)
  cost                  Cost of the interface (1..200000000)
  disable               Disable Spanning Tree on port
  edge                  Port is an edge port
  mode                  Interface mode (P2P or shared)
  no-root-port          Enable root-protect feature on this port
  priority              Interface priority (in increments of 16 - 0,16,..240)
```

You also define and set the location of the root bridge. In the default topology (Figure 6-29), Rum was chosen as the root bridge simply because it has the lowest bridge ID, due to having the lowest MAC address. In general, the root bridge should be your most powerful switch since much of the forwarding of the frame will transit this switch. Also, it makes sense to choose a bridge that has some significance for being the central point of the tree. This could be where the highest-speed interfaces connect or where the multiple intercampus or WAN links exist. In our topology, Vodkila would seem like a logical choice, as it is the largest switch due to its Virtual Chassis (VC) configuration and because it connects to the default gateway, which is Scotch.

To configure Vodkila to be the root bridge, simply set the bridge priority to a lower value than the default:

```
[edit]
lab@Vodkila# set protocols rstp bridge-priority 4k
```

After this change is committed, the first portion of the bridge ID changes to 4096 on the active ports:

```
lab@Vodkila> show spanning-tree interface

Spanning tree interface parameters for instance 0

Interface    Port ID   Designated   Designated       Port     State  Role
                       port ID      bridge ID        Cost
ge-0/0/0.0   128:513   128:513      4096.001f123dd280   20000  FWD    DESG
ge-0/0/3.0   128:516   128:516      4096.001f123dd280  200000  FWD    DESG
ge-0/0/7.0   128:520   128:520      4096.001f123dd280 2000000  FWD    DESG
ge-0/0/8.0   128:521   128:521      4096.001f123dd280   20000  FWD    DESG
ge-0/0/9.0   128:522   128:522      4096.001f123dd280  200000  FWD    DESG
ge-0/1/1.0   128:610   128:610      32768.001f123dd280  20000  BLK    DIS
ge-1/0/0.0   128:625   128:625      4096.001f123dd280   20000  FWD    DESG
ge-1/0/1.0   128:626   128:626      4096.001f123dd280   20000  FWD    DESG
ge-1/0/2.0   128:627   128:627      4096.001f123dd280   20000  FWD    DESG
ge-1/0/3.0   128:628   128:628      4096.001f123dd280  200000  FWD    DESG
ge-1/0/8.0   128:633   128:633      32768.001f123dd280  20000  BLK    DIS
```

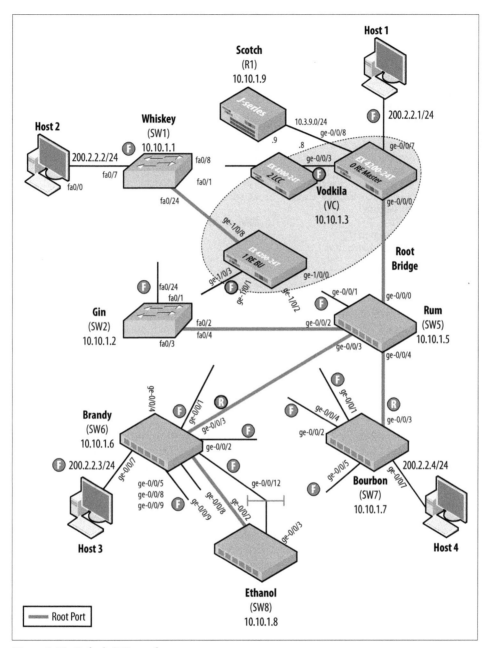

Figure 6-29. Default STP topology

Verify that Vodkila is the root bridge by checking that the root ID equals the bridge ID:

```
lab@Vodkila>

lab@Vodkila> show spanning-tree bridge
STP bridge parameters
Context ID                          : 0
Enabled protocol                    : RSTP
  Root ID                           : 4096.00:1f:12:3d:d2:80
  Hello time                        : 2 seconds
  Maximum age                       : 20 seconds
  Forward delay                     : 15 seconds
  Message age                       : 0
  Number of topology changes        : 13
  Time since last topology change   : 24 seconds
  Local parameters
    Bridge ID                       : 4096.00:1f:12:3d:d2:80
    Extended system ID              : 0
    Internal instance ID            : 0
```

When BPDUs Attack

It is possible that an edge port could receive BPDUs as a result of a misconfiguration or even an attack. These could be from a PC, a simple switch bought from your local electronics store, or an access switch. In these cases, the receipt of the BPDU could result in a spanning tree calculation, or worse, a root bridge recalculation. Two JUNOS features help guard against this issue: BPDU protection and root protection. In BPDU protection, the receipt of a BPDU will cause the port to transition to the blocking state and stop forwarding frames. The command to enable this is set ethernet-switching-options bpdu-block interface <interface name>. To verify whether the port is blocked, issue the show ethernet-switching interfaces command. In order to transition the interface out of the blocked state, either set a timer value (disable-timeout <seconds>) for the blocking and wait for the timeout to expire, or issue the clear ethernet-switching bpdu-error command.

Another alternative is to block the port only if it is receiving superior BPDUs on that port. This is the case when you have a small switch on which you may want to run STP but you don't want to allow it to become the root bridge, which could happen due to a misconfiguration or faulty software (sorry, electronics store). In order to ensure that a port cannot receive superior BPDUs, set the no-root-port command after the interface in [edit protocols rstp]. If the port receives a superior BPDU, it will transition into the blocking state and will not leave that state until it no longer receives superior BPDUs.

Spanning Tree Configuration Summary

We looked at how to configure Spanning Tree Protocol as well as Rapid Spanning Tree Protocol. Although RSTP solves many convergence issues, it does have a major problem when you are dividing your LAN segment into multiple VLANs. RSTP allows only a single instance of a spanning tree, causing all of your VLANs to traverse a single link. If you want to use multiple links, Multiple Spanning Tree Protocol, examined next, would be your solution.

Multiple Spanning Tree Protocol

RSTP solves many of STP's limitations, but it does have one major drawback, in that it is not "VLAN-aware." In other words, only one spanning tree is created for the entire LAN, and there is only a single path for all VLANs to use. This causes underutilization of links that are in the blocking state and results in no load balancing. MSTP (or 802.1s) provides for a new spanning tree instance per VLAN or per group of VLANs, allowing more links in the LAN to be utilized. MSTP was originally defined in IEEE 802.1s and later merged into IEEE 802.1Q-2003. That specification describes the operation of MSTP as follows:

> MSTP allows frames assigned to different VLANs to follow separate paths, each based on an independent Multiple Spanning Tree Instance (MSTI), within Multiple Spanning Tree (MST) Regions composed of LANs and or MST Bridges. These Regions and the other Bridges and LANs are connected into a single Common Spanning Tree (CST).

Figure 6-30 gives an example of the goal of MSTP; here, it allows certain VLANs to utilize the link from A to D1, while others utilize the link from A to D2. This would be a common topology if Switch A were an access switch.

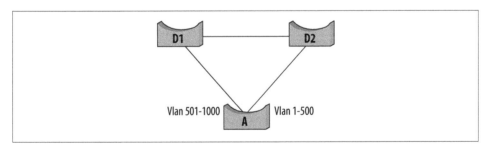

Figure 6-30. VLAN issue

 PVST+ was created to solve the multiple-STP issue on Cisco devices. However, PVST+ creates an instance per VLAN, which can have massive scaling issues as the number of VLANs grows. Consequently, Cisco also recommends the use of MSTP over PVST+.

MSTP allows VLANs 501–1000 to be classified into one instance, and VLANs 1–500 to be classified into another region, resulting in a setup such as the one in Figure 6-31. Notice that each instance is blocked on the opposite port due to the spanning tree.

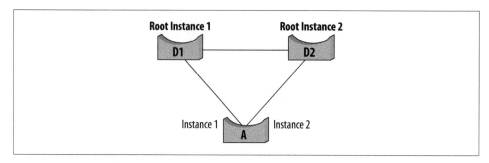

Figure 6-31. MSTP solution

MSTP groups switches together in an MSTP *region*. Figure 6-32 depicts an MSTP region. A region is defined by three parameters:

Region name
> User-defined name for the region.

Revision level
> User-defined value that defines the region.

Element table
> Provides mapping of VLANs to instances. This information is not actually sent out in the BPDU, but rather is a numerical digest that is generated from the local configuration.

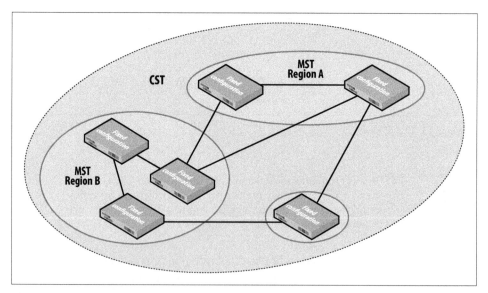

Figure 6-32. MSTP region

Each region defines the boundary for the MSTI BPDUs that will be sent out. Up to 64 regions can be configured. Each region is interconnected via a Common Spanning Tree (CST), shown in Figure 6-33. The CST interconnects MSTP regions or even devices that are simply running RSTP. Since the BPDUs that are sent out in each region are standard RSTP BPDUs with the MSTI messages placed at the end of the BPDU, standard RSTP switches will interpret the MSTP BPDUs as RSTP and will view the entire region as a single spanning tree instance. CST's role is to manage these regions and RSTP islands in order to ensure an accurate LAN topology. CST turns each MSTP region into a virtual bridge. So, from a logical point of view, it appears that there are multiple switches running STP, while in fact, physically they are the same switch. CST creates the "magic" by managing these logical separations.

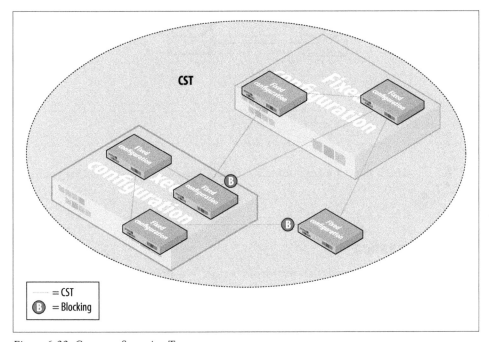

Figure 6-33. Common Spanning Tree

Within an MSTP region, standard RSTP operation applies. The only difference is that on a given link both ends of a link can send and receive BPDUs simultaneously. This is because, as shown in Figure 6-34, each bridge can be designated for one or more instances and needs to transmit BPDUs. In other words, STP changes its view from operating on a per-port basis to operating on a per-instance basis.

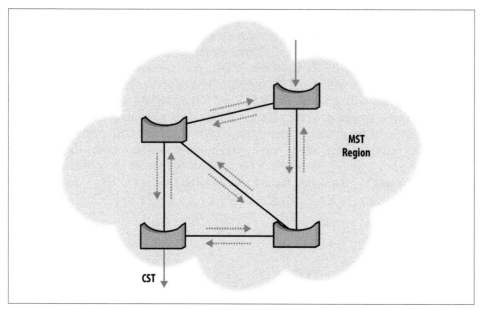

Figure 6-34. RSTP in an MSTP region

 MSTP allows 64 instances to be created. This was increased from an older value of 16.

MSTP Configuration

MSTP is configured under [edit protocols mstp]. We already discussed most of the configuration options elsewhere in this chapter:

```
lab@Bourbon# set protocols mstp ?
Possible completions:
  <[Enter]>              Execute this command
+ apply-groups           Groups from which to inherit configuration data
+ apply-groups-except    Don't inherit configuration data from these groups
  bpdu-block-on-edge     Block BPDU on all interfaces configured as edge (BPDU
                         Protect)
  bridge-priority        Priority of the bridge (in increments of 4k - 0,4k,8k,..60k)
  configuration-name     Configuration name (part of MST configuration identifier)
  disable                Disable MSTP
  forward-delay          Time spent in listening or learning state (4..30 seconds)
  hello-time             Time interval between configuration BPDUs (1..10 seconds)
> interface
  max-age                Maximum age of received protocol bpdu (6..40 seconds)
  max-hops               Maximum number of hops (1..255)
> msti                   Per-MSTI options
  revision-level         Revision level (part of MST configuration identifier)
```

```
>  traceoptions        Tracing options for debugging protocol operation
   |                    Pipe through a command
```

We are going to configure the example topology as shown in Figure 6-35. Brandy has three access ports, each with a single VLAN 10, 20, or 30. It also is connected to a host on port ge-0/0/7, which is untagged. The goal is to send all VLAN 30 traffic toward Rum, and all other traffic toward Vodkila.

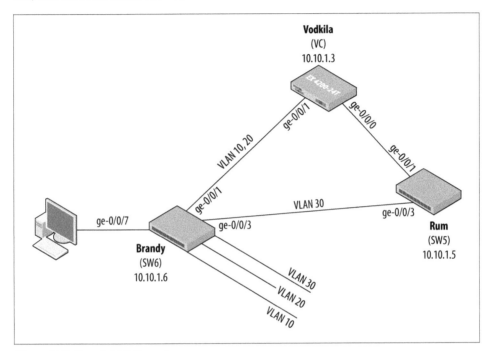

Figure 6-35. Sample MSTP topology

The VLANs are defined on Brandy:

```
lab@Brandy> show configuration vlans
v10 {
    vlan-id 10;
}
v20 {
    vlan-id 20;
}
v30 {
    vlan-id 30;
}
```

The interfaces of ge-0/0/1 and ge-0/0/3 are defined as trunk ports:

```
lab@Brandy> show configuration interfaces ge-0/0/1
unit 0 {
    family ethernet-switching {
        port-mode trunk;
```

```
        vlan {
            members [ v10 v20 ];
        }
    }
}

lab@Brandy> show configuration interfaces ge-0/0/3
unit 0 {
    family ethernet-switching {
        port-mode trunk;
        vlan {
            members v30;
        }
    }
}
```

The MSTP parameters are then defined. The first information that you must decide on is the revision level and configuration name. These values must be the same on each switch in order to identify the MSTP region and translate the incoming BPDUs:

```
lab@Brandy# show protocols mstp
configuration-name book-example;
revision-level 1;
msti 10 {
    vlan [ 10 20 ];
}
msti 30 {
    vlan 30;
}
```

After the configuration is committed, examine the MSTP configuration. Notice that proper VLANs assigned to MSTI 10 and 30 and the rest of the VLANs automatically mapped to the default MSTI of 0:

```
lab@Brandy> show spanning-tree mstp configuration
MSTP information
Context identifier    : 0
Region name           : book-example
Revision              : 1
Configuration digest  : 0x53973bbb358bdb2d6dcab806a189064f

MSTI      Member VLANs
   0 0-9,11-19,21-29,31-4094
  10 10,20
  30 30

  30 30
```

Then examine the spanning tree bridge parameters. Notice that Brandy is the root bridge for MSTI 10 and 30, but not for instance 0:

```
lab@Brandy> show spanning-tree bridge
STP bridge parameters
Context ID                        : 0
Enabled protocol                  : MSTP
```

```
STP bridge parameters for CIST
   Root ID                          : 32768.00:19:e2:56:ee:80
   Root cost                        : 20000
   Root port                        : ge-0/0/1.0
   CIST regional root               : 32768.00:1f:12:3d:d2:80
   CIST internal root cost          : 20000
   Hello time                       : 2 seconds
   Maximum age                      : 20 seconds
   Forward delay                    : 15 seconds
   Hop count                        : 19
   Message age                      : 1
   Number of topology changes       : 7
   Time since last topology change  : 250 seconds
   Local parameters
      Bridge ID                     : 32768.00:1f:12:30:96:80
      Extended system ID            : 0
      Internal instance ID          : 0

STP bridge parameters for MSTI 10
   MSTI regional root               : 32778.00:1f:12:30:96:80
   Hello time                       : 2 seconds
   Maximum age                      : 20 seconds
   Forward delay                    : 15 seconds
   Local parameters
      Bridge ID                     : 32778.00:1f:12:30:96:80
      Extended system ID            : 0
      Internal instance ID          : 1

STP bridge parameters for MSTI 30
   MSTI regional root               : 32798.00:1f:12:30:96:80
   Hello time                       : 2 seconds
   Maximum age                      : 20 seconds
   Forward delay                    : 15 seconds
   Local parameters
      Bridge ID                     : 32798.00:1f:12:30:96:80
      Extended system ID            : 0
      Internal instance ID          : 2
```

Looking at the forwarding states of the different spanning tree instances, notice that instance 10 contains the access ports for VLANs 10 and 20 as well as the trunk port. Instance 30 contains the access port for VLAN 30 and the trunk port. Instance 0 contains all the interfaces and is the CST that is tying the MSTP regions together. Instance 0 is sometimes called the internal spanning tree for the system and is a portion of the CST as a whole:

```
lab@Brandy> show spanning-tree interface

Spanning tree interface parameters for instance 0

Interface    Port ID   Designated    Designated          Port      State  Role
                       port ID       bridge ID           Cost
ge-0/0/1.0   128:514      128:626    32768.001f123dd280  20000     FWD    ROOT
ge-0/0/3.0   128:516      128:516    32768.001f12309680  20000     FWD    DESG
ge-0/0/5.0   128:518      128:518    32768.001f12309680  20000     FWD    DESG
ge-0/0/7.0   128:520      128:520    32768.001f12309680  2000000   FWD    DESG
```

```
ge-0/0/8.0    128:521      128:521  32768.001f12309680    20000  FWD   DESG
ge-0/0/9.0    128:522      128:522  32768.001f12309680    20000  FWD   DESG

Spanning tree interface parameters for instance 10

Interface    Port ID  Designated     Designated       Port   State  Role
                      port ID        bridge ID        Cost
ge-0/0/1.0   128:514      128:514  32778.001f12309680    20000  FWD   DESG
ge-0/0/5.0   128:518      128:518  32778.001f12309680    20000  FWD   DESG
ge-0/0/8.0   128:521      128:521  32778.001f12309680    20000  FWD   DESG

Spanning tree interface parameters for instance 30

Interface    Port ID  Designated     Designated       Port   State  Role
                      port ID        bridge ID        Cost
ge-0/0/3.0   128:516      128:516  32798.001f12309680    20000  FWD   DESG
ge-0/0/9.0   128:522      128:522  32798.001f12309680    20000  FWD   DESG
```

The actual root for the CST is Rum:

```
lab@Rum> show spanning-tree bridge
STP bridge parameters
Context ID                            : 0
Enabled protocol                      : MSTP

STP bridge parameters for CIST
  Root ID                             : 32768.00:19:e2:56:ee:80
  CIST regional root                  : 32768.00:19:e2:56:ee:80
  CIST internal root cost             : 0
  Hello time                          : 2 seconds
  Maximum age                         : 20 seconds
  Forward delay                       : 15 seconds
  Number of topology changes          : 5
  Time since last topology change     : 371 seconds
  Local parameters
    Bridge ID                         : 32768.00:19:e2:56:ee:80
    Extended system ID                : 0
    Internal instance ID              : 0

STP bridge parameters for MSTI 30
  MSTI regional root                  : 32798.00:19:e2:56:ee:80
  Hello time                          : 2 seconds
  Maximum age                         : 20 seconds
  Forward delay                       : 15 seconds
  Local parameters
    Bridge ID                         : 32798.00:19:e2:56:ee:80
    Extended system ID                : 0
    Internal instance ID              : 1
```

It is also interesting to look at Vodkila as we see the ports toward Rum—for instance, 10 and 30—in the master state. The master state is a port on the shortest path from the entire region to the common root bridge, connecting the MSTP region to the common root bridge:

```
lab@Vodkila> show spanning-tree interface
```

```
Spanning tree interface parameters for instance 0

Interface    Port ID   Designated   Designated          Port     State  Role
                       port ID      bridge ID           Cost
ge-0/0/0.0   128:513      128:513   32768.001f123dd280    20000   FWD    DESG
ge-0/0/3.0   128:516      128:516   32768.001f123dd280   200000   FWD    DESG
ge-0/0/7.0   128:520      128:520   32768.001f123dd280  2000000   FWD    DESG
ge-0/0/8.0   128:521      128:521   32768.001f123dd280    20000   FWD    DESG
ge-0/0/9.0   128:522      128:522   32768.001f123dd280   200000   FWD    DESG
ge-0/1/1.0   128:610      128:610   32768.001f123dd280    20000   BLK    DIS
ge-1/0/0.0   128:625      128:514   32768.0019e256ee80    20000   FWD    ROOT
ge-1/0/1.0   128:626      128:626   32768.001f123dd280    20000   FWD    DESG
ge-1/0/2.0   128:627      128:627   32768.001f123dd280    20000   FWD    DESG
ge-1/0/3.0   128:628      128:628   32768.001f123dd280   200000   FWD    DESG
ge-1/0/8.0   128:633      128:633   32768.001f123dd280    20000   BLK    DIS

Spanning tree interface parameters for instance 10

Interface    Port ID   Designated   Designated          Port     State  Role
                       port ID      bridge ID           Cost
ge-1/0/0.0   128:625      128:625   32778.001f123dd280    20000   FWD    MSTR
ge-1/0/1.0   128:626      128:514   32778.001f12309680    20000   FWD    ROOT

Spanning tree interface parameters for instance 30

Interface    Port ID   Designated   Designated          Port     State  Role
                       port ID      bridge ID           Cost
ge-1/0/0.0   128:625      128:625   32798.001f123dd280    20000   FWD    MSTR
ge-1/0/1.0   128:626      128:626   32798.001f123dd280    20000   FWD    DESG
```

MSTP Summary

In order to balance multiple VLANs across the links in your LAN, you should deploy
MSTP. MSTP is essentially multiple instances of STP tied together efficiently. When
diving into the details, the complexity of the protocol becomes clear; however, that
complexity is not a burden due to the automatic learning built into the specification
itself.

Redundant Trunk Groups

Another feature that can help in loops is a feature called Redundant Trunk Groups
(RTGs). Essentially, RTGs provide a fast and simplified Layer 2 failover mechanism,
without introducing the complexity of running STP. RTG allows for one physical (or
LAG) interface to back up another in case of a failure.

In the Cisco world, RTG is often referred to as FlexLink.

One of the most common situations in which to use RTG is when an access layer switch is dual-homed to the aggregation or distribution switch (Figure 6-36). In this scenario, configuration is needed only on the local Switch C, as Switch A and Switch B do not need to know that RTG is enabled. However, if a link is enabled for RTG, it cannot participate in STP nor does it need to, as you manually solved the issue that STP was trying to solve. If STP was allowed on interfaces running RTG, confusion could occur if the active link suddenly became blocked due to an STP calculation. JUNOS software will display a commit error if STP and RTG are attempted at the same time:

```
[edit ethernet-switching-options]
lab@Brandy# commit
error: XSTP : msti 0 STP and RTG cannot be enabled on the same interface ge-0/0/1.0
error: configuration check-out failed
```

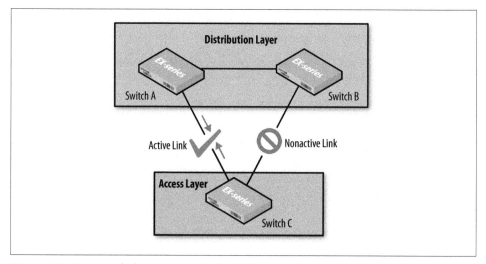

Figure 6-36. Common deployment scenario for RTG

RTG Configuration

RTG is configured under [edit ethernet-switching-options] and defines two interfaces under a group: a primary interface and a backup. JUNOS software allows up to 16 RTGs to be configured.

In Figure 6-37, Brandy is dual-homed to Vodkila and Rum. The ge-0/0/3 interface should be configured as primary and ge-0/0/1 should be configured as secondary:

```
[edit ethernet-switching-options]
lab@Brandy# show
redundant-trunk-group {
    group dual {
        interface ge-0/0/1.0;
        interface ge-0/0/3.0 {
            primary;
```

```
        }
      }
    }

[edit ethernet-switching-options]
lab@Brandy# commit and-quit
```

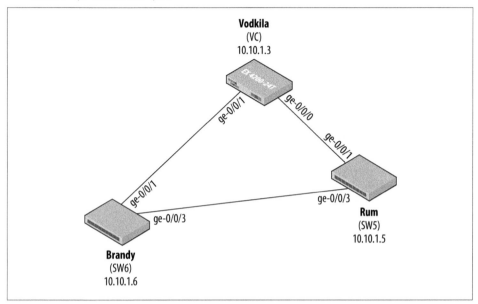

Figure 6-37. RTG example

Verify that RTG is working as configured:

```
lab@Brandy> show redundant-trunk-group group-name dual

Interface   State       Bandwidth   Time of last flap          Flap
                                                                count

ge-0/0/3.0  Up/Pri/Act  1000 Mbps   Never                         0
ge-0/0/1.0  Up          1000 Mbps   Never                         0
```

Notice that ge-0/0/3 is set to *primary* and *active* under the State tag. The ge-0/0/3
interface fails, and the ge-0/0/1 interface becomes the active interface:

```
lab@Brandy# run show redundant-trunk-group group-name dual

Interface   State     Bandwidth   Time of last flap                     Flap
                                                                        count

ge-0/0/3.0  Dwn/Pri   1000 Mbps   2008-08-06 10:36:20 UTC (00:00:22 ago)   1
ge-0/0/1.0  Up/Act    1000 Mbps   Never                                    0
```

Since the ge-0/0/3 interface is set as the primary interface, it has a *revertive capability*.
This means that if the primary interface fails and then returns, it will revert back to the
active link. In our example, ge-0/0/3 returns:

```
lab@Brandy# run show redundant-trunk-group group-name dual

Interface   State       Bandwidth   Time of last flap                   Flap
                                                                        count

ge-0/0/3.0  Up/Pri/Act  1000 Mbps   2008-08-06 10:37:08 UTC (00:00:01 ago)    2
ge-0/0/1.0  Up          1000 Mbps   Never                                     0
```

 Although a backup RTG interface will not forward traffic, Layer 2 management protocols such as Link Layer Discovery Protocol (LLDP) will still function across it.

It is possible to create a non-revertive RTG group by removing the primary flag. This takes some planning, as the software will always set the interface with the higher interface value to be the active link:

```
[edit ethernet-switching-options]
lab@Brandy# show
redundant-trunk-group {
    group dual {
        interface ge-0/0/1.0;
        interface ge-0/0/3.0;
    }
}
```

Without the primary knob, ge-0/0/3 becomes the active link as before, but with the absence of the primary flag in the State field:

```
lab@Brandy# run show redundant-trunk-group group-name dual

Interface   State   Bandwidth   Time of last flap               Flap
                                                                count

ge-0/0/3.0  Up/Act  1000 Mbps   Never                              0
ge-0/0/1.0  Up      1000 Mbps   Never                              0
```

When the failure occurs, ge-0/0/1 takes over as the active link:

```
lab@Brandy# run show redundant-trunk-group group-name dual

Interface   State   Bandwidth   Time of last flap                   Flap
                                                                    count

ge-0/0/3.0  Dwn     1000 Mbps   2008-08-06 10:46:28 UTC (00:00:03 ago)    1
ge-0/0/1.0  Up/Act  1000 Mbps   Never                                     0
```

However, when the ge-0/0/3 link comes back, the active link stays at ge-0/0/1. The only way to switch back to ge-0/0/3 is to temporarily disable ge-0/0/1 and initiate a switchover:

```
[edit]
lab@Brandy# run show redundant-trunk-group group-name dual
```

```
Interface   State    Bandwidth   Time of last flap                    Flap
                                                                      count

ge-0/0/3.0  Up       1000 Mbps   2008-08-06 10:47:07 UTC (00:00:03 ago)   2
ge-0/0/1.0  Up/Act   1000 Mbps   Never                                    0
```

RTG Summary

RTG is a simple alternative to the spanning tree and redundancy in simple topologies. It eliminates all the complex messaging of STP and makes the link decision a purely local matter. RTG is not designed for complex topologies, however, and is best used when an access switch is multihomed to two distribution switches.

Conclusion

This chapter examined Spanning Tree Protocol and its evolution, namely, the creation of the original Spanning Tree Protocol and the problems that it solved. It was evident pretty early on that some improvements needed to be made, and Rapid Spanning Tree Protocol was thus born. RSTP also had some limitations for VLANs and redundancy, so finally, Multiple Spanning Tree Protocol was created.

The spanning tree is not the only way to prevent loops and provide redundancy, however, so we also examined Redundant Trunk Groups.

Now that you have the Layer 2 functionality established, you can start looking at some Layer 3 functionality in the next chapters.

Chapter Review Questions

1. What is the primary purpose of Spanning Tree Protocol?
    a. Load-balance across multiple links
    b. Eliminate loops
    c. Learn neighbor switch information
    d. Create optimized paths throughout the LAN
2. Which bridge is the primary forwarder on a given link?
    a. Designated
    b. Primary
    c. Active
    d. Forwarder
3. By default, how is the root bridge chosen?
    a. Based on the number of total ports
    b. Port identifier

    c. Total port cost

    d. Bridge ID

4. Which two port states are valid STP states:

    a. `Blocking`

    b. `Waiting`

    c. `Listening`

    d. `Establishing`

5. Which command is issued to view the STP port states on an EX Series switch?

    a. `show stp`

    b. `show stp bridge`

    c. `show stp interface`

    d. `show stp ports`

6. True or false: STP provides subsecond failover time.

    a. True

    b. False

    c. There are no absolutes in the universe!

7. Which two new port types are introduced in RSTP?

    a. Backup

    b. Root

    c. Blocking

    d. Alternate

    e. Fast

8. Choose two items that RSTP changes from the original STP specification:

    a. More root bridges

    b. Fewer port states

    c. Different link costs

    d. Faster tree algorithm

9. Which command protects the switch from receiving superior BPDUs?

    a. `Bpdu-block`

    b. `inferior`

    c. `no-root-port`

    d. `edge port`

10. How many regions are allowed when configuring MSTP?

    a. 3

    b. 32

c. 64

d. 100

e. 128

11. What is the maximum number of root bridges that can be configured in MSTP?

   a. 1

   b. 64

   c. 65

   d. 100

12. Which command allows a primary link to revert back after a failover in RTG?

   a. `primary`

   b. `active`

   c. `revert`

   d. `master`

Chapter Review Answers

1. Answer: B. STP's primary goal is to create a loop-free topology. Answer D is close, as it is an optimized topology, but that is not the primary goal of STP.

2. Answer: A. The designated bridge is the device chosen to forward local traffic on a given link. This ensures that on a segment, a bridge does not receive a multicast or broadcast packet and reinject it into the LAN.

3. Answer: D. The root bridge is the bridge with the lowest bridge ID. Remember that the bridge ID consists of the MAC address and the priority field, which is user-configurable.

4. Answer: A, C. Blocking and listening are the only valid STP states in the list.

5. Answer: C. The only other valid command is answer B, but that command lists only local bridge parameters.

6. Answer: B. The failover times in STP could range as high as 30 seconds using default timers. The convergence time of STP is a known issue that has been corrected in RSTP.

7. Answer: A, D. The new ports in RSTP are the backup ports and the alternate ports. Both provide rapid switchover in case of a root or designated port failure.

8. Answer: B, C. RSTP actually decreases the number of port states an interface can be in. This allows for much quicker transition times. The port cost values were also adjusted in order to include modern high-speed ports.

9. Answer: C. This command will place the port in the blocking state if a superior BPDU is received.

10. Answer: C. Sixty-four regions are allowed in MSTP, with each region allowing multiple VLANs.

11. Answer: C. One root bridge is possible per MSTP region, as well as a single root bridge for the common spanning tree.

12. Answer: A. The primary knob causes that interface to be revertive. In other words, the switch will always use the primary interface to forward traffic if it is available.

## CHAPTER 7

Routing on the EX

Wait, routing in a Layer 2/LAN switching book! What gives?

EX switches run JUNOS software, and therefore inherit a rich legacy of carrier-grade routing support. EX routing support pretty much comes for free, which is to say IPv4 routing is included in the base EX chassis. There are no special "routing-enabled" EX software images, or added hardware requirements, or any special commands to turn on/enable IP routing. You simply have to decide that Layer 3 forwarding is desired, determine which routing protocols to use, and then configure the EX to route as you would any other JUNOS device.

In many network designs, the optimal solution combines Layer 2 switching at the edge with Layer 3 routing in the distribution and core layers. The built-in (rather than bolted-on) nature of EX routing support makes the deployment of such hybrid networks both simple and reliable, given the known stability of JUNOS routing and the respectable Layer 3 scaling limits supported by the EX platforms.

Large enterprise networks or service provider data centers may warrant the extended routing capabilities of a purpose-built routing platform such as a J Series, MX Series, or even M or T Series routing platform.

It should be noted that the goal here is to expose the reader to general EX routing capabilities and basic configurations and operational verification examples. A detailed discussion of IP routing and routing protocols, including aggregate and generated routes, multicast, OSPF, IS-IS, BGP, and complex routing policy, is outside the scope of this book. Readers interested in expanding on the topic of IP routing in a JUNOS environment are encouraged to consult the companion volume to this book, *JUNOS Enterprise Routing (http://oreilly.com/catalog/9780596514426/)*, by Doug Marschke and Harry Reynolds (O'Reilly).

The topics covered in this chapter include:

- EX routing overview
- Inter-VLAN routing and the Routed VLAN Interface

- Static routing
- RIP routing

EX Routing Overview

This section provides a high-level overview of key routing concepts, some JUNOS-specific, that you need to understand before moving into any particular routing scenario. Readers familiar with IP routing in a JUNOS environment are expected to be familiar with the material presented here; the review may be worthwhile, nonetheless.

What Is Routing?

Recall that Chapter 1 provided a general overview of bridging and routing concepts. Rather than reinvent that wheel, let's do a quick review of the key differences between bridging and routing. They are summarized in Table 7-1.

Table 7-1. Bridging versus routing summary

Bridge	Router
Forwards frames based on flat Media Access Control (MAC) address, limited scalability	Forwards packets based on variable length IP address (longest match), supports addressing hierarchy for global scalability
Transparent to end stations (shared subnet, no Time to Live [TTL] decrement)	End stations participate in routing via default gateway and local/non-local decision, TTL decrement at each hop
Floods Broadcast, Unknown, or Multicast (BUM) addresses	Filters BUM addresses
Prone to loops, uses Spanning Tree Protocol (STP) to block redundant paths, no load balancing, slow convergence	Tolerates loops, no blocking, load balancing over multiple paths, fast convergence
Uses bridging protocols to find and block loops for a single path	Uses routing protocols to find and select optimal paths; in some cases unequal cost load balancing is possible
Limited filtering/firewall/services	Rich set of filtering/firewall and enhanced services such as Dynamic Host Configuration Protocol (DHCP) or stateful firewall/deep packet inspection

Interior Gateway Protocol overview

Interior Gateway Protocols (IGPs) provide routing connectivity *within* the *interior* of a given routing domain (RD). An RD is defined as a set of routers under common administrative control that share a common routing protocol. Most multivendor IP networks use a standardized IGP to permit interoperability. Small networks tend to use Routing Information Protocol (RIP), while larger, more complicated designs favor the performance and scalability of a link state protocol such as Open Shortest Path First (OSPF).

In general, an IGP functions to advertise and learn network prefixes (routes) from neighboring routers. It uses this information to populate the routing table (RT) with entries for all sources advertising reachability for a given prefix. A route selection algorithm then selects the best (i.e., the shortest) path between the local router and each destination, and the associated next hop is pushed into the forwarding table (FT) to direct matching packets along that path.

When network conditions change, perhaps due to equipment failure or management activity, the IGP both generates and receives updates and recalculates a new best route to the affected destinations. Here, the concept of a "best" route is normally tied to a route metric, which is the criterion used to determine the relative path cost of a given route. Generally speaking, a route metric is significant only to the routing protocol it's associated with, and is meaningful only within a given RD. In some cases, a router may learn multiple paths to an identical destination from more than one routing protocol. Given that metric comparison between two different IGPs is meaningless, the selection of the best route between multiple routing sources is controlled by a route preference.

In addition to advertising internal network reachability, IGPs are often used to advertise routing information that is external to that IGP's RD through a process known as *route redistribution*. Route redistribution is performed via routing policy in JUNOS software.

EX Routing Capabilities

All EX platforms ship with support for basic IPv4 routing. While described as "basic," the EX's support of static, RIP, and OSPF routing, most users will agree, is quite complete, and more than they will ever need. An advanced routing license is needed to support Border Gateway Protocol (BGP) and the Intermediate System to Intermediate System (IS-IS) routing protocols; a single advanced routing license is used to unlock both features simultaneously.

Table 7-2 reflects EX Layer 3 routing capabilities as of the 9.3 JUNOS release. Note that the software demonstrated in this book is based on the 9.2 release, which means, for example, that IPv6 routing is not supported in the current EX test bed.

Table 7-2. EX Layer 3 feature support circa JUNOS 9.3

Layer 3 feature	License	Release	Comment
IPv4	No	9.0R2	IPv4 unicast and multicast offered in the initial 9.0R2 EX release.
Static routing	No	9.0R2	Often frowned upon, but static routes have their place.
OSPF	No	9.0R2	Link state (LS) IGP for IPv4. No support for Traffic Engineering (TE) extensions.
RIPv1 and v2	No	9.0R2	Distance-vector-based IPv4 unicast routing protocol.
IS-IS	Yes	9.0R2	LS IGP for IPv4, requires advanced license. No support for End System to Intermediate System (ES-IS), IPv6 multicast, or TE.
BGP/MBGP	Yes	9.0R2	Full BGP and Multiprotocol BGP support with advanced license.

Layer 3 feature	License	Release	Comment
SSM	No	9.0R2	Single-Source Multicast support.
PIM SM	No	9.0R2	Protocol Independent Multicast for IPv4 operating in Sparse mode.
PIM DM/SSM	No	9.3R2	Protocol Independent Multicast for IPv4 operating in Dense or Source-Specific mode.
IGMPv1 through v3	No	9.0R2	IPv4 host registration protocols for multicast.
IGMP snooping	No	9.1R1/9.2R1	Allows Layer 2 switches to become multicast-group-aware; snooping support on Routed VLAN Interface (RVI) added in 9.2R1.
Virtual router (VR)	No	9.3R2	Virtual router instances for "VRF-Lite" support (multiple VRs in a single CE router).
IPv6 unicast	No	9.3R2	IPv6 multicast support to follow in a later release.
OSPF3, RIPng, MBGP	No	9.3R2	IPv6 unicast routing support via OSPF3, RIPng, and MBGP.
Routed VLAN Interface	Yes	9.0R2/9.4 (8200)	RVI is like an IOS SVI, used to route/support IGMP snooping between virtual LANs (VLANs).

What's missing?

Even when you factor in the latest 9.3 capabilities, a number of Layer 3 protocols are supported in JUNOS routing platforms that are not supported in the EX JUNOS builds. Some of the more significant Layer 3 features not supported include:

- IPv6 multicast
- IP Security (IPSec)
- Multiprotocol Label Switching (MPLS) in all forms, including Label Distribution Protocol (LDP) and Resource Reservation Protocol (RSVP), TE, and by extension, Virtual Private LAN Service (VPLS), Layer 2/Layer 3 virtual private network (VPNs), and Layer 2 circuits
- Generic Routing Encapsulation (GRE) tunneling
- Logical routers
- Traffic sampling

The complete list of unsupported Layer 3 functionality for the 9.3 EX JUNOS release is available at *http://www.juniper.net/techpubs/en_US/junos9.3/topics/reference/general/ex-series-l3-protocols-not-supported.html*. Given the relatively recent release of the EX platform, and its switching rather than routing focus, the lack of complete Layer 3 parity with routing flavors of JUNOS is understandable. Additional Layer 3 capabilities are likely to be added as the product matures; as always, you should check Juniper's website at *http://www.juniper.net/products_and_services/switching.html* to confirm the latest feature and protocol support.

Layer 3 scaling limits

Everything has a limit, and there is no exception to this universal rule afforded to the EX Series. Generally speaking, the EX is designed to excel at Layer 2 filtering and forwarding, and as a result, some compromises have been made with respect to EX Layer 3 scaling, at least when compared to Juniper's purpose-built routing platforms. Then again, as most JUNOS routing platforms do not offer *any* bridging (the MX is the exception), the Layer 3 scaling limits for the EX are both understandable and respectable.

Table 7-3 summarizes Layer 3 scaling guidelines for EX switches.

Table 7-3. EX Layer 3 scaling

Feature	EX3200/4200	EX8200	Comment
IPv4 unicast routes	16K	512K	FT entries for IPv4, RT is limited by RAM
IPv4 multicast routes	2K	256K	FT entries for IPv4, RT is limited by RAM
IPv6 unicast routes	4K	128K	FT entries for IPv6, RT is limited by RAM
IPv6 multicast routes	512	64K	FT entries for IPv6, RT is limited by RAM
Multicast groups	2K	4K	Active multicast group addresses
BGP peers	128	128	BGP peering limited by memory and number of routes per peer
Address Resolution Protocol (ARP) entries	20K	64K	ARP table holds Layer 3 to Layer 2 address bindings

JUNOS Routing Concepts

This section details generic JUNOS routing concepts that you must understand before delving into any specific routing protocol examples. Many of these concepts exist in other routing products, usually with a different name to keep the innocent guessing.

Global route preference

A router always seeks to forward a packet based on the longest match, and the longest match always wins, regardless of how the longest matching route was learned. It's possible to learn routes of the same specificity from multiple sources. To break ties, each source is assigned a global preference. It can be said that the global preference determines the "goodness" of a routing source. Therefore, routes learned through local administrative action—for example, static routes—are more believable than the same routes learned through a routing protocol such as OSPF. In Cisco IOS, this concept is called *administrative distance*. Table 7-4 shows the default protocol preferences for JUNOS software.

Table 7-4. Global protocol preference values

Source	Purpose	Default preference
Local	Local IP address of the interface	0
Directly connected network	Subnet corresponding to the directly connected interface	0
Static	Static routes	5
RSVP	Routes learned from the Resource Reservation Protocol used in MPLS	7
LDP	Routes learned from the Label Distribution Protocol used in MPLS	9
OSPF internal route	OSPF internal routes, such as interfaces that are running OSPF	10
IS-IS Level 1 internal route	IS-IS Level 1 internal routes, such as interfaces that are running IS-IS	15
IS-IS Level 2 internal route	IS-IS Level 2 internal routes such as interfaces that are running IS-IS	18
Redirects	Routes from Internet Control Message Protocol (ICMP) redirect	30
Kernel	Routes learned via route socket from kernel	40
Simple Network Management Protocol (SNMP)	Routes installed by Network Management System	50
Router discovery	Routes installed by ICMP router discovery	55
RIP	Routes from Routing Information Protocol (IPv4)	100
RIPng	Routes from Routing Information Protocol (IPv6)	100
PIM	Routes from Protocol Independent Multicast	105
DVMRP	Routes from Distance Vector Multicast Routing Protocol	110
Aggregate	Aggregate and generated routes	130
OSPF AS external routes	Routes from Open Shortest Path First that have been redistributed into OSPF	150
IS-IS Level 1 external route	Routes from IS-IS Level 1 that have been redistributed into IS-IS	160
IS-IS Level 2 external route	Routes from IS-IS Level 1 that have been redistributed into IS-IS	165
BGP	Routes from Border Gateway Protocol	170

As with a route metric, numerically lower preference values are preferred. You can alter the default preference values through configuration. Altering global route preference should be considered only after careful thought and, better yet, testing—there can be unpredictable side effects.

Routing tables and RIB groups

JUNOS automatically creates a number of RTs that are used for a variety of purposes. In advanced routing scenarios, users can create their own RTs, either indirectly through the use of virtual routers or Layer 2/Layer 3 VPNs, or directly through the use of Routing Information Base (RIB) groups.

Generally speaking, each RT/RIB populates a designated portion of the FT. This creates a single FT that's partitioned based on each specific RT context. Packets are forwarded based on this RT context. The ability to maintain separate RTs and FTs is a key component of any VPN type of service.

You can view the contents of a particular RT using the command show route table <table name>. The general naming convention for RTs takes the form of the protocol family such as inet (Internet), followed by a period and a non-negative integer. Routing instance table names are somewhat the exception here, taking the form of instance-name.inet.0, where the first part consists of a user-assigned symbolic name, followed by the protocol family and table ID, which is inet.0 in this example.

The inet.0 table

The inet.0 table is the default unicast RT for the IPv4 protocol. This is the main RT used to store unicast routes such as interface local/direct, static, or dynamically learned routes. When you issue a show route command, all tables are listed chronologically, starting with inet.0:

```
lab@Rum> show route

inet.0: 3 destinations, 3 routes (3 active, 0 holddown, 0 hidden)
+ = Active Route, - = Last Active, * = Both

10.10.1.5/32        *[Direct/0] 2d 03:22:59
                     > via lo0.0
172.16.69.0/24      *[Direct/0] 2d 03:22:59
                     > via me0.0
172.16.69.5/32      *[Local/0] 2d 03:22:59
                      Local via me0.0

__juniper_private1__.inet.0: 4 destinations, 6 routes (1 active, 0 holddown, 3
hidden)
+ = Active Route, - = Last Active, * = Both

128.0.0.0/2         *[Direct/0] 2d 03:22:59
                     > via bme0.32768
                     [Direct/0] 2d 03:22:59
                     > via bme0.32768
                     [Direct/0] 2d 03:22:59
                     > via bme0.32768
```

In this example, you see that the inet.0 table has a total of three routes, all of which are active; two of the routes are from a Direct routing source, which means the route is directly connected and learned via IP address assignment to the related interface. This type of route is referred to as *connected* in IOS-speak. The third inet.0 entry is an automatically generated Local route, which is used to direct matching traffic to the local host (the routing engine, or RE). A local route represents the /32 local host addresses for the interface itself when a network's mask that is less than 32 bits long is specified. Here, the 172.16.69.5/32 route is the local host route resulting from the assignment of a 172.16.69.5/24 network address to an interface.

In this example, there are no statically defined or learned routes at present, but this will change shortly. The Direct routes stem from IP address assignment to the loopback and Out of Band (OoB) management interfaces. The /32 local route is automatically created for direct interface routes that represent a network rather than a host and are used to direct matching traffic to the local host itself. A network's routes will have a network mask length that is less than /32, whereas a local route's mask length is always /32.

Only active routes can be considered for packet forwarding. A route can be hidden because of problems installing the route—for example, the inability to resolve the route's next hop to a local forwarding next hop. You can use the hidden switch to display hidden routes; include the detail or extensive switch to see additional detail about any matching routes.

This example also illustrates the presence of a second RT named juniper_private. This table is automatically generated by JUNOS and is used for internal communications with chassis components. Here the table is used to provide communications between the three EX Packet Forwarding Engine (EX-PFE) complexes contained in the switch.

Routing policy

We demonstrate the JUNOS routing policy briefly later in this chapter, and we describe it in detail in Chapter 8. For now, it's important to note that routing policy is used to control the exchange and redistribution of routes among various routing sources according to your *administrative* policies and agreements. Policy gives you control over what routes are accepted for advertisement by a given routing protocol, and allows you to match on and manipulate or modify various route attributes such as metric, next hop, preference, and so on.

On Cisco Systems platforms, these types of functions are often performed through some combination of the redistribute command, distribute lists, or a route-map and its associated IP access lists. While different, to be sure, JUNOS software routing policy provides the same functionality with a consistent set of semantics/syntax, for all protocols, and all in one place!

Router ID and Autonomous System Number

Many routing protocols require that the source of routing information be uniquely identified using the concept of a router ID (RID). A RID normally takes the form of an IPv4 address, and in most cases does not have to be reachable to correctly function as a RID. JUNOS permits one RID per routing instance, and the same value is used by all protocols that require a RID in that instance (OSPF, OSPFv3, and BGP). The current best practice is to base the RID on the router's globally routable lo0 address, a process that happens automatically. You can override this behavior and explicitly configure a RID at the [edit routing-options] hierarchy.

An Autonomous System Number (ASN) is required for BGP operation; you cannot commit a BGP-related configuration without also defining the router's own ASN. In this regard, it can be said that the ASN is not really protocol-independent, but for whatever reason it is configured under [routing-options], rather than in the [protocols BGP] stanza itself. We do not demonstrate BGP routing in this book, so an ASN will not be necessary.

Summary of EX Routing Capabilities

This section reviewed key differences between bridging and routing, summarized the Layer 3 support offered by EX switches in the 9.3 release, and defined general JUNOS routing concepts that you need to understand before configuring and monitoring IP routing.

Given their JUNOS pedigree, EX switches arrived at the party with a rich and sophisticated set of Layer 3 features. The one JUNOS model means you can directly leverage routing experience on Juniper M, MX, or T Series platforms on your new EX switch, or vice versa, as the case may be.

The next section delves into the thick of it with a detailed look at the EX platform's Routed VLAN Interface (RVI), which will be used to link together the Sales and Admin VLAN islands we discussed at the end of the preceding chapter.

Inter-VLAN Routing

EX switches support the notion of a Routed VLAN Interface used to route IP traffic between VLANs. The RVI is a purely logical interface construct that does not require any additional interface hardware or cabling.

Figure 7-1 depicts the VLAN topology example, which is pretty much the exit state from Chapter 5.

Thinking back to Chapter 5, recall that things began with a single flat Layer 2 network that encompassed all four hosts. This result was a single logical IP subnet (LIS) shared among all hosts, which then had full IP layer connectivity.

Then, two VLANs were deployed, and continued connectivity *within* each VLAN was confirmed, along with a *lack* of communications for *inter-VLAN* traffic. The original LIS was intentionally not renumbered along with the VLAN deployment to help demonstrate the inherent isolation of VLANs. After all, stations on separate subnets *require* a router to communicate, and there was no router in Chapter 5. Therefore, leaving the stations on the same LIS gave them the maximum chance for continued connectivity, which, as noted, failed after VLANs were deployed.

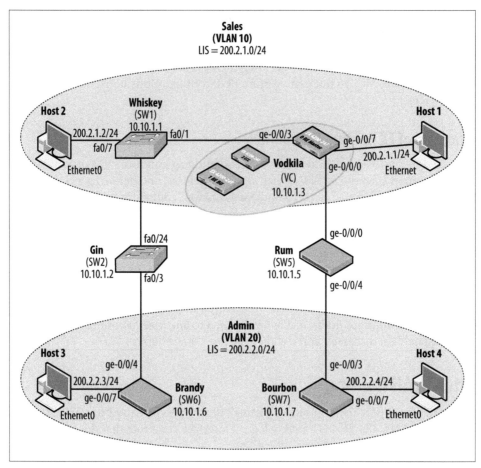

Figure 7-1. VLAN switching topology

The failure to communicate is because VLAN-aware switches segregate traffic on access links, permitting only traffic that is associated with the access link's VLAN to be sent out that port. With VLAN segregation successfully demonstrated, and in preparation for routing, the machines in the Sales VLAN are renumbered to be on the 200.2.1.0/24 network, while Host3 and Host4 remain on their original 200.2.2.0/24 network. The result is a distinct LIS per VLAN model, which represents the current best practice for aligning Layer 3 network numbering to Layer 2 broadcast domains. While the extra steps obviously did not matter from the viewpoint of the two VLANs and the Layer 2 network, a conventional IP network numbering scheme makes future deployment of Layer 3 routing services relatively painless.

The isolation of VLAN traffic at Layer 2 creates the need for Layer 3 forwarding (i.e., routing) for inter-VLAN traffic. The need for routing between VLANs offers the inherent advantage of tight control over what can be forwarded, and between which endpoints, given that routers can impose firewall and policy restrictions at Layers 3 and 4, and because routers can provide enhanced services at the IP layer, such as IPSec-based encryption, Network Address Translation (NAT), URL screening, and so on.

A Router on a Stick

Of course, the obvious downside to the need for a Layer 3 device to facilitate inter-VLAN communications is the need for the routing device itself. This need led to a somewhat novel solution known as a *router on a stick*, as shown in Figure 7-2.

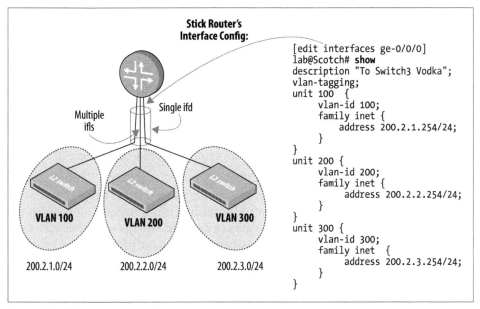

Figure 7-2. The router on a stick

Figure 7-2 shows three VLAN communities that are provided Layer 3 connectivity via a router that's attached to all three VLANs. Note that each VLAN community is assigned a unique IP subnet, and that the router is attached to it via a single physical interface (an interface device, or ifd in JUNOS); this is the stick part of the router-on-a-stick model. This single interface is in turn carved into multiple logical interfaces (subinterfaces, or interface logical, or ifl in JUNOS) through VLAN tagging. This logical partitioning is shown in the sample interface configuration stanza. Actually, deploying a router with a single interface, a device that historically was referred to as a "boat anchor" (or "doorstop"), is the genesis of the router-on-a-stick concept, a situation that also explains why it's sometimes called "lollipop" or "U-turn" routing.

Besides the cost of yet another lump of silicon, the packet-looping nature of the router on a stick can have performance and reliability impacts, especially when some old router that happened to be lying around is pressed into an extended tour of service—

only this time while perched in a somewhat ungainly fashion high up on a stick, as it were. A failure or performance bottleneck in this critical role affects all inter-VLAN traffic, making this "stick" more like a throne, at least as far as your company's intranet is concerned!

It's certainly possible to use a physically distinct interface per VLAN model. This approach can improve both reliability and performance, albeit oftentimes along with a non-linear increase in solution cost. This is because in most cases, the majority of a routing platform's cost stems from its interfaces rather than the router's chassis. When reliability is a concern, multiple stick routers running Virtual Router Redundancy Protocol (VRRP) is generally the best bet; such a design eliminates single points of failure in the Layer 3 access portion of the network. We demonstrate VRRP in a later section, and discuss it in detail in Chapter 11.

Enter the Routed VLAN Interface

While lollipop routing has a certain aesthetic appeal, it was not long before a sharp engineer realized that by adding some primitive Layer 3 functionality to the switch, a purely software-based interface construct could be implemented to provide inter-VLAN routing services without the need for an external routing device. Sort of a *virtual stick*, if you will.

An integrated VLAN routing capability is attractive because it serves to lower cost while increasing reliability. As long as the Layer 3 forwarding performance is on par with the device's Layer 2 capabilities, which isn't always the case, there is very little drawback to integrated routing; this is especially true when the device can provide Layer 3 services such as firewall filters or policing, in addition to the basic routing function.

In the Juniper architecture, the Layer 3 interface that interconnects to VLAN instances is called a Routed VLAN Interface (RVI). A similar function is provided on Cisco switches via a Switched Virtual Interface (SVI).

Full Layer 3 functionality

In keeping with Juniper tradition, EX switches provide a high-performance solution to the inter-VLAN routing problem. An EX forwards Layer 3 at wire rate, supports filtering and policing, and offers a rich set of routing policy options so that you can tailor Layer 3 access and interconnectivity based on your specific needs.

In contrast, an IOS device may support the basic SVI interconnect while lacking *full* IP routing capabilities; in other words, the full Enhanced Multilayer Software Image (EMI) may not be installed, yielding extremely limited Layer 3 capabilities. Stated differently, you can configure an SVI on an IOS build that does not permit the IP routing configuration statement, which is good, in that you can interconnect VLAN without a Layer 3-capable IOS version, but bad in that you wind up with limited Layer 3 functionality that may end up forcing a software upgrade or use of a standalone router anyway. All

EX versions have full `IP routing` (BGP/IS-IS advanced routing license notwithstanding), and these capabilities are available on the RVI, as well as any interfaces that are configured for Layer 3 operation.

Deploy an RVI

In this section, you'll configure an RVI on `Vodkila` to provide Layer 3 connectivity between stations in the Sales and Admin VLANs. Before making any changes, refer back to Figure 7-1, and again verify intra-VLAN connectivity for both the 200.2.1.0/24 and 200.2.2.0/24 subnets. Start at `Host1`:

```
Host1#ping 200.2.1.2

Type escape sequence to abort.
Sending 5, 100-byte ICMP Echos to 200.2.1.2, timeout is 2 seconds:
!!!!!
Success rate is 100 percent (5/5), round-trip min/avg/max = 4/5/8 ms
```

The `Host1` to `Host2` pings succeed, so move on to `Host3`:

```
Host3# ping 200.2.2.4

Type escape sequence to abort.
Sending 5, 100-byte ICMP Echos to 200.2.2.4, timeout is 2 seconds:
!!!!!
Success
```

The results confirm the expected IP connectivity within each logical IP subnet. Given that the current IP addressing does not accommodate direct communications between machines in different VLANs, you quickly reassign `Host1`'s IP address to place it on the LIS associated with the Admin VLAN. This allows a quick confirmation that, as before, VLAN-aware switching successfully prevents inter-VLAN communications, as evidenced by the ping failure:

```
Host1#conf t
Enter configuration commands, one per line.  End with CNTL/Z.
Host1(config)#inter ethernet 0
Host1(config-if)#ip add 200.2.2.1 255.255.255.0
Host1(config-if)#^Z
Host1#sho ip inter brie
Interface              IP-Address     OK? Method Status
Protocol
. . .
Ethernet0              200.2.2.1      YES manual up             up
. . .
Host1#ping 200.2.2.4

Type escape sequence to abort.
Sending 5, 100-byte ICMP Echos to 200.2.2.4, timeout is 2 seconds:
.....
Success rate is 0 percent (0/5)
Host1
```

Since we're satisfied with the baseline VLAN trunking behavior, Host1's IP address is reconfigured to reflect the Sales VLAN LIS:

```
Host1#show ip interface brief ethernet 0
Interface               IP-Address      OK? Method Status
Protocol
Ethernet0               200.2.1.1       YES manual up              up
```

Configure and test an RVI

Figure 7-3 details the RVI deployment plans for this scenario.

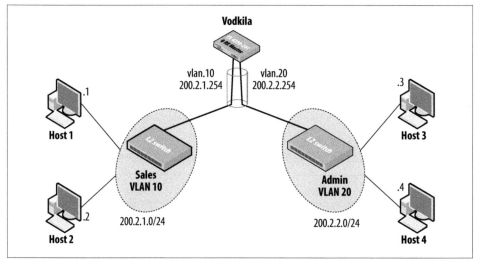

Figure 7-3. RVI test topology

In this example, the Vodkila switch is chosen to house the RVI function because of its somewhat central location, and the fact that it already has both VLANs defined—given that it's currently trunking both the Sales and Admin VLANs over to Whiskey. Also, as a Virtual Chassis (VC), Vodkila represents the most resilient node in the network. Figure 7-3 shows the two VLAN communities, along with their respective LISs.

The magic happens at Vodkila, where a virtual interface that must be named vlan is defined with two logical units. This example shows the units matching their respective VLANs; this is purely a best practice to help avoid confusion, given that there is no significance to a unit number other than it having to be unique under a given interface device. Each unit under the vlan interface is assigned a unique IP address, and a network mask that places it into the LIS associated with its respective VLAN. This example sets the VLAN interface host bits to the value .254, which is the highest usable host assignment in this case. This, or using the lowest available host IP address (1), is a common practice that helps make each LIS's default gateway easy to compute.

With the host addressing and switched VLAN infrastructure already confirmed, the only thing needed to complete this task is to configure the RVI at Vodkila. This is a two-step process: first, define the RVI itself, and then associate the RVI's units with the desired VLAN domains.

Begin by defining the vlan interface, as you would any other interface, at the [edit interfaces] hierarchy:

```
[edit]
lab@Vodkila# edit interfaces

[edit interfaces]
lab@Vodkila# set vlan unit 10 family inet address 200.2.1.254/24

[edit interfaces]
lab@Vodkila# set vlan unit 20 family inet address 200.2.2.254/24

[edit interfaces]
lab@Vodkila# show vlan
unit 10 {
    family inet {
        address 200.2.1.254/24;
    }
}
unit 20 {
    family inet {
        address 200.2.2.254/24;
    }
}
```

The changes are committed, and the instantiation of the vlan interface is confirmed:

```
[edit]
lab@Vodkila# run show interfaces vlan
Physical interface: vlan, Enabled, Physical link is Up
  Interface index: 128, SNMP ifIndex: 109
  Type: 33, Link-level type: 71, MTU: 1518
  Device flags   : Present Running
  Interface flags: SNMP-Traps
  Link flags     : None
  Current address: 00:1f:12:3d:b4:c0, Hardware address: 00:1f:12:3d:b4:c0
  Last flapped   : Never
    Input packets : 0
    Output packets: 0

  Logical interface vlan.10 (Index 65) (SNMP ifIndex 133)
    Flags: Link-Layer-Down SNMP-Traps 0x0 Encapsulation: Unspecified
    Input packets : 0
    Output packets: 0
    Protocol inet, MTU: 1500
      Flags: None
      Addresses, Flags: Dest-route-down Is-Preferred Is-Primary
        Destination: 200.2.1/24, Local: 200.2.1.254, Broadcast: 200.2.1.255

  Logical interface vlan.20 (Index 68) (SNMP ifIndex 137)
```

```
      Flags: Link-Layer-Down SNMP-Traps 0x0 Encapsulation: Unspecified
      Input packets : 0
      Output packets: 0
      Protocol inet, MTU: 1500
        Flags: None
        Addresses, Flags: Dest-route-down Is-Preferred Is-Primary
          Destination: 200.2.2/24, Local: 200.2.2.254, Broadcast: 200.2.2.255
```

Note that the interface device is reported as up, while both logical units are reported as down. This is the result of a lack of binding to a VLAN instance, and it's easily remedied with the l3-interface statement at the [edit vlans <vlan-name>] hierarchy:

```
[edit interfaces]
lab@Vodkila# top edit vlans

[edit vlans]
lab@Vodkila# set Sales_vlan l3-interface vlan.10

[edit vlans]
lab@Vodkila# set Admin_vlan l3-interface vlan.20
```

The configuration change is confirmed, and then committed:

```
[edit vlans]
lab@Vodkila# show
Admin_vlan {
    vlan-id 20;
    l3-interface vlan.20;
}
Sales_vlan {
    vlan-id 10;
    l3-interface vlan.10;
}
```

After the assignment to a VLAN, the RVI's units are confirmed to be operational:

```
[edit]
lab@Vodkila# run show interfaces vlan.10 detail
  Logical interface vlan.10 (Index 65) (SNMP ifIndex 133) (Generation 139)
    Flags: SNMP-Traps 0x0 Encapsulation: Unspecified
    Traffic statistics:
     Input  bytes  :                   0
     Output bytes  :                   0
     Input  packets:                   0
     Output packets:                   0
    IPv6 transit statistics:
     Input  bytes  :                   0
     Output bytes  :                   0
     Input  packets:                   0
     Output packets:                   0
    Local statistics:
     Input  bytes  :                   0
     Output bytes  :                   0
     Input  packets:                   0
     Output packets:                   0
    Protocol inet, MTU: 1500, Generation: 158, Route table: 0
      Flags: None
```

```
        Addresses, Flags: Is-Preferred Is-Primary
            Destination: 200.2.1/24, Local: 200.2.1.254, Broadcast: 200.2.1.255,
            Generation: 154
```

Note that currently the traffic statistics for the vlan.20 unit reflect that no traffic has been sent or received, which is expected at this time. The show vlans extensive command also confirms correct RVI-to-VLAN mapping:

```
[edit]
lab@Vodkila# run show vlans extensive
VLAN: Admin_vlan, Created at: Mon Jan 10 04:49:32 2005
802.1Q Tag: 20, Internal index: 2, Admin State: Enabled, Origin: Static
Protocol: Port Mode, Layer 3 interface: vlan.20 (UP)
IP addresses: 200.2.2.254/24
Number of interfaces: Tagged 2 (Active = 2), Untagged  0 (Active = 0)
        ge-0/0/0.0*, tagged, trunk
        ge-0/0/3.0*, tagged, trunk

VLAN: Sales_vlan, Created at: Mon Jan 10 04:49:32 2005
802.1Q Tag: 10, Internal index: 3, Admin State: Enabled, Origin: Static
Protocol: Port Mode, Layer 3 interface: vlan.10 (UP)
IP addresses: 200.2.1.254/24
Number of interfaces: Tagged 1 (Active = 1), Untagged  1 (Active = 1)
        ge-0/0/3.0*, tagged, trunk
        ge-0/0/7.0*, untagged, access
. . .
```

With the RVI interface defined and bound to the VLAN instances, you can test reachability to the RVI IP address from within each VLAN:

```
Host1#ping 200.2.1.254

Sending 5, 100-byte ICMP Echos to 200.2.1.254, timeout is 2 seconds:
!!!!!
Success rate is 100 percent (5/5), round-trip min/avg/max = 1/4/8 ms
Host1#
```

The result confirms that Host1 has IP-level connectivity to the RVI address assigned to the Sales VLAN. Host3 is also able to ping the 200.2.2.254 RVI address for the Admin VLAN (not shown). With VLAN-to-RVI connectivity confirmed within each VLAN, you move on to confirm inter-VLAN connectivity with a ping between Host1 and Host3:

```
Host1#ping 200.2.2.3

Type escape sequence to abort.
Sending 5, 100-byte ICMP Echos to 200.2.2.3, timeout is 2 seconds:
.....
Success rate is 0 percent (0/5)
Host1#
```

Unfortunately, there is no joy in Pingville for you.

But what could be wrong? Thinking back on the concepts of routing, you recall that, unlike a bridge, a router is not transparent, and an end station must participate in the act of routing by making a local versus remote delivery decision for each packet it generates. The need to consciously evoke the services of a router is an offshoot of the fact that routers do not flood unknown traffic, and they process broadcasts only for the purpose of local host delivery. In other words, a router tends to only process packets that are explicitly addressed to the router's MAC address (at Layer 2), but which are then found to have a destination IP address that does not match any locally assigned IP addresses. Stated differently, a router routes a packet that is *not* addressed to the router itself, when that packet is received in a frame that *is* addressed to the router's receiving interface.

What this boils down to is that Host1 needs to know a default gateway address in order to reach destinations that are not on its local IP subnet. A host generally learns its default router through administrative action, or via a dynamic mechanism such as DHCP. Given that the latter is currently not running in the network (although an EX can function as a DHCP server, this is not configured), and even if our "hosts" are actually Cisco routers, generally routers do not auto-configure a default gateway using DHCP. Hosts are assumed to be dumb, and therefore they need DHCP, whereas routers are assumed to be managed by smart folks like you. As a result, a router might *provide* the DHCP service, but it is rarely a client for one. For one reason, it could be bad having a pair of routers caught up in some auto-configuration race condition caused by them both accepting and then acting upon configuration parameters found in each other's DHCP messages.

A show ip route command on Host1 quickly identifies the problem:

```
Host1#show ip route 200.2.2.0
% Network not in table
```

The output makes it clear that there is no network route for the remote Admin LIS. Perhaps there is a default route....

```
Host1#show ip route
Codes: C - connected, S - static, I - IGRP, R - RIP, M - mobile, B - BGP
       D - EIGRP, EX - EIGRP external, O - OSPF, IA - OSPF inter area
       N1 - OSPF NSSA external type 1, N2 - OSPF NSSA external type 2
       E1 - OSPF external type 1, E2 - OSPF external type 2, E - EGP
       i - IS-IS, L1 - IS-IS level-1, L2 - IS-IS level-2, ia - IS-IS inter area
       * - candidate default, U - per-user static route, o - ODR
       P - periodic downloaded static route

Gateway of last resort is not set

C    200.2.1.0/24 is directly connected, Ethernet0
```

And there you have it. Host1 clearly has no network route to the target 200.2.2.0 subnet, nor does it have a default route to fall back on. To resolve this problem, you need to add a default gateway (a candidate route of last resort) to all four hosts. In each case, the next hop of the default route is the RVI address associated with that station's VLAN. This process is shown for Host1:

```
Host1#conf t
Host1(config)#ip route 0.0.0.0 0.0.0.0 200.2.1.254
Host1(config)#^Z
```

The static default route is confirmed to be present and pointing to the IP address associated with the vlan.10 unit:

```
Host1#show ip route
Codes: C - connected, S - static, I - IGRP, R - RIP, M - mobile, B - BGP
       D - EIGRP, EX - EIGRP external, O - OSPF, IA - OSPF inter area
       N1 - OSPF NSSA external type 1, N2 - OSPF NSSA external type 2
       E1 - OSPF external type 1, E2 - OSPF external type 2, E - EGP
       i - IS-IS, L1 - IS-IS level-1, L2 - IS-IS level-2, ia - IS-IS inter area
       * - candidate default, U - per-user static route, o - ODR
       P - periodic downloaded static route
Gateway of last resort is 200.2.1.254 to network 0.0.0.0

C    200.2.1.0/24 is directly connected, Ethernet0
S*   0.0.0.0/0 [1/0] via 200.2.1.254
```

Note how the default route at Host3 correctly points to the Admin VLAN's RVI address as its next hop. With all hosts now having a default route to the RVI unit serving their VLAN, we can expect Vodkila to route between the two subnets. All that is needed now is some inter-VLAN traffic for the RVI to route, and this is easily accommodated with a ping between Host1 and Host3:

```
Host1# ping 200.2.2.3

Type escape sequence to abort.
Sending 5, 100-byte ICMP Echos to 200.2.2.3, timeout is 2 seconds:
!!!!!
Success rate is 100 percent (5/5), round-trip min/avg/max = 4/9/32 ms
```

Awesome! The ping is successful, as is a ping between Host2 and Host4, which is not shown. A traceroute is performed as final confirmation that all is working to your Layer 3 plan:

```
Host1#traceroute 200.2.2.3

Type escape sequence to abort.
Tracing the route to 200.2.2.3

  1 200.2.1.254 0 msec 0 msec 4 msec
  2 200.2.2
```

The results confirm that one transit hop is needed to reach the target destination, and that the first hop is indeed the default gateway for the 200.2.1.0/24 subnet. A confirmation of non-zero packet counts on the RVI completes the confirmation of the inter-VLAN routing scenario:

```
[edit]
lab@Vodkila# run show interfaces vlan.10
  Logical interface vlan.10 (Index 65) (SNMP ifIndex 133)
    Flags: SNMP-Traps 0x0 Encapsulation: Unspecified
    Input packets : 21
```

```
    Output packets: 11
  Protocol inet, MTU: 1500
    Flags: None
    Addresses, Flags: Is-Preferred Is-Primary
      Destination: 200.2.1/24, Local: 200.2.1.254, Broadcast: 200.2.1.255
```

Use VRRP with an RVI

As noted previously, one advantage to the JUNOS RVI construct is the ability to leverage the existing JUNOS Layer 3 infrastructure, which includes firewall filters and VRRP. VRRP is a standardized form of Cisco's Hot Standby Routing Protocol (HSRP), and is designed to provide redundancy in the first hop, where hosts tend to use a single default gateway address to reach all remote destinations. In this case, having multiple routers alone does not solve the redundancy issue, as the failure of the currently used next hop requires that an updated default route that points to the backup gateway be pushed out to all hosts. Later, when service is restored, the default gateway change has to be backed out to allow normal forwarding over the primary router to resume.

Note that routers *run* VRRP, but they do not use the related Virtual IP (VIP) as a next hop. This is because, unlike hosts, a router normally runs a routing protocol that allows it to learn of changes in the forwarding topology, and to react accordingly. Aside from the VRRP messages themselves, and ARP replies that result from ARP requests to the VIP, the router generates all of its local control plane traffic from the real, rather than virtual, IP address.

In this example, the goal is to add redundancy to the default gateway serving VLAN 20 (Admin), such that the loss of any one EX switch does not prevent stations from reaching their default gateway, and ostensibly the rest of the routed world. Figure 7-4 outlines a quick VRRP *proof of concept* test.

This is called a proof of concept because given the network topology, it does not make much sense to have Rum serve as the Admin VLAN's backup default gateway, since Rum must transit Vodkila to provide connectivity between Host3 and Host4 in the Admin VLAN. As a result, losing Vodkila leaves a broken Admin VLAN, in that Host4 can still reach the VLAN's default gateway, but not much else. In this example, you can believe that an additional link is added to Rum and serves to connect it to the rest of the Layer 2/Layer 3 network. Both switches are assigned the same VRRP group number and a shared VIP address. Note that the shared VIP address is "owned" by the current VRRP master. Each RVI is also assigned a unique IP address. Given that no switch owns the VIP address itself in this example, the accept-data keyword is added to allow responses to pings targeted to the VIP. Lastly, Vodkila is assigned a higher VRRP priority so that it will win a mastership election against Rum. The preempt keyword ensures that Vodkila will actually overthrow an operational master, rather than just win the next election. Together these options provide determinism by ensuring that Vodkila is always the active VRRP master whenever it's operational.

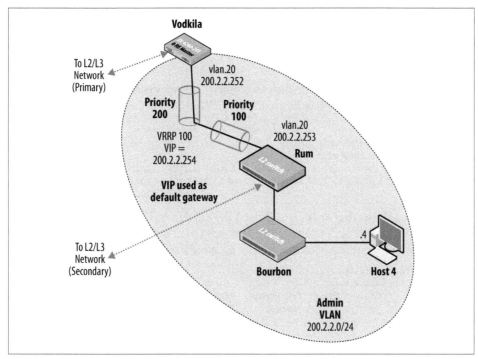

Figure 7-4. VRRP and RVI integration

Here's the VRRP configuration for Vodkila:

```
[edit interfaces vlan unit 20]
lab@Vodkila# show
family inet {
    address 200.2.2.252/24 {
        vrrp-group 20 {
            virtual-address 200.2.2.254;
            priority 200;
            preempt;
            accept-data;
        }
    }
}
```

An RVI is now defined at Rum, and similar VRRP settings are added; note the lower priority and the duplicated (shared) VIP:

```
[edit interfaces vlan]
lab@Rum# show
unit 20 {
    family inet {
        address 200.2.2.253/24 {
            vrrp-group 20 {
                virtual-address 200.2.2.254;
                priority 100;
```

```
                accept-data;
            }
        }
    }
}
```

Use the various show vrrp commands to confirm proper VRRP operation. You begin with confirmation that Vodkila is master of VRRP group 20, and that it is using the correct VIP:

```
[edit interfaces vlan unit 20]
lab@Vodkila# run show vrrp
Interface    State  Group  VR state  Timer       Type  Address
vlan.20      up        20  master    A  0.351    lcl   200.2.2.252
                                                 vip   200.2.2.254
```

The display confirms that Vodkila is indeed the current VIP master, and that the VIP is 200.2.2.254, which is the value still configured as a default gateway in Host3 and Host4. You quickly confirm that Rum admits it's a backup for this VRRP group:

```
[edit interfaces vlan]
lab@Rum# run show vrrp
Interface    State  Group  VR state  Timer       Type  Address
vlan.20      up        20  backup    D  2.987    lcl   200.2.2.253
                                                 vip   200.2.2.254
                                                 mas   200.2.2.252
```

The output also confirms that Rum believes the current VIP master is indeed the owner of IP address 200.2.2.252, which is again our good friend Vodkila. As a final verification, remove the RVI to the Admin VLAN binding at Vodkila:

```
[edit]
lab@Vodkila# delete vlans Admin_vlan l3-interface

[edit]
lab@Vodkila# commit
```

And after a few moments, check back on Rum to confirm that it has seized the VIP as master of the VRRP group:

```
[edit interfaces vlan]
lab@Rum# run show vrrp
Interface    State  Group  VR state  Timer       Type  Address
vlan.20      up        20  master    A  0.455    lcl   200.2.2.253
                                                 vip   200.2.2.254
```

The output confirms that the loss of the previous VRRP master resulted in promotion of Rum to active VRRP master. As a result, Rum now owns the VIP, and is responsible for forwarding traffic sent to that address; that is, Rum is now the active default gateway for the Admin VLAN. You confirm continued connectivity to the VIP, now provided by Rum's RVI. Recall that a successful ping to a VIP requires the accept-data keyword:

```
Host3#ping 200.2.2.254

Type escape sequence to abort.
```

```
Sending 5, 100-byte ICMP Echos to 200.2.2.254, timeout is 2 seconds:
!!!!!
Success rate is 100 percent (5/5), round-trip min/avg/max = 4/4/4 ms
Host3#ping 200.2.2.253

Type escape sequence to abort.
Sending 5, 100-byte ICMP Echos to 200.2.2.253, timeout is 2 seconds:
!!!!!
Success rate is 100 percent (5/5), round-trip min/avg/max = 1/3/4 ms
Host3#
```

Satisfied that VRRP failover worked correctly, you bring the old master back online to confirm the desired mastership preemption behavior:

```
[edit]
lab@Vodkila# rollback 1
load complete

[edit]
lab@Vodkila# commit
```

And a few seconds later, back at Rum, the VRRP mastership state again reflects the starting conditions, showing that the revertive VRRP mastership process worked:

```
[edit interfaces vlan]
lab@Rum# run show vrrp
Interface    State  Group   VR state   Timer      Type   Address
vlan.20      up        20   backup     D  2.940   lcl    200.2.2.253
                                                  vip    200.2.2.254
                                                  mas    200.2.2.252
```

Before moving on, you again confirm that end stations in the Admin VLAN are able to reach the VIP, except this time it's Vodkila, the current VRRP master, that's answering the pings:

```
Host3#ping 200.2.2.254

Type escape sequence to abort.
Sending 5, 100-byte ICMP Echos to 200.2.2.254, timeout is 2 seconds:
!!!!!
Success rate is 100 percent (5/5), round-trip min/avg/max = 1/3/4 ms
```

Once again, the ping to the VIP is successful. The results confirm that you can combine an RVI with existing JUNOS support for VRRP to achieve significant availability/reliability gains for the first hop of a routed VLAN infrastructure. Satisfied that an EX vlan interface can leverage the JUNOS Layer 3 infrastructure, the VRRP-related changes are backed out of Vodkila and Rum.

Restricting RVI Communications

Routers like to route; it's what they do. Network addresses assigned to RVI units appear as directly connected networks in the RT. Because a router does not need any routing protocols, or static routes, to route among its directly connected networks, the use of routing policy does not assist when the goal is to limit Layer 3 connectivity among a set of VLANs that share units on the vlan interface.

For example, assume that two new VLANs are added to the test topology. As soon as their respective RVI units are defined and bound to the new VLAN instances, the default behavior is to provide full IP layer connectivity among *all four* VLANs, unless you choose to restrict (or police) this traffic using a Layer 3 firewall filter. We describe filters in detail in Chapter 8; here we demonstrate a quick RVI and firewall filter scenario to reinforce this important concept.

RVI and Layer 3 filters

In this section, the goal is to deploy a Layer 3 firewall filter to restrict inter-VLAN communications between Host2 and Host4, while permitting communications between all other stations. Refer back to Figure 7-3 as needed for the topology specifics.

There are many ways to achieve such a goal using a firewall filter. Given the need to restrict communications based on a host pairing, the use of a /32 (host) address match condition seems a reasonable approach. Note that this filter will have no effect on Layer 2 communications for stations within the same VLAN. A Layer 2 filter is needed to affect inter-VLAN communications.

Before making any changes, communication is again confirmed between Host2 and both Admin VLAN hosts:

```
Host2#ping 200.2.2.4

Type escape sequence to abort.
Sending 5, 100-byte ICMP Echos to 200.2.2.4, timeout is 2 seconds:
.!!!!
Success rate is 80 percent (4/5), round-trip min/avg/max = 4/6/8 ms
Host2#ping 200.2.2.3

Type escape sequence to abort.
Sending 5, 100-byte ICMP Echos to 200.2.2.3, timeout is 2 seconds:
.!!!!
Success rate is 80 percent (4/5), round-trip min/avg/max = 4/6/8 ms
Host2#
```

Both pings are a success, proving there are unrestricted communications between the two VLANs. A sample source-address-based firewall filter is crafted and displayed:

```
[edit firewall family inet filter no_host2_to_host3]
lab@Vodkila# show
term 1 {
    from {
        source-address {
            200.2.1.2/32;
        }
        destination-address {
            200.2.2.3/32;
        }
```

```
        }
        then {
            count host2_blocked-to_host3;
            discard;
        }
    }
    term else {
        then accept;
    }
```

In this example, term 1 does what's needed as far as matching on and then restricting only those packets sent by Host2 to Host3. In this case, a counter is also added to aid in filter confirmation. The else term is critical, as the default behavior for a JUNOS filter is an implicit deny all; omitting this term results in a complete blockage of *all* inter-VLAN traffic, which is not the goal here. You must apply the filter to the Layer 3-enabled vlan interface to press it into service. The filter can be applied to either or both units, and can be applied in either or both the input and output directions. In this case, it seems logical to apply the no_host2_to_host3 filter only as input, and only on the vlan unit associated with the Sales VLAN.

The filter is applied to vlan.10 as an input filter, and the result is displayed:

```
[edit interfaces vlan unit 10]
lab@Vodkila# set family inet filter input no_host2_to_host3

[edit interfaces vlan unit 10]
lab@Vodkila# show
family inet {
    filter {
        input no_host2_to_host3;
    }
    address 200.2.1.254/24;
}
```

After committing the change, the host2_blocked-to_host3 firewall counter is displayed with a show firewall command:

```
[edit]
lab@Vodkila# run show firewall

Filter: no_host2_to_host3
Counters:
Name                          Bytes            Packets
host2_blocked-to_host3            0                  0
```

The 0 count is expected, given the lack of inter-VLAN Layer 3 stimulus. The pings from Host2 to Host3 and Host4 are repeated:

```
Host2#ping 200.2.2.3

Type escape sequence to abort.
Sending 5, 100-byte ICMP Echos to 200.2.2.3, timeout is 2 seconds:
.....
Success rate is 0 percent (0/5)
Host2#ping 200.2.2.4
```

```
Type escape sequence to abort.
Sending 5, 100-byte ICMP Echos to 200.2.2.4, timeout is 2 seconds:
!!!!!
Success rate is 100 percent (5/5), round-trip min/avg/max = 4/6/8 ms
Host2#
```

The results are as expected: Host2 can still chat with Host4, but not so much with Host3. The firewall counter is again displayed to verify that the discard counter matches the test traffic generated during the failed ping attempt:

```
[edit]
lab@Vodkila# run show firewall

Filter: no_host2_to_host3
Counters:
Name                              Bytes              Packets
host2_blocked-to_host3             610                 5
```

The output reflects five filtered packets, which correctly matches the number of ping attempts that failed. Before calling this a success, it's a good idea to confirm that Host1 continues to enjoy unfettered access to both Admin hosts:

```
Host1#ping 200.2.2.3

Type escape sequence to abort.
Sending 5, 100-byte ICMP Echos to 200.2.2.3, timeout is 2 seconds:
!!!!!
Success rate is 100 percent (5/5), round-trip min/avg/max = 8/12/32 ms
Host1#ping 200.2.2.4

Type escape sequence to abort.
Sending 5, 100-byte ICMP Echos to 200.2.2.4, timeout is 2 seconds:
!!!!!
Success rate is 100 percent (5/5), round-trip min/avg/max = 4/5/8 ms
Host1#
```

Both pings succeed, thereby confirming that the filter is not affecting Host1, as per design. As this was only a quick Layer 3 RVI-to-firewall filter integration proof of concept, the filter is removed to prepare the test bed for the static routing section:

```
[edit]
lab@Vodkila# delete firewall

[edit]
lab@Vodkila# delete interfaces vlan unit 10 family inet filter
```

The change is committed at Vodkila, and full Layer 3 connectivity between both the Sales and Admin VLANs is again confirmed (not shown for brevity).

RVI Summary

The RVI construct is an important EX feature that eliminates the need for any external Layer 3 devices when you wish to route traffic between VLANs. Because the vlan

---

interface is virtual, or software-based, you can provide Layer 3 connectivity without burning any real interface hardware.

Because all EX versions support full routing, you can easily combine an RVI with the rich set of existing JUNOS Layer 3 features, such as VRRP, firewall filters, policers, and routing policy.

An EX switch can support only one vlan interface, but this interface can have multiple logical units that are bound to VLAN instances as needed. Note that by default, full Layer 3 connectivity is provided over the RVI. You can restrict communications over the RVI using a Layer 3 filter. Stated differently, you can define only one vlan interface, and all of its child units are considered directly connected IP networks, and packets are freely routed between all such subnets by default.

Static Routing

Although the use of static routing is often frowned upon and considered bad form, there are many practical applications for static routes, along with their aggregate/generated counterparts.

Static routing suffers from a general lack of dynamism (although Bidirectional Forwarding Detection [BFD] can mitigate this issue), which often leads to loss of connectivity or inefficient forwarding during network outages due to their static, *nailed-up* nature. Static routes can quickly become maintenance and administration burdens for networks that have frequent adds, moves, or changes. With that said, static routing is often used at the network edge to support attachment to stub networks, which, given their single point of entry/egress, are well suited to the simplicity of a static route.

Next Hop Types

Static routes support various next hop types, some that actually forward traffic and others that *black-hole* matching packets. Here are the specifics for each type of next hop:

Discard

A discard next hop results in the silent discard of matching traffic. *Silent* here refers to the fact that no ICMP error message is generated back to the source of the packet. You normally choose a discard next hop when the goal is to advertise a single aggregate that represents a group of prefixes, with the expectation that any traffic attracted by the aggregate route will longest-match against one of the more specific routes, and therefore will be forwarded according to the related next hop, rather than reject or discard the next hop of the aggregate route itself.

The use of discard is currently a best practice when advertising an aggregate because the generation of ICMP error messages can consume system resources, and may end up bombarding an innocent third party; for instance, in the case of spoofed source addressing as part of a distributed denial of service (DDoS) attack.

Reject

A reject next hop results in the generation of an ICMP error message reporting an unreachable destination for matching traffic. This is the default next hop type of an aggregated route and for a generated route when it has no contributors.

Forwarding

A forwarding next hop is used to move traffic to a downstream node, and is typically specified as the IP address of a directly connected device. Matching traffic is then forwarded to the specified next hop. On a multi-access network such as a LAN, this involves the resolution of the IP address to a link layer address through ARP or some form of static mapping. When directing traffic over a point-to-point (P-to-P) interface, the next hop can be specified as an interface name; however, LAN interface types require an IP address next hop due to their multipoint nature.

Forwarding next hop qualifiers

When defining a static route with a forwarding next hop, you can use qualifiers that influence how the next hop is resolved and handled. Specifically:

`resolve`

The `resolve` keyword allows you to define an indirect next hop for a static route, which is to say an IP forwarding address that does not resolve to a directly connected interface route. For example, you could specify a static route that points to a downstream neighbor's loopback address. In this case, matching traffic will result in a recursive lookup against the specified (lo0) next hop to select a directly connected forwarding next hop. If a parallel connection exists, the failure of the currently used link results in a new recursive lookup and selection of the remaining link for packet forwarding.

`qualified-next-hop`

The `qualified-next-hop` keyword allows you to define a single static route with a list of next hops that are individually qualified with a preference. In operation, the most preferred qualified next hop that is operational (i.e., the next hop can be resolved and the interface is operational) is used. When that next hop is no longer usable, the next-best-qualified next hop is selected. That is to say, when the primary link is down, the router selects the next preferred next hop, which in this example maps to a low-speed backup facility.

Route Attributes and Flags

When you define a static route, you can include various route attributes such as Autonomous System (AS) path, BGP community, route tag, metric, and so forth. These attributes may or may not come into play when the route is later redistributed into a specific routing protocol. For example, OSPF has no notion of a BGP community or AS path, and therefore these attributes are not injected into OSPF, despite being attached to the route's static definition. The route attributes can be defined individually for each route, or as part of a default template that is inherited by all related routes, unless specifically overwritten by a competing attribute.

You can also attach to a static route flags that control various aspects of how the route is handled or operates. For example, the no-advertise flag prevents the associated route from being exported into routing protocols, even when the policy configuration would otherwise select that route for redistribution. You can display the list of available route attributes and flags with the command-line interface (CLI) ? feature:

```
[edit routing-options]
lab@Rum# set static route 10/8 ?
Possible completions:
  active                  Remove inactive route from forwarding table
+ apply-groups            Groups from which to inherit configuration data
+ apply-groups-except     Don't inherit configuration data from these groups
> as-path                 Autonomous system path
> bfd-liveness-detection  Bidirectional Forwarding Detection (BFD) options
> color                   Color (preference) value
> color2                  Color (preference) value 2
+ community               BGP community identifier
  discard                 Drop packets to destination; send no ICMP unreachables
  install                 Install route into forwarding table
> metric                  Metric value
> metric2                 Metric value 2
> metric3                 Metric value 3
> metric4                 Metric value 4
+ next-hop                Next hop to destination
  next-table              Next hop to another table
  no-install              Don't install route into forwarding table
  no-readvertise          Don't mark route as eligible to be readvertised
  no-resolve              Don't allow resolution of indirectly connected next hops
  no-retain               Don't always keep route in forwarding table
  passive                 Retain inactive route in forwarding table
> preference              Preference value
> preference2             Preference value 2
> qualified-next-hop      Next hop with qualifiers
  readvertise             Mark route as eligible to be readvertised
  receive                 Install a receive route for the destination
  reject                  Drop packets to destination; send ICMP unreachables
  resolve                 Allow resolution of indirectly connected next hops
  retain                  Always keep route in forwarding table
> tag                     Tag string
> tag2                    Tag string 2
[edit routing-options]
lab@Rum# set static route 10/8
```

The reader is encouraged to consult JUNOS software documentation at *http://www .juniper.net/techpubs/software/junos/junos92/swconfig-routing/configuring-static -routes.html#id-10352299* for a detailed description of the various attributes and flags that can be attached to static routes.

Floating Static Routes

A *floating static route* is nothing more than a static route that has a modified preference, causing it to be less preferred than a dynamically learned copy. The defaults cause a static route to always be preferred over a dynamic route. A floating static route is often used to provide backup in the event of a network or protocol malfunction. When all is operating normally, the static route remains idle because the dynamically learned routing is preferred. When routing protocol disruption results in the loss of a learned route, the previously inactive static route becomes active.

The following code sample creates a floating static route by assigning a modified preference that makes the route *less* preferred than an OSPF internal route, which has a default preference of 10. Note that the configuration statements relating to static routing live in the [edit routing-options] hierarchy:

```
[edit routing-options]
lab@Rum# show
static {
    route 0.0.0.0/0 {
        next-hop 172.16.1.1;
        preference 11;
    }
}
```

The CLI's display set function conveniently confirms the syntax used to create the static route:

```
[edit routing-options]
lab@Rum# show | display set
set routing-options static route 0.0.0.0/0 next-hop 172.16.1.1
set routing-options static route 0.0.0.0/0 preference 11
```

The successful creation of the floating static route is confirmed with the operational mode show route command, in this case filtering the output to only those routes learned through the *static* protocol:

```
[edit routing-options]
lab@Rum# run show route 200.0.0.0

inet.0: 12 destinations, 12 routes (12 active, 0 holddown, 0 hidden)
+ = Active Route, - = Last Active, * = Both

0.0.0.0/0          *[Static/11] 00:00:06
                    > to 172.16.1.1 via fe-0/0/0.412
```

EX Static Routing Scenario

It's time to put your newfound static routing knowledge to work by configuring static routing on Vodkila to provide Internet access for the IP hosts in the Sales and Admin VLANs. Figure 7-5 details the EX static routing topology.

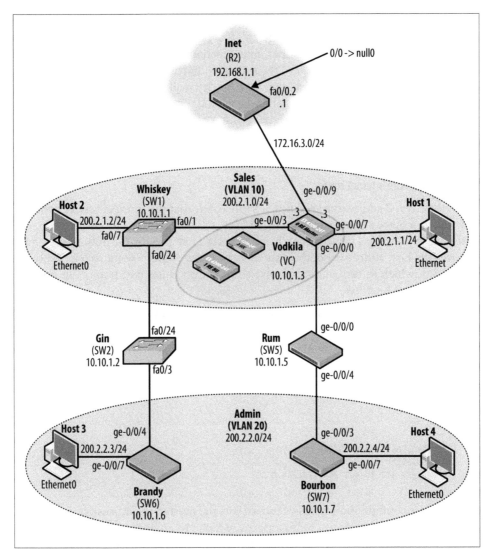

Figure 7-5. EX static routing

The topology shown in Figure 7-5 is well suited to static routing; because of the single ingress and egress points between Vodkila and the Inet router, there is little to gain from the ability to dynamically select the best next hop. Nope, the single link either works, or does not, and no amount of RIP or OSPF will make things better.

The design goal is to provide external access to the host in the Sales and Admin VLANs. Here this means that all Layer 3 entities should be able to ping the 192.168.1.1 loopback and 172.16.3.1 interface addresses associated with the Inet router when this traffic is sourced from any internal IP address. The Inet router has a default route that points to `null0`; this route is intended to represent the other 4 billion or so possible IPv4 addresses on the planet. Note that the default IOS behavior for packets that match such a route is to generate an ICMP destination unreachable error message. While possibly bad form—from the preservation of the router's control plane during a DDoS attack perspective—such behavior assists you here, in that you can expect an explicit error message when attempting to ping any non-local (and therefore non-defined) addresses. Receipt of the error validates the internal routing of external traffic to the `Inet` router as well as its ability to route return traffic back to that source; error message or not, it's still an IP packet that is sent back, after all.

Static routing in the Internet router

Given the Layer 3 addressing specifics, it's safe to say that the Internet service provider (ISP) needs to have a static route for 10.10.1/24, which covers the loopback addresses of the Layer 3-enabled devices `Scotch` and `Vodkila`, and also for the 200.2.0/22 prefix to cover the host addresses. Both of these static routes should point to a next hop of 172.16.3.3, which in turn should resolve to the direct route that points out the Inet router's `fa0/0.2` interface.

Here are the relevant portions of the `Inet` router's configuration:

```
interface Loopback0
 ip address 192.168.1.1 255.255.255.255
!
. . .
interface FastEthernet0/0
 no ip address
 duplex auto
 speed auto
!
interface FastEthernet0/0.2
 encapsulation dot1Q 70
 ip address 172.16.3.1 255.255.255.0
!
```

The interface configuration is basic, and assigns the needed IP addresses for the loopback and Ethernet interfaces. Subinterfaces, which are analogous to logical units in JUNOS, are used to support VLAN tagging. VLAN tagging is used here to allow the same physical interface to be used multiple times, as subinterfaces, to the various other devices in the test topology that may connect to it at some point. The result is a Layer 3 interface that is configured for VLAN trunking with each VLAN treated as a separate IP interface:

```
. . .
ip route 0.0.0.0 0.0.0.0 Null0
ip route 10.10.1.0 255.255.255.0 172.16.3.3
ip route 200.2.0.0 255.255.252.0 172.16.3.3
. . .
```

The static routing section defines the three static routes, described earlier. Note that the "/22" mask length for the 200.2.0 supernet is entered using a somewhat dated dotted decimal format, one that forces the conversion of the /22 mask length into binary (1111 1100), and from there on to decimal, which in this case is a 252. What fun.

The presence of a VLAN switch ensures that the Inet router's end of the Ethernet link is up. The lack of Layer 2 keepalives or BFD static routing support at the Cisco end means we can expect all to look fine and dandy, even though the EX's ge-0/0/9 interface is not even configured yet:

```
Inet-Rtr#show ip route
Codes: C - connected, S - static, I - IGRP, R - RIP, M - mobile, B - BGP
       D - EIGRP, EX - EIGRP external, O - OSPF, IA - OSPF inter area
       N1 - OSPF NSSA external type 1, N2 - OSPF NSSA external type 2
       E1 - OSPF external type 1, E2 - OSPF external type 2, E - EGP
       i - IS-IS, su - IS-IS summary, L1 - IS-IS level-1, L2 - IS-IS level-2
       ia - IS-IS inter area, * - candidate default, U - per-user static route
       o - ODR, P - periodic downloaded static route

Gateway of last resort is 0.0.0.0 to network 0.0.0.0

     172.16.0.0/24 is subnetted, 2 subnets
C       172.16.9.0 is directly connected, FastEthernet0/0.1
C       172.16.3.0 is directly connected, FastEthernet0/0.2
     10.0.0.0/24 is subnetted, 1 subnets
S       10.10.1.0 [1/0] via 172.16.3.3
     192.168.1.0/32 is subnetted, 1 subnets
C       192.168.1.1 is directly connected, Loopback0
S*   0.0.0.0/0 is directly connected, Null0
S    200.2.0.0/22 [1/0] via 172.16.3.3
```

Note that both static routes point to 172.16.3.3 as the next hop address. When a packet matches against one of these static routes, an important concept called *recursion* kicks in. Recursion refers to the process of having to refer *back* to the RT to locate a forwarding next hop based on the IP address found during a previous route lookup; a forwarding next hop always involves an egress interface, and in the case of multipoint interfaces, a next hop IP address and the need to resolve the underlying Layer 2 address (the Destination MAC, or DMAC) for that IP address.

The output confirms that packets matching one of the static routes should be sent to 172.16.3.3. The recursive route lookup on 172.16.3.3 in turn resolves to a Connected 172.16.3.0/24 route that points to the FastEthernet0/0.2 egress interface subinterface that the Inet router uses to reach Scotch. At this point, all that is needed to correctly route packets matching these static routes is an ARP exchange so that the Inet router can learn the MAC address of Scotch's ge-0/0/8 interface. When combined with knowledge of the correct egress interface, the DMAC address is all that is needed for the Inet router to correctly send packets to Scotch. The expected ARP entry is confirmed at the Inet router:

```
R2#show ip arp 172.16.9.9
Protocol  Address      Age (min)  Hardware Addr   Type   Interface
Internet  172.16.9.9       2      001d.b50e.9601  ARPA   FastEthernet0/0.1
R2#
```

However, as expected, despite all being well with the Inet router's static routes, Inet-router-initiated pings to Vodkila's end of the access link fail:

```
Inet-Rtr#ping 172.16.3.3

Type escape sequence to abort.
Sending 5, 100-byte ICMP Echos to 172.16.3.3, timeout is 2 seconds:
.....
Success rate is 0 percent (0/5)
Inet-Rtr#
```

Looks like it's time to configure some static routing JUNOS-style on the EX!

EX static routing

Things begin on EX switch Vodkila with the configuration of its Layer 3 access link to the Inet router. The VLAN switch between the two devices is configured with the Inet router end as a trunk to allow the transport of tagged packets. The Vodkila end is set as an access link, and the result is that untagged traffic is sent and received at the Vodkila end, while tagged traffic is sent and received at the Inet router end. Tagging is not needed at Vodkila, as its one Layer 3 interface needs to attach to only one remote Layer 3 device, while the Inet router may connect to multiple devices and therefore needs multipoint capability:

```
[edit]
lab@Vodkila# edit interfaces ge-0/0/9

[edit interfaces ge-0/0/9]
lab@Vodkila# set unit 0 family inet address 172.16.3.3/24
```

After activating the Layer 3 configuration, you again try to ping from the Inet router; there is no point in adding a static route if basic IP forwarding is not working between the two:

```
Inet-Rtr#ping 172.16.3.3

Type escape sequence to abort.
Sending 5, 100-byte ICMP Echos to 172.16.3.3, timeout is 2 seconds:
.!!!!
Success rate is 80 percent (4/5), round-trip min/avg/max = 1/1/4 ms
Inet-Rtr#
```

Much better! With basic IP connectivity confirmed, you move on to the static route definitions. Before adding any static routes, test external connectivity at Host1:

```
Host1#traceroute 30.1.1.1

Type escape sequence to abort.
Tracing the route to 30.1.1.1
  1 200.2.1.254 0 msec 0 msec 4 msec
  2 * * *
  3 * * *
  . . .
 29 * * *
```

```
   30  *   *   *
Host1#
```

As expected, the traceroute fails at the subnet's default gateway. Things should improve when the default route is added to Vodkila. Note again that the host sends its non-local traffic to the default gateway, which is the vlan.10 interface and its 200.2.1.254 address in this example. Currently, Vodkila has no route for the destination, and so, by default, it performs a silent discard, resulting in no error message being received.

It seems that to get things flowing, so to speak, all that is needed at Vodkila is a simple static default route pointing out the ge-0/0/09 interface toward 172.16.3.1. The static default route is defined, and the modified configuration is displayed:

```
[edit routing-options]
lab@Vodkila# set static route 0/0 next-hop 172.16.3.1

[edit routing-options]
lab@Vodkila# show
static {
    route 0.0.0.0/0 next-hop 172.16.3.1;
}
```

As noted previously, the lack of Layer 2 Operational Administration and Maintenance (OAM)/keepalives, coupled with the lack of any routing protocols (this is static, after all), means it is very easy for connectivity problems to go undetected. In effect, the static route is usable as long as the local device's interface status remains up. Not only will data plane issues go undetected as a result, but the presence of the VLAN switch also means that a complete failure at one end (fiber cut) does not result in any Physical layer abnormalities, alarms, or error indications at the other end. A lack of visibility into the peer's forwarding/operational status is one of the significant drawbacks to using static routing, after all.

While it is true that this particular topology affords only one egress link, and therefore a dynamic routing protocol is overkill, there is something to be said for at least knowing when your one access link is broken, as this can simplify fault isolation and allow you to get a complaint lodged all the faster.

EX switches support the BFD protocol, which is useful in just this type of environment because it provides a protocol-agnostic keepalive mechanism that lives to rapidly detect faults in the forwarding plane, for whatever reason, and in this example it hides the static route when the BFD session fails. Sounds good, but the older IOS-based router used in the Inet role does not support BFD. Neither of the devices currently support Ethernet OAM, which takes that option off the table as well.

The change is committed, and the presence of an active, default route is confirmed:

```
[edit routing-options]
lab@Vodkila# run show route protocol static

inet.0: 10 destinations, 10 routes (10 active, 0 holddown, 0 hidden)
+ = Active Route, - = Last Active, * = Both
```

```
0.0.0.0/0              *[Static/5] 00:09:51
                       > to 172.16.3.1 via ge-0/0/9.0
```

The output looks good, in that the route is not hidden and shows the correct forwarding interface and IP next hop. It's easy to test the default route's efficacy with a ping to some non-existent address, as the default route should dispatch such a packet to the Inet router, where it meets the cold embrace of null0:

```
[edit routing-options]
lab@Vodkila# run ping 30.1.1.1
PING 30.1.1.1 (30.1.1.1): 56 data bytes
36 bytes from 172.16.3.1: Destination Host Unreachable
Vr HL TOS  Len   ID Flg  off TTL Pro  cks      Src       Dst
 4  5  00 0054 026c   0 0000  3f  01 ab28 172.16.3.3  30.1.1.1

. . .

^C
--- 30.1.1.1 ping statistics ---
2 packets transmitted, 0 packets received, 100% packet loss
```

Again, the ping failure is expected and is considered a success by virtue of receiving the resulting error message from 172.16.3.1. The test is repeated, but this time traffic is sourced from the switch's loopback address. This is a critical step, because it proves the Internet router has a route back to a non-directly connected address, thus proving its end of the static routing solution is working. The previous ping was sourced from the 172.16.3.3 interface address, and therefore tested only direct routing:

```
[edit routing-options]
lab@Vodkila# run ping 30.1.1.1 source 10.10.1.3
PING 30.1.1.1 (30.1.1.1): 56 data bytes
36 bytes from 172.16.3.1: Destination Host Unreachable
Vr HL TOS  Len   ID Flg  off TTL Pro  cks      Src       Dst
 4  5  00 0054 0299   0 0000  3f  01 4f02 10.10.1.3  30.1.1.1
. . .
^C
--- 30.1.1.1 ping statistics ---
2 packets transmitted, 0 packets received, 100% packet loss
```

The error message is again received, confirming that static routing is correctly configured in the Inet router. To complete validation, you test external reachability from Host1. In this case, the IP traffic must first be switched over the Sales VLAN to reach the RVI, at which point the packet enters the Layer 3 processing stream, where the routing process matches against the static (default) route and directs the packet to the Inet router:

```
Host1#ping 30.1.1.1

Type escape sequence to abort.
Sending 5, 100-byte ICMP Echos to 30.1.1.1, timeout is 2 seconds:
U.U.U
Success rate is 0 percent (0/5)
Host1#
```

IP traffic sent by a host router to a remote network address will be addressed to the device's default gateway at the MAC layer. The EX switch determines whether it should perform network-level processing on a packet versus the far simpler Layer 2 switching by determining whether a frame is addressed to the switch's local MAC address. Traffic that is switched within the same VLAN is directly delivered to the target host's MAC address, meaning such traffic is never addressed to the EX switch itself.

Traffic that is sent to a remote IP network is directly addressed to a default router. In the case of an RVI as a default router, this traffic will be addressed to the switch itself, and therefore will enter Layer 3 processing.

The ping generated by Host1 is again sent to a non-existent address. Once again, the return of an error message is expected, and again proves all is working per design; the Us are IOS's way of indicating that a destination *U*nreachable packet was received, by the way. Again, the receipt of this error is good, as it proves the packet was correctly routed to the Inet router, and that a response could be correctly routed back to the packet's source. The last ping validates that a non-locally attached host can be correctly VLAN-switched to the RVI and then correctly routed. In this case, a traceroute is performed from Host3, which is attached to Brandy and has to be trunked over to Vodkila, to the loopback address of the Inet router:

```
Host3#traceroute 192.168.1.1

Type escape sequence to abort.
Tracing the route to 192.168.1.1

  1 200.2.2.254 4 msec 0 msec 4 msec
  2 172.16.3.1 4 msec 4 msec *
Host3#
```

The success of the traceroute confirms that RVI-based static routing is working. Who says this Layer 3 stuff is hard?

Static Routing Summary

This section demonstrated a typical static routing scenario that provides Internet access among a set of LAN hosts that in turn access Layer 3 services through an RVI. Static routing often gets a bad rap. While there are definite shortcomings that become more pronounced as the level of network "meshiness" grows, there are times when a static route is a technically sound choice.

Despite the lack of operational visibility, when there is a single way out, or in, or when you are interfacing to another network that's under different administrative control, the complexity of a dynamic routing protocol is oftentimes just not worth it. Here the complexity is in the configuration and possible troubleshooting of such a protocol, but also in the ongoing processing burden within the router (which on some routers can be a big problem), in addition to the complexities of working out peering arrangements with other ASs, and so on.

Yep, there are times when a static route is the right tool for the job. The next section demonstrates the use of the RIP routing protocol in a slightly more complicated topology, where there is a need for intelligent next hop selection brought about by redundant access links.

RIP Routing

Routing Information Protocol (RIP) is a venerable old workhorse, having been deployed way back in the early 1970s to support Xerox Networking Services. The protocol was later updated to support IP routing and was defined in RFC 1058, back in 1988. RIP Version 2 (RIPv2) was originally defined in RFC 1388 (1993) and is currently specified in RFC 2453 (1998).

RIP is a distance vector (DV), or Bellman-Ford, routing protocol that suffers from many performance and functional shortcomings when compared to newer IGP options, such as OSPF, which is based on LS shortest-path algorithms. That being said, RIP was updated with a version 2, and later with IPv6 support via RIPng, and it is supported in virtually all IP routing gear, in part because of its inherent simplicity and long implementation history. Being easy to implement is great; this inherent simplicity also results in frugal use of router CPU and memory resources, which can be a significant issue in low-end or older routing equipment. When compared to LS protocols, RIP is also found to be somewhat intuitive and easy to understand. This translates to easier deployment and support activities by *normal* folks—you know, those who may have a life outside of IP networking, and therefore do not "dine on data and snack on RFCs."

As noted in the static routing section, there is a time when a simple nail and hammer, that is, a static route, is the best solution. While you may gain an ego boost by solving the same problem with glue, welding, and perhaps even a high-tech hook and loop fastener technology, such a solution is not always cost-effective. RIP is well suited to relatively small internetworks (15 or fewer hops) that are relatively stable, and which don't demand subsecond reconvergence when the odd failure does occur.

In the end, there is no "right" answer when it comes to the best IGP choice for *your* network. Too many variables need to be factored. Can you run multi-area OSPF on a two-router internetwork? Sure. Should you? Probably not. Most enterprises will likely deploy the EX with no routing; they are marketed as switches, after all. Those that do enable routing will likely use whatever IGP they are already running. If there is no routing in the current network, it's likely that when routing is first added, a simple protocol such as RIP will be chosen. A data center customer, on the other hand, likely has an extensive routed infrastructure in place, and has numerous trained IT technicians who will have no problem deploying more complicated protocols. The somewhat demanding needs of a modern data center may well benefit from the added performance and scalability gains found with the more complicated LS routing option.

RIP Overview

RIP is classified as a DV routing protocol because it advertises reachability information in the form of distance/vector pairs, which is to say, each route is represented as a cost (distance) to reach a given prefix (vector) tuple. DV routing protocols typically exchange entire RTs among their set of directly connected peers on a periodic basis. Figure 7-6 displays key operational characteristics of RIP.

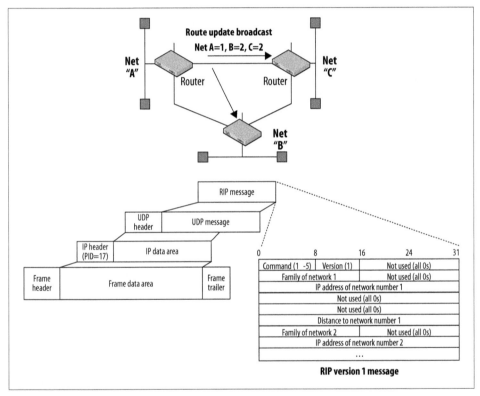

Figure 7-6. RIP operational overview

While Figure 7-6 is admittedly busy, it's composed of three parts that can easily be tackled. The top of the figure shows a stylized network running a DV protocol such as RIP. Each router connects to a LAN segment with a unique network number. As with most DV protocols, RIP operates through the periodic exchange of the local router's route table (with a few exceptions for stability, as described shortly), among all of its directly attached neighbors—a case of telling all your friends everything you know, every so often. Here Router 1 advertises its cost to reach the three networks it knows about. Note that its local network has a metric of 1, while the other two networks each have a cost of 2, indicating that it and one additional router must be crossed to reach these networks. Each router receives the updates sent by its peers and then performs a best-route selection process. The trophy goes to the route with the lowest metric, which is typically based on hop count, and that route is installed into the RT. Only the best routes are installed in the RT, and only these routes are subsequently selected for re-advertisement to other peers.

The lower-left portion of Figure 7-6 shows a RIP message's IP/User Datagram Protocol (UDP)-based encapsulation, as defined in the RFC. RIP uses UDP port 520, and v1 messages are sent as IP broadcast (all 1s), while v2 uses the reserved 224.0.0.9 multicast address. The lower-right portion of the figure shows the RIPv1 message format; v2 uses a compatible format, but includes a network mask for subnet support and can also support authentication. A v2 speaker can interoperate with a v1 implementation, but at the cost of forcing the fallback of all speakers to the lesser set of v1 capabilities.

The behavior just described, although direct and easy to understand, leads to many of the disadvantages associated with DV routing protocols. Specifically:

- Increased network bandwidth consumption stemming from the periodic exchange of potentially large RTs, even during periods of network stability. This can be a significant issue when routers connect over low-speed or usage-based network services.

- Slow network convergence, and as a result, a propensity to produce routing loops when reconverging around network failures. To alleviate (but not eliminate) the potential for routing loops, mechanisms such as split horizon, poisoned reverse, route hold downs, and triggered updates are generally implemented. These stability features come at the cost of prolonging convergence.

- Association (usually) with crude route metrics. The typical metric (cost) for DV protocols is a simple hop count, which is a crude measure of actual path cost, to say the least. For example, most users realize far better performance when crossing several routers interconnected by Gigabit Ethernet links, as opposed to half as many routers connected over low-speed serial interfaces.

To help illustrate what is meant by *slow to converge*, consider that the protocol's architects ultimately defined a *hop count* (the number of routers that need to be crossed to reach a destination) of 16 to be infinity! Setting infinity to a rather low value was needed because in some conditions, RIP can converge only by cycling through a series of route exchanges between neighbors, with each such iteration increasing the route's cost by one, until the condition is cleared by the metric reaching infinity and both ends finally agree that the route is not reachable. With the default 30-second update frequency, this condition is aptly named a *slow count to infinity*, and is shown in Figure 7-7.

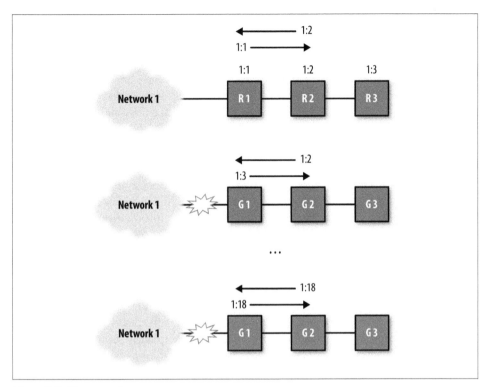

Figure 7-7. Slow counting

Things begin at the top, where all is well, and the three routers have converged on a loop-free topology that is optimized on hop count. At this time, R1's link to Network 1 is up, resulting in its advertisement to R2 with a cost of 1; in this example, R2 finds this to be the best route, and it installs Network 1 in its RT with the advertised costs plus its cost (1) to reach R1. Here R2 is not running split horizon (explained shortly), which results in it later readvertising Network 1 back to R1, now with its cost of 2. Because R1 *still* has a better cost of 1 for this network, all is well.

Later on, in the middle of the figure, R1 loses its link to Network 1. In this example, R1 does not implement triggered updates (described shortly), so it plans to let R2 and the rest of its friends know of this change at its convenience—say, at the next scheduled update. By default, this is every 30 seconds. In this example, R2 happens to send its advertisement back to R1 first, and this is where things go bad.

Because R1 currently has *no* route to Network 1 (it has a local metric of 16), it now appears as though R2 has a better route. R1 is fooled, and incorrectly believes that it can reach Network 1 via R2 at a cost of 3 (the advertised plus local metric to reach R2), and so installs the route to Network 1 with a next hop pointing back to R2. Thus, a loop is born—R2, you'll recall, is still pointing its copy of the route back at R1. At the next update, R3 advertises its new metric, R2 updates its table/costs, and the process repeats as the route is again advertised back to R1. This process continues until the route is finally poisoned by reaching a hop count of 16, which for RIP equals infinity, causing the route to be considered unreachable.

Over the years, problems such as the slow count to infinity have been solved by various tweaks to RIP's operation. The next section explores these tweaks, all of which are implemented in JUNOS to give you the best RIP experience possible.

RIP stability and performance tweaks

Hold downs serve to increase stability, at the expense of rapid convergence, by preventing installation of a route with a reachable metric, after that same route was recently marked as unreachable (cost = 16) by the local router. This behavior helps to prevent loops by preventing the local router from installing route information for a route that was *originally* advertised by the local router, and which is now being *readvertised* by another neighbor. It's assumed that the slow count to infinity will complete before the hold down expires, after which the router will be able to install the route using the lowest advertised cost.

Split horizon prevents the advertisement of routing information back over the interface from which it was learned, and *poisoned reverse* alters this rule to allow readvertisement back out to the learning interface, as long as the cost is explicitly set to infinity: a case of "I can reach this destination, NOT!" This helps to avoid loops by making it clear to any receiving routers that they should not use the advertising router as a next hop for the prefix in question. This behavior is designed to avoid the need for a slow count to infinity that might otherwise occur because the explicit indication that "I cannot reach destination X" is less likely to lead to misunderstandings when compared to the absence of information associated with split horizon. To prevent unnecessary bandwidth waste that stems from bothering to advertise a prefix that you cannot reach, most RIP implementations use split horizon, except when a route is marked as unreachable, at which point it is advertised with a poisoned metric for some number of update intervals (typically three).

Triggered updates allow a router to generate event-driven as well as ongoing periodic updates, serving to expedite the rate of convergence as changes propagate quickly. When combined with hold downs and split horizon, a RIP network can be said to receive bad news quickly while good news travels slowly.

RIP and RIPv2

Although the original RIP version still works and is currently supported on Juniper Networks routers, it's assumed that readers of this book will consider deploying only RIP Version 2. Although the basic operation and configuration are the same, several important benefits are associated with RIPv2 and no real drawbacks (considering that virtually all modern routers support both versions and that RIPv2 messages can be made backward compatible with v1 routers, albeit while losing the benefits of RIPv2 for those v1 nodes).

RIPv2's support of Variable Length Subnet Masking/Classless Inter-Domain Routing (VLSM/CIDR), combined with its ability to authenticate routing exchanges, has resulted in new life for our old friend RIP (pun intended). Table 7-5 provides a summary comparison of the two RIP versions.

Table 7-5. Comparing characteristics and capabilities of RIP and RIPv2

Characteristic	RIP	RIPv2
Metric	Hop count (16 max)	Hop count (16 max)
Updates/hold down/route timeout	30/120/180 seconds	30/120/180 seconds
Max prefixes per message	25	25 (24 when authentication is used)
Authentication	None	Plain text or Message Digest 5 (MD5)
Broadcast/multicast	Broadcast to all nodes using all 1s, RIP-capable or not	Multicast only to RIPv2-capable routers using 224.0.0.9 (broadcast mode is configurable)
Support for VLSM/CIDR	No, only classful routing is supported (no netmask in updates)	Yes
Route tagging	No	Yes (useful for tracking a route's source, i.e., internal versus external)

RIP Deployment Scenario

In this section, our goal is to deploy RIP on an EX as part of a switched access-to-routed distribution layer scenario. Figure 7-8 provides the RIP topology details.

While similar to the static routing case, this topology is designed to reflect a typical Layer 2 switched access infrastructure that connects to a distribution layer through a Layer 3 routing service. Here, the existing access layer is composed of our two favorite VLANs, Sales and Admin, along with their respective hosts. This example brings a J Series router named Scotch into play, where it provides redundant access to the distribution layer. The presence of alternate paths helps to justify the use of a routing protocol, as opposed to simple static routing. As before, R2 has a default route pointing to null0 to simulate the rest of the network's destinations.

Scotch is configured to run RIPv2 to R2, a decision made to help provide an access link keepalive mechanism given the presence of a VLAN switch, and to ensure that when things are down, the affected access layer's routes are withdrawn from the rest of the network. This behavior ensures that the router does not attract data from remote parts of the network only to have it discarded once it reaches the distribution layer router.

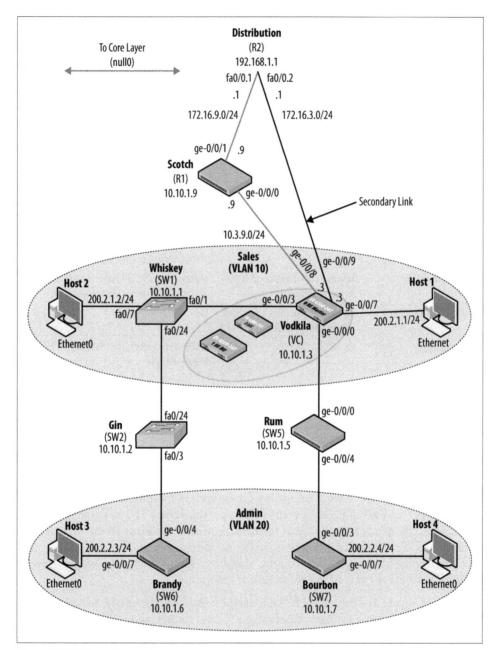

Figure 7-8. RIP routing topology

In this example, Vodkila is configured with two static routes representing the addressing space of the access layer network, and the required policy to have RIP advertise these routes to both of its upstream peers.

An aggregate route could have been used. Such a route needs a contributing route to be active. In this example, the aggregates would be activated by one of the locally connected routes, thereby effectively nailing them up, so the routes may as well be static. More complicated scenarios that would benefit from an aggregate route are beyond the scope of this chapter.

In this example, one of the two static routes is an aggregate in the sense that it represents multiple Class C networks using a mask that is less than 24 bits in length. In JUNOS, an aggregate route is more a type of next hop and does not imply super- or subnetting.

In the reverse direction, the distribution layer router is configured to advertise a default route to both access layer peers in order to provide them with reachability to non-local destinations. An import policy is used to restrict RIP to accepting *only* a default route from the distribution layer peer, a strategy planned to protect the router from abnormally large RTs in the event of misconfiguration or other abnormal events in remote portions of the network.

To make things interesting, your challenge in this example is to ensure that traffic is forwarded through the J Series router when that link is operational, despite this path having more hops and therefore being less preferred by RIP. Such a restriction may stem from a need for some service, such as sampling or stateful firewall, which is currently not available on the EX platform.

When there is a problem with the J Series path, traffic should flow directly from Vodkila to R2, given RIP's predilection for paths with fewer hops. This type of control of forwarding paths, and *not blocking* so much as simply *not choosing* to use redundant paths, is what separates routing from bridging—and for that matter the "packets from the frames," so to speak.

Configure RIP

We'll start with the configuration of RIP on Vodkila. The distribution layer router is already configured per the requirements of this example. If things do not work, you can assume it's due to misconfiguration in the access layer. For completeness, the relevant portions of the IOS-based device's configuration are shown:

```
. . .
interface Loopback0
 ip address 192.168.1.1 255.255.255.255
!
interface FastEthernet0/0
 no ip address
 duplex auto
 speed auto
!
interface FastEthernet0/0.1
 encapsulation dot1Q 69
```

```
  ip address 172.16.9.1 255.255.255.0
!
interface FastEthernet0/0.2
  encapsulation dot1Q 70
  ip address 172.16.3.1 255.255.255.0
!
router rip
  version 2
  redistribute static
  network 172.16.0.0
  no auto-summary
!
ip classless
ip route 0.0.0.0 0.0.0.0 Null0
. . .
```

The summary of this IOS configuration is that it created a VLAN-tagged interface with two logical units, each assigned an IP address, and RIP is configured to run over both subinterfaces.

 As noted previously, the IOS interface configuration makes use of VLAN-based subinterfaces, which are analogous to logical units in JU-NOS, to allow the same physical interface to be used multiple times with differing logical connectivity. This topology uses a one-to-many model with regard to a single Internet router that connects to multiple discrete devices. The use of VLAN tagging facilitates the desired logical connectivity from a single interface, and is limited to the Internet router side of the topology. As a result, the remote end of the Inet to Scotch link does not use VLAN tagging and has a single logical unit defined.

RIP is set to redistribute static routes (a default route in this example), and RIPv2 is selected with no auto-summary to permit class routing/subnetting.

 If IOS to JUNOS RIP, OSPF, EIGRP, and Layer 3 services interoperation and integration is your bag, consult the companion volume in this series, *JUNOS Enterprise Routing*.

Because no protocol can overcome a broken physical layer or misconfigured IP address, start with configuration of the Layer 3 interfaces at Vodkila and Scotch. The interfaces at Scotch are shown, and proper IP communication is confirmed:

```
[edit]
lab@Scotch# show interfaces
ge-0/0/0 {
    description "To Switch3 Vodka";
    unit 0 {
        family inet {
            address 10.3.9.9/24;
        }
```

```
            }
        }
    ge-0/0/1 {
        description "To R2 Core"
        unit 0 {
            family inet {
                address 172.16.9.9/24;
            }
        }
    }
    ge-0/0/3 {
        description "OoB Interface";
        unit 0 {
            family inet {
                address 172.16.69.9/24;
            }
        }
    }
    lo0 {
        unit 0 {
            family inet {
                address 10.10.1.9/32;
            }
        }
    }
}
```

And now the pings to both of Scotch's peers:

```
[edit]
lab@Scotch# run ping 10.3.9.3
PING 10.3.9.3 (10.3.9.3): 56 data bytes
64 bytes from 10.3.9.3: icmp_seq=0 ttl=64 time=603.526 ms
64 bytes from 10.3.9.3: icmp_seq=1 ttl=64 time=1.403 ms
64 bytes from 10.3.9.3: icmp_seq=2 ttl=64 time=1.387 ms
^C
--- 10.3.9.3 ping statistics ---
3 packets transmitted, 3 packets received, 0% packet loss
round-trip min/avg/max/stddev = 1.387/202.105/603.526/283.847 ms

[edit]
lab@Scotch# run ping 172.16.9.1
PING 172.16.9.1 (172.16.9.1): 56 data bytes
64 bytes from 172.16.9.1: icmp_seq=0 ttl=255 time=3.865 ms
64 bytes from 172.16.9.1: icmp_seq=1 ttl=255 time=2.385 ms
^C
--- 172.16.9.1 ping statistics ---
2 packets transmitted, 2 packets received, 0% packet loss
round-trip min/avg/max/stddev = 2.385/3.125/3.865/0.740 ms
```

And the same at Vodkila:

```
[edit interfaces]
lab@Vodkila# show ge-0/0/8
unit 0 {
    family inet {
        address 10.3.9.3/24;
    }
}
```

```
    }

[edit interfaces]
lab@Vodkila# show ge-0/0/9
unit 0 {
    family inet {
        address 172.16.3.3/24;
    }
}

[edit interfaces]
lab@Vodkila# run ping 172.16.3.1
PING 172.16.3.1 (172.16.3.1): 56 data bytes
64 bytes from 172.16.3.1: icmp_seq=0 ttl=255 time=1.390 ms
64 bytes from 172.16.3.1: icmp_seq=1 ttl=255 time=1.217 ms
^C
--- 172.16.3.1 ping statistics ---
2 packets transmitted, 2 packets received, 0% packet loss
round-trip min/avg/max/stddev = 1.217/1.304/1.390/0.086 ms
```

The pings to directly connected neighbors succeed, confirming that nothing is stopping RIP now, except, of course, its glaring absence. RIP is configured at the [edit protocols rip] hierarchy, where the general options are shown:

```
[edit protocols rip]
lab@Vodkila# set ?
Possible completions:
+ apply-groups         Groups from which to inherit configuration data
+ apply-groups-except  Don't inherit configuration data from these groups
  authentication-key   Authentication key (password)
  authentication-type  Authentication type
  check-zero           Check reserved fields on incoming RIPv2 packets
> graceful-restart     RIP graceful restart options
> group                Instance configuration
  holddown             Hold-down time (10..180 seconds)
+ import               Import policy
  message-size         Number of route entries per update message (25..255)
  metric-in            Metric value to add to incoming routes (1..15)
  no-check-zero        Don't check reserved fields on incoming RIPv2 packets
> receive              Configure RIP receive options
> rib-group            Routing table group for importing RIP routes
  route-timeout        Delay before routes time out (30..360 seconds)
> send                 Configure RIP send options
> traceoptions         Trace options for RIP
  update-interval      Interval between regular route updates (10..60 seconds)
[edit protocols rip]
lab@Vodkila# set
```

It should be apparent that many aspects of RIP are configurable within JUNOS software. Some options are global, such as the authentication key/type and import/export policy, which means they apply to all groups (unless negated by a more specific group setting, if available). Other parameters can be specified only at a subsequent hierarchy. For example, a neighbor can be defined only within a group. You can quickly explore the options available under send and receive using the CLI's ? help utility:

```
[edit protocols rip]
lab@ Vodkila# set send ?
Possible completions:
  broadcast              Broadcast RIPv2 packets (RIPv1 compatible)
  multicast              Multicast RIPv2 packets
  none                   Do not send RIP updates
  version-1              Broadcast RIPv1 packets
. . .
lab@Vodkila# set receive ?
Possible completions:
  both                   Accept both RIPv1 and RIPv2 packets
  none                   Do not receive RIP packets
  version-1              Accept RIPv1 packets only
  version-2              Accept only RIPv2 packets
. . .
```

It's apparent from the display that the send and receive settings globally control the RIP version, and you can also tell whether multicast (default for v2) or broadcast packets are sent. It just so happens that these same settings can also be specified on a per-neighbor (interface) basis—as with all JUNOS software, a more-specific group-level configuration hierarchy setting always overrides a less-specific global value. Now for a quick look at the options available under a group—this is where you define RIP neighbors in the form of interface names that will run RIP:

```
lab@Vodkila# set group test ?
Possible completions:
   <[Enter]>               Execute this command
 + apply-groups            Groups from which to inherit configuration data
 + apply-groups-except  Don't inherit configuration data from these groups
 > bfd-liveness-detection  Bidirectional Forwarding Detection options
 + export                  Export policy
 + import                  Import policy
   metric-out              Default metric of exported routes (1..15)
 > neighbor                Neighbor configuration
   preference              Preference of routes learned by this group
   route-timeout           Delay before routes time out (30..360 seconds)
   update-interval         Interval between regular route updates (10..60 seconds)
   |
```

Configuration options found at the neighbor level include the import or export keyword, which is used to apply routing policy to receive or transmit route updates, respectively. Note that when applied at the neighbor level, any globally defined import or export policies are negated. The router runs *either* the global or the group policy, never both, and the router always chooses the most specific application—a neighbor level is more specific than a global level, of course.

The terminology of groups and neighbors may seem a bit confusing at first, given the way RIP is configured in IOS. JUNOS software is optimized when routing peers with similar export policies are placed into the same group. As a result, even if you have only one peer, that neighbor needs to belong to a RIP group. Also, the term *neighbor* here actually means *interface*, given that RIP messages are not unicast to specific machines, but instead are broadcast or multicast to all RIP speakers on a given link. This means that specifying a single neighbor in the form of a multi-access interface results in RIP communications with *all* RIP-capable routers on that LAN segment.

Also note that in IOS, the `neighbor` keyword is used to define *unicast* RIP exchanges (when combined with the `passive` statement) among a subset of RIP speakers on a given interface. There is no keyword equivalent for this non-standard RIP operation in JUNOS, but firewall filters and/or policy can be used to constrain routing exchanges for similar effect.

Vodkila's RIP configuration

Vodkila's RIP process is configured in accordance with the operational guidelines set forth for this scenario. Analysis begins with a `routing-options` stanza and static route definition:

```
[edit]
lab@Vodkila# show routing-options
static {
    route 200.2.0.0/22 reject;
}
```

The static route for the 200.2.0/22 supernet catches subnets in the range of 200.2.0 through 200.2.3, and therefore encompasses the host subnets in the access layer with a bit of room for later expansion. Analysis continues with the RIP stanza:

```
[edit]
lab@Vodkila# show protocols rip
group distribution {
    export rip_export_add;
    neighbor ge-0/0/9.0 {
        metric-in 3;
        import default_only;
    }
}
group Scotch {
    export rip_export;
    neighbor ge-0/0/8.0;
}
```

Two RIP groups are used because in this example there is a need for export policy differences between the two RIP neighbors. Recall that in the JUNOS implementation, RIP export is available only at the group level, while import policy can be applied at the global, group, or neighbor level. Each group defines a single RIP neighbor, which is somewhat of a misnomer, in that it's really identifying a RIP-enabled *interface*; RIP operates multipoint, and one such interface may therefore advertise and receive from *all* RIP speakers attached to the direct subnet. Multiple RIP-enabled interfaces can be defined within a single group only when all such interfaces share a common export policy.

The distribution group's export policy increases the default RIP metric to ensure that R2 prefers Scotch's version of these routes. This group is set to perform a similar metric addition to received RIP routes, in this case adding 3 rather than the default 1 to all such routes. This operation could also be accomplished with import policy. The approach taken here allows a common import policy to be shared among both groups.

The distribution group makes use of a neighbor-level import policy named default_only; given there is only one neighbor, this policy could have been applied at the group level with the same effect. The policy accepts only a RIP-learned default and is applied *only* to RIP exchanges received from R2. The Scotch group uses the default RIP import policy, which accepts all sane RIP routes. This is done to ensure that both Scotch and Vodkila maintain RIP-learned loopback address reachability in the event that both RIP peerings to R2 are lost, which in turn causes the loss of the default route.

Here is Vodkila's routing policy:

```
[edit]
lab@Vodkila# show policy-options
policy-statement default_only {
    term permit{
        from {
            protocol rip;
            route-filter 0.0.0.0/0 exact;
        }
        then accept;
    }
    term deny {
        then reject;
    }
}
```

The default_only import policy accepts only a RIP-learned default route. This is sufficient to provide Vodkila with routing to all external destinations, including Scotch's loopback address, in the event that the RIP session to Scotch is disrupted. Due to the metric manipulation, it's expected that Vodkila will use the default learned via Scotch, and therefore will route egress traffic over the primary link when it's operational:

```
policy-statement rip_export {
    term static {
        from {
            protocol static;
            route-filter 200.2.0.0/22 exact;
        }
        then accept;
    }
```

```
        term direct {
            from {
                protocol direct;
                route-filter 10.10.1.3/32 exact;
            }
            then accept;
        }
        term readvertise_rip {
            from protocol rip;
            then accept;
        }
        term deny {
            then reject;
        }
    }
```

The `rip_export` policy's first term advertises the access layer's 200.2.0/22 aggregate route and the direct route for the local loopback address, and readvertises any RIP-learned routes. This last bit may seem odd, as this is the default behavior in most RIP implementations. In JUNOS RIP, the default RIP export policy is to reject all, *even* RIP-learned, routes. The `readvertise_rip` term causes JUNOS to act like other RIP routers, and is used here to ensure maximum connectivity in the face of failures. If Scotch's upstream link fails, this term allows Vodkila to readvertise the 10.10.1.9 loopback address, which is learned via RIP, on to the distribution router:

```
policy-statement rip_export_add {
    term static {
        from {
            protocol static;
            route-filter 200.2.0.0/22 exact;
        }
        then {
            metric add 3;
            accept;
        }
    }
    term direct {
        from {
            protocol direct;
            route-filter 10.10.1.3/32 exact;
        }
        then {
            metric add 3;
            accept;
        }
    }
    term readvertise_rip {
        from protocol rip;
        then {
            metric add 3;
            accept;
        }
    }
    term deny {
```

```
        then reject;
    }
}
```

The `rip_export_add` policy is applied as export-only to the R2 peering. It functions as the `export_rip`, except it artificially raises the advertised metric by adding 3. This poisons the secondary link from the perspective of R2; as a result, it will be used only when the primary link is down.

 This is not your father's RIP. As observed in this section, the JUNOS RIP implementation is a bit more complicated than what you may be familiar with in IOS. In IOS, a simple `router rip` statement followed by a `network` statement or two is all that is needed. The choice of default reject export policy in JUNOS means that, in almost all cases, you will need to define a RIP export policy that accepts RIP routes for export just to get a JUNOS device to behave like any other RIP router. Folks often miss this and scratch their heads when things don't initially work as expected.

The configuration at Scotch is simpler, in that no metric manipulation is needed. The ability to use a common export policy for the RIP-learned default route *and* the local direct loopback address allows you to meet the stated goals with the definition of a single RIP group:

```
[edit]
lab@Scotch# show protocols rip
group rip_peers {
    export rip_export;
    neighbor ge-0/0/0.0;
    neighbor ge-0/0/1.0 {
        import default_only;
    }
}
```

Both RIP neighbors share the `rip_export` policy, as noted, but the neighbor-level application of the `default_only` policy means it is applied to the R2 peering only. This allows Scotch to import RIP routes that originate at Vodkila, in addition to its RIP-learned default route, in the event of disruption to the peering session between Scotch and R2. The related policies are displayed:

```
[edit]
lab@Scotch# show policy-options
policy-statement default_only {
    term 1 {
        from {
            protocol rip;
            route-filter 0.0.0.0/0 exact;
        }
        then accept;
    }
    term 2 {
```

```
        then reject;
    }
}
policy-statement rip_export {
    term readvertise_rip {
        from protocol rip;
        then accept;
    }
    term direct {
        from {
            protocol direct;
            route-filter 10.10.1.9/32 exact;
        }
        then accept;
    }
    term deny {
        then reject;
    }
}
```

Verify RIP

With RIP configured, it's time to confirm that all is well. We focus on Vodkila, as this is an *EX* book, but the common JUNOS software also running on J Series router Scotch means the same commands and techniques apply equally there. You begin with the obligatory display of which show rip operational mode commands are available at the CLI:

```
[edit]
lab@Vodkila# run show rip ?
Possible completions:
  general-statistics   Show RIP general statistics
  neighbor             Show RIP interfaces
  statistics           Show RIP statistics
[edit]
```

Confirming that RIP is running on the right interfaces is easy with a show rip neighbor command:

```
[edit]
lab@Vodkila# run show rip neighbor
                  Source        Destination   Send   Receive  In
Neighbor   State  Address       Address       Mode   Mode     Met
--------   -----  -------       -----------   ----   -------  ---
ge-0/0/8.0    Up  10.3.9.3      224.0.0.9     mcast  both      1
ge-0/0/9.0    Up  172.16.3.3    224.0.0.9     mcast  both      3
```

The show rip statistics command also indicates general operational health as a function of messages being sent and received, versus any error message counts, and so on:

```
[edit]
lab@Vodkila# run show rip statistics
RIPv2 info: port 520; holddown 120s.
    rts learned  rts held down  rqsts dropped  resps dropped
```

```
            2              0              0              0

ge-0/0/8.0:  2 routes learned; 2 routes advertised; timeout 180s; update interval 30s
Counter                       Total   Last 5 min  Last minute
-------                   -----------  ----------- -----------
Updates Sent                    235          10            2
Triggered Updates Sent            7           0            0
Responses Sent                    0           0            0
Bad Messages                      0           0            0
RIPv1 Updates Received            0           0            0
RIPv1 Bad Route Entries           0           0            0
RIPv1 Updates Ignored             0           0            0
RIPv2 Updates Received          412          20            4
RIPv2 Bad Route Entries           0           0            0
RIPv2 Updates Ignored             0           0            0
Authentication Failures           0           0            0
RIP Requests Received             4           0            0
RIP Requests Ignored              0           0            0
```

Sooner or later, routing protocol verification invariably comes around to looking at the RT. The RT is, after all, the end result of a routing protocol's operation. A show route protocol rip command is an easy way to filter the RT to display only RIP-learned routes:

```
[edit]
lab@Vodkila# run show route protocol rip

inet.0: 15 destinations, 15 routes (15 active, 0 holddown, 0 hidden)
+ = Active Route, - = Last Active, * = Both

0.0.0.0/0          *[RIP/100] 01:33:07, metric 3, tag 0
                    > to 10.3.9.9 via ge-0/0/8.0
10.10.1.9/32       *[RIP/100] 00:09:29, metric 2, tag 0
                    > to 10.3.9.9 via ge-0/0/8.0
224.0.0.9/32       *[RIP/100] 00:09:40, metric 1
                      MultiRecv

__juniper_private1__.inet.0: 4 destinations, 6 routes (1 active, 0 holddown, 3 hidden)
```

The output shows that Vodkila has learned two routes through RIP. The 224.0.0.9 entry is the multicast route entry (MultiRecv) used to facilitate reception of RIPv2's multicast route updates. The entry is created by the routing process and is not advertised (or learned) in any RIP updates. As hoped, Vodkila has both a default route and a route to Scotch's loopback address, and, just as important, both point to the primary link via its link to Scotch. Note that the metric value for these routes reflects the local router's cost; the cost advertised to other RIP speakers will normally be one higher. The 0/0 route in this case is received with a hop count of 2, reflecting the need to transit two routers in order to reach its origination point at R2. When installed in the RT, Vodkila adds its costs to reach that neighbor, a 1 in this case, resulting in the metric value 3 that is shown.

Recall that the RIP configuration at Vodkila adds 3, rather than the default 1, to the updates received over its ge-0/0/9 interface. The metric for the default routes learned over this interface is therefore the received cost (1), plus 3 for the metric addition, for a total metric of 4. This explains why the secondary path has not been chosen.

You can use the show route receive protocol rip <remote-neighbor-ip> command to display active routes that are received on a particular interface from a particular RIP source. This helps to confirm that Scotch is in fact advertising two RIP routes to Vodkila. Note that the neighbor argument to this command for RIP is a remote neighbor IP address:

```
[edit]
lab@Vodkila# run show route receive-protocol rip 10.3.9.9

inet.0: 15 destinations, 15 routes (15 active, 0 holddown, 0 hidden)
+ = Active Route, - = Last Active, * = Both

0.0.0.0/0          *[RIP/100] 00:00:40, metric 3, tag 0
                    > to 10.3.9.9 via ge-0/0/8.0
10.10.1.9/32       *[RIP/100] 00:23:43, metric 2, tag 0
                    > to 10.3.9.9 via ge-0/0/8.0
```

This command has a send counterpart in the form of a show route advertising protocol rip <local-ip-address> command. Note that in this context the neighbor argument is the local interface's IP address, rather than its name. Vodkila is confirmed to be sending RIP routes to both of its upstream neighbors:

```
[edit]
lab@Vodkila# run show route advertising-protocol rip 10.3.9.3

inet.0: 15 destinations, 15 routes (15 active, 0 holddown, 0 hidden)
+ = Active Route, - = Last Active, * = Both

10.10.1.3/32       *[Direct/0] 1d 05:00:14
                    > via lo0.0
200.2.0.0/22       *[Static/5] 02:32:37
                      Reject
```

The advertisements sent out to the 10.3.9.3 interface confirm that Vodkila is advertising the access network summary and its loopback address, and that it's obeying split horizon by *not* advertising the default route back to the speaker from which it was learned:

```
[edit]
lab@Vodkila# run show route advertising-protocol rip 172.16.3.3

inet.0: 15 destinations, 15 routes (15 active, 0 holddown, 0 hidden)
+ = Active Route, - = Last Active, * = Both

0.0.0.0/0          *[RIP/100] 00:04:02, metric 3, tag 0
                    > to 10.3.9.9 via ge-0/0/8.0
10.10.1.3/32       *[Direct/0] 1d 05:00:16
                    > via lo0.0
```

```
10.10.1.9/32        *[RIP/100] 00:27:05, metric 2, tag 0
                     > to 10.3.9.9 via ge-0/0/8.0
200.2.0.0/22        *[Static/5] 02:32:39
                     Reject
```

In contrast, the advertisements to the 172.16.3.3 interface show that Vodkila is readvertising its (RIP-learned) default route to R2, but note the increased hop count that ensures that R2 will not use it (its static default has a lower [more] administrative distance [preference] anyway, so the RIP metric does not even matter in this case). Note that the command output shows the *local route metric*, rather than what is actually advertised. Recall that here the metric sent to R2 will have 3 added due to the configuration.

Because Vodkila prefers the routes learned via the Scotch peering, no RIP routes are installed over its ge-0/0/9 interface. As a result, the show route receive-protocol rip 172.16.3.1 command does not return any output:

```
[edit protocols rip]
lab@Vodkila# run show route receive-protocol rip 172.16.3.1

inet.0: 15 destinations, 15 routes (15 active, 0 holddown, 0 hidden)
```

Be careful here, as this makes it easy to incorrectly believe there is a RIP malfunction resulting in R2's failure to advertise the expected routes to the local router. Tracing, which is akin to debugging in IOS, is added to Vodkila to gain added insight as to whether RIP is correctly operating over the R2 peering, given that so far it has been hard to tell:

```
[edit protocols rip]
lab@Vodkila# show traceoptions
file rip;
flag error detail;
flag packets detail;
flag update detail;
```

The rip trace log is monitored with a monitor start rip command:

```
[edit protocols rip]
lab@Vodkila# run monitor start rip
```

And like clockwork, RIP starts firing off its scheduled updates, resulting in send and receive activity:

```
[edit protocols rip]
lab@Vodkila#
*** rip ***
Jan 12 09:40:14.185692 Preparing to send RIPv2 updates on nbr ge-0/0/8.0, group:
Scotch.
Jan 12 09:40:14.185921 Update job: sending 20 msgs; nbr: ge-0/0/8.0; group: Scotch;
msgp: 0x2567c00.
Jan 12 09:40:14.185996  nbr ge-0/0/8.0; msgp 0x2567c00.
Jan 12 09:40:14.186025              10.10.1.3/0xffffffff: tag 0, nh
0.0.0.0, met 1.
Jan 12 09:40:14.186047              200.2.0.0/0xfffffc00: tag 0, nh
```

```
                       0.0.0.0, met 1.
                       Jan 12 09:40:14.186069              sending msg 0x2567c04, 2 rtes
                       Jan 12 09:40:14.186465 Update job done for nbr ge-0/0/8.0 group: Scotch
```

The first activity is an update message sent to Scotch over the ge-0/0/8 interface. It clearly shows the 200.2.0/22 and 10.10.1.3 prefixes (routes) being advertised as expected and confirmed with previous commands. Note that the metric that is actually being advertised is now shown, which, in this update, is 1 (the default):

```
                       Jan 12 09:40:20.449687 Preparing to send RIPv2 updates on nbr ge-0/0/9.0, group:
                       distribution.
                       Jan 12 09:40:20.449918 Update job: sending 20 msgs; nbr: ge-0/0/9.0; group:
                       distribution; msgp: 0x2567a00.
                       Jan 12 09:40:20.449944   nbr ge-0/0/9.0; msgp 0x2567a00.
                       Jan 12 09:40:20.449970              200.2.0.0/0xffffffc00: tag 0, nh
                       0.0.0.0, met 3.
                       Jan 12 09:40:20.449992              10.10.1.3/0xffffffff: tag 0, nh
                       0.0.0.0, met 3.
                       Jan 12 09:40:20.450014              10.10.1.9/0xffffffff: tag 0, nh
                       0.0.0.0, met 5.
                       Jan 12 09:40:20.450036              0.0.0.0/0x00000000: tag 0, nh
                       0.0.0.0, met 6.
                       Jan 12 09:40:20.450057              sending msg 0x2567a04, 4 rtes
                       Jan 12 09:40:20.450445 Update job done for nbr ge-0/0/9.0 group: distribution
```

The next message is a RIP update sent to R2 over the ge-0/0/9 interface. Note that the locally originated routes have a metric of 3, and the routes being readvertised are set to 5. This confirms the correct operation of the metric add 3 logic in the rip_export_add export policy:

```
                       Jan 12 09:40:30.577936 received response: sender 172.16.3.1, command 2, version 2,
                       mbz: 0; 5 routes.
                       Jan 12 09:40:30.578035              0.0.0.0/0x00000000: tag 0, nh          0.0.0.0,
                       met 1.
                       Jan 12 09:40:30.578093              10.10.1.3/0xffffffff: tag 0, nh        0.0.0.0,
                       met 3.
                       Jan 12 09:40:30.578132              10.10.1.9/0xffffffff: tag 0, nh        0.0.0.0,
                       met 2.
                       Jan 12 09:40:30.578231              172.16.9.0/0xffffff00: tag 0, nh       0.0.0.0,
                       met 1.
                       Jan 12 09:40:30.578266              200.2.0.0/0xffffffc00: tag 0, nh       0.0.0.0,
                       met 3.
```

The next message is a received RIP update from the R2 router. This is what we have been waiting for, as it's the first proof so far that it's correctly advertising the default route to Vodkila over the secondary link. The tracing represents what came off the wire, and therefore shows the received metric. Recall that this neighbor has the metric-in option set to 3 to poison the secondary route, unless the primary fails and it's the only option left, of course.

With things looking good at Vodkila, the tracing is removed, and attention shifts to Scotch, where the route table is displayed:

```
[edit]
lab@Scotch# run show route protocol rip

inet.0: 11 destinations, 11 routes (11 active, 0 holddown, 0 hidden)
+ = Active Route, - = Last Active, * = Both

0.0.0.0/0          *[RIP/100] 00:26:09, metric 2, tag 0
                    > to 172.16.9.1 via ge-0/0/1.0
10.10.1.3/32       *[RIP/100] 01:03:49, metric 2, tag 0
                    > to 10.3.9.3 via ge-0/0/0.0
200.2.0.0/22       *[RIP/100] 01:03:49, metric 2, tag 0
                    > to 10.3.9.3 via ge-0/0/0.0
224.0.0.9/32       *[RIP/100] 00:26:17, metric 1
                       MultiRecv
```

Things also look good here. Note that the RIP-learned default route is pointing out the primary ge-0/0/1 path toward R2, in addition to the two RIP routes known to originate at Vodkila. As this is a convenient maintenance window, a quick failover test is initiated by downing the primary link at Scotch:

```
[edit]
lab@Scotch# deactivate interfaces ge-0/0/1
```

After committing the change, you note that it takes longer than expected for the router with the locally downed interface to remove the default RIP route, which is observed to still be pointing out a down interface:

```
[edit]
lab@Scotch# run show route protocol rip

inet.0: 9 destinations, 9 routes (8 active, 1 holddown, 0 hidden)
+ = Active Route, - = Last Active, * = Both

0.0.0.0/0          [RIP/100] 00:00:05, metric 2, tag 0
                    > to 172.16.9.1 via ge-0/0/1.0
. . .
```

You note that the route's age shows it was refreshed recently, as in some five seconds ago. Also of note is the presence of a hold-down route. Intrigued, you add the extensive switch to see what additional details you can glean:

```
[edit]
lab@Scotch# run show route protocol rip extensive

inet.0: 9 destinations, 9 routes (8 active, 1 holddown, 0 hidden)
0.0.0.0/0 (1 entry, 1 announced)
TSI:
RIP route tag 0; poison reverse, holddown metric 16, ends in 114 secs
     nbr ge-0/0/0.0: 0.0.0.0/0.0.0.0, met: 16, nh: 0.0.0.0
         RIP     Preference: 100
                 Next hop type: Router, Next hop index: 529
                 Next-hop reference count: 1
                 Next hop: 172.16.9.1 via ge-0/0/1.0, selected
                 State: <Delete Int>
                 Age: 6  Metric: 2      Tag: 0
```

```
Task: RIPv2
Announcement bits (1): 1-RIPv2
AS path: I
Route learned from 172.16.9.1 has expired
```

The added detail combined with RIP operational knowledge sheds light on what at first seems like odd behavior. The router was not asleep at the switch, and is aware of the down interface and resulting useless next hop. As a result, the route is set to expired (without waiting for a timeout), and the router has placed it into poison reverse state. This sets the metric to infinity in an attempt to expedite the removal of the local router as a next hop for this destination in all the downstream nodes that will listen. The route is retained as active in the RT to prevent the chance of a later route update, one with a *higher metric*, being installed in its absence. Such a route could represent a loop because a higher metric can imply that the advertising router is using the local router as its next hop. During this time, a new route with a better metric than the current hold-down route can be installed, but the old route will still be advertised with a poisoned metric until the timeout expires.

Later yet, a similar condition is noted at Vodkila, which is the recipient of all that poison reverse love Scotch is generating:

```
[edit protocols rip]
lab@Vodkila# run show route protocol rip

inet.0: 15 destinations, 15 routes (14 active, 1 holddown, 0 hidden)
+ = Active Route, - = Last Active, * = Both

0.0.0.0/0          [RIP/100] 00:02:00, metric 3, tag 0
                   > to 10.3.9.9 via ge-0/0/8.0
. . .
```

Notice that some two minutes have now elapsed since the last update for the default route, which is also in a hold-down state, in this case because of the poison reverse actions at Scotch that result in continued advertisements of the route with a poisoned metric (16) of 120 seconds. Recall that poison reverse helps to ensure that other routers have no misconceptions about their being able to reach that destination through the advertising router, while a hold down helps to guard against loops by not installing a new next hop for a deleted router until enough time has expired to ensure that the old route will have aged out of all speakers. After the route has been poisoned for a while, and the hold down has expired, it's assumed that it is safe to begin listening for the next best route to that destination. Within 30 seconds, someone will advertise such a route, assuming one still exists, and the network will reconverge.

This "poor" performance is all part and parcel of RIP's DV nature. The use of unreliable periodic route exchange means that several updates must be lost before the death of an upstream neighbor is detected. The default timeout in JUNOS is 180 seconds. Triggered updates and poison reverse help to move bad news along, but even then hold-down timers are needed to help guard against loops, and in JUNOS the default hold down is four update periods or 120 seconds. This means that in a worst-case scenario, it will take 180 seconds to time out a route, then another 120 seconds for hold down. Ouch. The good news is that most of these delays occur when routes are removed, not when they are added. So, upon restoration, things tend to heal pretty quickly. It can truly be said that in RIP, bad news travels slowly and good news quickly. And we did state that RIP is not the fastest horse in the stable, right?

Just as you're beginning to worry, the protocol finally converges:

```
[edit protocols rip]
lab@Vodkila# run show route protocol rip

inet.0: 15 destinations, 15 routes (15 active, 0 holddown, 0 hidden)
+ = Active Route, - = Last Active, * = Both

0.0.0.0/0          *[RIP/100] 00:00:46, metric 4, tag 0
                    > to 172.16.3.1 via ge-0/0/9.0
. . .
```

The display confirms a RIP-learned default that now points out the secondary link at Vodkila. Note that the displayed metric of 4, which, combined with previous tracing that showed a received metric of 1, finally provides confirmation of the metric-in knob's functionality. Before restoring the link, a few traceroutes are performed to confirm routing over the secondary link:

```
R2#traceroute 200.2.2.3

Type escape sequence to abort.
Tracing the route to 200.2.2.3

  1 172.16.3.3 0 msec 0 msec 0 msec
  2 200.2.2.3 4 msec 4 msec *
R2#
```

The trace from R2 to Host3 confirms that during primary failures, inbound traffic takes the secondary link. The result also confirms the R2 routes to the access layer's 200.2.0/22 subnets:

```
Host3#traceroute 192.168.1.1

Type escape sequence to abort.
Tracing the route to 192.168.1.1

  1 200.2.2.254 24 msec 0 msec 4 msec
  2 172.16.3.1 4 msec 4 msec *
Host3#
```

And the traceroute from Host3 to R2's loopback also succeeds, thereby confirming that default outbound routing is working over the secondary link during a primary outage. The primary link is restored:

```
[edit]
lab@Scotch# rollback 1
load complete
```

And a short time later, since good news travels faster than bad in the land of RIP, all route tables are observed to have converged back onto the primary link:

```
[edit protocols rip]
lab@Vodkila# run show route protocol rip

inet.0: 15 destinations, 15 routes (15 active, 0 holddown, 0 hidden)
+ = Active Route, - = Last Active, * = Both

0.0.0.0/0          *[RIP/100] 00:00:08, metric 3, tag 0
                    > to 10.3.9.9 via ge-0/0/8.0
. . .
```

The traceroutes are repeated, and they confirm routing symmetry over the primary link:

```
R2#traceroute 200.2.2.3

Type escape sequence to abort.
Tracing the route to 200.2.2.3

  1 172.16.9.9 0 msec 4 msec 4 msec
  2 10.3.9.3 0 msec 0 msec 0 msec
  3 200.2.2.3 4 msec 4 msec *
R2#
Host3#traceroute 192.168.1.1

Type escape sequence to abort.
Tracing the route to 192.168.1.1

  1 200.2.2.254 4 msec 0 msec 4 msec
  2 10.3.9.9 8 msec 4 msec 8 msec
  3 172.16.9.1 4 msec 4 msec *
Host3#
```

The results confirm the extra hop and the IP addresses of the primary link in both directions. This completes the validation of the EX RIP deployment lab.

Routing protocol interaction can be complex and sometimes has unanticipated behaviors. It's always a good idea to test failover scenarios as part of routine maintenance to ensure that things do what you expect. There is nothing worse than learning your design is fatally flawed *during* an actual network outage. The design demonstrated here was revertive, in that the goal was to use the primary link whenever it was available. Some designs prefer to stick on a backup link until manual intervention decides the primary link is stable enough to switch back to. In this example, correct failover—and, later, the revertive switch back to the primary link—were both confirmed.

RIP Summary

RIP is not glamorous, but it's well vetted and easy to understand, and it's hard to find a router that does not support it. In this fast-moving business, it's easy to fall victim to networking one-upmanship, and to feel as though you must be running with the latest version of Internet draft [protocol name here] in order to stake your claim to living on the bleeding edge.

There are times when RIP is just a good fit, and while another, far more complicated protocol can also work, it's best to avoid protocol envy and IT peer pressure and make the decision that is right for your network.

Conclusion

EX platforms provide a rich set of IP routing right out of the box, with no "routing-enabled" images or messy software trains to wade through.

Routing is one of those things that is hard to get right. At a small scale, the situation may be forgiving, but history shows that many switch vendors tried to become core class routers—and they have all failed. With the EX, however, you have a reverse case of carrier-class, Internet core-proven IP routing that happens to ship with your Ethernet switch because it's already in JUNOS. While the EXs have plenty of room to mature and grow into more Layer 2 features, they kind of had Layer 3 covered from day one.

The impressive scaling capabilities of the EX platforms and the common JUNOS image mean that folks familiar with routing in JUNOS software will be able to hit the ground running with an EX under each arm; you may be able to put off buying those router upgrades and just let the EX do it all!

Note that while this chapter provides demonstrations of static and RIP routing, the EX platforms are also capable of full-blown OSPF, BGP, and IS-IS, as well as IPv6 routing using RIPng or OSPF3. Even though IS-IS and BGP currently require an advanced routing license, that's still less expensive, and likely far more reliable, than adding a standalone router for Layer 3 handling.

Chapter Review Questions

1. Which is true regarding routing?
   a. End stations are not aware of routers
   b. End stations participate in routing and intentionally evoke routing services when needed
   c. Routing is based on a flat addressing space
   d. Bridging is based on a hierarchical address space
2. Stations in different VLANs can communicate:
   a. Directly when Layer 2 ACLs permit
   b. Indirectly, through transparent bridging
   c. Directly via a router
   d. Indirectly via a router

3. True or false: when a station in one VLAN attempts to reach a destination in another VLAN, the ARP is sent to the target station directly.

4. You can provide inter-VLAN routing on the EX with:
    a. A switched VLAN interface
    b. A routed VLAN interface
    c. A tunnel interface that is placed into both VLANs
    d. Two interfaces with an external cable

5. Four VLANs are defined, and each is served by a vlan interface unit. Which is true?
    a. By default, no inter-VLAN communications are possible
    b. By default, all stations will have inter-VLAN connectivity at Layer 3
    c. Policy is needed to permit communications
    d. Firewall filters must be defined with an accept action for inter-VLAN communications to succeed

6. What is the difference between a discard and a reject next hop on a static route?
    a. Discard silently tosses traffic, while reject generates an error message
    b. Discard generates an error message, while reject tosses traffic silently
    c. Discard black-holes traffic, while reject prevents the route from being readvertised in another protocol
    d. Neither is valid; a static route must point to a forwarding next hop

7. Where can export policy be applied in the JUNOS RIP implementation?
    a. Globally
    b. Group level
    c. Neighbor level
    d. All of the above

8. What is the default RIP export policy?
    a. Accept all sane RIP routes
    b. Accept all direct interface routes on which the protocol is running
    c. Reject all
    d. Both A and B

9. What command displays RIP routes received from a given neighbor?
    a. show route advertising-protocol rip <neighbor>
    b. show route receiving-protocol rip <neighbor>
    c. show route protocol rip
    d. show ip rip route

10. What networks are covered by the prefix 200.0.0.0/22?

a. 200.0.0.0 through 200.0.0.22

b. 200.0.0 through 200.0.22

c. 200.0.0.0 through 200.0.3.0

d. None of the above; a Class C network must have a mask of 24 bits or longer

11. What is the purpose of a subnet mask (prefix length)?

a. The mask identifies the network portion of the address, which is used to route

b. The mask identifies the host portion of the address, which is used to route

c. The mask indicates when a portion of the network space is redefined to evoke subnetting

d. The mask indicates when a portion of the host space is redefined to evoke supernetting (CIDR)

12. A RIP route is not displayed with a `show route receiving-protocol rip` command at one end, yet the remote end shows the route in its `show route advertising-protocol rip` command. What could account for this?

a. Another route is preferred, causing the update to be ignored

b. Import policy is filtering the route

c. You may be entering the wrong neighbor argument, which is easy to do and returns null

d. All of the above

Chapter Review Answers

1. Answer: B. Routing is not transparent. End stations participate in the decision to use a router when they decide a destination is not local. Routing is based on hierarchical addressing.

2. Answer: D. Routing is needed to interconnect stations in different VLANs. Routing is an indirect delivery model. Layer 2 filters can be used to restrict communications with this same VLAN; Layer 3 filters are normally used to filter inter-VLAN traffic.

3. Answer: False. When attempting to reach a station on a different link, the ARP is sent to the next hop, which is a router for indirect delivery. Within the same VLAN, the ARP is sent to the target address itself.

4. Answer: B. The RVI is used for internal routing among VLANs. The SVI is in IOS, and while two interfaces with an external cable could work, they are not required.

5. Answer: B. All units on a VLAN interface are seen as directly connected routes. Layer 3 filters are needed to prevent routing among these direct networks.

6. Answer: A. In JUNOS, discard is a silent discard, while reject generates an error message. Both are valid next hops for a static route.

7. Answer: B. RIP in JUNOS accepts the **export** keyword only at the group level. Import policy can be applied at all three levels.

8. Answer: C. By default, RIP advertises/readvertises nothing, not even active RIP routes. Odd, isn't it?

9. Answer: B. The `show route advertising-protocol` form displays routes sent out of a RIP interface. The `show route protocol rip` command does display RIP routes, but may not make it obvious from which RIP neighbor they were received.

10. Answer: C. In classful form, a class network address is assumed to have a /24 mask. Supernetting and subnetting allow the mask length to be decreased or increased, respectively. The /22 indicates only 24 bits of network prefix, making option A incorrect. Option B is incorrect because with only the least significant two bits of the third octet available for use, only four decimal combinations are possible, and these are 0–3.

11. Answer: A. The mask identifies the network and host portions of the address. Routing is done on the network portion. A subnet is when a classful mask is extended, while a supernet reduces classful mask length to form CIDR blocks (aggregates).

12. Answer: D. Any of the stated conditions could result in a route being omitted from the `show route receiving-protocol rip` command. It's easy to get confused as to what the neighbor value should be. For the `advertising-protocol` form, the neighbor is the local end's IP address:

```
[edit]
lab@Vodkila# run show route advertising-protocol rip ?
Possible completions:
  <neighbor>          IP address of neighbor (local for RIP and RIPng)
[edit]
lab@Vodkila# run show route advertising-protocol rip
```

For the `receiving-protocol` form, the neighbor is the remote neighbor's IP address:

```
[edit]
lab@Vodkila# run show route receive-protocol rip ?
Possible completions:
  <peer>              IP address of neighbor
[edit]
lab@Vodkila# run show route receive-protocol rip
```

The CLI help strings are a useful way to remember the neighbor syntax differences between the advertising and receiving forms of the command; recall that specifying an incorrect neighbor simply returns an empty list with no compliant, which sometimes hides the fact that the wrong command syntax was used, and this can in turn lead your fault isolation down the wrong path.

Routing Policy and Firewall Filters

Routing policy and firewall filters serve very different purposes, but because they share a common structure, they can be grouped together. A shot of tequila can be used to celebrate a birthday or a team win, or to block out the birthday or team loss. Either way, liquor is a common tool (or language) that can lead to very different results.

Once you understand the framework of policy language, you can switch between policies and filters seamlessly, concentrating on the application needed and not the syntax.

This chapter is divided into two sections. The first half discusses policy and includes the following topics:

- Policy overview and import and export policy
- Policy components (terms, match conditions, actions, and policy chains)
- Route filters
- Testing and monitoring policy
- Policy case study

The second half of this chapter discusses firewall filters:

- Firewall filter overview
- Firewall filter processing
- Firewall filter monitoring

Routing Policy

This section details JUNOS software routing policy operation and configuration. The actual application of policy to solve some specific networking requirement is generally left to the protocol to which the policy is applied. These particular applications are discussed in various sections throughout the book. You configure policy-related options and statements at the [edit policy-options] hierarchy. Routing policy and firewall filters have a similar syntax in JUNOS software. The first deals with routes in the control plane, whereas the second deals with packets in the data plane.

What Is Routing Policy, and When Do I Need One?

Simply put, routing policy is used to:

- Control what routes are installed into the routing table (RT) for possible selection as an active route
- Control what routes are exported from the RT, and into which protocols
- Alter attributes of routes, either at reception or at the time of advertisement to other peers

Given that routing policy is used to control the reception and transmission of routing information and to alter route attributes, it's safe to say that you need routing policy when the default policy does not meet your requirements.

The specifics of the various default policies are covered later, but to provide an example, consider that by default, directly connected routes are not advertised into any routing protocol, which of course includes Routing Information Protocol (RIP). If your goal is to get direct routes advertised into RIP, the default policy obviously does not meet your needs and a custom policy must be written and applied to achieve your goal of redistributing direct routes into RIP.

Where and How Is Policy Applied?

You can apply policy in one of two places: at import or at export. Generally speaking, you use a command of the form set protocols `<protocol-name>` import to apply an import policy, or set protocols `<protocol-name>` export to apply an export policy. Figure 8-1 illustrates this concept.

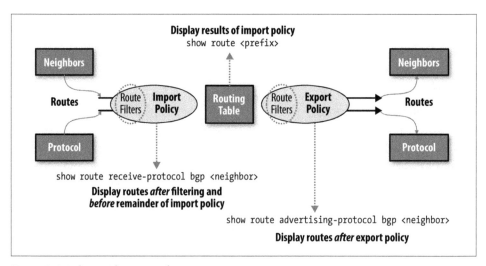

Figure 8-1. Policy application and monitoring points

Figure 8-1 shows routes being received through some protocol, and how import policy serves to filter and adjust route attributes before they are copied into the RT. In contrast, export policy comes into play when routes are being selected from the RT for inclusion in transmitted route updates. Once again, the export policy serves to filter and adjust route attributes to meet the specific needs of the networking environment.

It is worth noting that distance vector (DV) protocols such as RIP and Border Gateway Protocol (BGP) actually support the notion of received and transmitted routes.

 We do not discuss BGP in this book because it is not a common protocol deployment on an EX.

For information on BGP, please see *JUNOS Enterprise Routing (http://oreilly.com/catalog/9780596514426/)*.

These protocols support the show route receiving-protocol <protocol-name> <neighbor-address> and show route advertising-protocol <protocol> <neighbor-address> commands, which are very useful when troubleshooting or analyzing policy operation. Figure 8-1 shows how the receiving-protocol form of the command is used to display routes *after* route filtering, but *before* attribute manipulation. In contrast, the advertising-protocol form of the command is executed after all export policy operations, to include route filtering and attribute modification. You simply issue a show route <prefix> command to display a route as it exists in the RT, which will include any modified attribute resulting from import policy operations.

Applying policy to link state routing protocols

Link state (LS) protocols such as Open Shortest Path First (OSPF) and Intermediate System to Intermediate System (IS-IS) do not send and receive routes directly. Instead, they flood link state advertisement (LSA) packets, which are used to build a topological database, from which each router computes an RT. As such, LS protocols do not support much in the way of import policy. OSPF import policies that prevent installation of external routes into the RT are supported, but have very limited application. In general, it is best to know the rules and limitations of your intended routing protocols *before* making any leaps or assumptions!

If you wish to filter LSAs, protocol-specific mechanisms are required to ensure that LS database consistency is maintained. Chapter 7 covers the concepts of stub areas and LSA filtering.

You can apply export policy to an LS protocol to affect route redistribution, but the external route is still flooded in an LSA rather than being sent outright; the result is that the show route receiving-protocol and show route advertising-protocol commands are not effective when dealing with LS protocols.

When you apply policy to an LS protocol, you do so globally, which is to say the policy is not applied to particular interfaces or areas. In the case of OSPF, you apply export policy at the [edit protocol ospf] hierarchy:

```
[edit protocols ospf]
lab@Rum# show
export test_export; ## 'test_export' is not defined
```

The command-line interface (CLI) warning provides a nice reminder that the related test_export policy does not yet exist. Because the presence (or absence) of a policy can have a dramatic effect on overall network operation, you will not be able to commit a configuration with this type of omission. You can define a policy that is never applied, but once applied, the policy must exist before you can commit the changes.

Applying policy to RIP

RIP supports the application of import and export policy, and supports application of *import* policies at different hierarchies. Export policies are allowed only at the group level, however.

Focusing on import policies for the moment, you can apply a policy at one of three different hierarchies—global, group, or neighbor. The following code snippet provides an example of this concept:

```
[edit protocols bgp]
lab@RUM# show
import global_import;
group rip-example {
    import internal_import;
    neighbor ge-0/0/0.0 {
        import neighbor_ge-0/0/0_import;
    }
    neighbor ge-0/0/1.0;
}
group other {
    neighbor ge-0/0/2.0;
}
```

In this example, a policy named global_import is applied at the global level, another policy named internal_import is applied at the group level, and yet a third policy named neighbor_ge-0/0/0_import is applied at the neighbor level.

A key point, and one that is often misunderstood and that can lead to problems, is that in such a configuration *only* the most explicit policy is applied. A neighbor-level policy is more explicit than a group-level policy, which in turn is more explicit than a global policy. Hence, neighbor ge-0/0/0 is subjected only to the neighbor_ge-0/0/0_import policy, whereas neighbor ge-0/0/1, lacking anything more specific, is subjected only to the internal_import policy. Meanwhile, neighbor ge-0/0/2 in group other has no group- or neighbor-level policy, so it uses the global_import policy.

The use of the neighbor may seem confusing at first, as RIP does not have explicit neighbors, unlike other routing protocols. Instead, RIP sends messages with a destination IP address of a broadcast packet or multicast packet. So, in this example, the neighbor refers to the local interface in which these messages will be sent, which could contain one or many devices.

What if you need to have neighbor ge-0/0/0 perform the function of all three policies? Simple—you could write and apply a new neighbor-level policy that encompasses the functions of the other three, or simply apply all three *existing* policies, as a chain, to neighbor ge-0/0/0. Note the use of brackets in the following command to open a set of values; if desired, each policy can be specified individually:

```
[edit protocols rip group rip-example]
lab@RUM# set neighbor ge-0/0/0.0 export [global-import
internal_import]

[edit protocols rip]
lab@RUM# show group rip-example neighbor ge-0/0/0.0
export [ neighbor_ge-0/0/0_import global_import internal_import];
```

As with access control lists (ACLs) or firewall filters, chained policy statements are evaluated in a specific left-to-right order, and only up to the point when a route is either accepted or rejected. As a result, you must consider factors such as whether a policy makes use of a match-all deny term at its end, which is common for a standalone policy. However, when applied at the front of a policy chain, the match-all aspect of such a policy prevents route processing by any remaining policies.

To help illustrate this point, consider two policies, one named deny, which denies all, and another named accept, which accepts all. Given the nature of the two policies, you will see a dramatic difference between the two policy chains, even though they are composed of the same parts:

```
export [accept deny];
export [deny accept];
```

Here, the first policy chain results in *all* routes being *accepted*, whereas the reverse application results in *all* routes being *denied*. You can use the CLI's insert feature to rearrange the order of applied policies, or simply delete and reapply the policies to get the order needed. Note that a newly applied policy always takes the leftmost place in a policy chain, where it becomes the first in line for route evaluation.

We covered a few critical points here, so much so that they bear repeating, in another form. The first point is that when multiple policies are applied at different CLI hierarchies for the same protocol, only the most specific application is evaluated, to the exclusion of other, less-specific policy applications. Second, a given route is evaluated against a chain of policies starting with the leftmost policy, up until the route meets a terminating action of either accept or reject. This leads to ordering sensitivity of both terms within a policy, and for policies when they are chained together.

Although these points always seem to make sense when you are learning them, they are somehow easily forgotten during switch configuration, when two policies that individually worked as expected suddenly break when they are combined, or when you mistakenly believe that a neighbor-level policy is combined with a global or group-level policy, only to find that your policy behavior is not what you anticipated.

Policy Components

Generally speaking, a policy statement consists of one or more named terms, each consisting of two parts: a from statement that defines a set of match criteria, and a corresponding then statement that specifies the set of actions to be performed for matching traffic. It is possible to create a policy with a single term, in which case the term can be unnamed, such as in these two examples:

```
[edit policy-options]
lab@RUM# show
policy-statement explicit_term {
    term 1 {
        from protocol direct;
        then accept;
    }
}
policy-statement implict_term {
    from protocol direct;
    then accept;
}
```

The two policy statements perform identical functions: both have a match criterion of direct, and both have an associated action of accept. The explicit term format is generally preferred, because new terms can be added without the need to redefine the existing term. Note that any new terms are added to the end of the policy statement, as shown here, where, oddly enough, a new term named new is added to the explicit_term policy statement:

```
[edit policy-options]
lab@RUM# set policy-statement explicit_term term new from protocol
direct

[edit policy-options]
lab@RUM# set policy-statement explicit_term term new then reject
```

```
[edit policy-options]
lab@RUM# show policy-statement explicit_term
term 1 {
    from protocol direct;
    then accept;
}
term new {
    from protocol direct;
    then reject;
}
```

As with policy chains, term ordering within a policy is significant. In the example, explicit_term policy, term 1, and term new are diametrically opposed, with one accepting and the other denying the same set of direct routes. Although making little practical sense, it does afford the opportunity to demonstrate term resequencing with the insert function:

```
[edit policy-options]
lab@RUM# edit policy-statement explicit_term

[edit policy-options policy-statement explicit_term]
lab@RUM# insert term new before term 1

[edit policy-options policy-statement explicit_term]
lab@RUM# show
term new {
    from protocol direct;
    then reject;
}
term 1 {
    from protocol direct;
    then accept;
}
```

There is no practical limit to the number of terms that can be specified in a single policy, or how many policies can be chained together.

Logical OR and AND functions within terms

It's possible to define a term with multiple match criteria defined under a single from statement. For a match to occur, all of the from conditions must be met, which is a logical AND. However, for a specific match type, such as protocol, you can specify multiple values, in which case each protocol match condition functions as a logical OR. Consider this example:

```
[edit policy-options]
lab@RUM# show
policy-statement test {
    term 1 {
        from {
            protocol [ ospf rip ]; ##logical OR within brackets
            interface ge-0/0/0.0; ## logical AND with other match
criteria
```

```
        }
      then next term;
    }
  }
```

In this case, a match will occur when a route is learned over the ge-0/0/0 interface *and* is learned from OSPF *or* RIP.

Policy Match Criteria and Actions

JUNOS software policy provides a rich set of criteria you can match against, and an equally rich set of actions that can be performed as a result of a match. The various match and action functions are well documented, so the goal here is not to re-create the wheel by rehashing each option—as noted at the beginning of this chapter, our objective is to acquaint you with a box of tools; later chapters will provide specific examples of those tools being used.

Policy match criteria

The list of available match criteria is long in the JUNOS Software 9.2 release:

```
[edit]
lab@Vodkila# set policy-options policy-statement test term 1 from ?
Possible completions:
  aggregate-contributor  Match more specifics of an aggregate
+ apply-groups           Groups from which to inherit configuration data
+ apply-groups-except    Don't inherit configuration data from these groups
  area                   OSPF area identifier
+ as-path                Name of AS path regular expression (BGP only)
+ as-path-group          Name of AS path group (BGP only)
  color                  Color (preference) value
  color2                 Color (preference) value 2
+ community              BGP community
+ condition              Condition to match on
> external               External route
  family
  instance               Routing protocol instance
+ interface              Interface name or address
  level                  IS-IS level
  local-preference       Local preference associated with a route
  metric                 Metric value
  metric2                Metric value 2
  metric3                Metric value 3
  metric4                Metric value 4
> multicast-scope        Multicast scope to match
+ neighbor               Neighboring router
+ next-hop               Next-hop router
  next-hop-type          Next-hop type
  origin                 BGP origin attribute
+ policy                 Name of policy to evaluate
  preference             Preference value
  preference2            Preference value 2
> prefix-list            List of prefix-lists of routes to match
```

```
  > prefix-list-filter   List of prefix-list-filters to match
  + protocol             Protocol from which route was learned
    rib                  Routing table
  > route-filter         List of routes to match
    route-type           Route type
  > source-address-filter  List of source addresses to match
  + tag                  Tag string
    tag2                 Tag string 2
```

The key takeaway here is that you can match on things such as interface, protocol, route tag, next hop, metric, source address, area, and so on. Route filtering based on prefix and mask length is performed with the route-filter keyword. There is significant power (and complexity) in route filtering, and it is covered in its own section later in this chapter.

Policy actions

When a match occurs, a wide range of actions are available:

```
[edit policy-options]
lab@RUM# set policy-statement test term 1 then ?
Possible completions:
  accept                 Accept a route
+ apply-groups           Groups from which to inherit configuration data
+ apply-groups-except    Don't inherit configuration data from these groups
> as-path-expand         Prepend AS numbers prior to adding local-as (BGP only)
  as-path-prepend        Prepend AS numbers to an AS path (BGP only)
  class                  Set class-of-service parameters
> color                  Color (preference) value
> color2                 Color (preference) value 2
> community              BGP community properties associated with a route
  cos-next-hop-map       Set CoS-based next-hop map in forwarding table
  damping                Define BGP route flap damping parameters
  default-action         Set default policy action
  destination-class      Set destination class in forwarding table
> external               External route
  forwarding-class       Set source or destination class in forwarding table
> install-nexthop        Choose the next hop to be used for forwarding
> load-balance           Type of load balancing in forwarding table
> local-preference       Local preference associated with a route
> metric                 Metric value
> metric2                Metric value 2
> metric3                Metric value 3
> metric4                Metric value 4
  next                   Skip to next policy or term
> next-hop               Set the address of the next-hop router
  origin                 BGP path origin
> preference             Preference value
> preference2            Preference value 2
  reject                 Reject a route
  source-class           Set source class in forwarding table
> tag                    Tag string
> tag2                   Tag string 2
  trace                  Log matches to a trace file
```

Actions include changing the route color (internal tiebreaker), specifying the metric, altering a packet's forwarding class, adding a route tag, and so forth. Key actions include accept and reject, which are termination actions. The next keyword allows you to skip to the next term or policy in the chain, and is useful for shunting routes from one term or policy into another.

Route Filters

The ability to match on specific routes to accept or reject them, or to modify some attribute, is a critical aspect of virtually any networking scenario. The majority of JUNOS software routing policy strikes most users as intuitive and logical, given the easy-to-follow *if, then* construct of policy syntax translated into from/then in the actual JUNOS CLI syntax.

The exception always seems to be route filtering, because to truly understand how this is performed in JUNOS software, you must first understand the binary radix tree nature of the route lookup table, and how the binary tree is used in conjunction with route filters.

Binary trees

Binary trees have been used in computer science for several decades as a way to quickly locate a desired bit of information. In the case of route lookup, the goal is to quickly find the longest match for some prefix, with the corresponding next hop being the information that is sought. The Juniper Networks implementation of a binary tree is called the J-Tree, and it forms the basis of both route lookup and policy-based route filtering. Figure 8-2 shows the root of a binary tree, along with a few of its branches.

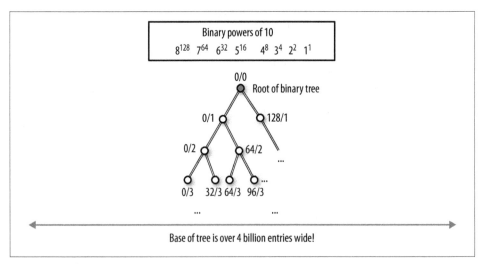

Figure 8-2. A binary tree

Figure 8-2 shows a "binary to powers of a decimal" chart, to help in understanding the structure of the J-Tree. For example, the binary sequence 0100 000 equates to a decimal 64, whereas 0110 0000 codes a decimal 96. In this example, bit 8, which has the decimal power of 128, represents the second set of nodes from the top of the tree. The top of the tree represents no bit, and the first pair of nodes down represents a test of the Most Significant Bit (MSB), which is bit 8 in this example, as either 0 (0) or 1 (128).

The binary tree is based on nodes that test the state of a particular bit that makes up the 32-bit IP address or route prefix. The bit being tested is indicated by the related prefix (mask) length. For example, the top of the tree is testing no bits, as indicated by the /0 prefix length. All prefixes match when you do not bother to test any bits, so the top of the tree effectively represents a default route, which is to say when no other patterns match you are guaranteed to match the first node—whether such a match actually results in forwarding depends on whether a default route has been installed, but that is another story.

The tree branches to the left when a given bit is a 0, and branches to the right for a 1. As a result, the first two nodes below the root represent the state of the MSB in the most significant byte, which is either a 0 or a 1. If it is a 0, you have a 0/1 match, which codes a decimal 0. If that bit is a 1, you have a 1/1 match, which codes a decimal 128. Each node then branches out, based on the test of the next bit, until you reach the bottom of the tree, representing a test of all 32 bits, which is *sometimes* necessary when doing a route lookup or route filter that is based on a /32 prefix length.

In actual operation, the J-Tree is optimized, and can quickly jump to a longest match when other portions of the tree are eliminated. In fact, it could be said that the act of finding a longest match against a binary tree is not so much finding what you seek as it is quickly eliminating all that cannot be what you want, and then simply looking at what is left. By way of example, a 32-bit IP address can take more than 4 billion combinations. However, half of these (2 billion) will have a 0 in the high-order bit position, whereas the other half will have a 1. By simply testing the status of one bit you have effectively eliminated one-half of the tree as not possible to match. With each subsequent bit test eliminating one-half of the remaining possibilities, you quickly arrive at a node that does not match the prefix being evaluated—in which case, you back up one node and that is the longest match for this prefix.

Route filters and match types

When you configure a route filter, you specify a starting prefix and initial prefix length, and then include a match type to indicate whether routes with prefixes longer than the initial value should be considered as matching. Put another way, a route filter is based on a match against the specified prefix bits, as based on the provided mask, in addition to the overall mask length of the prefix being evaluated. As such, it can be said that a Juniper route filter cares as much about the prefix *length* as it does the *prefix itself*.

Figure 8-3 illustrates the supported `route-filter` match types in the context of a J-Tree; we mentioned before, and state here again, that you cannot effectively use route filters if you do not first understand the operation of the J-Tree. This is especially true for the **through** match type, which 99.9% of the time is applied incorrectly, and therefore does *not* do what the operator wants.

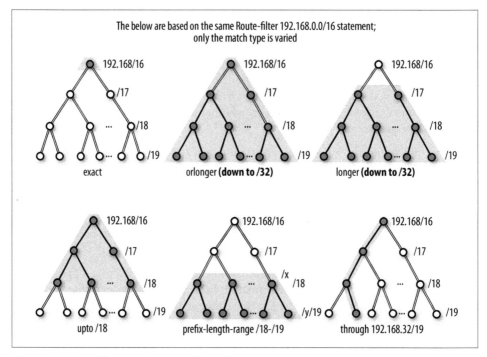

Figure 8-3. Route filter match types and the J-Tree

Figure 8-3 is based on a portion of the J-Tree that represents route 192.168/16. Entries below the starting node all share the same high-order 16 bits of 192.168, but differ from the root prefix in that they have longer mask lengths, as shown by the two nodes below the first, each of which is testing bit 17, therefore indicating a /17 mask length.

Each route filter match type is described against the corresponding portion of the figure:

exact

> The **exact** match type is just what it sounds like. To match with **exact**, both of the initial prefix bits must match, and the prefix length must be equal to the value specified. If the prefix bits do not match, or if the prefix length is either shorter or longer, the exact match type does not match. Figure 8-3 shows that a `route-filter` 192.168.0.0/16 **exact** matches only on that node of the J-Tree, to the exclusion of all others.

orlonger

> The orlonger match type matches the specified prefix and initial mask length, and matches on prefixes with longer mask lengths when they share the same high-order 16 bits, as indicated by the specified prefix. In this example, the result is a match against 192.168.0.0/16 itself, as well as 192.168.0/17 and 192.168.128/17 and all longer mask lengths, up to /32.

longer

> The longer match type excludes the exact match and catches all routes with the same prefix bits, but only when their masks are longer than the prefix length specified. The difference between orlonger and longer is shown in Figure 8-3. Here, the latter excludes the exact match, which is prefix 192.168.0.0/16 in this case.

upto

> The upto match type matches against the initial prefix and mask length, as well as matching prefixes with masks that are longer than the initial value, upto the ending mask length value. In the example, the initial prefix of 192.168.0.0/16 matches, as well as all other 192.168 prefixes that have mask lengths upto the specified value, which is 18 in this example. Therefore, 192.168.192/18 will match, whereas 192.168.1/24 does not.

prefix-length-range

> The prefix-length-range match type matches against routes with the same prefix as specified in the initial mask length, but only when the associated mask falls between the starting and ending values. The result is that the exact match is excluded, whereas routes with the same high-order prefix bits, but masks that fall within the specified range, are accepted. This match type is especially useful when the goal is to filter the route based on mask length alone, which is a common policy within service provider networks, and many refuse to carry routes with masks longer than 28 in an effort to keep RT size manageable. To prevent installing any route with a mask length longer than /28, you can use a route-filter 0/0 prefix-length-range /28-/32 reject statement. Because the initial prefix length is 0, all prefix values match, making the decision to reject one that is based strictly on mask length.

 It's worth noting that route-filter syntax supports a short form of action linking, in which the related then action can be specified directly on the route-filter line. Functionally, there is no difference between the short form and adding an explicit then action.

through

> The through match type is generally misunderstood, and rarely works the way folks think it will. This is not to say that it is broken, but it has led to this somewhat humorous rule of thumb: "When you are thinking of using through, think again." Often when people use through, what they really want is more of an upto or

prefix-length-range type of match; this rule of thumb is intended to warn users that in most cases, through is not what they really want, and that the decision to use it should be carefully thought through (pardon the pun).

A through match type matches the initial prefix and mask length exactly, as well as the ending prefix and mask length, and matches on the *contiguous* set of nodes between the two points. The through match type was originally offered to meet a corner case, in which a customer was found to be using 32 exact matches, all based on some form of a default route. Although a true default is 0/0, the customer wanted to ensure that they did not install any 0.0.0.0 prefixes, regardless of mask length. So, rather than a 0/0 exact, 0/1 exact, 0/2 exact … 0/32 exact, the through match type was created to allow the same effect with a single 0/0 through 0/32 statement. This matches the top of the tree, all the way down the left side to the very bottom, and all contiguous points in between.

In Figure 8-3, the through match type is specified as 192.168.0.0/16 through 192.168.32.0/19. The line shows the sequence of contiguous matches between the two points, which in this case includes 192.168.0.0/16, 192.168.0.0/17, 192.168.0.0/18, and 192.168.32.0/19. Now ask yourself, and be honest: is this what you expected a 192.168/16 through 192.168.32/19 to match?

Longest match wins, but may not…

As with routing in general, route filter processing is based on finding a longest match, and then performing the action associated with that match. There are cases where this behavior may lead to unexpected behavior because users do not always take into account the consequences of different match types. Recall that the longest match function is based on the high-order prefix bits, whereas the match type focuses more on mask length. Consider this route-filter example, and what will happen when route 200.0.67.0/24 is evaluated against it:

```
[edit policy-options policy-statement test_me]
user@host# show
from {
    route-filter 200.0.0.0/16 longer reject;
    route-filter 200.0.67.0/24 longer;
    route-filter 200.0.0.0/8 orlonger accept;
}
then {
    metric 10;
    accept;
}
```

The question is, will route 200.0.67.0/24 match this term, and if so, is it accepted, is it rejected, or does it have its metric set to 10 before being accepted? Think carefully, and consider how longest matching is performed, along with how the match type comes into play.

If you answered, "The route does not match, and is neither accepted nor rejected, and no metric modification is made," give yourself a well-deserved pat on the back. It's quite all right if you answered differently—this little tidbit alone may well justify the expenditure for this book (you did pay for this book, right?). The key here is that the longest match, as based on specified prefix, is against the second `route-filter` statement—here the first 24 bits of the prefix do in fact match 200.0.67, which is more exact than either 200/8 or 200.0/16. However, the longest match in this example has a match type of `longer`, meaning that only a route with a mask length of /25 through /32 with the 24 high-order bits set to 200.0.67 is considered to match.

Because this route has a mask length that is equal to the value specified, it does not match. A given route is only evaluated against the longest match in a given term. This is to say that if the longest match ends up not really matching, as shown in this example, other `route-filter` statements within that same term are not evaluated. Instead, the route falls through to the next term; policy; or, lacking any of those, default policy for the routing protocol in question.

Default Policies

The last hurdle in understanding JUNOS software policy is to be familiar with the default policy associated with each protocol used in your network. Understanding the default policy is important because it ultimately decides the fate of any route that is not matched against in your user-defined policy. Some operators rely on the default policy to do something, and others prefer to ensure that their policy is written to match on *all* possible routes, which means the default policy is negated because it never gets a chance to come into play.

OSPF default policy

The default import policy for OSPF is to accept all routes learned through that protocol. JUNOS software releases support explicit import policy, but *only* to filter external routes from being installed into the RT. Such an import policy does not filter external route LSAs from the database, however.

The default LS export policy is to reject everything. LSA flooding is not affected by export policy, and is used to convey routing in an indirect manner in an LS protocol. The result of this flooding is the advertisement of local interfaces that are enabled to run OSPF, as well as the readvertisement (flooding) of LSAs received from other routers.

IS-IS default policy

The default import policy for IS-IS is to accept all routes learned through that protocol. Unlike OSPF, there is currently no option to apply any import policy.

The export policy in IS-IS determines which IP information is contained in the link state packet (LSP). By default, all IP information on interfaces enabled for IS-IS is sent and all other IP information is rejected. If the default policy were explicitly written, it would contain two terms; the first term matches on configured IS-IS interfaces and accepts them, and the second term matches all and rejects.

This is a minor but important difference between OSPF and IS-IS.

RIP default policy

The default RIP import policy is to accept all received RIP routes that pass a sanity check. In contrast, the default export policy is to advertise no routes. None, zip, nada, zilch. Not even RIP-learned routes are advertised with the default RIP export policy. Although it may be an odd choice of default behavior, the net effect is that for any practical RIP deployment, you will need to create and apply a custom export policy to readvertise RIP-learned routes to other RIP speakers.

BGP default policy

For completeness, let's take a look at BGP default policy. The default BGP import policy is to accept all received BGP routes that pass a sanity check—for example, those routes that do not have an Autonomous System (AS) loop, as indicated by the AS path attribute.

The default BGP export policy is to readvertise all learned BGP routes to all BGP speakers, while obeying protocol-specific rules that prohibit one Internal BGP (IBGP) speaker from readvertising routes learned from another IBGP speaker, unless it is functioning as a route reflector.

Testing and Monitoring Policy

Congratulations. You have made it to this point, and therefore you now possess an in-depth and practical understanding of routing policy. This section explores some advanced policy concepts, some of which are quite interesting, but rarely used. The use of regular expressions (regexes) is treated as an advanced topic, but differs from the remaining topics because the use of AS path or community regex matching is somewhat common, especially in large networks such as those operated by service providers.

Testing policy results

Making a mistake in a `route-filter` statement can have a dramatic impact on network stability, security, and overall operation. For example, consider the operator who does not notice that in the following policy example (appropriately called whoops), rather than then accept being added to term 1, as intended, the accept action was mistakenly added as part of a final, *unnamed* term. Because this term has no from statement, it matches on *all* possible routes and routing sources!

```
[edit policy-options]
lab@Ethanol# show policy-statement whoops
term 1 {
    from {
        route-filter 0.0.0.0/0 prefix-length-range /8-/24;
    }
}
then accept; ###this action is part of an unnamed match all term!
```

Applying a broken policy such as this in a production network could result in network meltdown when all routes, rather than the expected subset, are suddenly advertised within your network.

JUNOS software offers a test policy feature that is designed to avoid this type of problem. You use the test command to filter routes through the identified policy to determine which routes are accepted (those displayed) versus rejected.

The test policy command is primarily useful for route-filter testing. You cannot test route redistribution policies, because the default policy for a policy test is to *accept all* protocol sources. Thus, a given route-filter policy might match against static routes, but the same policy when applied to RIP may *not* result in the advertisement of the same static routes. This is because the default policy for RIP does not accept static routes, whereas the default for the test policy did. As an example, consider this policy:

```
[edit policy-options]
lab@Ethanol# show policy-statement test_route_filter
term 1 {
    from {
        route-filter 0.0.0.0/2 orlonger;
    }
    then next policy;
}
term 2 {
    then reject;
}
```

With the test_route_filter policy shown, the test policy command will match on and accept static, direct, OSPF, and RIP, as well as routes that match the route filter (routes in the range of 0 to 63), while the same policy applied as an import policy to RIP results in the receipt of only RIP routes that match the filter. Again, this is because the matching routes are not explicitly accepted by the test_route_filter policy in this example, and therefore are subjected to the default policy for RIP.

There are a number of OSPF routes on Ethanol. The test_route_filter policy is run against a route that does not fall in the 0/2 or longer range:

```
lab@Ethanol> test policy test_route_filter 10.3.5.0/24

inet.0: 27 destinations, 27 routes (27 active, 0 holddown, 0 hidden)
+ = Active Route, - = Last Active, * = Both

10.3.5.0/24        *[OSPF/10] 00:07:36, metric 2
                    > to 172.16.69.5 via me0.0
```

```
Policy test_route_filter: 1 prefix accepted, 0 prefix rejected
```

The result confirms that a prefix outside the range of 0 to 63 is rejected:

```
lab@Ethanol> test policy test_route_filter 200.4.7/24

Policy test_route_filter: 0 prefix accepted, 1 prefix rejected
```

This result confirms that a prefix inside the range of 0 to 63 is accepted. To test against all possible routes, use 0/0:

```
lab@Ethanol> test policy test_route_filter 0/0

inet.0: 27 destinations, 27 routes (27 active, 0 holddown, 0 hidden)
+ = Active Route, - = Last Active, * = Both

10.1.3.0/24        *[OSPF/10] 00:00:43, metric 3
                      to 172.16.69.5 via me0.0
                      to 172.16.69.6 via me0.0
                    > to 172.16.69.7 via me0.0
                      to 172.16.69.9 via me0.0
                      to 10.6.8.6 via ge-0/0/2.0
                      to 10.7.8.7 via ge-0/0/3.0
10.2.6.0/24        *[OSPF/10] 00:02:22, metric 2
                      to 172.16.69.6 via me0.0
                    > to 10.6.8.6 via ge-0/0/2.0
10.3.5.0/24        *[OSPF/10] 00:02:22, metric 2
                    > to 172.16.69.5 via me0.0
10.3.9.0/24        *[OSPF/10] 00:02:22, metric 2
                    > to 172.16.69.9 via me0.0
10.4.5.0/24        *[OSPF/10] 00:02:22, metric 2
                    > to 172.16.69.5 via me0.0
10.4.6.0/24        *[OSPF/10] 00:02:22, metric 2
                      to 172.16.69.6 via me0.0
                    > to 10.6.8.6 via ge-0/0/2.0
10.4.7.0/24        *[OSPF/10] 00:00:43, metric 2
                    > to 172.16.69.7 via me0.0
                      to 10.7.8.7 via ge-0/0/3.0
10.5.6.0/24        *[OSPF/10] 00:02:22, metric 2
                    > to 172.16.69.5 via me0.0
10.5.7.0/24        *[OSPF/10] 00:00:43, metric 2
                    > to 172.16.69.5 via me0.0
                      to 172.16.69.7 via me0.0
                      to 10.7.8.7 via ge-0/0/3.0
10.6.7.0/24        *[OSPF/10] 00:00:43, metric 2
                      to 172.16.69.6 via me0.0
                      to 172.16.69.7 via me0.0
                      to 10.6.8.6 via ge-0/0/2.0
                    > to 10.7.8.7 via ge-0/0/3.0
10.6.8.0/24        *[Direct/0] 00:02:31
                    > via ge-0/0/2.0
10.6.8.8/32        *[Local/0] 00:02:31
                      Local via ge-0/0/2.0
10.7.8.0/24        *[Direct/0] 00:02:31
                    > via ge-0/0/3.0
```

```
10.7.8.8/32          *[Local/0] 00:02:31
                        Local via ge-0/0/3.0
10.10.1.3/32         *[OSPF/10] 00:00:43, metric 2
                      > to 172.16.69.5 via me0.0
                        to 172.16.69.6 via me0.0
                        to 172.16.69.7 via me0.0
                        to 172.16.69.9 via me0.0
                        to 10.6.8.6 via ge-0/0/2.0
                        to 10.7.8.7 via ge-0/0/3.0
10.10.1.5/32         *[OSPF/10] 00:02:22, metric 1
                      > to 172.16.69.5 via me0.0
10.10.1.6/32         *[OSPF/10] 00:02:22, metric 1
                      > to 172.16.69.6 via me0.0
                        to 10.6.8.6 via ge-0/0/2.0
10.10.1.7/32         *[OSPF/10] 00:00:43, metric 1
                      > to 172.16.69.7 via me0.0
                        to 10.7.8.7 via ge-0/0/3.0
10.10.1.8/32         *[Direct/0] 00:13:04
                      > via lo0.0
10.10.1.9/32         *[OSPF/10] 00:02:22, metric 1
                      > to 172.16.69.9 via me0.0

Policy test_route_filter: 20 prefix accepted, 8 prefix rejected
```

The output confirms that both direct and OSPF routes are matching the
route-filter in the test_route_filter policy. Note again that the policy being tested
does not have an explicit accept action, and instead uses the next policy for matching
routes; the acceptance in this case is the result of the default accept-all policy for the
test policy. This same policy applied to RIP will advertise no routes that match the filter
(since by default RIP does not send out any routes), unless you add an explicit accept
action to the first term.

Policy tracing

Another useful tool in troubleshooting policy is to turn on *policy tracing*, which logs
routes as they match a policy and a term. Since this could result in a large number of
routes and lots of log entries to filter through, you should use it only as a temporary
tool. First, set the trace option in the then statement in your policy:

```
lab@Ethanol# show policy-options
policy-statement send-connected {
    term 2 {
        from {
            route-filter 172.16.69.0/24 exact
                {                   trace;
                                    reject;
}
        }
    }
    term 1 {
        from protocol direct;
        then {
            trace;
```

```
            accept;
        }
    }
}
```

Second, turn `traceoptions` on under `routing-options` and specify a flag of `trace`. In this example, the result is sent to a file called *policy-trace*:

```
lab@Ethanol# top show routing-options
traceoptions {
    file policy-trace;
    flag policy;
}
```

Lastly, examine the result of your log by issuing a `show log policy-trace` or, in real time, a `monitor start policy-trace`:

```
Aug  6 12:53:02.574334 export: Dest 0.0.0.0 proto RIP
Aug  6 12:53:02.574411 policy_match_qual_or: Qualifier proto Sense: 0
Aug  6 12:53:02.574458 export: Dest 10.6.8.0 proto Direct
Aug  6 12:53:02.574495 policy_match_qual_or: Qualifier proto Sense: 1
Aug  6 12:53:02.574549 policy_export_trace: Prefix 10.6.8.0/24 term 1 --> accept
Aug  6 12:53:02.574602 export: Dest 10.7.8.0 proto Direct
Aug  6 12:53:02.574638 policy_match_qual_or: Qualifier proto Sense: 1
Aug  6 12:53:02.574674 policy_export_trace: Prefix 10.7.8.0/24 term 1 --> accept
Aug  6 12:53:02.574726 export: Dest 10.10.1.8 proto Direct
Aug  6 12:53:02.574765 policy_match_qual_or: Qualifier proto Sense: 1
Aug  6 12:53:02.574802 policy_export_trace: Prefix 10.10.1.8/32 term 1 --> accept
Aug  6 12:53:02.574844 export: Dest 172.16.69.0 proto Direct
Aug  6 12:53:02.574887 policy_export_trace: Prefix 172.16.69.0/24 term unnamed -->
reject
```

The output shows the policy accepting three prefixes and rejecting one.

Policy Case Study

Let's see some basic policy in action. In Figure 8-4 we show a topology that is running OSPF and RIP. OSPF is running on the J Series router, as well as every switch except Ethanol. RIP is running between Ethanol and Bourbon. The goals for this topology are simple:

- All switches should see Ethanol's local interface networks, except the local management network.
- All switches should receive a default route for Internet connectivity that is being originated from Scotch in OSPF.

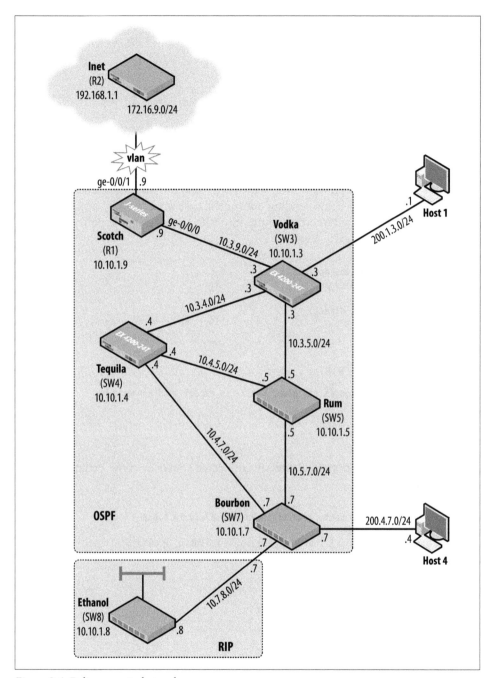

Figure 8-4. Policy case study topology

First, let's make sure that Ethanol is advertising the correct routes via RIP. Examining Ethanol shows that RIP is indeed running, but that routes are not actually being advertised:

```
lab@Ethanol# run show rip neighbor
                     Source        Destination   Send    Receive   In
Neighbor    State    Address       Address       Mode    Mode      Met
--------    -----    -------       -----------   ----    -------   ---
ge-0/0/3.0  Up       10.7.8.8      224.0.0.9     mcast   both      1

[edit]
lab@Ethanol# run show route advertising-protocol rip 10.7.8.8
```

This is due to the default policy of RIP, which is to send out no routes unless explicitly told to do so. In order to begin sending out routes, first write a policy on Ethanol that sends the locally connected interface routes into RIP. Here's a simple policy to accomplish this:

```
[edit]
lab@Ethanol# show policy-options
policy-statement send-connected {
    term 1 {
        from protocol direct;
        then accept;
    }
}
```

Then apply the policy to RIP:

```
lab@Ethanol# set protocols rip group small-rip export send-connected

[edit]
lab@Ethanol# commit
commit complete
```

Examine the result and note that directly connected routes are now being advertised via RIP:

```
[edit]
lab@Ethanol# run show route advertising-protocol rip 10.7.8.8

inet.0: 8 destinations, 8 routes (8 active, 0 holddown, 0 hidden)
+ = Active Route, - = Last Active, * = Both

10.6.8.0/24          *[Direct/0] 00:22:20
                      > via ge-0/0/2.0
10.10.1.8/32         *[Direct/0] 00:32:53
                      > via lo0.0
172.16.69.0/24       *[Direct/0] 00:32:53
                      > via me0.0
```

However, one out-of-place subnet is still appearing—the management network—so an additional term is added to reject that subnet:

```
lab@Ethanol# set term 2 from route-filter 172.16.69/24 exact reject
[edit policy-options policy-statement send-connected]
lab@Ethanol# commit
commit complete
```

After waiting awhile, and applying the policy, the route still appears to be advertised:

```
[edit policy-options policy-statement send-connected]
lab@Ethanol# run show route advertising-protocol rip 10.7.8.8

inet.0: 8 destinations, 8 routes (8 active, 0 holddown, 0 hidden)
+ = Active Route, - = Last Active, * = Both

10.6.8.0/24        *[Direct/0] 00:24:27
                    > via ge-0/0/2.0
10.10.1.8/32       *[Direct/0] 00:35:00
                    > via lo0.0
172.16.69.0/24     *[Direct/0] 00:35:00
                    > via me0.0
```

Perhaps something is astray in the policy... View the policy on Ethanol again:

```
[edit policy-options policy-statement send-connected]
lab@Ethanol# show
term 1 {
    from protocol direct;
    then accept;
}
term 2 {
    from {
        route-filter 172.16.69.0/24 exact reject;
    }
}
```

Recall that once a route reaches a terminating action, it is no longer processed. In this case, the me0 subnet is matching the more global first term, and therefore never reaches the second term. A simple rearrangement fixes that:

```
[edit policy-options policy-statement send-connected]
lab@Ethanol# insert term 2 before term 1

[edit policy-options policy-statement send-connected]
lab@Ethanol# commit

commit complete
[edit policy-options policy-statement send-connected]
lab@Ethanol# show
term 2 {
    from {
        route-filter 172.16.69.0/24 exact reject;
    }
}
term 1 {
    from protocol direct;
    then accept;
}
```

After waiting for poison reverse to finish (see the sidebar "Poison Reverse in Action" for a discussion of poison reverse), the management network is no longer advertised:

```
lab@Ethanol# run show route advertising-protocol rip 10.7.8.8

inet.0: 8 destinations, 8 routes (8 active, 0 holddown, 0 hidden)
+ = Active Route, - = Last Active, * = Both

10.6.8.0/24         *[Direct/0] 00:27:37
                     > via ge-0/0/2.0
10.10.1.8/32        *[Direct/0] 00:38:10
                     > via lo0.0
```

Poison Reverse in Action

When changing a policy in RIP, remember that RIP "poisons a route" by setting the metric to infinity (16) for several route advertisements. So, when an export policy rejects a route that was previously accepted for RIP, it may take several minutes for the entry to be removed from the **advertising** command. Here the route is shown on **Ethanol**:

```
[edit policy-options policy-statement send-connected]
lab@Ethanol# run show route advertising-protocol rip
10.7.8.8

inet.0: 8 destinations, 8 routes (8 active, 0
holddown, 0 hidden)
+ = Active Route, - = Last Active, * = Both

10.6.8.0/24         *[Direct/0] 00:25:15
                     > via ge-0/0/2.0
10.10.1.8/32        *[Direct/0] 00:35:48
                     > via lo0.0
172.16.69.0/24      *[Direct/0] 00:35:48
                     > via me0.0
```

However, the route is not shown in the receive direction on the neighboring switch. This is due to the fact that JUNOS actually tosses away routes with a metric of infinity:

```
lab@Bourbon> show route receive-protocol rip 10.7.8.8

inet.0: 33 destinations, 35 routes (33 active, 0
holddown, 0 hidden)
+ = Active Route, - = Last Active, * = Both

10.6.8.0/24         [RIP/100] 00:07:11, metric 2, tag
0
                     > to 10.7.8.8 via ge-0/0/5.0
10.10.1.8/32        *[RIP/100] 00:07:11, metric 2, tag 0
                     > to 10.7.8.8 via ge-0/0/5.0
```

As additional verification, the route also is not in the route table:

```
lab@Bourbon> show route 172.16.69.0/24

inet.0: 33 destinations, 35 routes (33 active, 0
holddown, 0 hidden)
+ = Active Route, - = Last Active, * = Both
```

```
172.16.69.0/24      *[Direct/O] 2d 04:16:52
                     > via me0.0
                     [OSPF/10] 00:12:27, metric 2
                     > to 10.4.7.4 via ge-0/0/1.0
                       to 10.6.7.6 via ge-0/0/2.0
                       to 10.5.7.5 via ge-0/0/3.0
172.16.69.7/32      *[Local/0] 2d 07:30:27
                      Local via me0.0
```

This is poison reverse in action; after about one and a half minutes, the route no longer is advertised with an infinity metric:

```
lab@Ethanol# run show route advertising-protocol rip
10.7.8.8

inet.0: 8 destinations, 8 routes (8 active,
0 holddown, 0 hidden)
+ = Active Route, - = Last Active, * = Both

10.6.8.0/24      *[Direct/O] 00:27:37
                  > via ge-0/0/2.0
10.10.1.8/32     *[Direct/O] 00:38:10
                  > via lo0.0
```

Next, Bourbon must take the routes it receives from RIP and redistribute them into OSPF. Bourbon also applies a large metric to these routes in order to avoid any better "back door" options:

```
[edit policy-options policy-statement send-connected term 2]

lab@Bourbon# show policy-options
policy-statement rip-ospf {
    term 1 {
        from protocol rip;
        then {
            metric 100;
            accept;
        }
    }
}

lab@Bourbon# set protocols ospf export rip-ospf

[edit]
lab@Bourbon# commit
commit complete
```

The OSPF database now shows these routes as Type 5 LSAs:

```
lab@Bourbon# run show ospf database external extensive
    OSPF AS SCOPE link state database
 Type       ID       Adv Rtr          Seq      Age Opt  Cksum Len
 Extern  0.0.0.0  10.10.1.9        0x80000001  858 0x22 0x197b  36
   mask 0.0.0.0
   Topology default (ID 0)
     Type: 2, Metric: 0, Fwd addr: 0.0.0.0, Tag: 0.0.0.0
```

```
   Aging timer 00:45:42
   Installed 00:14:16 ago, expires in 00:45:42, sent 00:14:16 ago
   Last changed 00:14:16 ago, Change count: 1
Extern *10.10.1.8  10.10.1.7        0x80000001    69  0x22 0xba5a  36
   mask 255.255.255.255
   Topology default (ID 0)
     Type: 2, Metric: 100, Fwd addr: 0.0.0.0, Tag: 0.0.0.0
   Gen timer 00:48:51
   Aging timer 00:58:51
   Installed 00:01:09 ago, expires in 00:58:51, sent 00:01:09 ago
   Last changed 00:01:09 ago, Change count: 1, Ours
```

Notice that only the 10.10.1.8/24 route is sent, as the direct route between Bourbon and Ethanol will be more preferred than the RIP routes making the 10.6.8/24 an inactive RIP route. Only active routes can be exported:

```
[edit]
lab@Bourbon# run show route protocol rip

inet.0: 33 destinations, 35 routes (33 active, 0 holddown, 0 hidden)
+ = Active Route, - = Last Active, * = Both

10.6.8.0/24          [RIP/100] 00:10:21, metric 2, tag 0
                    > to 10.7.8.8 via ge-0/0/5.0
10.10.1.8/32        *[RIP/100] 00:10:21, metric 2, tag 0
                    > to 10.7.8.8 via ge-0/0/5.0
224.0.0.9/32        *[RIP/100] 00:00:40, metric 1
                        MultiRecv
```

Lastly, a policy is written to send a default route to Ethanol for Internet connectivity. This default route is already present on Bourbon, so all that is required is a policy to match on that route and send it into RIP:

```
policy-statement send-default {
    term 1 {
        from {
            protocol ospf;
            route-filter 0.0.0.0/0 exact accept;
        }
    }
}
[edit]
lab@Bourbon# set protocols rip group smallrip export send-default
```

Verify that the route exists on Ethanol:

```
[edit]
lab@Ethanol# run show route 0/0 exact

inet.0: 9 destinations, 9 routes (9 active, 0 holddown, 0 hidden)
+ = Active Route, - = Last Active, * = Both

0.0.0.0/0            *[RIP/100] 00:01:56, metric 2, tag 0
                    > to 10.7.8.7 via ge-0/0/3.0
```

Routing Policy Summary

This section detailed JUNOS software routing policy. The policy framework provides a consistent and easy-to-fathom environment for all of your route exchange and attribute manipulation needs. Although route filters and the whole J-Tree thing can be a bit daunting when first encountered, the overall logic of a JUNOS policy is easy to follow, and the consistent way in which it is applied to routing protocols makes network administration that much easier. Rather than a collection of network statements, default-information-originate statements, distribute lists, route maps, and so on, with Juniper policy you create and advertise a static route into OSPF, BGP, or RIP using the same approach and the same syntax.

This section also covered the commands and procedures used to monitor and debug the operation of your import and export policies.

Firewall Filters

In order to protect the switch and the network, packet filters can be deployed to allow only certain traffic into the switch's control plane (routing engine, or RE) or to transmit the switch out of a Packet Forwarding Engine (PFE) interface. These filters have different names on each switch OS, but they operate in the same stateless manner. On a Cisco device these filters are called *access lists*, and on a Juniper switch they are called *firewall filters*. These filters look very similar to the policy we discussed in the previous section; however, firewall filters operate on the actual data forwarding plane. Table 8-1 gives a comparison of the two features.

Table 8-1. Firewall filters versus routing policies

Feature	Firewall filter	Routing policy
Operates in...	Forwarding plane	Control plane
Match keyword	from	from
Action keyword	then	then
Match attributes	Packet fields	Route attributes
Default action	Discard	Depends on default policy
Applied to...	Interfaces	Routing protocols/tables
Named terms required	Yes	No
Chains allowed	Yes	Yes
Absence of from	Match all	Match all

Firewall filter syntax takes a human-friendly, intuitive form:

```
firewall {
    family inet {
        filter filter-1 {
```

```
        term term-1 {
            from {
                protocol tcp;
                destination-port telnet;
            }
            then {
                accept;
            }
        }
    }
  }
}
```

This filter matches on Telnet traffic and accepts the packets. As you can see, the syntax is very similar to a routing policy with the match conditions in the from term and the actions specified in a then term.

In the EX Series switches, firewall filters are implemented in hardware for faster processing and are stored in the TCAM.

Types of Filters

The EX Series switches support a variety of filters. The syntax for these filters looks the same, but the place of application is slightly different.

The following firewall filter types are supported for EX Series switches:

Port (Layer 2) firewall filter
Port firewall filters apply to Layer 2 switch ports. You can apply port firewall filters only in the ingress direction on a physical port.

VLAN firewall filter
Virtual LAN (VLAN) firewall filters provide access control for packets that enter a VLAN, are bridged within a VLAN, and leave a VLAN. You can apply VLAN firewall filters in both ingress and egress directions on a VLAN. VLAN firewall filters are applied to all packets that are forwarded to or forwarded from the VLAN.

Router (Layer 3) firewall filter
You can apply a router firewall filter in both ingress and egress directions on Layer 3 (routed) interfaces and Routed VLAN Interfaces (RVIs). You can also apply a router firewall filter in the ingress direction on the loopback interface.

You can configure and apply no more than one firewall filter per port, VLAN, or router interface, per direction (input or output). At the time of this writing, the number of filter terms within a filter is 2,048, which should be more than enough for today's networks.

Firewall filters are not supported on aggregated Ethernet (AE) interfaces as of JUNOS 9.2.

Egress firewall filters do not affect the flow of locally generated control packets from the RE. In other words, this traffic bypasses an egress firewall filter.

Of course, there will be times when multiple filter types are applied. Figure 8-5 displays the general processing order. However, it does vary somewhat depending on whether a Layer 2 frame or a Layer 3 packet is being processed, as Layer 2 frames are not processed by Layer 3 filters. (No square pegs in round holes.)

For Layer 2 (bridged) unicast packets, the following firewall filter processing points apply, in the order shown:

1. Ingress port firewall filter
2. Ingress VLAN firewall filter
3. Egress VLAN firewall filter

For Layer 3 (routed and multilayer-switched) unicast packets, the following firewall filter processing points apply in the order shown:

1. Ingress port firewall filter
2. Ingress VLAN firewall filter
3. Ingress router firewall filter
4. Egress router firewall filter
5. Egress VLAN firewall filter

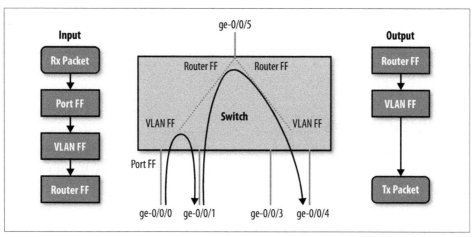

Figure 8-5. Firewall filter processing

A router firewall filter will *not* process switched packets in the same VLAN.

Filter Term Processing

Similar to a policy, a filter is made up of multiple terms, and each term is examined in the order listed. If there is a match in a term and there is a terminating action, no other term is examined (see Figure 8-6). Terminating actions include:

`accept`
   Allows the packet through the filter

`discard`
   Silently discards the packet

`reject`
   Discards the packet with an Internet Control Message Protocol (ICMP) error message (default: administratively prohibited)

Action modifier
   Any action modifier, such as `log`, `count`, `syslog`, and so forth

> The presence of an action modifier such as `count` without an explicit `accept`, `discard`, or `reject` will result in a default action of `accept`. If the desired action is to discard or reject the packet, it must be explicitly configured.

If the packet does not match any terms in the filter, it is discarded.

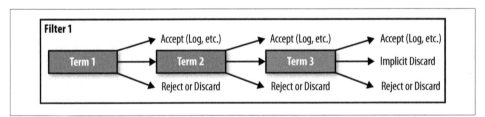

Figure 8-6. Filter processing

Filter Match Conditions

When examining the possible match conditions, the general rule of thumb is that if it is a field in the IP, Transmission Control Protocol (TCP), User Datagram Protocol (UDP), or ICMP header, it is probably a potential match. For Layer 3 filters, the possible match conditions are:

```
lab@Bourbon# set firewall family inet filter foo term 1 from ?
Possible completions:
+ apply-groups          Groups from which to inherit configuration data
+ apply-groups-except   Don't inherit configuration data from these groups
> destination-address   Match IP destination address
+ destination-port      Match TCP/UDP destination port
```

```
  + dscp                    Match Differentiated Services (DiffServ) code point
    fragment-flags          Match fragment flags (in symbolic or hex formats)
  + icmp-code               Match ICMP message code
  + icmp-type               Match ICMP message type
  > interface               Match interface name
  + ip-options              Match IP options
    is-fragment             Match if packet is a fragment
  + packet-length           Match packet length
  + precedence              Match IP precedence value
  + protocol                Match IP protocol type
  > source-address          Match IP source address
  + source-port             Match TCP/UDP source port
    tcp-flags               Match TCP flags (in symbolic or hex formats)
    tcp-initial             Match initial packet of a TCP connection
  + ttl                     Match IP ttl type
  + ttl-except              Do not match IP ttl type
```

For Layer 2 filters, many of the same match conditions are available, in addition to Layer 2 options such as Media Access Control (MAC) addresses, 802.1Q bits, Ether-Types, and VLAN tags:

```
lab@Bourbon# set family ethernet-switching filter foo term 1 from ?
Possible completions:
  + apply-groups           Groups from which to inherit configuration data
  + apply-groups-except    Don't inherit configuration data from these groups
  > destination-address    Match IP destination address
  > destination-mac-address  Match MAC destination address
  + destination-port       Match TCP/UDP destination port
  + dot1q-tag              Match Dot1Q Tag Value
  + dot1q-user-priority    Match Dot1Q user priority
  + dscp                   Match Differentiated Services (DiffServ) code point
  + ether-type             Match Ethernet Type
    fragment-flags         Match fragment flags (in symbolic or hex formats) -
(Ingress only)
  + icmp-code              Match ICMP message code
  + icmp-type              Match ICMP message type
  > interface              Match interface name
    is-fragment            Match if packet is a fragment
  + precedence             Match IP precedence value
  + protocol               Match IP protocol type
  > source-address         Match IP source address
  > source-mac-address     Match MAC source address
  + source-port            Match TCP/UDP source port
    tcp-flags              Match TCP flags (in symbolic or hex formats) - (Ingress
only)
    tcp-initial            Match initial packet of a TCP connection - (Ingress only)
  + vlan                   Match Vlan Id or Name
```

The match conditions fall into three general categories: numeric, address, and bit field matches (Table 8-2).

Table 8-2. General match conditions

Numeric match	Address match	Bit field match
Protocol fields	Source address	IP options
Port numbers	Destination address	TCP flags
Class of service (CoS) fields	Source-prefix lists	IP fragmentation
ICMP type codes	Destination-prefix lists	Time to Live (TTL)

A term can have zero or many match conditions specified. The absence of a from state-ment creates a match-all condition, whereas multiple match conditions are treated as a logical AND or OR, depending on common versus uncommon match conditions. A common match is treated as a logical OR, which the router groups together in square brackets. The filter example matches on TCP or UDP packets:

```
filter example {
    term common {
        from {
            protocol [ tcp udp ];
        }
    }
}
```

An uncommon match is treated as a logical AND. These logical ANDs and ORs can be combined in the same term with limitless possibilities. Adding to the example, the following filter matches on TCP or UDP packets and source port 123:

```
filter example {
    term common {
        from {
            protocol [ tcp udp ];
            source-port 123;
        }
    }
}
```

Also, numeric matches such as port or protocol values can take either the numeric match or the more user-friendly keywords. For example, the first and second terms of the filter called same are equivalent, but the second term is written in a more efficient and user-friendly method. JUNOS software does not auto-convert the numerical num-bers into the names for you. The number remains the way it is configured:

```
firewall {
    filter same {
        term numbers {
            from {
                protocol 6;
                source-port 23;
            }
            then accept;
        }
        term user-friendly {
```

```
            from {
                protocol tcp;
                port telnet;
            }
            then accept;
        }
    }
}
```

 Bit field matching such as IP options and TCP flags also support numeric values or more user-friendly terms. In these cases, the numeric support must be written in hex format, so a TCP flag match for SYN packets could be written with the keyword syn or the value 0x2. No reason to break out the hex converter—make life easy and use the keywords!

Filter Actions

Besides the terminating actions already discussed (accept, discard, reject), other common action modifiers include:

analyzer <analyzer name>
> Mirrors port traffic to a specified destination port or VLAN that is connected to a protocol analyzer application. Mirroring copies all packets seen on one switch port to a network monitoring connection on another switch port. The analyzer-name and mirroring parameters must be configured under [edit ethernet-switching-actions analyzer].
>
> You can specify mirroring for ingress ports, VLANs, and router firewall filters only.

count <counter name>
> Counts the total number of packets and bytes that match a term. Counters can be viewed with the show firewall command.

policer
> Rate-limits traffic based on bandwidth and burst size limits (discussed later in this chapter).

forwarding-class
> Sends packets to a forwarding class, which maps to a queue.

Applying a Filter

The final step after writing the filter is actually applying it to the interface. Filters can be applied to either transit or non-transit traffic. To apply a filter to transit traffic, apply the filter to any PFE interface as either an input or output filter.

Remember, a filter can be applied in one of three places: the port level, VLAN level, or logical unit level. Also, only a single filter can ever be applied per direction.

Applying a filter at the port level

To apply a filter at the port level, apply the filter to the interface under family ethernet-switching. A port-level filter can be applied only in the input direction:

```
lab@Bourbon# set interface ge-0/0/2 unit 0 family ethernet-switching filter ?
Possible completions:
+ apply-groups          Groups from which to inherit configuration data
+ apply-groups-except   Don't inherit configuration data from these groups
  input                 Name of filter applied to received packets
```

Applying a filter at the VLAN level

To apply a filter at the VLAN level, apply the filter as an input or output filter under the named VLAN. This applies the same filter to all interfaces that contain that VLAN:

```
[edit]
lab@Bourbon# set vlans Admin_vlan filter ?
Possible completions:
+ apply-groups          Groups from which to inherit configuration data
+ apply-groups-except   Don't inherit configuration data from these groups
  input                 Name of filter applied to received packets
  output                Name of filter applied to transmitted packets
```

Applying a filter at the Layer 3 level

To apply a filter at the Layer 3 level, apply the filter under [edit interfaces] on a per-logical-unit basis. For firewall filters applied to Layer 3 routed interfaces, the family address type must be inet. This filter could be an input and output filter, and the logical unit could be untagged or tagged:

```
lab@Bourbon# set interfaces ge-0/0/2 unit 0 family inet filter ?
Possible completions:
+ apply-groups          Groups from which to inherit configuration data
+ apply-groups-except   Don't inherit configuration data from these groups
  input                 Name of filter applied to received packets
  output                Name of filter applied to transmitted packets
```

In order to protect traffic to the switch itself, a filter can be applied to the loopback interface (Figure 8-7). Local traffic is any packet that is destined to the switch itself, such as routing protocols, ICMP, SSH, and other management protocols.

Transit Filter Case Study

To illustrate the filters, we will use the example topology in Figure 8-8. Two VLANs are defined:

VLAN 10
> For internal employees with subnet 200.2.2/24

VLAN 20
> For guests on the network with subnet 66.66.66/24

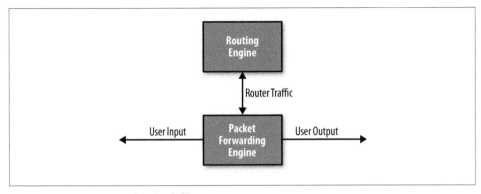

Figure 8-7. Transit versus loopback filters

There is a single Internet connection via Vodkila, which acts as the default gateway for all VLANs.

The goal is to protect the network as a whole, as well as provide different access for a guest versus an employee.

Layer 3 filter

First, we are going to apply the filter for traffic coming from the Internet on Vodkila. Before we begin typing away on the switch, though, we must write down the goals of the filter. In this case study, all outbound traffic from the network to the Internet is allowed, while some traffic will be filtered inbound. The goals are as follows:

- TCP connections are only allowed to be initiated outbound to the Internet.
- TCP fragments are allowed.
- UDP packets should be allowed inbound for traceroutes and return traffic for outbound UDP connections.
- Ping and traceroute are allowed outbound.
- Traceroute is allowed inbound.

Let's examine the filter called internet-in, and match each term with our five goals. First, we define our first term to deny TCP sessions inbound that are destined for internal subnets. We also count these packets to a counter called deny-i-tcp:

```
lab@RUM# show firewall family inet
term deny-inbound-tcp {
    from {
        destination-address {
            200.2.2.0/24;
            66.66.66.0/24;
        }
        protocol tcp;
        tcp-initial;
    }
```

```
        then {
            count deny-i-tcp;
            discard;
        }
    }
```

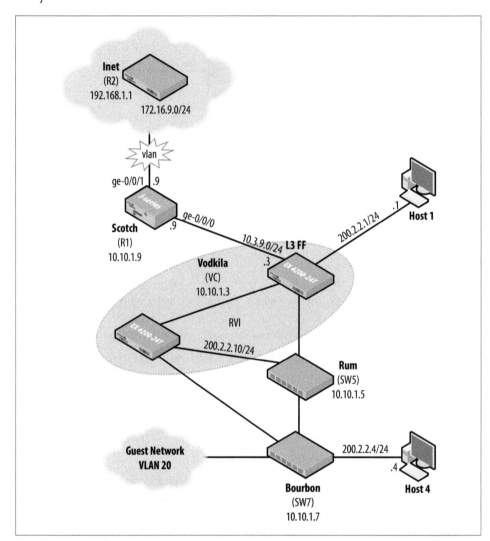

Figure 8-8. Example filter topology

Next, we create a term that accepts TCP packets from internal subnets. Since the TCP connections inbound were already denied by the previous term, this effectively allows outbound TCP connections. The packets are also counted:

```
term allow-outbound-tcp {
    from {
```

```
            destination-address {
                200.2.2.0/24;
                66.66.66.0/24;
            }
            protocol tcp;
        }
        then {
            count allow-o-tcp;
            accept;
        }
    }
```

TCP fragments are allowed in the next term and counted. The `is-fragment` keyword matches on all fragments except the first fragment:

```
    term allow-tcp-frags {
        from {
            is-fragment;
            protocol tcp;
        }
        then {
            count tcp-frags;
            accept;
        }
    }
```

Next, internal subnets are allowed to receive incoming UDP packets that are not fragments, and the packets are counted. This step allows return traffic for outbound UDP sessions, as well as inbound traceroute packets that use UDP inbound:

```
    term allow-udp {
        from {
        destination-address {
            200.2.2.0/24;
            66.66.66.0/24;
        }
            protocol udp;
        }
        then {
            count count-udp;
            accept;
        }
    }
```

Finally, ping and traceroute are allowed outbound and counted. Since this is an input filter, the return traffic is actually being allowed in for both ping (echo replies) and traceroute (time exceeds messages). Additionally, unreachable messages are allowed in for any possible outbound error responses:

```
    term allow-some-icmp-outbound {
        from {
        destination-address {
            200.2.2.0/24;
            66.66.66.0/24;
        }
            protocol icmp;
```

```
            icmp-type [ echo-reply time-exceeded unreachable ];
        }
        then {
            count icmp;
            accept;
        }
    }
```

A final deny term is added at the end of the filter. Although this is the default behavior for a filter, it is explicitly configured here to allow denied traffic to be counted:

```
term denied-traffic {
    then {
        count denied;
        discard;
    }
}
```

The filter is applied as an input filter on Vodkila on the interface toward the router Scotch:

```
lab@Vodkila# set unit 0 family inet filter input internet-in
[edit interfaces ge-0/0/8]
lab@Vodkila# show
unit 0 {
    family inet {
        filter {
            input internet-in;
        }
        address 10.3.9.8/24;
    }
}

[edit interfaces ge-0/0/8]
lab@Vodkila# commit
fpc1:
configuration check succeeds
fpc0:
commit complete
fpc2:
commit complete
fpc1:
commit complete
```

But after we commit, users begin to complain about losing Internet connectivity. A quick look at the counter shows that traffic is being denied:

```
lab@Vodkila# run show firewall

Filter: internet-in
Counters:
Name                            Bytes              Packets
allow-o-tcp                         0                    0
count-udp                           0                    0
denied                            516                    6
deny-i-tcp                          0                    0
```

```
icmp                                          0                    0
tcp-frags                                     0                    0
```

After some analysis in the logs, we notice that OSPF went down between Vodkila and Scotch about 40 seconds after the filter was applied as a result of OSPF dead-timer expiration. We neglected to allow OSPF in the filter, which is vital to the operation of the network. So, let's add a term to allow OSPF:

```
lab@Vodkila# set term allow-ospf from protocol ospf

[edit firewall family inet filter internet-in]
lab@Vodkila# set term allow-ospf then accept
```

We must view the filter before committing to make sure everything is correct. Remember that new terms are placed at the end of the filter. In this case, the term allow-ospf has been placed at the end of the filter, which would never be used due to the catchall term deny that is also counting all other traffic:

```
[edit firewall family inet filter internet-in]
lab@Vodkila# show
term deny-inbound-tcp {
    from {
        destination-address {
            200.2.2.0/24;
            66.66.66.0/24;
        }
        protocol tcp;
        tcp-initial;
    }
    then {
        count deny-i-tcp;
        discard;
    }
}
term allow-outbound-tcp {
    from {
        destination-address {
            200.2.2.0/24;
            66.66.66.0/24;
        }
        protocol tcp;
    }
    then {
        count allow-o-tcp;
        accept;
    }
}
term allow-tcp-frags {
    from {
        is-fragment;
        protocol tcp;
    }
    then {
        count tcp-frags;
        accept;
```

```
        }
    }
    term allow-udp {
        from {
            destination-address {
                200.2.2.0/24;
                66.66.66.0/24;
            }
            protocol udp;
        }
        then {
            count count-udp;
            accept;
        }
    }
    term allow-some-icmp-outbound {
        from {
            destination-address {
                200.2.2.0/24;
                66.66.66.0/24;
            }
            protocol icmp;
            icmp-type [ echo-reply time-exceeded unreachable ];
        }
        then {
            count icmp;
            accept;
        }
    }
    term denied-traffic {
        then {
            count denied;
            discard;
        }
    }
    term allow-ospf {
        from {
            protocol ospf;
        }
        then accept;
    }
```

Move the new term higher in the chain:

```
lab@Vodkila# insert term allow-ospf before term denied-traffic

lab@Vodkila# show
term deny-inbound-tcp {
    from {
        destination-address {
            200.2.2.0/24;
            66.66.66.0/24;
        }
        protocol tcp;
        tcp-initial;
    }
}
```

```
        then {
            count deny-i-tcp;
            discard;
        }
    }
    term allow-outbound-tcp {
        from {
            destination-address {
                200.2.2.0/24;
                66.66.66.0/24;
            }
            protocol tcp;
        }
        then {
            count allow-o-tcp;
            accept;
        }
    }
    term allow-tcp-frags {
        from {
            is-fragment;
            protocol tcp;
        }
        then {
            count tcp-frags;
            accept;
        }
    }
    term allow-udp {
        from {
            destination-address {
                200.2.2.0/24;
                66.66.66.0/24;
            }
            protocol udp;
        }
        then {
            count count-udp;
            accept;
        }
    }
    term allow-some-icmp-outbound {
        from {
            destination-address {
                200.2.2.0/24;
                66.66.66.0/24;
            }
            protocol icmp;
            icmp-type [ echo-reply time-exceeded unreachable ];
        }
        then {
            count icmp;
            accept;
        }
    }
}
```

```
term allow-ospf {
    from {
        protocol ospf;
    }
    then accept;
}
term denied-traffic {
    then {
        count denied;
        discard;
    }
}
```

```
[edit firewall family inet filter internet-in]
lab@Vodkila# commit
```

Generate some test traffic, first from the internal network to the Internet (2.2.2.2) from Host4. We start with a ping and traceroute:

```
Host4#ping 2.2.2.2

Type escape sequence to abort.
Sending 5, 100-byte ICMP Echos to 2.2.2.2, timeout is 2 seconds:
!!!!!
Success rate is 100 percent (5/5), round-trip min/avg/max = 1/12/48 ms
Host4#trace 2.2.2.2

Type escape sequence to abort.
Tracing the route to 2.2.2.2

  1 200.2.2.10 0 msec 4 msec 4 msec
  2 2.2.2.2 4 msec 4 msec 8 msec

Host4#trace 2.2.2.2

Type escape sequence to abort.
Tracing the route to 2.2.2.2

  1 200.2.2.10 0 msec 0 msec 4 msec
  2 2.2.2.2 8 msec 40 msec 20 msec
```

then test that TCP connections can be initiated from the Internet via a telnet command:

```
Host4#telnet 2.2.2.2
Trying 2.2.2.2 ... Open

Scotch (ttyp0)

login: lab
Password:

--- JUNOS 9.0R1.10 built 2008-02-14 03:13:25 UTC
lab@Scotch> exit

[Connection to 2.2.2.2 closed by foreign host]
```

Finally, we'll run similar tests from Scotch toward the internal network; 200.2.2.10 is an address on Vodkila:

```
lab@Scotch> ping 200.2.2.10
PING 200.2.2.10 (200.2.2.10): 56 data bytes
c^C
--- 200.2.2.10 ping statistics ---
4 packets transmitted, 0 packets received, 100% packet loss

lab@Scotch> traceroute 200.2.2.10
traceroute to 200.2.2.10 (200.2.2.10), 30 hops max, 40 byte packets
 1  200.2.2.10 (200.2.2.10)  1.330 ms  3.675 ms  1.988 ms

lab@Scotch> traceroute 200.2.2.1
traceroute to 200.2.2.1 (200.2.2.1), 30 hops max, 40 byte packets
 1  10.3.9.8 (10.3.9.8)  1.955 ms  1.995 ms  1.972 ms
 2  200.2.2.1 (200.2.2.1)  5.579 ms *  28.569 ms

lab@Scotch> telnet 200.2.2.10
Trying 200.2.2.10...
telnet: connect to address 200.2.2.10: Operation timed out
telnet: Unable to connect to remote host
```

Verify that the traffic is hitting the filter by examining the various counters that have been set up in the filter with the count action. These counters can be viewed by issuing the show firewall command:

```
lab@Vodkila# run show firewall

Filter: internet-in
Counters:
Name                            Bytes              Packets
allow-o-tcp                      1904                   28
count-udp                         576                    9
denied                            408                    4
deny-i-tcp                        642                    9
icmp                              812                    8
tcp-frags                           0                    0
```

VLAN filters

Create some rules for the employee VLAN (10) and the guest VLAN (20). The criteria are as follows:

- All employee-to-employee traffic is allowed.
- Employee-to-Internet traffic is counted and analyzed.
- Guests are allowed to communicate with other guests, but not with employees.
- Guests are allowed web access but not access to peer-to-peer applications.

These filters will be placed on multiple switches, but for the sake of brevity we'll only look at the ones on Bourbon. The first filter that is written allows guests to communicate with other guests, while allowing web access but not peer-to-peer application access.

This is implemented as a multiterm policy. The first term allows communication to the 66.66.66.0/24 subnet, which is the subnet set aside for the guest network. The second term, allow-int-no-peer, allows traffic to be sent to a particular MAC address. This MAC address is the gateway MAC for the PCs, which is Vodkila. All other traffic is denied:

```
lab@Bourbon# show firewall
family ethernet-switching {
    filter guest-vlan {
        term allow-employee {
            from {
                destination-address {
                    66.66.66.0/24;
                }
            }
            then accept;
        }
        term allow-int-no-peer {
            from {
                destination-mac-address {
                    00:1f:12:3d:d2:80;
                }
            }
            then accept;
        }
    }
}
```

Apply this filter to the VLAN Guest on Bourbon. This is applied as an input filter to the interface directly connected to the Guest network:

```
lab@Bourbon# set vlans Guest_vlan filter input guest-vlan
```

Next, let's configure the employee VLAN filter. This filter allows all traffic within the employee subnet of 200.2.2.0/24, as well as web traffic, to the Internet. Internal traffic is simply accepted, but web traffic is counted and sent to an analyzer called web-traffic. The analyzer parameters are configured under [edit ethernet-switching-options]:

```
filter employee-vlan {
    term accept-corp {
        from {
            destination-address {
                200.2.2.0/24;
            }
        }
        then accept;
    }
    term monitor-internet {
        from {
            protocol tcp;
            destination-port 80;
        }
        then {
```

```
            analyzer web-traffic;
            count web-traffic;
            accept;
        }
    }
```

Apply this filter to the employee VLAN:

```
lab@Bourbon# set vlans Employee_vlan filter output employee_vlan
```

Case Study: Loopback Filters

Our next task is to secure traffic destined for the switches themselves. The goals of this case study are to allow:

- OSPF traffic
- SSH from the MGT network of 172.16.9/24
- VRRP packets
- Ping and traceroute
- Domain Name System (DNS) replies
- Simple Network Management Protocol (SNMP) and Network Time Protocol (NTP) traffic

This filter should be applied to every switch in the network, but is shown only on Bourbon for the sake of brevity.

The filter protect-switch is created with the first term, allowing SSH traffic to *and* from the switch matching on either the source or destination port:

```
filter protect-switch {
  term allow-ssh {
    from {
        source-address {
            172.16.9.0/24;
        }
        protocol tcp;
        source-port ssh;
        destination-port ssh;
    }
    then accept;
  }
```

Then we create a term to allow for OSPF packets:

```
            term allow-ospf {
                from {
                    protocol ospf;
                }
                then accept;
            }
```

and we allow VRRP traffic:

```
term allow-vrrp {
    from {
        protocol vrrp;
    }
    then accept;
}
```

Let's not forget DNS replies. Since these are stateless, we filter the return traffic so that DNS resolution is allowed in:

```
term dns-replies {
    from {
        protocol udp;
        source-port 53;
    }
    then accept;
}
```

SNMP is allowed:

```
term snmp {
    from {
        protocol udp;
        source-port [ snmp snmptrap ];
    }
    then accept;
}
```

and UDP packets with a TTL of 1, in order for traceroute to operate:

```
term traceroute {
    from {
        protocol udp;
        ttl 1;
    }
    then accept;
}
```

We allow pings, traceroutes, and error messages:

```
term allow-icmp {
    from {
        protocol icmp;
        icmp-type [ echo-request echo-reply time-exceeded
        unreachable ];
    }
    then accept;
}
```

and also NTP:

```
term allow-ntp {
    from {
        source-address 172.16.69/24
        protocol udp;
        source-port ntp;
```

```
        }
        then accept;
    }
```

Lastly, we will use a term that denies all other traffic (the default), but allows the denied traffic to be counted as well as logged to a syslog file:

```
    term match-denied {
        then {
            count bad-packets;
            discard;
        }
    }
  }
}
```

We then apply the filter to the loopback interface as an input filter. Even though it is just a single filter, it is added as a list for future expansion:

```
lab@Bourbon#    set interface lo0.0 family inet filter input
protect-switch
```

This is a good point at which to dust off the commit confirmed to make sure the filter does not break the current network or, worse yet, lock us out of the router:

```
[edit]
lab@Bourbon# commit confirmed
commit confirmed will be automatically rolled back in 10 minutes unless
confirmed
commit complete

# commit confirmed will be rolled back in 10 minutes
[edit]
lab@Bourbon# commit
commit complete
```

> During the writing of this book, PR 402722 was created with the following issue: "In a Virtual Chassis configuration, applying a family inet firewall filter to the loopback (lo0) interface might cause communication problems between the member switches of the Virtual Chassis configuration." This caused the VC to lose membership, and at the time of this writing, the fix is a software upgrade to JUNOS 9.3.

Policers

In order to rate-limit traffic entering an interface, a *policer* can be deployed. The policers implemented in the Juniper switch are token-based, and use the IP packet to limit based on bandwidth and bursts. The bandwidth is measured as the average number of bits over a one-second interval (Figure 8-9). The burst size is the number of bytes that can exceed the bandwidth constraints.

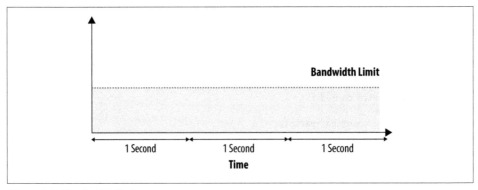

Figure 8-9. Bandwidth limit

 A maximum of 512 policers can be configured for port firewall filters, *and* a maximum of 512 policers can be configured for VLAN and Layer 3 firewall filters.

The burst size is what implements the "token-based" behavior of the policer. The burst size sets the initial and maximum sizes of a bucket in bytes (tokens) that are accessed each time data needs to be sent. As a packet is sent, the bucket bytes (tokens) are removed from the bucket. If there are not enough tokens to send the packet, the packet will be policed. The bucket is then replenished at the bandwidth rate.

In Figure 8-10, a packet is sent that bursts above the bandwidth limit but is still sent, as there are enough tokens in the "bucket." After the packet is sent, the number of tokens is decreased based on the packet size.

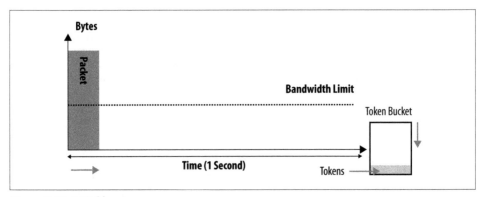

Figure 8-10. Initial burst

Sometime later, another packet needs to be sent that is also above the bandwidth limit. Since there are no longer enough tokens left in the bucket, the packet is policed (Figure 8-11).

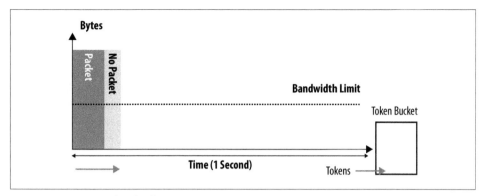

Figure 8-11. Empty token bucket

As time goes by, the bucket replenishes at a rate equal to the bandwidth limit. When a new packet arrives, it can be sent, as now tokens are available in the bucket. This process continues over a one-second interval, and the end result is a rate equal to the bandwidth limit (Figure 8-12).

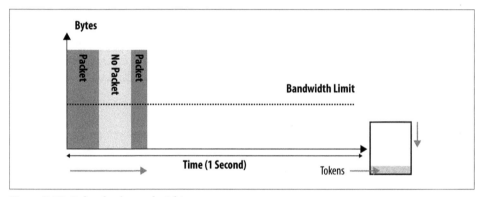

Figure 8-12. Token bucket replenishing

Burst-size-limit mystery

The setting of the burst size is a mystery for many operators. Set this value too low, and potentially all packets will be policed. Set the value too high, and no packets will be policed. A good rule of thumb is that the burst size should never be lower than 10 times the maximum transmission unit (MTU). The recommended value is to set the amount of traffic that can be sent over the interface in five milliseconds. So, if your interface is a fast Ethernet interface, the minimum is 15,000 bytes (10 × 1,500) and the recommended value would be 62,500 bytes. This value is derived by finding the number of bytes per millisecond and then multiplying that value by five. First convert 100 Mbps to bytes by dividing by 8. The result is 12.5 bytes per second. Then multiply it by 1,000 to get bytes per millisecond, and multiply that result by 5 (12,500 bytes/ms × 5).

Policer actions

Once the policer limits have been configured, we must choose an action if a packet is received that exceeds the policer limits. Two types of policing are available: *soft* policing and *hard* policing. Hard policing specifies that the packet will be dropped if it exceeds the policer's traffic profile. Soft policing simply marks or reclassifies the packets, which could change the probability of the packets being dropped at the egress interface during times of congestion. Soft policing is implemented by either setting the packet loss priority (PLP) on the packet or placing the packet into a different forwarding class.

Configuring and applying policers

Policers are configured under the [edit firewall] level. The policer will be named and then the burst size applied in bytes/second, the bandwidth limit in bits/second, or the percentage of interface bandwidth set along with the policer action. For example:

```
policer simple {
    if-exceeding {
        bandwidth-limit 50m;
        burst-size-limit 15k;
    }
    then discard;
}
```

Once the policer is configured, it must be applied to an interface by referencing the policer name in the firewall filter. If the policer is referenced in a filter, specific types of traffic can be policed, as the entire toolkit of filter actions are allowed.

Policer example

In this example, we will examine a policer that limits the guest network's Internet traffic to 1 Mb with a 5 KB burst.

First, the policer is defined under the firewall level:

```
lab@Bourbon# show firewall

policer limit-guest {
    if-exceeding {
        bandwidth-limit 1m;
        burst-size-limit 5k;
    }
```

Then a filter is created to match on the MAC address of the default gateway. No other traffic is rate-limited. Recall that traffic will also be filtered by the VLAN filters that have been applied in previous sections:

```
filter limit-guest {
    term limit-all-traffic {
        from {
            destination-mac-address {
                00:1f:12:3d:d2:80;
            }
```

```
            }
            then {
                accept;
                policer limit-guest;
            }
        }
        term accept-other {
            then accept;
        }
    }
}
```

Apply the filter to the interface as a port-based filter. Port-based filters can be applied only in the input direction:

```
lab@Bock# show interfaces ge-0/0/12

ge-0/0/12 {
    unit 0 {
        family ethernet-switching {
            vlan {
                members 20;
            }
            filter {
                input limit-guest;
            }
        }
    }
}
```

To verify the policer and the counters, issue the show policer command:

```
lab@Bourbon# run show policer

Filter: employee-vlan

Filter: guest-vlan

Filter: limit-guest
Policers:
Name                                          Packets
limit-guest-limit-all-traffic                       0
```

One difficulty is determining how much traffic the policer is allowing in order to evaluate whether the exceeding parameters are too large or small. This can be accomplished using policer counters, interface statistics, and a little math. First, determine the byte-per-packet size viewed by the policer by dividing the bytes by the number of packets seen by the policer counter. Then multiply the egress rate in packets per second by the per-packet size, and 8 bits to get the bytes per second.

For example, say the policer counter claimed that 1,406,950 bytes and 18,494 packets have exceeded the policer. This example calculates an average per-packet size of 76 bytes (1,406,950 / 18,494). Via the show interfaces command, the interface rate is determined to be 203 pps. So, 203 pps times 76 bytes/packet times 8 bits per second will provide the bytes per second of 123,424, which should be close to the configured bandwidth rate.

Storm Control and Rate Limiting

We already examined traffic storms in Chapter 5. Recall that a traffic storm could be from a loop due to a bad design, which could cause poor or even disrupted network service. EX Series switches have two methods for alleviating these storms:

- Send all unknown unicast traffic to a particular VLAN for processing.
- Rate-limit broadcast and unknown traffic to a threshold and drop any traffic that exceeds this threshold.

Both of these are configured under [edit ethernet-switching-options]. For example, the following configuration sets a broadcast and unknown unicast threshold to 70% of the ge-0/0/7 interface. By default, both broadcast and unknown unicasts match this threshold, but we can exempt either type of traffic from rate limiting by using the no-broadcast or no-unknown-unicast statement:

```
lab@Vodkila# set ethernet-switching-options storm-control interface ge-0/0/7 level
70
```

Filters and Policers Summary

This section looked at filters and policers for the switch. Filters can be applied at a variety of levels in the switch: port level, VLAN level, or router Layer 3 level. Although only a single filter can be applied per direction, multiple types of filters can be applied at the same time. Policers are tied into filters because they are referenced as a filter action. Multiple policers can be referenced in the same filter, which allows for per-application filtering.

Conclusion

Although policy and firewall filters operate on different planes (control versus forwarding), they share a common syntax and language. Once a language is understood, a policy and a filter can be written and implemented quite easily. A policy is used to control the flow of routing information in your network, and a firewall filter is used to control the flow of data packets. This is a very powerful tool on the switches; a misconfigured policy or filter could lead to very painful results for yourself, your colleagues, and especially your boss.

Chapter Review Questions

1. You have configured RIP between three routers connected in a serial chain, but no RIP routes are being learned. Which policy results in full RIP connectivity for all direct routes?

    a. A RIP import policy of the form:

    ```
    term 1 {
        from protocol [ rip direct ];
        then accept;
    }
    ```

    b. A RIP export policy of the form:

    ```
    term 1 {
        from protocol [ rip direct ];
        then accept;
    }
    ```

    c. A RIP import policy of the form:

    ```
    term 1 {
        from protocol direct;
        then accept;
    }
    ```

    d. A RIP export policy of the form:

    ```
    term 1 {
        from protocol direct;
        then accept;
    }
    ```

2. What happens when the static route 192.168.10/24 is evaluated by this policy?

    ```
    [edit policy-options policy-statement test]
    lab@RUM# show
    term 1 {
        from {
            protocol rip;
            route-filter 192.168.0.0/16 orlonger reject;
            route-filter 192.168.10.0/24 exact {
                metric 10;
                accept;
            }
        }
    }
    ```

    a. Nothing, because no match occurs

    b. The route is longest-matched against the first route filter and rejected

    c. The route is longest-matched against the second route filter and has its metric set to 10

    d. Both B and C

3. What type of import policy can you apply to OSPF?

    a. None; LS protocols do not support the notion of import policies because the policy breaks database consistency

    b. You can apply policy to filter certain LSA types, such as AS externals, to create a stub area

    c. Import policy for OSPF can be used only to filter AS external LSAs from being flooded

    d. Import policy for OSPF can be used to prevent installation of AS external routes into the RT, but has no effect on flooding

4. What is the default action of the `test policy` command?

    a. `accept`

    b. `reject`

    c. Depends on the default policy

    d. None

5. Choose three types of filters that can be configured in an EX3200:

    a. Port

    b. Layer 3

    c. VLAN group

    d. VLAN

    e. Queue

6. How many terms can be configured in a filter?

    a. 1,024

    b. 512

    c. 768

    d. 2,048

7. What kind of filter would be written to protect control traffic destined for the switch?

    a. A filter applied to the default VLAN

    b. A filter applied to the native VLAN

    c. A filter applied to the management interface

    d. A filter applied to the loopback interface

8. Choose the correct order of filter processing for a Layer 2 packet:

    a. Input VLAN, input port, output VLAN

    b. Input port, input VLAN, output VLAN, output port

    c. Input port, input VLAN, output VLAN

    d. Input VLAN, output VLAN, output port

9. Which JUNOS software command shows filter counters?

   a. `show filter`

   b. `show firewall`

   c. `show counter`

   d. `show filter counters`

10. True or false: a policer can be applied directly to the interface.

11. Which JUNOS software commands show counters for packets that exceed a policer? (Choose two.)

    a. `show firewall`

    b. `show counter`

    c. `show policer`

    d. `show policer counters`

12. True or false: Juniper EX Series switches can limit broadcast storms.

Chapter Review Answers

1. Answer: B. The default RIP import policy accepts RIP routes. To send direct routes, you need the direct protocol, and to readvertise RIP-learned routes, you need the RIP protocol. The default RIP export policy is to reject all.

2. Answer: A. A static route can never match from a protocol RIP condition, so it does not match the term. There is a logical AND for distinct conditions such as `route-filter` and `protocol` when listed under the same statement.

3. Answer: D. You cannot use policy to control LSA flooding. Import policy simply allows route filtering from the link state database (LSDB) to the RT.

4. Answer: A. The default action of the `test policy` command is to accept the route. For testing purposes, it makes sense to specify an action of reject when creating the policy.

5. Answer: A, B, D. Port, VLAN, and Layer 3 filters can be configured on an EX Series switch.

6. Answer: D. A total of 2,048 terms can be configured on a single filter.

7. Answer: D. In order to protect the switch itself, a filter should be applied to the loopback interface. This allows traffic coming in from any interface destined to the switch to be matched.

8. Answer: C. Port filters are evaluated first, then VLAN filters. There are no egress port filters.

9. Answer: B. The `show firewall` command displays all counters in all filters.

10. Answer: False. In an EX Series switch, policers must be referenced in a firewall filter, and then the filter is applied to the interface.

11. Answer: A, C. The `show firewall` and `show policer` commands will both show the counter associated with a given policer.

12. Answer: True. The switch can be configured to either rate-limit broadcast or unknown unicast packets or redirect these packets to a VLAN or interface.

Port Security and Access Control

Security is a serious concern for any modern network. Juniper Networks EX switches facilitate a rich set of security architectures through a range of Layer 2 security features that allow you to harden and secure the switched portions of your network. By adding JUNOS Layer 3 security features as described in Chapter 8, you can use the same EX switches to secure the routed portions of your network. The ability to combine Layer 2 and Layer 3 security features within the same box without a substantial impact to performance is a significant benefit provided by the JUNOS heritage enjoyed by the EX line.

The topics covered in this chapter include:

- Layer 2 security overview
- Media Access Control (MAC) limiting, Dynamic Host Configuration Protocol (DHCP) snooping, and dynamic ARP inspection (DAI)
- IEEE 802.1X port security

Layer 2 Security Overview

Layer 2 networks can present some unique security challenges, especially for those who are already familiar with IP technologies and common security approaches used for IP-based networks. IP security tends to *begin* at Layer 3, and carry on into the upper transport and even application layers to provide deep packet inspection and related services such as firewall and Network Address Translation/Port Address Translation (NAT/PAT). In contrast, a Layer 2 network can transport any number of upper-layer protocols, which may or may not include IP. Further, by definition a Layer 2 device may not even be able to inspect the contents of the frames that are transported. Such inspection brings that device into the realm of Layer 3 processing, a domain typically associated with the role of a router or security appliance.

While Layer 2 networks can be multiprotocol, the reality is that most modern networks are IP-centric; this IP *assumption* allows some Layer 2 devices to offer features that rely on peeking into IP-packet exchanges. This snooping provides a switch with some insight into key IP state, which in turn results in the ability to prevent certain IP and Address Resolution Protocol (ARP) spoofing attacks. While IP-based features are nice, it must be stressed that true Layer 2 security can make no assumptions about the upper-layer protocols, and must therefore be performed at Layer 2. In the case of Ethernet, this involves MAC frame processing, MAC learning, and the use of IEEE 802.1X, which encompasses a suite of Layer 2-specific authentication mechanisms built upon an extensible protocol framework.

Because EX switches have built-in (rather than bolted-on) IP routing, it should be no surprise to see that they offer a range of IP-based security features in addition to support for Layer 2-centric mechanisms such as 802.1X. This chapter focuses on key Layer 2 security features, which include a few value-added IP packet-snooping-based capabilities.

The rich set of EX security features makes it possible to deploy a robust internetwork that requires little, if any, external hardware. Some networks demand additional protection or sophisticated packet services that are currently not available on EX platforms. Enhanced security support is available on Juniper Networks routing platforms such as the J, M, and MX Series, which can provide stateful packet inspection, IP Security (IPSec), and NAT services. These topics are described in detail in the companion volume to this book, *JUNOS Enterprise Routing*, by Doug Marschke and Harry Reynolds. It should be noted that Juniper also offers an extensive range of security-focused solutions that include intrusion detection and prevention/antivirus (IDP/AV) control, Secure Sockets Layer (SSL)-based virtual private networks (VPNs), and Integrated Security Gateway (ISG) products.

EX Layer 2 Security Support

EX platforms support a number of Layer 2 and IP-enabled Layer 3 security features, and the list grows longer with each new release. In the 9.1 release used as the basis of this book, the primary Layer 2 security capabilities are:

MAC limiting

Switches filter and forward based on the MAC address. Unlike a hierarchical Layer 3 address, the flat MAC address structure does not permit aggregation or information hiding. This means that all 48 bits of each MAC address are significant, a fact that forces switches to maintain (per-VLAN) MAC tables for every active source in the broadcast domain. As all storage is finite, having to learn too many MAC addresses can lead to premature aging of previously learned MAC entries, which in turn leads to unnecessary flooding of related traffic when the specific MAC address is no longer known. MAC limiting caps how many MAC addresses can be stored in a given virtual LAN (VLAN) instance. This protects the MAC storage resources of other VLANs from a malicious user, who potentially lives on another VLAN, and isolates any resulting flooding to the problem VLAN rather than affecting all VLANs on the switch.

DHCP snooping

> DHCP is a widely used server-based/stateful protocol that allows for the automatic configuration of parameters needed for an IP machine to successfully communicate, both on and off the local network. DHCP snooping enables an EX to eavesdrop on DHCP exchanges to prevent unauthorized DHCP servers and ensure that each host in fact uses the IP addressing assigned to it during the DCHP process.

Dynamic ARP inspection

> ARP is used in Ethernet networks to bind a Layer 3 IP address to a corresponding Layer 2 MAC (hardware) address. A station can spoof ARP traffic by sending a gratuitous response with its own MAC address and another station's IP address. The result is that the switch forwards IP traffic intended for the legitimate destination to the spoofing station. This situation allows for a range of *man-in-the-middle* attacks. DAI works with DCHP snooping to ensure that ARP messages reflect the station's Source MAC (SMAC) and the source IP address matches that were assigned through DHCP. This process prevents ARP cache poisoning, thereby preventing many attack vectors.

802.1X

> The IEEE 802.1X standard defines a port-based Network Access Control (NAC) protocol. Here, the term *port* refers to a single point of attachment to a LAN network (i.e., a switch port). 802.1X is based on Extensible Authentication Protocol (EAP), which, true to its moniker, is designed to permit easy extensions to address new technologies or new security requirements. EX switches support the 802.1X authenticator role and offer a range of security options, such as support for unresponsive MAC addresses, firewall filter assignment, or a guest VLAN for unauthorized users.

This section provided an overview of EX security features that can be used to harden a Layer 2 switch infrastructure from denial of service (DoS) attacks and security threats, as well as to authenticate users and limit network access based on their authorization levels. We will demonstrate these concepts in the following sections in a variety of networking scenarios.

MAC Limiting, DHCP, and ARP

This section describes EX support for MAC limits, DHCP snooping, and DAI. It should be noted that the configuration examples and subsequent verification steps are based on a lab topology that differs from that used elsewhere in this book. This is because of the need for a RADIUS server with support for 802.1X extensions, and the desire for a command-line 802.1X supplicant (client) to help demonstrate details of the protocol's operation. Figure 9-1 details the Layer 2 security topology.

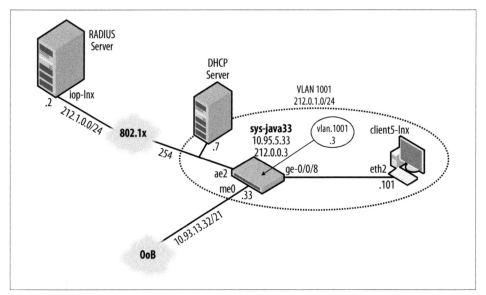

Figure 9-1. Layer 2 security topology

Of note in Figure 9-1 is the use of a Linux-based client and a RADIUS/DCHP server named client5-lnx and iop-lnx, respectively. In this case, the prefix iop is not really significant and simply differentiates this machine from other "operations" servers such as hop-lnx. The server is running Red Hat Enterprise Linux 5 with DHCP Server v3.0.5. The server also runs FreeRADIUS version 1.1.7, which offers support for 802.1X via support for EAP extensions. The client also runs Enterprise Linux 5, and makes use of the Open1X group's Xsupplicant version 1.2.8 package for its supplicant functionality. The client also runs the Internet Systems Consortium DHCP Client v3.0.5 package for DHCP operation.

The Device Under Test (DUT) is an EX3200-24P platform called sys-java33 that's running JUNOS Software Release 9.5R1. The majority of the functionality demonstrated in this chapter is also supported in the 9.1R1 release used as the basis for this book. In the time it took to prepare this material, Juniper Networks began shipping the 9.5 release, and that happened to be the version found running in the 802.1X test bed. Having no good reason to downgrade, we simply left it in place. Besides, newer releases should result in fewer interoperability issues and greater stability in general, two traits that prove beneficial later when testing 802.1X.

The EX switch is already configured to be part of a larger test network; the portions of its configuration that are relevant to the upcoming Layer 2 security labs are detailed here.

An me0-based Out of Band (OoB) management network is in place to support Telnet, FTP, and so on. The OoB management network simply works, and we will not discuss it further here. The switch is assigned two loopback addresses in the form of 10.93.5.33

and 212.0.0.3; the 10.93.5.33 address is marked as the primary address. We mention the loopback addresses here for the sake of completeness, but they have no direct bearing on the security activities of this chapter.

The switch is configured with a VLAN 1001, named **v1001**, which consists of the **ge-0/0/8** access link and the **ae2** trunk interface. Here's the access and trunk interface configuration:

```
[edit]
lab@sys-java33# show interfaces ge-0/0/8
unit 0 {
    family ethernet-switching {
        vlan {
            members v1001;
        }
    }
}
[edit]
lab@sys-java33# show interfaces ge-0/0/10
ether-options {
    802.3ad ae2;
}

[edit]
lab@sys-java33# show interfaces ge-0/0/11
ether-options {
    802.3ad ae2;
}

[edit]
lab@sys-java33# show interfaces ae2
unit 0 {
    family ethernet-switching {
        port-mode trunk;
        vlan {
            members [ 1001-1010 1012-1030 ];
        }
    }
}

[edit]
lab@sys-java33# show chassis aggregated-devices
ethernet {
    device-count 4;
}
```

Although we are concerned only with VLAN 1001 in this case, the example shows how a VLAN range, or set of ranges, can be specified.

The VLAN range used in this example may not work with some Cisco devices because these VLANs are internally reserved for Fiber Distributed Data Interface (FDDI) and Token Ring support. This is not an interop demonstration, so the VLAN ranges used here do not present an issue.

The operational status of the aggregated Ethernet (AE) interface is confirmed:

```
[edit]
lab@sys-java33# run show interfaces ae2 detail
Physical interface: ae2, Enabled, Physical link is Up
  Interface index: 159, SNMP ifIndex: 165, Generation: 163
  Link-level type: Ethernet, MTU: 1514, Speed: 2000mbps, BPDU Error: None, MAC-
REWRITE Error: None, Loopback: Disabled,
  Source filtering: Disabled, Flow control: Disabled, Minimum links needed: 1,
Minimum bandwidth needed: 0
  Device flags   : Present Running
  Interface flags: SNMP-Traps Internal: 0x0
  Current address: 00:19:e2:50:dd:02, Hardware address: 00:19:e2:50:dd:02
  Last flapped   : 2009-03-31 15:22:41 PDT (20:47:50 ago)
  Statistics last cleared: Never
  Traffic statistics:
   Input  bytes  :          50542449725                 18640 bps
   Output bytes  :         657792487705              62293336 bps
   Input  packets:            463148780                    34 pps
   Output packets:           1284806332                 15209 pps
   IPv6 transit statistics:
   Input  bytes  :                    0
   Output bytes  :                    0
   Input  packets:                    0
   Output packets:                    0

  Logical interface ae2.0 (Index 128) (SNMP ifIndex 187) (Generation 234)
   Flags: SNMP-Traps 0x0 Encapsulation: ENET2
   Statistics      Packets       pps         Bytes         bps
   Bundle:
      Input :            0         0             0           0
      Output:        45941         0       8098484           0
   Link:
    ge-0/0/10.0
      Input :            0         0             0           0
      Output:         2498         0        611927           0
    ge-0/0/11.0
      Input :            0         0             0           0
      Output:         2498         0        611927           0
   Marker Statistics:   Marker Rx   Resp Tx   Unknown Rx   Illegal Rx
    ge-0/0/10.0               0         0            0            0
    ge-0/0/11.0               0         0            0            0
   Protocol eth-switch, Generation: 248, Route table: 0
    Flags: Trunk-Mode
```

The output confirms that the **ae2** interface is operational, and that it's made up of two link members, **ge-0/0/10** and **ge-0/0/11**. The VLAN configuration is displayed next:

```
[edit]
lab@sys-java33# show vlans v1001
vlan-id 1001;
interface {
    ge-0/0/9.0;
}
l3-interface vlan.1001;
```

Note that v1001 is bound to VLAN tag 1001, and that a Routed VLAN Interface (RVI) is bound to the Layer 2 instance. Recall that an RVI provides Layer 3 services for a VLAN, as detailed in Chapter 8. Note that the interface ge-0/0/9 declaration under the VLAN definition is an alternative way to bind interfaces to a VLAN; this capability started in Release 9.3 and is functionally equivalent to the vlan members statement specified at the [edit interfaces <interface-name> unit <unit-number> family ethernet-switching] hierarchy.

Here is the related RVI configuration:

```
[edit]
lab@sys-java33# show interfaces vlan.1001
family inet {
    address 212.0.1.3/24;
}
```

The 212.0.1.0/24 logical IP subnet (LIS) is confirmed to match Figure 9-1, shown earlier. Note that both client5-lnx and the VLAN's RVI share an LIS, which is good, because otherwise a router would be needed to facilitate communications between the two IP endpoints. Given that the RVI *is the router* for this VLAN, it has to be on the same LIS as its clients.

The VLAN's operational status is confirmed:

```
[edit]
lab@sys-java33# run show vlans v1001 detail
VLAN: v1001, 802.1Q Tag: 1001, Admin State: Enabled
  Primary IP: 212.0.1.3/24
Number of interfaces: 9 (Active = 4)
  Untagged interfaces: ge-0/0/8.0*, ge-0/0/9.0*, ge-0/0/16.0, ge-0/0/17.0
  Tagged interfaces: ae0.0*, ae1.0, ae2.0*, ae3.0, ge-0/0/18.0
```

The output confirms that a VLAN v1001 is defined and uses 212.0.1.3 for its RVI, that ge-0/0/8 is an untagged access port, and that ae2 is a (tagged) trunk port.

With all apparently set up and ready to go, final verification comes when IP reachability within VLAN 1001 is confirmed. Start with IP connectivity between the client and the VLAN's RVI:

```
[root@client5-lnx ~]# ping 212.0.1.3
PING 212.0.1.3 (212.0.1.3) 56(84) bytes of data.
64 bytes from 212.0.1.3: icmp_seq=1 ttl=64 time=10.9 ms
64 bytes from 212.0.1.3: icmp_seq=2 ttl=64 time=0.769 ms
```

Great, the client is able to reach the VLAN's RVI. It's most likely that the first ping has a longer response time due to the need for an ARP transition, but JUNOS does not attempt to prioritize ping replies, so you can expect significant skew in ping response times anyway. As a result, you expect that the client's MAC has now been learned in the VLAN's context, and your expectation is confirmed:

```
[edit]
lab@sys-java33# run show ethernet-switching table vlan 1001 | match ge-0/0/8
  v1001                00:50:8b:6f:60:3a Learn          0 ge-0/0/8.0
```

Attention is now directed to how the RVI in turn routes toward remote destinations (e.g., the server). In this case, a simple static route allows sys-java33 to route packets to the remote server:

```
[edit]
lab@sys-java33# show routing-options static
route 212.0.0.0/8 {
    next-hop 212.0.1.254;
    no-readvertise;
}
```

The no-readvertise flag prevents this route from being later readvertised into some routing protocol. This flag is normally used for OoB-related routes, as such routes are normally not advertised beyond the local machine. It could be said that in this example a semi-OoB network is deployed for the purposes of performing 802.1X-based authentication.

Based on the static route, you conclude that the next hop for any packet destined to the 212/8 network is a gateway (router) with address 212.0.1.254. Layer 3 reachability to this forwarding next hop is verified from sys-java33:

```
[edit]
lab@sys-java33# run ping 212.0.1.254
PING 212.0.1.254 (212.0.1.254): 56 data bytes
64 bytes from 212.0.1.254: icmp_seq=0 ttl=64 time=14.033 ms
64 bytes from 212.0.1.254: icmp_seq=1 ttl=64 time=0.876 ms
^C
--- 212.0.1.254 ping statistics ---
2 packets transmitted, 2 packets received, 0% packet loss
round-trip min/avg/max/stddev = 0.876/7.454/14.033/6.579 ms
```

The ping is successful, and given that Ethernet-based pings first require a successful ARP exchange, you expect that the MAC address associated with 212.0.1.254 has now been learned in the context of VLAN 1001:

```
[edit]
lab@sys-java33# run show arp no-resolve | match .254
00:12:f2:21:cf:00 10.93.15.254    me0.0          none
00:00:5e:00:01:01 212.0.1.254     vlan.1001      none
```

Forwarding from VLAN 1001's RVI to the RADIUS server is verified:

```
[edit]
lab@sys-java33# run traceroute 212.1.0.2 no-resolve
traceroute to 212.1.0.2 (212.1.0.2), 30 hops max, 40 byte packets
 1  212.0.1.7  1.240 ms  0.964 ms  1.008 ms
 2  212.1.6.2  1.346 ms  0.975 ms  1.079 ms
```

```
     3  212.1.4.2  1.197 ms  3.132 ms  1.009 ms
     4  212.1.0.2  0.780 ms  1.769 ms  0.680 ms
```

With reachability working correctly, the correct learning of the MAC address that's associated with the 212.0.1.254 next hop is verified:

```
[edit]
lab@sys-java33# run show ethernet-switching table vlan v1001
Ethernet-switching table: 4 unicast entries
  VLAN          MAC address        Type       Age Interfaces
  v1001         *                  Flood        - All-members
  v1001         00:00:5e:00:01:01  Learn        0 ae2.0
  v1001         00:19:e2:50:dd:00  Static       - Router
  v1001         00:19:e2:50:f8:40  Learn        0 ae2.0
  v1001         00:21:59:f7:b4:00  Learn        0 ae2.0
```

The display shows that the EX's routing function has determined that it can forward Layer 3 routed traffic destined for 212/8 to a next hop address of 212.0.1.254 using the MAC address 00:00:5e:00:01:01 by forwarding the resulting frame out its ae2 interface. Note that as an Ethernet switch, the switching table is based strictly upon MAC addresses; there is no reference to IP, which is good because some LANs may still run non-IP protocols. The Router entry represents the presence of a Layer 3 interface (RVI) and the related MAC address is confirmed to be owned by the chassis:

```
[edit]
lab@sys-java33# run show chassis mac-addresses
    FPC 0   MAC address information:
      Public base address    00:19:e2:50:dd:00
      Public count           64
```

The ability to use the same interfaces for simultaneous Layer 2 switching and Layer 3 routing makes the EX a multilayer switch and is the primary motivation for the RVI construct. Recall that with a multilayer switch, the decision to switch versus route is based on MAC address; frames that are sent to the switches' own MAC address are sent to the Layer 3 processing engine (i.e., the VLAN's RVI), while all others are switched at Layer 2, thus saving several complex processing steps.

The confirmation of IP-level reachability within VLAN 1001 for the client to the RVI and for the RVI to the RADIUS server completes verification of the network's operational baseline. Subsequent sections modify this baseline to demonstrate key Layer 2 security features.

MAC Limiting

As mentioned earlier, MAC limiting is an invaluable tool in your Layer 2 security arsenal. By limiting the number of MAC addresses that can be learned within a VLAN, you can prevent exploits that seek to fill switch tables until they overflow, causing previously learned addresses to be forgotten in favor of more recent MAC activity. The usual goal of such an exploit is to force the switch to fall back to a flooding behavior for valid MAC addresses, which are (now) unknown because of table overflow. The performance impact of flooding can be bad enough; when you factor in potential security and DoS threats, it's clear that protecting a switched infrastructure from MAC flooding exploits is a good practice.

You configure MAC limits on a per-port basis only; there is no VLAN-wide global value on the number of MACs that can be learned. You can use the `interface all` keyword to define a global MAC limit that is applied to all interfaces on the switch. You can then exempt specific interfaces by specifying the desired MAC limit and action along with the interface's name. JUNOS will always perform the most specific match action, and an interface's name is always more specific than the keyword `all`.

EX switches support a limit on total MAC count, as well as the ability to limit based on an explicit list of allowed MAC addresses via the `allowed-mac` keyword. Managing a list of permissible MAC addresses can be a burden, but in some cases an automated provisioning system is used to maintain the MAC database and to push changes out to the switch.

Limiting MAC moves

EX platforms also provide control over the number of MAC moves that are permitted in a one-second window. A MAC move is defined as a MAC address being learned on one interface, followed by the receipt of a frame with that same SMAC on a different port. In such cases, the switch's control plane is interrupted to update the relevant Ethernet switching table, which should always reflect the last interface on which a given MAC address was seen.

MAC move limits are intended to protect the switches' control plane resources from exhaustion stemming from having to process abnormally high rates of MAC learn events and potential resource consumption as the switch attempts to keep its switching tables up-to-date. High MAC move rates can be the result of an intentional attack or a network malfunction that results in a loop.

Unlike the per-port setting of MAC limits, you define MAC move limits on a per-VLAN basis. This makes sense, as a MAC move by definition involves at least two ports, hence the need to control the aggregate number of MAC moves with the context of a specific VLAN. Note that on an AE port, MAC learning is based on the `ae` interface number, not the individual member links, to ensure that MAC moves are not falsely sensed when the same MAC address appears on multiple member interfaces.

MAC limit actions

By default, MAC limiting is not enabled on EX switches. When you configure MAC limiting, you choose what action is taken when a port exceeds its MAC learning or move limit. MAC limit exceeding actions include:

MAC limiting is disabled by default, which makes none the default limiting action. You can specify the none action explicitly for a given interface to exempt it from global actions evoked with an `interface all` statement. In the following code, interface ge-0/0/8 is not subject to the limits and actions associated with the `interface all` statement:

```
[edit]
lab@sys-java33# show ethernet-switching-options
secure-access-port {
    interface ge-0/0/8.0 {
        mac-limit 4294967295 action log;
    }
    interface all {
        mac-limit 500 action shutdown;
    }
}
```

log

The log action results in a syslog warning that is printed to the *messages* file. This action only warns and does not actually prevent the MAC address from being learned or moved.

drop

The drop action drops the MAC learning or move event and generates a syslog message that warns of the event. This action is good for protection of the switch but can result in loss of connectivity for legitimate users when the number of devices grows and the configuration is not updated to reflect the increased MAC count. Here, the syslog entry can assist in troubleshooting what can otherwise be a tricky problem. Note that in some cases, conventional MAC aging may result in a previously broken device suddenly starting to work as its MAC is no longer dropped (or is now allowed to move), just about the time some new user begins to complain of similar symptoms.

shutdown

The shutdown action is the most Draconian measure that can be taken. Upon exceeding the limit, the port is shut down and a syslog message is generated. The heavy-handed nature of this option is both good and bad. This option is prone to isolating legitimate users when thresholds are not correctly managed. At the same time, it sure makes troubleshooting what happened a fair bit simpler, as the port must be administratively enabled with a `clear ethernet-switching table interface <interface-name>` command before operation can resume.

Deploy and verify MAC limiting

The goal of the following example scenario is to configure MAC limiting on sys-java33 to meet these criteria:

- Allow 10 MACs on the ge-0/0/8

- Upon exceeding this limit, log an error message but allow additional MACs to be learned

The operational requirements make it clear that you need to use the MAC limiting feature's log action, as only this option satisfies the second criterion of allowing excess MACs to be learned while still generating the required error message upon exceeding the threshold.

MAC limiting is configured at the [edit ethernet-switching-options secure-access-port] hierarchy. Here are the options available at this hierarchy:

```
[edit ethernet-switching-options secure-access-port]
lab@sys-java33# set ?
Possible completions:
+ apply-groups         Groups from which to inherit configuration data
+ apply-groups-except  Don't inherit configuration data from these groups
> dhcp-snooping-file   Configure DHCP snooping persistence file, write-interval and
timeout
> interface            Configure access port security for this interface
> vlan                 Configure access port security for this VLAN
```

Of interest here are the interface and vlan subhierarchies. Thinking back, you recall that MAC limiting is set on a per-port basis and that MAC move limits are defined per VLAN. Given the goal of MAC limiting, it's clear that you need to deal with interface-level settings for the ge-0/0/8 interface, so you position yourself at that hierarchy, and again display the available options:

```
[edit ethernet-switching-options secure-access-port]
lab@sys-java33# edit interface ge-0/0/8

[edit ethernet-switching-options secure-access-port interface ge-0/0/8.0]
lab@sys-java33# set ?
Possible completions:
+ allowed-mac          Allowed MAC address on this interface
+ apply-groups         Groups from which to inherit configuration data
+ apply-groups-except  Don't inherit configuration data from these groups
  dhcp-trusted         Make this interface trusted for DHCP
> mac-limit            Number of MAC addresses allowed on this interface
  no-allowed-mac-log   Do not log violation of allowed MAC on this interface
  no-dhcp-trusted      Don't make this interface trusted for DHCP
> static-ip            Static IP address configuration
[edit ethernet-switching-options secure-access-port interface ge-0/0/8.0]
lab@sys-java33# set
```

A number of interface-level security options are available. Most are self-explanatory, or will be demonstrated at some point in this chapter; move on to configure the needed MAC limit behavior:

```
[edit ethernet-switching-options secure-access-port interface ge-0/0/8.0]
lab@sys-java33# set mac-limit ?
Possible completions:
  <limit>              Number of MAC addresses allowed on this interface
(1..4294967295)
  action               Action to take if limit is exceeded
```

```
[edit ethernet-switching-options secure-access-port interface ge-0/0/8.0]
lab@sys-java33# set mac-limit 10 action ?
Possible completions:
  drop                  Drop the packet and log it
  log                   Log a message
  none                  Take no action
  shutdown              Shut down the interface
```

The command-line interface (CLI) help function makes the needed syntax seem pretty clear, so it's entered:

```
[edit ethernet-switching-options secure-access-port interface ge-0/0/8.0]
lab@sys-java33# set mac-limit 10 action log
```

And the resulting configuration is displayed, and committed (not shown):

```
[edit ethernet-switching-options secure-access-port interface ge-0/0/8.0]
lab@sys-java33# show
mac-limit 10 action log;

[edit ethernet-switching-options secure-access-port interface ge-0/0/8.0]
lab@sys-java33#
```

Before generating any test traffic, display the current state of VLAN 1001 MAC learning:

```
[edit]
lab@sys-java33# run show ethernet-switching table vlan 1001
Ethernet-switching table: 4 unicast entries
  VLAN              MAC address        Type        Age Interfaces
  v1001             *                  Flood         - All-members
  v1001             00:00:5e:00:01:01  Learn         0 ae2.0
  v1001             00:19:e2:50:dd:00  Static        - Router
  v1001             00:21:59:f7:b4:00  Learn         0 ae2.0
  v1001             00:50:8b:6f:60:3a  Learn         0 ge-0/0/8.0
```

The output confirms that the v1001 learning table has fewer than 10 MAC entries, and that a single entry has been learned for the ge-0/0/8 port attached to client5-lnx. So far, so good. Enable syslog monitoring to watch for any changes in real time:

```
[edit]
lab@sys-java33# run monitor start messages
```

With all set on the EX, the macof utility is used at client5-lnx to generate some random MACs—20, to be exact:

```
[root@client5-lnx ~]# macof -i eth2 -n 20
f9:c:62:2:28:85 dd:d3:7a:23:bf:59 0.0.0.0.12564 > 0.0.0.0.13373: S
1492723364:1492723364(0) win 512
d1:28:d3:2e:d:6f e1:dd:31:3a:f7:7f 0.0.0.0.1514 > 0.0.0.0.29728: S
128837449:128837449(0) win 512
. . .
5a:d9:ce:1f:14:49 8a:69:2b:15:5d:cf 0.0.0.0.16072 > 0.0.0.0.25606: S
633015079:633015079(0) win 512
a4:bd:9a:5a:ff:67 97:3f:13:62:28:d1 0.0.0.0.13225 > 0.0.0.0.22271: S
1237001916:1237001916(0) win 512
[root@client5-lnx ~]#
```

With the MAC storm (or was that more of a sprinkle?) over, focus shifts back to sys-java33, where you expect to see a log warning reporting the excess number of MACs:

```
[edit]
lab@sys-java33#
*** messages ***
Apr  1 16:43:47  sys-java33 eswd[2809]: ESWD_MAC_LIMIT_EXCEEDED: MAC limit (10)
exceeded at ge-0/0/8.0
```

The expected log warning is confirmed, which is a good sign that things are proceeding to plan. Because the log action does not preclude the learning of excess MACs, you examine the resulting switching table to confirm that all MACs have in fact been learned:

```
[edit]
lab@sys-java33# run show ethernet-switching table vlan 1001 | match ge-
  v1001          00:50:8b:6f:60:3a Learn          52 ge-0/0/8.0
  v1001          1a:70:ed:6c:60:ab Learn          44 ge-0/0/8.0
  v1001          1e:68:34:22:c5:94 Learn          45 ge-0/0/8.0
  v1001          42:fa:2d:28:a1:fc Learn          43 ge-0/0/8.0
  v1001          5a:d9:ce:1f:14:49 Learn           0 ge-0/0/8.0
  v1001          62:8e:a4:28:4a:e4 Learn           0 ge-0/0/8.0
  v1001          6c:9c:a4:71:dc:9e Learn          40 ge-0/0/8.0
  v1001          7c:ed:a0:49:68:64 Learn           0 ge-0/0/8.0
  v1001          a2:06:91:03:32:2a Learn          31 ge-0/0/8.0
  v1001          a4:bd:9a:5a:ff:67 Learn           0 ge-0/0/8.0
  v1001          b6:06:8d:4d:09:88 Learn          26 ge-0/0/8.0
  v1001          ca:5f:35:35:a4:b1 Learn          33 ge-0/0/8.0
  v1001          ce:0e:21:57:b0:84 Learn          23 ge-0/0/8.0
  v1001          d2:0f:26:2f:bf:ee Learn          38 ge-0/0/8.0
  v1001          d2:90:78:6b:43:3e Learn           0 ge-0/0/8.0
  v1001          d4:a0:a2:74:55:0f Learn          35 ge-0/0/8.0
  v1001          e2:40:23:7e:93:a2 Learn           0 ge-0/0/8.0
  v1001          ec:ad:c1:42:7c:de Learn          46 ge-0/0/8.0
  v1001          ee:b4:c1:18:e8:8c Learn          33 ge-0/0/8.0
  v1001          fa:c3:19:70:09:c5 Learn          42 ge-0/0/8.0
  v1001          fc:05:83:2c:56:ef Learn           0 ge-0/0/8.0
```

The display is piped to the CLI match function so that only lines with ge- are displayed. This makes it easy to confirm that all of the spoofed MACs have been learned, which is in keeping with the scenario's requirements. The CLI count function is used to help save space in subsequent demonstrations. Here, quick work is made of the need to confirm that the expected count of 21 MAC addresses have been learned on VLAN 1001's ge-0/0/8 interface:

```
[edit]
lab@sys-java33# run show ethernet-switching table vlan 1001 | match ge- | count
Count: 21 lines
```

Having confirmed the log action, you quickly alter the configuration to drop excess MACs:

```
[edit ethernet-switching-options secure-access-port interface ge-0/0/8.0]
lab@sys-java33# set mac-limit action drop
```

```
[edit ethernet-switching-options secure-access-port interface ge-0/0/8.0]
lab@sys-java33# show | compare
[edit ethernet-switching-options secure-access-port interface ge-0/0/8.0]
-  mac-limit 10 action log;
+  mac-limit 10 action drop;
```

The Ethernet switching table is cleared:

```
[edit ethernet-switching-options secure-access-port interface ge-0/0/8.0]
lab@sys-java33# run clear ethernet-switching table vlan 1001
```

and the MAC flood is repeated (not shown):

```
[edit ethernet-switching-options secure-access-port interface ge-0/0/8.0]
Apr  1 17:03:38  sys-java33 eswd[2809]: ESWD_MAC_LIMIT_DROP: MAC limit (10)
exceeded at ge-0/0/8.0: dropping the packet
```

In addition to the log message, the real proof is in the MAC pudding, so to speak, and comes with the confirmation that only 10 MAC addresses have been learned on the ge-0/0/8 interface:

```
[edit ethernet-switching-options secure-access-port interface ge-0/0/8.0]
lab@sys-java33# run show ethernet-switching table vlan 1001 | match ge- | count
Count: 10 lines
```

With the drop action confirmed, the configuration is again modified, this time to use the shutdown action:

```
[edit ethernet-switching-options secure-access-port interface ge-0/0/8.0]
lab@sys-java33# set mac-limit action shutdown
```

```
[edit ethernet-switching-options secure-access-port interface ge-0/0/8.0]
lab@sys-java33# show | compare
[edit ethernet-switching-options secure-access-port interface ge-0/0/8.0]
-  mac-limit 10 action drop;
+  mac-limit 10 action shutdown;
```

After repeating the MAC flood, a log warning is noted:

```
[edit ethernet-switching-options secure-access-port interface ge-0/0/8.0]
lab@sys-java33# Apr  1 17:13:01  sys-java33 eswd[2809]: ESWD_MAC_LIMIT_BLOCK: MAC
limit (10) exceeded at ge-0/0/8.0: shutting down the interface
Apr  1 17:13:01  sys-java33 fpc0 MRVL-L2,mrvl_brg_port_stg_entry_set(),9771:Exact
msti rt state for IFL ge-0/0/8, stg_index 0,stg_port_state 1, oper_type 1
Apr  1 17:13:01  sys-java33 fpc0 MRVL-L2,mrvl_brg_port_stg_entry_set(),9771:Exact
msti rt state for IFL ge-0/0/8, stg_index 1,stg_port_state 1, oper_type 1
```

and the offending interface is confirmed to be shut down with a show ethernet-switching interface ge-0/0/8 command:

```
[edit ethernet-switching-options secure-access-port interface ge-0/0/8.0]
lab@sys-java33# run show ethernet-switching interface ge-0/0/8
Interface   State   VLAN members        Blocking
ge-0/0/8.0  up      v1001               blocked - MAC limit exceeded
```

Note that using the wrong form of this command, which includes the table keyword, results in the display of the interface's VLAN association without any indication that it's shut down:

```
[edit ethernet-switching-options secure-access-port interface ge-0/0/8.0]
lab@sys-java33# run show ethernet-switching table interface ge-0/0/8
Ethernet-switching table: 0 unicast entries
  VLAN             MAC address      Type      Age Interfaces
  v1001            *                Flood       - All-members
```

Once shut down, all the interface's learned MAC addresses are flushed from the table; serves it right for misbehaving, I say:

```
[edit ethernet-switching-options secure-access-port interface ge-0/0/8.0]
lab@sys-java33# run show ethernet-switching table vlan 1001
Ethernet-switching table: 3 unicast entries
  VLAN             MAC address      Type      Age Interfaces
  v1001            *                Flood       - All-members
  v1001            00:00:5e:00:01:01 Learn      0 ae2.0
  v1001            00:19:e2:50:dd:00 Static      - Router
  v1001            00:19:e2:50:f8:40 Learn      0 ae2.0
```

Use the clear ethernet-switching table interface <interface-name> command to enable an interface that's been disabled due to MAC limiting:

```
[edit ethernet-switching-options secure-access-port interface ge-0/0/8.0]
lab@sys-java33# run clear ethernet-switching table interface ge-0/0/8
```

Currently, there is no feature to automatically reenable a port after some period of time; administrative intervention is required. Because the log is still being monitored, a few seconds later an indication that the interface has returned to an operational state appears:

```
[edit ethernet-switching-options secure-access-port interface ge-0/0/8.0]
lab@sys-java33# Apr  1 17:13:38  sys-java33 fpc0 MRVL-
L2,mrvl_brg_port_stg_entry_set(),9771:Exact msti rt state for IFL ge-0/0/8,
stg_index 0,stg_port_state 4, oper_type 1
Apr  1 17:13:38  sys-java33 fpc0 MRVL-L2,mrvl_brg_port_stg_entry_set(),9771:Exact
msti rt state for IFL ge-0/0/8, stg_index 1,stg_port_state 4, oper_type 1
```

This concludes the MAC limiting case study; you can remove the related MAC limits from the configuration to prepare for the next section:

```
[edit ethernet-switching-options secure-access-port]
lab@sys-java33# delete interface ge-0/0/8
```

DHCP Snooping and ARP Inspection

DHCP is primarily defined in RFC 2131, but many other RFCs further elaborate on its operation or enhance its functionality. DHCP is designed to facilitate auto-configuration of network parameters in IP-based machines. In addition to providing a network address, mask, and default gateway, DHCP can also provide the boot image name and location for diskless PCs; its extensible nature permits support of various vendor- or application-specific extensions such as those needed for support of Microsoft's NetBIOS over TCP/IP transport.

Complete coverage of DHCP and its numerous protocol extensions is outside the scope of this book; our goal is to demonstrate DHCP snooping and other DHCP-related security enhancements supported by EX Series switches. To that end, Figure 9-2 provides a high-level overview of typical DHCP operation.

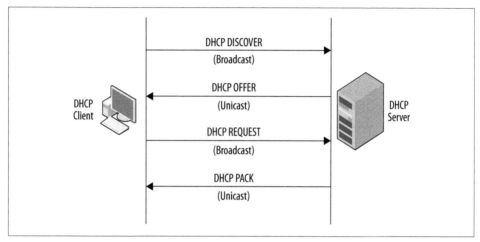

Figure 9-2. A typical DHCP exchange

Figure 9-2 begins with an IP-based host generating a DHCPDISCOVER message. This message is broadcast and therefore flooded through a Layer 2 domain, and is intended to dynamically discover the presence of one or more DHCP servers. Each server is then unicast back a DHCPOFFER, which presents the client with a proposed set of configuration parameters. The client chooses the preferred server based on some local policy, such as the first response received, and then broadcasts back a DHCPREQUEST message indicating the parameters (and server) it would like to use. The final step is the return of a unicast DHCPACK, which confirms the assignment of the related parameters and provides a lease duration during which they remain valid.

While highly useful and the norm in most of today's networks, DHCP's use of broadcast combined with lack of authentication and the exchange of network parameters in clear text opens it to a range of security threats. These range from brute force DoS attacks that attempt to flood servers with too many (or malformed) requests, to deployment of a bootleg server that provides bogus or conflicting information, all the way to sophisticated man-in-the-middle-type threats that center on a station eavesdropping on assigned DHCP parameters so that it may usurp their use and masquerade as the legitimate owner in an attempt to intercept traffic intended for another device.

ARP is a necessary part of IP operation on multipoint links. This is because of the independence of the 32-bit Layer 3 IP addresses versus the 48-bit MAC addresses used at the Link layer. ARP exploits are used to affect the redirection of IP traffic to a man-in-the-middle or sniffer device by sending an unsolicited ARP response, in which the black-hat station seeks to update a legitimate IP-to-MAC binding by substituting its own MAC address. The result is that IP traffic intended for the legitimate end station is now sent to the unicast MAC address of the attacker's machine.

Securing DHCP and ARP

EX switches support a range of security features specifically targeted at mitigating DHCP, ARP, and IP address-spoofing-related security threats. These include:

DHCP snooping
> The DHCP snooping feature filters and blocks ingress DHCP server messages on untrusted ports. The trusted DHCP server feature prevents users from setting up unauthorized DHCP servers. It also builds and maintains an IP address/MAC address binding database through monitoring of DHCP exchanges. The resulting DHCP snooping database is then used for a variety of related security features, such as IP Source Guard and DAI.
>
> DHCP snooping includes support for DHCP option 82. DHCP relay agents use this option to identify the client's point of attachment (port) and the relay agent's MAC address to the DHCP server. The server can in turn use this information to help formulate its response or to perform statistical tracking. When option 82 is enabled, the EX inserts option 82 information into DHCP messages it sends (or relays when in Layer 3 mode) to DHCP servers. The reply is expected to contain the same option 82 information, which the EX then strips before the response is sent to the client, thus making the option transparent to the DHCP user. Option 82 can increase security by allowing DHCP servers to generate alarms, or customize their response based on the ingress interface name or MAC address of the ingress EX switch. For example, all DHCP requests stemming from a "guest access port" might result in IP address assignment from some specific pool that is later filtered from private portions of the network.

Dynamic ARP inspection
> The DAI feature prevents ARP spoofing attacks by comparing ARP requests and replies against entries stored in the DHCP snooping database. ARP packets that do not match values in the snooping database are filtered.

IP Source Guard
> The IP Source Guard feature limits the effects of IP address spoofing attacks by validating the source IP address in packets received on an untrusted access interface against the DHCP snooping database. When the source IP address and SMAC address are found to be valid, the packet is accepted. In the case of a mismatch in either the source IP or SMAC address, the packet is discarded.

A key principle in the EX implementation of DHCP and ARP inspection is the concept of trusted and untrusted interfaces. By default, all access links are untrusted, and therefore, when enabled, all DHCP and ARP packets on these ports are inspected and are allowed to pass only when they match the snooping database. In contrast, trunk ports are trusted and don't perform DHCP or ARP inspection, allowing all such packets to pass unhindered.

Deploy DHCP snooping and ARP inspection

In this section, you will configure and validate the DCHP snooping and DAI features on an EX switch. Refer back to Figure 9-1 as needed for details on the test topology. Your goals are to:

- Ensure that DHCP responses are not accepted on access interface ge-0/0/8
- Configure VLAN v1001 for:
  — DHCP snooping with option 82 support using the interface name for the circuit ID, the switch name for the remote ID, and a vendor ID of Juniper_EX
  — DAI to prevent ARP spoofing
  — IP Source Guard to prevent IP spoofing

The set of configuration tasks is pretty straightforward. Knowing that DHCP snooping/ option 82, ARP inspection, and IP Source Guard are configured on a per-VLAN basis, while DHCP/ARP trusted versus untrusted is an interface-level setting, helps to get things started. Note that, by default, all trunk ports are considered DHCP trusted and all access ports are considered untrusted once DHCP snooping is enabled with the examine-dhcp keyword. You can then use the dhcp-trusted or no-dhcp-trusted keyword on a per-interface basis in order to alter the defaults.

Recall that a previously issued show vlan v1001 command confirmed that the ae2 interface is set to trunking mode, and that the ge-0/0/8 interface is operating in untagged access mode. Given the defaults, and the knowledge that ge-0/0/8 is operating in access mode, the first requirement is met by simply enabling DHCP snooping. The following configuration enables the functionality required in this scenario:

```
[edit]
lab@sys-java33# show ethernet-switching-options
secure-access-port {
    vlan v1001 {
        arp-inspection;
        examine-dhcp;
        ip-source-guard;
        dhcp-option82 {
            circuit-id;
            remote-id {
                prefix hostname;
            }
            vendor-id {
                Juniper_EX;
```

```
                }
            }
        }
    }
```

Notice that there is no configuration stanza for the ge-0/0/8 access interface. This is because access interfaces are set to DHCP untrusted by default, which makes an explicit setting unnecessary. Also note that there is no MAC limiting in this scenario, which is also a port-level setting and yet another reason for no explicit access port configuration in this example.

DHCP snooping is enabled for VLAN v1001 via the examine-dhcp keyword. Once enabled, the related DAI and IP Source Guard functionality can be turned on. Neither has any configuration options, so you simply need to enable or disable the features as desired; both features require a DHCP snooping database to work, making DHCP snooping a prerequisite to their support.

By leaving the circuit-id option blank, the default coding of the interface name/VLAN name is sent, and the remote-id option is set to specify the local host name per requirements. Likewise, the vendor-id is coded with the required string variable; note that when no argument is specified the default string is Juniper. With DHCP snooping and option 82 configured per the requirements, and with DAI and IP Source Guard enabled, you commit the changes and move on to verification.

Confirm DHCP snooping and ARP inspection

Start the verification process by firing off a DHCP request from client5-lnx. Normally the DHCP client is set to run at startup in the background, but here it's evoked in the foreground from a root shell to allow analysis of its operation. You start by displaying the current state of client5-lnx's eth2 interface:

```
[root@client5-lnx ~]# ifconfig eth2
eth2      Link encap:Ethernet  HWaddr 00:50:8B:6F:60:3A
          BROADCAST ALLMULTI MULTICAST  MTU:1500  Metric:1
          RX packets:2627393558 errors:0 dropped:0 overruns:11355539 frame:11355539
          TX packets:29883855 errors:0 dropped:0 overruns:0 carrier:0
          collisions:0 txqueuelen:1000
          RX bytes:1968164841 (1.8 GiB)  TX bytes:1261916105 (1.1 GiB)
```

Of note here is the lack of an UP indication, and the lack of any IP addresses. Seems like a job for DHCP! To that end, the dhclient program is activated at client5-lnx on the eth2 interface:

```
[root@client5-lnx ~]# dhclient eth2 -d
Internet Systems Consortium DHCP Client V3.0.5-RedHat
Copyright 2004-2006 Internet Systems Consortium.
All rights reserved.
For info, please visit http://www.isc.org/sw/dhcp/

Listening on LPF/eth2/00:50:8b:6f:60:3a
Sending on  LPF/eth2/00:50:8b:6f:60:3a
Sending on  Socket/fallback
```

```
DHCPREQUEST on eth2 to 255.255.255.255 port 67
DHCPREQUEST on eth2 to 255.255.255.255 port 67
DHCPACK from 212.0.1.7
bound to 212.0.1.101 -- renewal in 33929 seconds.
```

The display confirms that a server was located after the second DHCPREQUEST message, and that the parameters offered were acceptable, as confirmed by receipt of a DHCPACK from a DHCP server with IP address 212.0.1.7. The message confirms that the client has been assigned IP address 212.0.1.101, which is verified with a repeat of the ifconfig eth2 command:

```
[root@client5-lnx ~]# ifconfig eth2
eth2      Link encap:Ethernet  HWaddr 00:50:8B:6F:60:3A
          inet addr:212.0.1.101  Bcast:212.0.1.255  Mask:255.255.255.0
          inet6 addr: fe80::250:8bff:fe6f:603a/64 Scope:Link
          UP BROADCAST RUNNING ALLMULTI MULTICAST  MTU:1500  Metric:1
          RX packets:2627393673 errors:0 dropped:0 overruns:11355539 frame:11355539
          TX packets:29883886 errors:0 dropped:0 overruns:0 carrier:0
          collisions:0 txqueuelen:1000
          RX bytes:1968175337 (1.8 GiB)  TX bytes:1261925037 (1.1 GiB)
```

The highlights confirm the expected IP address assignment and that the interface is now UP. The successful result shows that the DCHP protocol itself is working, which is a good sign that the DHCP server actually supports option 82. A DHCP server that does not support the option does not echo it back, causing DHCP failures as the returned DHCP responses are typically not conveyed onto the client in such cases.

DHCP snooping should have built a database entry for the successful DHCP transaction between client5-lnx and the server at 201.0.1.7. Use the show dhcp snooping binding CLI command to verify current bindings:

```
[edit]
lab@sys-java33# run show dhcp snooping binding
DHCP Snooping Information:
MAC address        IP address      Lease (seconds) Type     VLAN    Interface
00:50:8B:6F:60:3A  212.0.1.101              85899 dynamic  v1001   ge-0/0/8.0
```

As expected, the database contains an entry for client5-lnx's eth2 MAC address and its newly assigned IP address. Displaying the results of DAI is just as easy with the show arp inspection statistics command:

```
[edit]
lab@sys-java33# run show arp inspection statistics
Interface      Packets received      ARP inspection pass  ARP inspection failed
    ae0               0                        0                     0
    . . .
  ge-0/0/8            1                        1                     0
  . . .
```

The output is rather basic, but clearly indicates that all of the ARP responses seen on access link ge-0/0/8 have been valid, which is to say that the reported MAC and IP source addresses do in fact match the contents of the snooping database. To test this, the arpspoof utility is fired up at client5-lnx in an attempt to have it masquerade as the DHCP server by having traffic that's intended for 212.0.1.7 directed to its own MAC address:

```
[root@client5-lnx ~]# arpspoof -i eth2 212.0.1.7
0:50:8b:6f:60:3a ff:ff:ff:ff:ff:ff 0806 42: arp reply 212.0.1.7 is-at
0:50:8b:6f:60:3a
0:50:8b:6f:60:3a ff:ff:ff:ff:ff:ff 0806 42: arp reply 212.0.1.7 is-at
0:50:8b:6f:60:3a
. . .
```

After sending a few such Gratuitous ARP replies, we check to see whether DAI has prevented ARP cache poisoning by blocking the flooding of this bogus ARP message to the rest of the hosts in the VLAN:

```
[edit]
regress@sys-java33# run show arp inspection statistics | match "(inspection|ge-
0/0/8)"
Interface    Packets received    ARP inspection pass  ARP inspection failed
   ge-0/0/8                 9                      2                      7
```

As expected, we now see that some seven packets have failed DAI, and as a result were dropped to foil the attempt to intercept traffic intended for the DHCP server. Use the `show ip-source-guard` command to view the IP source address database:

```
[edit]
lab@sys-java33# run show ip-source-guard
IP source guard information:
Interface    Tag   IP Address      MAC Address        VLAN
ge-0/0/8.0   0     212.0.1.101     00:50:8B:6F:60:3A  v1001
```

The display confirms that only packets with a source IP of 212.0.1.101 are permitted to ingress on port `ge-0/0/8`. This information is again based on DHCP snooping. To show that it works, a new IP is assigned to `client5-lnx`'s `eth2` interface, and the ping to the DHCP server is repeated:

```
[root@client5-lnx ~]# ifconfig eth2 211.0.1.102 netmask 255.255.255.0
[root@client5-lnx ~]# ifconfig eth2

eth2      Link encap:Ethernet  HWaddr 00:50:8B:6F:60:3A
          inet addr:211.0.1.102  Bcast:211.0.1.255  Mask:255.255.255.0
          inet6 addr: fe80::250:8bff:fe6f:603a/64 Scope:Link
          UP BROADCAST RUNNING ALLMULTI MULTICAST  MTU:1500  Metric:1
          RX packets:2627396876 errors:0 dropped:0 overruns:11355539 frame:11355539
          TX packets:29883926 errors:0 dropped:0 overruns:0 carrier:0
          collisions:0 txqueuelen:1000
          RX bytes:1968451198 (1.8 GiB)  TX bytes:1261929466 (1.1 GiB)

[root@client5-lnx ~]# ping 212.0.1.7
PING 212.0.1.7 (212.0.1.7) 56(84) bytes of data.

--- 212.0.1.7 ping statistics ---
5 packets transmitted, 0 received, 100% packet loss, time 3998ms
```

As expected, the pings fail, ostensibly due to an ingress discard that results from an invalid IP source address. Oddly, after the `ip-source-guard` feature is removed, the ping is still not allowed:

```
[edit ethernet-switching-options secure-access-port]
lab@sys-java33# delete vlan v1001 ip-source-guard

[edit ethernet-switching-options secure-access-port]
lab@sys-java33# commit
```

As noted, back at the client, the spoofed source address pings are found to still be failing:

```
[root@client5-lnx ~]# ping 212.0.1.7
. . .
From 212.0.1.102 icmp_seq=65 Destination Host Unreachable
From 212.0.1.102 icmp_seq=67 Destination Host Unreachable
. . .
```

Pondering the failure, you recall that before the client can send an IP packet to the destination, it must first perform an ARP. It's likely that while changing the interface's IP address, any related ARP entries were flushed, thereby requiring a new ARP exchange, and the ARP is filtered due to its use of an invalid source IP. So, while the removal of `ip-source-guard` does technically permit the transmission of IP datagrams on the ge-0/0/8 link with a source address *other* than 212.0.1.101, ARP packets are *still* subject to DAI, and they are not allowed to pass when either the source IP or SMAC address does not match the snooping database. The theory is confirmed by displaying ARP inspection statistics:

```
[edit]
lab@sys-java33# run show arp inspection statistics
Interface     Packets received    ARP inspection pass  ARP inspection failed
     ae0                    27                     27                      0
. .
  ge-0/0/8                 109                      4                    105
```

As suspected, the high failed count reflects that the ongoing ARP attempts generated by client5-lnx are in fact being filtered. Having confirmed ARP inspection, DAI is removed:

```
[edit]
lab@sys-java33# delete ethernet-switching-options secure-access-port vlan v1001
arp-inspection
```

And as predicted, client5-lnx is now able to ping a remote machine using an IP address that was not assigned via DHCP:

```
. .
64 bytes from 212.0.1.7: icmp_seq=70 ttl=64 time=2.00 ms
64 bytes from 212.0.1.7: icmp_seq=71 ttl=64 time=0.404 ms
. . .
```

These results conclude the confirmation of DHCP snooping and ARP inspection on the EX platform.

MAC Limiting, DHCP, and ARP Summary

This section detailed various port-level security features that mitigate many common attack vectors such as MAC table overflows and IP address spoofing. The next section addresses port-based authentication using the IEEE 802.1X standard.

The best security design is always multilayered. By combining port-level authentication with port security, you get additional protection. After all, now you have to be successfully authorized before you're even able to attempt some nefarious activity.

IEEE 802.1X Port-Based Authentication

The IEEE 802.1X standard defines port-based NAC. In English, this means the protocol authenticates users on a per-switch port (or Wireless Access Point [WAP]) basis, allowing access for valid users and effectively disabling the port when authentication fails. The 802.1X standard relies on EAP for its heavy lifting; EAP is currently defined in RFC 3748. 802.1X is most often associated with WAPs, for the obvious reason that a wireless infrastructure, by its very nature, opens itself up to any and all takers, and hence may want to authenticate users before allowing them in. That being said, there is no reason that what is good for a wireless network cannot also be a benefit for a wired infrastructure. For example, you may have wall jacks that are in an unsecured area in a public meeting room that is shared by internal users and external guests, and you would like to offer intranet and Internet access to the former, but only Internet access to the latter.

802.1X does not replace other security technologies. 802.1X works with port security features such as DHCP snooping, DAI, and MAC limiting to guard against DoS attacks and spoofing.

Terminology and Basic Operation

Before diving into the 802.1X configuration and verification lab, let's review some basic terminology and operational concepts. Figure 9-3 illustrates basic 802.1X concepts and EAP operation.

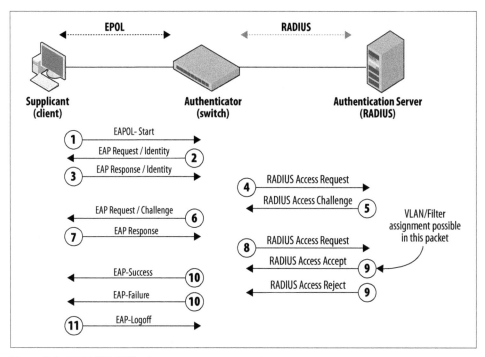

Figure 9-3. IEEE 802.1X basics

An 802.1X authentication system contains three basic components:

Supplicant

> While somewhat arcane, *supplicant* is the official term for a client that seeks au-thentication through 802.1X. The supplicant is a "humble petitioner: somebody who makes a humble and sincere appeal to a person who has the power to grant the request." Given this definition, *supplicant* does in fact seem to be a befitting term, albeit a bit on the obscure side. The supplicant can be responsive or non-responsive. A responsive supplicant is one that is 802.1X-enabled and provides authentication credentials—specifically, a username and password for EAP MD5, or a username and client certificates for EAP-TLS, EAP-TTLS, and EAP-PEAP. A non-responsive supplicant typically does not support 802.1X, and therefore does not initiate or react to any EAP stimulation. You must authenticate non-responsive supplicants using a MAC-based authentication method, or disable 802.1X and forgo simple authorization.

Authenticator port access entity

> This is the official IEEE term for the authenticator, which is typically a switch or WAP. The term *authenticator* is used to describe this role for the sake of brevity. The primary role of the authenticator is to block all traffic to and from supplicants until they are authenticated. The authenticator will pass gathered information to

an authentication server; per the 802.1X specification, this must be a RADIUS server.

Authentication server

The authentication server contains the user database that's used to make authentication decisions. The database normally contains credentials information for each supplicant that is expected to connect to the network, and is used to compare the credentials supplied by the supplicant. Access is granted only when the credentials match.

 Currently, EX Series switches support only RADIUS-based authentication and accounting servers for 802.1X.

A Word on 802.1X Clients

Windows XP ships with 802.1X supplicant support. You enable and configure the client by selecting the network adapter's properties, and then clicking on the Authentication tab, as shown in Figure 9-4.

It is common to have to specifically seek out and install an 802.1X client in other operating systems. For example, the Linux platform used as a client when developing this material was loaded with the Xsupplicant package from Open1X.org, available at *http://open1x.sourceforge.net/*.

Figure 9-3, seen earlier, shows the functional components of 802.1X authentication, and details the typical EAP and RADIUS packet exchanges used for authentication. Note that between the supplicant and the authenticator, only EAP messages are exchanged, and likewise between the authenticator and the authentication server, only RADIUS messages are exchanged.

In step 1, the client initiates authentication through transmission of an EAPOL-Start message. Alternatively, the switch can also initiate authentication when it receives a data packet from a currently unauthorized client. At this time, the authenticator sets the port to the initialized state. In this state, only 802.1X traffic is allowed; all other traffic is blocked at the Data Link layer.

At step 2, the authenticator sends an EAP Request for the supplicant's identity/credentials. The supplicant obliges at step 3 with an EAP Response message. The authenticator bundles the supplicant's credentials into a RADIUS Access Request message, which is then sent to the authentication server at step 4.

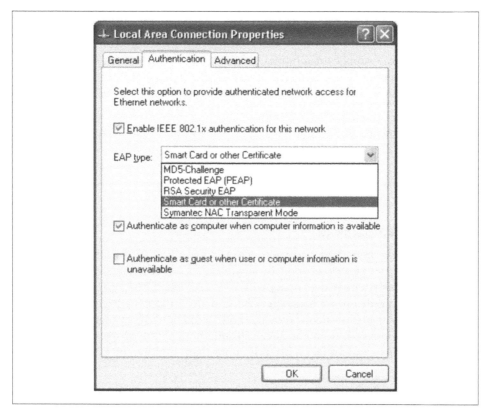

Figure 9-4. The Windows XP 802.c1X client

The authentication server then accepts or rejects the access request. When the access request is accepted, the authentication server sends a RADIUS Access Challenge, which is converted to an EAP Request (for a challenge response) by the authenticator, as shown in steps 5 and 6. The supplicant returns the challenge response, in EAP format, which is then relayed on to the RADIUS server in another RADIUS Access message, shown in steps 7 and 8.

When the supplicant's response meets the authentication server's expectations, a RADIUS Access Accept message is generated, or else access is rejected and the port continues to block all but EAP traffic. This process is shown in steps 9 and 10.

It should be noted that the RADIUS server can return a dynamically assigned VLAN and/or a firewall filter in its Access Accept message through support for vendor-specific attributes (VSAs), as described shortly. This capability allows a supplicant to begin authentication in one VLAN, and wind up in another based on the results of the authentication. If the supplicant meets the challenge, the authenticator sets the port to the authorized state and normal traffic is then accepted to pass through the port. If the authentication server rejects the RADIUS Access Request, the authenticator sets the port to the unauthorized state, blocking all traffic.

Step 11 shows the supplicant sending an EAP-Logoff upon disconnection from the network. In response, the authenticator returns the port to the unauthorized state, again blocking all non-EAP traffic as it patiently awaits its next supplicant.

The configurable `quiet-period` determines how long the switch waits before retrying an authentication attempt after it has failed `retries` times. The default is three retry attempts before entering a 60-second quiet period, a behavior designed to prevent excessive control plane resource consumption as a result of having to process too many EAP requests, as well as to thwart brute force hacking techniques, which in theory will grow bored and move on to a faster target.

Extensible Authentication Protocol

EAP is an authentication framework rather than a specific authentication mechanism. As such, it provides generic support for any number of authentication schemes, the operation of which is treated as opaque data by EAP itself. EAP provides a common set of functions needed to exchange packets, and also supports a negotiation method to determine which EAP authentication method should be used between the supplicant and the authenticator. EAP methods can include both standards-based and vendor-proprietary approaches. In fact, currently more than 40 different EAP methods are defined!

The EAP methods supported on EX switches are all IETF-standards-based, and are detailed in Table 9-1.

Table 9-1. Supported EAP methods

EAP method	What is exchanged
EAP-MD5	EAP-MD5 exchanges the MD5 hashed username and password, defined in RFC 3748. It can be vulnerable to dictionary attacks and does not offer support for mutual authentication or encryption key generation.
EAP-TLS	EAP Transport Layer Security (TLS) uses PKI-based authentication for strong security. TLS is a new name for what was formerly called Secure Sockets Layer (SSL). EAP-TLS is defined in RFC 2716. This widely supported method uses client- (and server-) side certificates for both strong authentication and dynamic encryption key generation. The need for a client-side certificate can complicate configuration. EAP-TLS is the most common EAP method in use today.
EAP-TTLS	The Tunneled Transport Layer Security (TTLS) method extends TLS by not requiring a client-side certificate. This protocol was co-developed by Funk Software (later acquired by Juniper) and Certicom. Like TLS, it offers strong security and never sends user credentials in clear text.
EAP-PEAP	The Protected Extensible Authentication Protocol (PEAP) method was jointly developed by Cisco Systems, Microsoft, and RSA Security. It's similar to TTLS, in that a client-side certificate is not required to establish a secure tunnel between the supplicant and the authentication server. There are several versions of PEAP. EX switches support PEAPv0/EAP-MSCHAPv2 as defined in (expired draft) draft-kamath-pppext-peapv0-00.

JUNOS 802.1X Feature Support

EX switches support a wide range of 802.1X options. This section describes the most common features and their typical usage.

Administrative modes

The administrative mode describes how a port behaves in the 802.1X context. A port can be in one of three 802.1 administrative states:

Automatic
> In automatic mode, the supplicant must be authenticated before access is granted. This is the default setting for an 802.1X-enabled port on an EX. A port operates in automatic mode when you enable it for 802.1X authentication by listing it under the [edit protocols dot1x] stanza.

Force authorized
> This mode forces a port to be in the authorized state at all times. In JUNOS, you can configure this mode explicitly with a `disable` statement at the [edit protocols dot1x authenticator interface <interface-name>] hierarchy. Alternatively, you can obtain force authorized behavior by simply not enabling 802.1X on that interface to begin with. This mode can be used to support unresponsive hosts when a MAC-based authentication approach is not desired.

Force unauthorized
> This mode forces a port to be in the unauthorized state at all times, thereby disabling the port for user traffic.
>
> With JUNOS, you do not configure this mode explicitly. You obtain force unauthorized behavior by disabling the port itself using a `set interfaces <interface-name> disable` configuration mode statement.

> When you disable a port, the related status LED is extinguished. You can view interface status LEDs directly when local to the switch, or remotely with a `show chassis lcd` command.

Supplicant modes

A supplicant can be authenticated in single mode, single-secure mode, or multiple mode. The difference between each mode is significant and must be understood to effectively deploy 802.1X on EX switches. You configure the supplicant mode for a given interface at the [edit protocols dot1x authenticator interface <interface-name>] hierarchy using the `supplicant` keyword.

Single
> The single mode authenticates only the first supplicant. All other supplicants that connect later to the port are allowed full access without any further authentication.

They effectively "piggyback" on the first supplicant's authentication. Single is the default supplicant mode.

Single-secure

The single-secure mode allows only one supplicant to connect to the port. No other supplicant is allowed to connect until the first supplicant logs out. This mode prevents the piggybacking of other clients upon the first authentication that occurs in single mode.

Multiple

Multiple mode provides fine-grained access control by allowing multiple supplicants per port while requiring that each such supplicant be individually authenticated.

Additional capabilities

EX switches offer some value-added features that relate to 802.1X authentication:

Guest VLAN

The guest VLAN feature provides limited network access for supplicants that fail 802.1X authentication. This feature is often used to provide external/Internet access while blocking intranet access for mobile users that are visiting a network location. Authentication can fail due to a non-responsive host that does not answer EAP queries, or because the provided credentials do not match those in the authentication server, resulting in a reject response.

Dynamic VLAN

With this feature, a supplicant's VLAN can be dynamically assigned/altered as part of the authentication process. The assigned VLAN must already exist on the local switch, however.

Dynamic firewall filters

This feature assigns a `family ethernet-switching` firewall to the authenticated port based on the supplied user credentials.

MAC-based authentication

This feature is designed to support non-responsive hosts, which is another term for machines that are not 802.1X-capable. MAC-based authentication bypasses 802.1X authentication for these hosts and can use either a local list of permitted MACs or RADIUS-based MAC authentication. MAC-based authentication can be used exclusively, or in addition to supported EAP methods for a responsive host.

Dynamic changes to a user session

This feature allows an administrator to terminate an established session, perhaps due to violations of the Acceptable Use Policy (AUP). The feature is enabled through support for the RADIUS Disconnect Message, as defined in RFC 3576. In-operation changes made to a supported RADIUS server result in a Disconnect Request being sent to the authenticator, which forces reauthentication of the affected user.

Support for VoIP

802.1X-enabled Voice over IP (VoIP) devices can be assigned a voice VLAN ID and class of service (CoS) forwarding class marker settings as part of the RADIUS Accept message. The tagging and marking parameters are then conveyed to the phone using Link Layer Discovery Protocol-Media Endpoint Discovery (LLDP-MED), as described in Chapter 10.

A VoIP device that lacks 802.1X support can rely on an 802.1X-1X-capable device that's connected to its data port to perform authentication and thereby enable the switch port. While such a method does not automatically ensure correct VLAN tagging and CoS marking for the VoIP traffic, it does allow voice traffic to pass to and from the phone.

RADIUS accounting

This feature sends accounting information to a RADIUS accounting server whenever a subscriber logs in or logs out. This feature is based on RFC 2866, "RADIUS Accounting."

Periodic reauthentication

This feature forces previously authenticated clients to reauthenticate after the configured period of time. This feature can be disabled with a `no-reauthentication` statement.

Non-responsive host/MAC-based authentication

You can configure MAC-based authentication while excluding all other forms, or as a fallback when EAP is not supported. In this mode, the username and password are set to the client's MAC address, requiring that a matching entry be defined on the authentication server. By adding `restrict` to the `mac-radius` statement, you force MAC-based authentication exclusively. This eliminates the 90-second delays that must normally elapse before the switch assumes that the host is non-responsive to EAP and falls back to a MAC-based approach.

Server fallback

In addition to allowing a primary and secondary authentication server, you can also define how the switch handles a RADIUS timeout. Such a timeout occurs when all RADIUS servers, or the communications path to reach them, are down. You configure the desired timeout or reject action on a per-interface basis at the [edit protocols dot1x authenticator interface *<interface-name>*] hierarchy using the `server-fail` and `server-reject-vlan` keywords, respectively. Actions include `permit`, `deny`, and `use-cached`, with the latter used to support reauthentication only; new users are denied.

Vendor-specific attributes

VSAs are the mechanism behind support for dynamic VLAN and firewall filter assignment. VSAs are *defined centrally* on the authentication server but are implemented in *distributed fashion* on each authenticator. By housing the VSAs on a central authentication server, you eliminate the need for configuring the same attributes, in the form of firewall filters, on every switch to which a given supplicant may connect.

Deploy and Verify 802.1X

In this section, you will configure and validate 802.1X authentication using both the EAP-MD5 and server-based MAC address methods. Once again, refer back to Figure 9-1 for topology details as needed.

The configuration criteria are as follows:

- Configure an access profile to support RADIUS authentication for dot1x users.
- Enable dot1x authentication on ge-0/0/8 to support EAP-MD5.
- Ensure that RADIUS requests originate from the 212.0.1.3 address.
- Ensure that only one supplicant can be active while blocking traffic from all other sources.
- Force the client to reauthenticate every 180 seconds.

You may assume that the RADIUS server iop-lnx has been correctly configured to support 802.1X, and that the supplicant software is correctly installed on client5-lnx. For completeness, key portions of the RADIUS server and supplicant's configuration are shown.

RADIUS server configuration

The configuration files are housed in the *usr/local/etc/raddb* directory. The *clients.conf* file contains an entry for Network Access Server (NAS), sys-java33:

```
. . .
client 212.0.1.3 {
        secret        = juniper
        shortname     = sys-java33
}
. . .
```

Please try to keep the hard-to-guess password a secret; strong passwords are such a pain to remember, after all. The EAP-MD5 user's credentials are configured in the *users* file:

```
. . .
md5user01 Auth-Type := EAP, User-Password == "md5user01"
          Tunnel-Type = VLAN,
          Tunnel-Medium-Type = IEEE-802,
          Tunnel-Private-Group-Id = 1001,
#         Juniper-Switching-Filter = "Match Destination-mac 00:00:00:00:11:22 Action
allow"
          Filter-Id = "f1"
. . .
```

The configuration includes the expected username and password pair, as these are the only credentials used in the EAP-MD5 method. This example also includes VSAs to identify the client's VLAN, which is 1001 in this example, and to dynamically build or apply an existing firewall filter via the `Juniper-Switching-Filter` or `Filter-Id` VSA, respectively. Using VSAs, it's possible to have the supplicant begin authentication in one VLAN with one set of filter rules, and as a result of successful authentication be reassigned to a new VLAN and a new set of filter rules.

The commented `Juniper-Switching-Filter` line provides sample syntax for a set of rules that are sent by the RADIUS server to the authenticator, where they are executed to dynamically build and apply the related filter. Alternatively, you can simply return the name of a preexisting filter, as is the case with the `f1` filter name in conjunction with the `Filter-Id` VSA in this example. Note again that for `Filter-Id` to be successful a `family ether-switching` filter with a matching name must already be configured on the authenticator.

VSAs and the Juniper Dictionary

For VSA support, you must have a supported RADIUS server, and you must configure it with the string definitions contained in a file called *dictionary.juniper*. Details are available at *http://www.juniper.net/techpubs/en_US/junos9.4/topics/task/configuration/802-1x-vsa-configuring-cli.html*. A sample *dictionary.juniper* file that supports VLAN, filter ID, and dynamic switching filters is shown; you can paste this information into a *dictionary.juniper* file of your own creation:

```
[root@iop-lnx2 ~]# cat
/usr/local/share/freeradius/dictionary.juniper
# -*- text -*-
#
#   dictionary.juniper
#
#       As posted to the list by Eric Kilfoil
<ekilfoil@uslec.net>
#
# Version:      $Id: dictionary.juniper,v 1.2.6.1 2005/11/30
22:17:25 aland Exp $
#

VENDOR          Juniper                         2636

BEGIN-VENDOR    Juniper

ATTRIBUTE       Juniper-Local-User-Name         1
string
ATTRIBUTE       Juniper-Allow-Commands          2
string
ATTRIBUTE       Juniper-Deny-Commands           3
string
ATTRIBUTE       Juniper-Allow-Configuration     4
string
ATTRIBUTE       Juniper-Deny-Configuration      5
string
```

```
ATTRIBUTE       Juniper-Switching-Filter            48
string

END-VENDOR      Juniper
```

EAP-MD5 supplicant configuration

With the relevant portions of the RADIUS server analyzed, the supplicant's configuration is displayed:

```
[root@client5-lnx xsupplicant-conf]# cat md5user01.conf
# This is an example configuration file for xsupplicant versions after 0.8b.
. . .
network_list = all
logfile = /var/log/xsupplicant.log
default
{
  identity = md5user01

  eap-md5 {
      username = md5user01
      password = ABCabc123
  }
}
```

The contents of the *md5user01.conf* file are pretty straightforward. A logfile location is specified for help in debugging, as the client is normally run in daemon (background) mode, and the desired EAP method, which is eap-md5, is specified. As you might expect, both the username and password are also specified in the configuration file. The network_list all statement allows the client to cache results for all networks it may encounter, and is primarily intended for roaming among a set of WAPs.

Configure RADIUS parameters

With details of the server and client settings established, you have enough to get cracking at the EX configuration that is supposed to make all of this happen. Begin with the definition of the RADIUS server and local NAS attributes that facilitate communications between the authenticator and the authentication server. Configure these parameters at the [edit access] hierarchy. The configuration options are displayed:

```
[edit access]
lab@sys-java33# set ?
Possible completions:
> address-pool          Address pool
+ apply-groups          Groups from which to inherit configuration data
+ apply-groups-except   Don't inherit configuration data from these groups
> group-profile         Group profile to use for this client
> profile               Set of attributes that define access
> radius-options        RADIUS options
> radius-server         RADIUS server configuration
[edit access]
```

Here the primary areas of concern are the `radius-server` stanza, where you define the specifics needed to access a given server, and the `profile` stanza, where you create an authentication profile that in turn evokes one or more of the servers you have defined. In the 9.5 release, the `radius-options` stanza is used to specify the `revert-interval`, which controls how long the local NAS waits before reverting back to a primary RADIUS server that previously was deemed to be unreachable or unresponsive. This functionality is not needed here; therefore, focus is on the `radius-server` stanza.

A working RADIUS definition is shown:

```
[edit access]
lab@sys-java33# show radius-server
212.1.0.2 {
    secret "$9$ZwDHmz39O1hfT1hSr8LGDi"; ## SECRET-DATA
    retry 5;
    source-address 212.0.1.3;
}
```

The main point to note here is the specification of the RADIUS server via its IP address, along with the corresponding secret used to authenticate the local NAS to that server. In this case, the MD5 hash of the password `juniper` is displayed; JUNOS never displays passwords in clear text, but you can load such a password into another configuration in its MD5 hashed form. Note that this password matches the one in the RADIUS server for this authenticator (NAS) shown previously; so far, so good.

The `retry` setting determines how many requests are sent before the server is considered unresponsive. The `source-address` statement is used to ensure that all RADIUS requests originate from the 212.0.1.3 address, which is assigned to the `vlan.1001` RVI. Note that a source address setting can be critical for proper operation. The RADIUS server is typically configured with a list of allowed NAS clients in the form of their source addresses, so sourcing requests from the wrong address makes it appear that the packet originated from an unauthorized NAS, resulting in the inability to authenticate users.

You move on to configure an authentication profile named `auth` that defines RADIUS as the accounting and authorization method:

```
[edit access]
lab@sys-java33# show profile auth
authentication-order radius;
radius {
    authentication-server 212.1.0.2;
    accounting-server 212.1.0.2;
}
accounting {
    order radius;
}
```

The `auth` profile also specifies a list of authentication and, if desired, accounting servers. Each server is tried in the order listed, according to the specifics defined for that server at the `[edit access radius-server]` hierarchy, as described previously.

Currently, EX platforms require RADIUS for 802.1X authentication and authorization. Additional server types may be supported in future releases.

With the authentication server housework now complete, you can move on to 802.1X configuration proper.

Configure 802.1X authenticator properties

802.1X authenticator properties are configured at the [edit protocols dot1x] hierarchy. Once again, here are the primary configuration options:

```
[edit protocols dot1x]
lab@sys-java33# set ?
Possible completions:
+ apply-groups          Groups from which to inherit configuration data
+ apply-groups-except   Don't inherit configuration data from these groups
> authenticator         802.1X authenticator options
> traceoptions          Trace options for 802.1X
[edit protocols dot1x]
```

The options allow you to configure 802.1X authenticator properties and tracing for the same. The authenticator options are displayed next:

```
[edit protocols dot1x]
lab@sys-java33# set authenticator ?
Possible completions:
  <[Enter]>             Execute this command
+ apply-groups          Groups from which to inherit configuration data
+ apply-groups-except   Don't inherit configuration data from these groups
  authentication-profile-name  Access profile name to use for authentication
> interface             802.1X interface specific options
> static                Static MAC configuration needed to bypass 802.1X
  |                     Pipe through a command
[edit protocols dot1x]
```

Here, the options include the authentication-profile-name, which is used to link to an authentication profile defined at the [edit access] hierarchy; the interface stanza, which is where specific 802.1X parameters are set; and the static hierarchy, where you configure a list of MAC addresses that are allowed to bypass 802.1X authentication. In this example, you link to the already defined auth profile, and 802.1X is enabled on the ge-0/0/8 interface. A configuration that should meet the current criteria is shown:

```
[edit protocols dot1x]
lab@sys-java33# show
authenticator {
    authentication-profile-name auth;
    interface {
        ge-0/0/8.0 {
            supplicant single-secure;
            reauthentication 180;
        }
```

```
        }
    }
```

The authenticator settings specify the interface and the authentication profile that are used for supplicants on this interface. This example sets the interface to single-secure mode, in keeping with the requirements that only one active client is permitted at any time. The default mode is single, which allows multiple clients' MAC addresses as long as at least one of them has been successfully authenticated.

Worth noting is the total lack of any EAP-method-specific settings. Such settings are not needed because the EX automatically senses and reacts to the supplicant's chosen method, assuming it's one of the methods supported, of course. Given that the EAP-MD5 method is supported, things should just work.

Verify 802.1X authentication

With the authentication server and authenticator configuration in place, it's time to fire up the supplicant and see what happens. In this case, the supplicant is evoked to run on the eth2 interface, to use the *md5user01.conf* configuration file, and to run in the foreground via the **-i**, **-c**, and **-f** flags, respectively:

```
[root@client5-lnx xsupplicant-conf]# xsupplicant -i eth2 -c md5user01.conf -f
Couldn't get encryption capabilites!
No configuration information for network "(null)" found. Using default.
Failed to authenticate eth2
Failed to authenticate eth2
. . .
```

Bum deal, and certainly not the most auspicious of beginnings for you and all things 802.1X. Given the supplicant's output, it's clear that there is a failure to authenticate. Hoping for more details, the 802.1X authentication status is displayed on the EX using a show dot1x interface command:

```
[edit]
lab@sys-java33# run show dot1x interface
802.1X Information:
Interface     Role          State          MAC address     User
ge-0/0/8.0    Authenticator Authenticating
```

The display confirms that ge-0/0/8 is in fact configured to run 802.1X, and that it's currently trying to authenticate the user. That's something, right? Sometime later the command is repeated to obtain current status:

```
[edit]
lab@sys-java33# run show dot1x interface
802.1X Information:
Interface     Role          State     MAC address         User
ge-0/0/8.0    Authenticator Held      00:50:8B:6F:60:3A   md5user01
```

While still not successfully authenticated, the display has been updated to reflect that the interface has transitioned to the held state. This means that the number of retry attempts, three by default, have expired without successful authentication. The interface remains in this state until the quiet-period elapses, at which point the authenticator again attempts to authorize the user. It's not all bad, however. There is some indication that EAP is working correctly between the EX and the supplicant, by virtue of the display correctly being updated to reflect the EAP client's name and MAC address. Use the detail switch to see additional information:

```
[edit]
lab@sys-java33# run show dot1x interface detail
ge-0/0/8.0
  Role: Authenticator
  Administrative state: Auto
  Supplicant mode: Single-Secure
  Number of retries: 3
  Quiet period: 60 seconds
  Transmit period: 30 seconds
  Mac Radius: Disabled
  Mac Radius Restrict: Disabled
  Reauthentication: Enabled
  Configured Reauthentication interval: 180 seconds
  Supplicant timeout: 30 seconds
  Server timeout: 30 seconds
  Maximum EAPOL requests: 2
  Guest VLAN member: <not configured>
  Number of connected supplicants: 1
    Supplicant: md5user01, 00:50:8B:6F:60:3A
      Operational state: Held
      Authentcation method: Radius
      Authenticated VLAN: configured
      Session Reauth interval: 0 seconds
      Reauthentication due in 0 seconds
```

The details confirm various aspects of 802.1X operation on the interface. The highlights call out key operational characteristics such as the need for periodic reauthorization and a single-secure operating mode. The Auto setting for the administration status simply indicates that 802.1X is enabled for all supported EAP modes. While useful, the details do not shed any light on the current malfunction. To help isolate where things are going astray, 802.1X tracing is added to the configuration:

```
[edit protocols dot1x]
lab@sys-java33# show traceoptions
file dot1x;
flag state;
flag dot1x-debug;
flag eapol;
```

After committing the changes, a monitor start dot1x command is issued to begin monitoring the trace file. A short time later you observe the following trace output. Some comments are interspersed to help break it up and shed light on what it all means:

```
[edit protocols dot1x]
lab@sys-java33# run monitor start dot1x

. . .
Apr  4 14:36:20.929129  EAPOL packet received on interface ge-0/0/8.0
Apr  4 14:36:20.929193  Creating background job to process EAPOL frame
```

```
Apr  4 14:36:20.929300  Entering background job to process received EAPOL frames
Apr  4 14:36:20.929340  Invoking state machine for frame received on interface ge-
0/0/8
Apr  4 14:36:20.929371  Received an EAPOL Frame...

Apr  4 14:36:20.929422  Frame is targetted to this machine...

Apr  4 14:36:20.929661  EAPOL Frame Received on Port: 104 !!!

Apr  4 14:36:20.929728  AuthHandleInEapFrame: Received MAC based Eap Frame

Apr  4 14:36:20.929782  Session Node for MAC: -508b6f-603a- Port: 104 obtained ...

Apr  4 14:36:20.929927  ASM Called with Event: RXRESPID, and State: Connecting

Apr  4 14:36:20.929978  for Port: 104, MAC: 508b6f - 603a

Apr  4 14:36:20.930023  Id: 4, SessionNode: 1d3f800

Apr  4 14:36:20.930068  ASM: Inside PnacAuthAsmRxrespConnecting

Apr  4 14:36:20.930124  TMR: Timer is deleted

Apr  4 14:36:20.930171  ASM moved to state: AUTHENTICATING !!

Apr  4 14:36:20.930221  BSM Called with Event: AUTHSTART, and State: Idle

Apr  4 14:36:20.930268  for Port: 104, MAC: 508b6f-603a

Apr  4 14:36:20.930313  Id: 4, SessionNode: 1d3f800

Apr  4 14:36:20.930375  TMR: Timer is started

Apr  4 14:36:20.930422  BSM moved to state: RESPONSE !!

Apr  4 14:36:20.930507  ASIF: Transferring Server-data to Auth Server for the user,
md5user01.

Apr  4 14:36:20.930573  SessId: 802.1x81d8001e strlen: 14
Apr  4 14:36:20.930711  Queuing message to auth client to validate mac address
0:50:8b:6f:60:3a, user md5user01 on interface ge-0/0/8.0
Apr  4 14:36:20.930818  ASIF: Radius REQUEST_ID: 3f
Apr  4 14:36:20.930853  ASIF: Tx of Server-data to Auth Server succeeded
```

In the first block, you see confirmation that an authorization process begins with receipt of an EAP Over LAN (EAPOL) frame from the client. The software tracks the related interface using the logical interface index, which is 104 in this example, as confirmed in this snippet from a show interfaces ge-0/0/8 command:

```
. . .
Logical interface ge-0/0/8.0 (Index 104) (SNMP ifIndex 159)
    Flags: SNMP-Traps 0x0 Encapsulation: ENET2
```

The trace goes on to show that the Authentication State Machine (ASM) moves into the authenticating state, and as a result constructs a message that is queued to be sent to the RADIUS server. The trace also confirms the expected username and interface name for this example. The trace output continues:

```
. . .
Apr  4 14:36:20.934947  Received Access-Challenge authentication message
Apr  4 14:36:20.934984  Invoking state machine for authentication response for mac
address 00:50:8B:6F:60:3A
Apr  4 14:36:20.935106  on intf ge-0/0/8.0

Apr  4 14:36:20.935166  ASIF: Handing over Server frame to Authenticator

Apr  4 14:36:20.935214  AUTH: Handling Server Frame

Apr  4 14:36:20.935263  SessNode got from SessIdtbl for Id 5 is : 1d3f800, Port: 104

Apr  4 14:36:20.935310  Code = 1, Id = 5, Len = 22

Apr  4 14:36:20.935385  BSM Called with Event: AREQ_RCVD, and State: Response

Apr  4 14:36:20.935439  for Port: 104, MAC: 508b6f-603a

Apr  4 14:36:20.935802  Id: 4, SessionNode: 1d3f800

Apr  4 14:36:20.935920  TMR: Timer is deleted

Apr  4 14:36:20.936004  Queuing EAPOL frame to be transmitted out on interface ge-
0/0/8
Apr  4 14:36:20.936093  EAP Req Frame Sent !!!

Apr  4 14:36:20.936163  TMR: Timer is started

Apr  4 14:36:20.936210  BSM moved to state: REQUEST !!
```

The next block shows that the authenticator received an Access Challenge Response from the RADIUS server, and that the challenge was sent as an EAP message to the supplicant over ge-0/0/8. Seeing the reply from the RADIUS server tells you that the authenticator to authentication server communications path is working, which also confirms the source-address and shared secret particulars. As a result, you can conclude that the issue *does not* relate to RADIUS server communications, which tells you a lot about what the problem *isn't*. The trace output continues:

```
. . .
Apr  4 14:36:20.937771  EAPOL packet received on interface ge-0/0/8.0
Apr  4 14:36:20.937837  Creating background job to process EAPOL frame
Apr  4 14:36:20.937944  Entering background job to process received EAPOL frames
Apr  4 14:36:20.937985  Invoking state machine for frame received on interface
ge-0/0/8

Apr  4 14:36:20.938016  Received an EAPOL Frame...

Apr  4 14:36:20.938065  Frame is targetted to this machine...

Apr  4 14:36:20.938112  EAPOL Frame Received on Port: 104 !!!
```

```
Apr  4 14:36:20.938165  AuthHandleInEapFrame: Received MAC based Eap Frame

Apr  4 14:36:20.938217  Session Node for MAC: -508b6f-603a- Port: 104 obtained ...

Apr  4 14:36:20.938269  BSM Called with Event: RXRESP, and State: Request

Apr  4 14:36:20.938392  for Port: 104, MAC: 508b6f-603a

Apr  4 14:36:20.938440  Id: 5, SessionNode: 1d3f800

Apr  4 14:36:20.938540  TMR: Timer is deleted

Apr  4 14:36:20.938608  TMR: Timer is started

Apr  4 14:36:20.938656  BSM moved to state: RESPONSE !!

Apr  4 14:36:20.938707  ASIF: Transferring Server-data to Auth Server for the user,
md5user01.

Apr  4 14:36:20.938767 SessId: 802.1x81d8001e strlen: 14
Apr  4 14:36:20.938904 Queuing message to auth client to validate mac address
0:50:8b:6f:60:3a, user md5user01 on interface ge-0/0/8.0
Apr  4 14:36:20.939017  ASIF: Radius REQUEST_ID: 40
Apr  4 14:36:20.939053  ASIF: Tx of Server-data to Auth Server succeeded
```

This block of trace confirms that the supplicants generated a Challenge Response, as indicated by the receipt of an EAP packet on the ge-0/0/8 interface. The trace also shows the authenticator packaging the supplicant's response into a RADIUS message that is successfully sent to the authentication server. Again, so far, so good. And the tracing continues:

```
. . .
Apr  4 14:36:23.944683 process_auth_reply len:2696
Apr  4 14:36:23.944717 No VLAN attributes configured
Apr  4 14:36:23.944747 Received Access-Reject authentication message
Apr  4 14:36:23.944780 pnac_apply_access_reject_vlan portnum:104
Apr  4 14:36:23.944824 Invoking state machine for authentication response for mac
address 00:50:8B:6F:60:3A
Apr  4 14:36:23.944855  on intf ge-0/0/8.0

Apr  4 14:36:23.944913  ASIF: Handing over Server frame to Authenticator

Apr  4 14:36:23.944962  AUTH: Handling Server Frame

Apr  4 14:36:23.945011  SessNode got from SessIdtbl for Id 5 is : 1d3f800, Port: 104

Apr  4 14:36:23.945140  Code = 4, Id = 5, Len = 4

Apr  4 14:36:23.945192  BSM Called with Event: AFAIL_RCVD, and State: Response

Apr  4 14:36:23.945240  for Port: 104, MAC: 508b6f-603a

Apr  4 14:36:23.945286  Id: 5, SessionNode: 1d3f800

Apr  4 14:36:23.945346  TMR: Timer is deleted
```

```
Apr  4 14:36:23.945775  Queuing EAPOL frame to be transmitted out on interface ge-0/0/8

Apr  4 14:36:23.945882  EAP Frame Sent with code: 4 !!!

Apr  4 14:36:23.945935  BSM moved to state: FAIL !!

Apr  4 14:36:23.945988  BSM moved to state: IDLE !!
```

The last bit of trace output confirms that the authenticator received a negative reply from the RADIUS server. This reply is then conveyed to the supplicant as an EAP code 4 packet, to indicate the failure. Based on the trace, it seems that nothing is amiss in the basic EAP to RADIUS and RADIUS to EAP conversion process, which validates the RADIUS settings and general EAP operation on the access link. Given that there is no protocol malfunction to blame, double-check the credentials specified in the *md5user01.conf* configuration file:

```
. . .
eap-md5 {
     username = md5user01
#     username = "HURRICANE-TEST\md5user01"
     password = ABCabc123
  }
}
```

The credentials, when compared to the server's configuration, make you want to blush:

```
. . .
md5user01 Auth-Type := EAP, User-Password == "md5user01"
          Tunnel-Type = VLAN,
          Tunnel-Medium-Type = IEEE-802,
. . .
```

The two passwords do not match! The error is corrected in the supplicant's configuration and authentication is retried:

```
[root@client5-lnx xsupplicant-conf]# xsupplicant -i eth2 -c md5user01.conf -f
Couldn't get encryption capabilites!
No configuration information for network "(null)" found.  Using default.
Successfully authenticated eth2
```

The supplicant confirms that authentication has succeeded. This is confirmed back at the authenticator with a `show dot1x interface` command:

```
[edit protocols dot1x]
lab@sys-java33# run show dot1x interface
802.1X Information:
Interface     Role          State          MAC address        User
ge-0/0/8.0    Authenticator Authenticated  00:50:8B:6F:60:3A  md5user01
```

As expected, the EX confirms that the supplicant with a MAC address of 00:50:8B:6F:60:3A is in the authenticated state. Adding the `detail` switch confirms additional state, such as the need to reauthenticate and the assignment of a user VLAN and dynamic filter:

```
[edit]
lab@sys-java33# run show dot1x interface detail ge-0/0/8.0
ge-0/0/8.0
  Role: Authenticator
  Administrative state: Auto
  Supplicant mode: Single
  Number of retries: 3
  Quiet period: 60 seconds
  Transmit period: 30 seconds
  Mac Radius: Disabled
  Mac Radius Restrict: Disabled
  Reauthentication: Enabled
  Configured Reauthentication interval: 300 seconds
  Supplicant timeout: 30 seconds
  Server timeout: 30 seconds
  Maximum EAPOL requests: 2
  Guest VLAN member: <not configured>
  Number of connected supplicants: 1
    Supplicant: md5user01, 00:50:8B:6F:60:3A
      Operational state: Authenticated
      Authentcation method: Radius
      Authenticated VLAN: v1001
      Dynamic Filter: f1
      Session Reauth interval: 300 seconds
      Reauthentication due in 291 seconds
```

Looking back at the interface configuration, you note that it was already set for VLAN v1001 membership:

```
[edit]
lab@sys-java33# show interfaces ge-0/0/8
unit 0 {
    family ethernet-switching {
        vlan {
            members v1001;
        }
    }
}
```

As such, there was no VLAN reassignment as a result of the authentication in the case of this user, but such reassignment is possible. Lastly, verify that a firewall filter named f1 has been dynamically applied to the interface with a show dot1x firewall command:

```
[edit protocols dot1x]
lab@sys-java33# run show dot1x firewall
Filter name: dot1x_ge-0/0/8
Counters:
Name                                     Bytes          Packets
c1__dot1x_ge-0/0/8_00508b6f603a-f1-t1      0                0
```

The output confirms that filter assignment VSA worked to dynamically apply the pre-defined f1 firewall filter to the interface as a result of user authentication. The packet counts are zero because the related firewall's match conditions have not been met. The goal here is to demonstrate filter assignment based on authentication, so the actual filter particulars do not matter much.

```
[edit]
lab@sys-java33# show firewall
family ethernet-switching {
    filter f1 {
        term t1 {
            from {
                destination-address {
                    212.0.1.254/32;
                }
            }
            then {
                accept;
                forwarding-class pc;
                loss-priority low;
                count c1;
            }
        }
        term t2 {
            then accept;
        }
    }
}
. . .
```

Note that VSA assigned filters must belong to the `ethernet-switching` family because they are ultimately applied to an interface operating in Layer 2 mode. These results confirm that 802.1X authentication is working, and they complete the verification of the IEEE 802.1X EAP-MD5-based authentication lab. The next section goes on to demonstrate configuration of MAC-based RADIUS authentication for a non-responsive host.

Configure MAC-based RADIUS authentication

In this section, you will deploy 802.1X authentication on an EX switch to support RADIUS-based MAC authentication for a host that does support the 802.1X supplicant role. The resulting configuration must meet these requirements:

- Ensure that only MAC-based authentication is permitted; no other EAP method should be possible.
- RADIUS-based authentication must be used.
- When authenticated, place the user in VLAN 1001.

Figure 9-5 shows the slightly modified test topology used for this lab.

The topology change involves the use of `eth1` at `client5-lnx` and the `ge-0/0/9` interface at EX switch `sys-java33`, and a new IP subnet value as the original 212.0.1.0/24 subnet is in use on the client's `eth2` interface (not shown). Figure 9-5 also documents the MAC address for the `eth1` interface as 00:30:48:8D:7B:53.

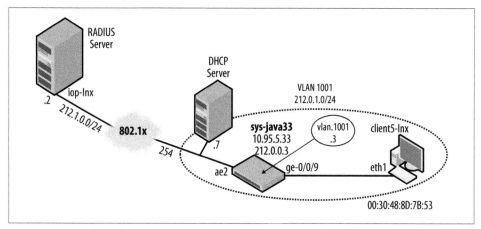

Figure 9-5. MAC-based authentication topology

The operational requirements seem simple enough. Recall that there are several ways to deal with a non-responsive host. You could simply disable 802.1X on that interface, or add the host's MAC address as being allowed to bypass 802.1X using the `static` keyword at the `[edit protocols dot1x authenticator]` hierarchy. However, these methods fail to meet the desired behavior of using the RADIUS server for the authentication; the former negates all authentication, while the latter uses a local MAC database on the switch itself to decide when a host can bypass 802.1X, and therefore both fail to achieve the server-based authentication that is required here.

MAC-based authentication makes use of RADIUS, but technically is not a defined EAP method and therefore does not require EAP in the supplicant. Unlike standard EAP methods, you must specifically enable MAC-based RADIUS authentication when desired using the `mac-radius` statement. MAC-based RADIUS authentication is not exclusive; EAP authentication exchanges are still permitted unless specifically disabled.

MAC-based authentication is normally used for hosts that do not support port 802.1X. An 802.1X-1X-capable client can cause problems with MAC-based RADIUS authentication. It's a good practice to explicitly disable any 802.1X client software on machines that you intend to authenticate via MAC address. In testing with the 9.5 release, it was found that when (a failing) EAP client was left enabled on the supplicant, the switch did not attempt to fall back to attempt a MAC-based approach, which would have succeeded if it were not fouled up by the ongoing EAP-MD5 authentication failure.

The requirements also state that you must limit the interface to MAC-based authentication *only*. Given that the first requirement forces you to enable 802.1X, and recalling that, by default, the EX will automatically react to any supported EAP method, this means you must disable EAP with the `restrict` keyword used in conjunction with `mac-radius`. This combination limits the supplicant to RADIUS-based MAC authentication only, and therefore meets the stated goals. Before making any changes, the configuration for the `ge-0/0/9` interface is displayed. Of note is the absence of any VLAN membership statement. As per requirements, the VLAN will be assigned as a result of successful authentication:

```
[edit]
lab@sys-java33# show interfaces ge-0/0/9
unit 0 {
    family ethernet-switching;
}
```

The same RADIUS server parameters are used in this scenario, so no changes are needed to the [`edit access`] hierarchy. The RADIUS server's */usr/local/etc/raddb/users* file is updated with the information needed for MAC authorization, shown here:

```
0030488d7b53 Auth-Type := EAP, User-Password == "0030488d7b53"
             Tunnel-Type = VLAN,
             Tunnel-Medium-Type = IEEE-802,
             Tunnel-Private-Group-Id = "1001"
```

Note that both the username and password are coded with the client's MAC address (in lowercase), and that the VSA `Tunnel-Private-Group-Id` correctly specifies VLAN ID 1001. The local definition of this VLAN is confirmed on the authenticator:

```
[edit]
lab@sys-java33# show vlans v1001
vlan-id 1001;
l3-interface vlan.1001;
```

Next, the supplicant's `eth1` interface is assigned an IP address and set to be operational via the up flag. The interface's newly assigned IP and burned-in MAC address are then confirmed, for the record:

```
[root@client5-lnx xsupplicant-conf]# ifconfig eth1 200.0.0.1 netmask 255.255.255.0
up

[root@client5-lnx xsupplicant-conf]# ifconfig eth1
eth1      Link encap:Ethernet  HWaddr 00:30:48:8D:7B:53
          inet addr:200.0.0.1  Bcast:200.0.0.255  Mask:255.255.255.0
          inet6 addr: fe80::230:48ff:fe8d:7b53/64 Scope:Link
          UP BROADCAST RUNNING ALLMULTI MULTICAST  MTU:1500  Metric:1
          RX packets:123 errors:0 dropped:0 overruns:0 frame:0
          TX packets:24 errors:0 dropped:0 overruns:0 carrier:0
          collisions:0 txqueuelen:1000
          RX bytes:10901 (10.6 KiB)  TX bytes:7317 (7.1 KiB)
          Interrupt:177
```

While the client displays its MAC address using uppercase letters for HEX values A–F, it was found that the RADIUS server was case-sensitive and failed to authenticate the user when the MAC address was entered with uppercase letters.

With client5-lnx's interface particulars confirmed, the server's user definition updated with the supplicant's eth1 MAC address, and the prerequisite configuration confirmed in sys-java33, you're ready to configure MAC-based authentication on the EX. The resulting dot1x stanza is displayed:

```
[edit]
lab@sys-java33# show protocols dot1x
authenticator {
    authentication-profile-name auth;
    interface {
        ge-0/0/8.0 {
            reauthentication 180;
        }
        ge-0/0/9.0 {
            supplicant multiple;
            mac-radius {
                restrict;
            }
        }
    }
}
```

The highlighted area calls out the delta. Interface ge-0/0/0 is set to support MAC-based RADIUS, as required, and is restricted to only MAC-based RADIUS via the restrict keyword. This prevents other forms of EAP authentication from working, as per requirements. In this case, the supplicant mode is set to multiple, which means that more than one client can be supported, but given the other settings each will require its own MAC-based authentication. While this behavior is not specifically called out, the requirements do not state that once one client MAC is authenticated all others should get a free pass, which is the default (single) supplicant behavior.

After committing the changes, traffic is generated at the client. The target address does not exist, so ping failures are expected. The resulting ARP traffic uses the client's SMAC address and is sufficient to trigger authentication:

```
[root@client5-lnx xsupplicant-conf]# ping 200.0.0.2
PING 200.0.0.2 (200.0.0.2) 56(84) bytes of data.
...

^c
--- 200.0.0.2 ping statistics ---
4 packets transmitted, 0 received, 100% packet loss, time 2999ms
```

Recall that in testing it was found that the EX stopped at the first RADIUS reject and did not attempt alternative authentication schemes. As a result, leaving an EAP-MD5 supplicant with improper authentication running on a port enabled for *both* EAP and MAC-based authentication resulted in EAP failure (as expected) and a disabled interface. The MAC approach was never attempted until EAP was disabled on the client.

With traffic stimulus in effect, the authentication status is verified:

```
[edit]
lab@sys-java33# run show dot1x interface
802.1X Information:
Interface    Role           State          MAC address         User
ge-0/0/15.0  Authenticator  Connecting
ge-0/0/17.0  Authenticator  Connecting
ge-0/0/8.0   Authenticator  Connecting
ge-0/0/9.0   Authenticator  Authenticated  00:30:48:8D:7B:53   0030488d7b53
```

The display confirms successful MAC-based authentication, as indicated by a username that equals the MAC address. A `show vlans` command confirms the resulting dynamic VLAN assignment:

```
[edit]
lab@sys-java33# run show vlans 1001
Name          Tag    Interfaces
v1001         1001
                     ae0.0*, ae1.0, ae2.0*, ae3.0, ge-0/0/8.0*, ge-0/0/9.0*, ge-
0/0/16.0, ge-0/0/17.0, ge-0/0/18.0
```

If you suspect a malfunction and want to start things fresh, issue a `restart dot1x-protocol` operational mode command to restart the 802.1X daemon. Use this with caution, as this globally resets all sessions. Alternatively, use the `clear dot1x interface <interface-name>` command to clear targeted sessions. Both of these commands reset hold timers and are useful for expediting reauthentication attempts after some change is made and you are impatient to see whether it worked. A `restart` example is provided:

```
[edit protocols dot1x authenticator interface ge-0/0/8.0]
lab@sys-java33# run show dot1x interface
802.1X Information:
Interface    Role           State          MAC address         User
. . .
ge-0/09.0    Authenticator  Authenticated  00:30:48:8D:7B:53   0030488d7b53

[edit protocols dot1x authenticator interface ge-0/0/8.0]
lab@sys-java33# run restart dot1x-protocol
Port based Network Access Control started, pid 7085

[edit protocols dot1x authenticator interface ge-0/0/8.0]
lab@sys-java33# run show dot1x interface
802.1X Information:
Interface    Role           State          MAC address         User
```

```
. . .
ge-0/0/9.0    Authenticator  Connecting
```

This example shows that as a result of an 802.1X daemon restart, a formerly authenticated client is transitioned back to the connection state, which forces a reauthentication attempt.

The successful RADIUS-based authentication of a non-responsive host concludes the 802.1X configuration and verification lab.

802.1X Port-Based Authentication Summary

EX platforms provide wide-ranging support for the IEEE 802.1X standard and offer both port- and MAC-level access control to a switched network. In addition to various EAP methods, you can support non-802.1X hosts using either local or RADIUS-based MAC authentication.

Support is offered for single or multiple supplicants per port, with the ability to individually authenticate each or allow others to ride on the coattails of the first authorization. EX switches also support VSAs for dynamic firewall or VLAN assignment, to include a guest VLAN concept that safely partitions users who fail authentication into a specific VLAN with limited access.

Conclusion

EX switches offer a variety of Layer 2 security and port-level access controls. These features help to ensure that only authorized users can access secured portions of your network, and also guard against common attacks such as unauthorized DHCP services, ARP poisoning, and IP address spoofing. When combined with Layer 2 security, built-in Layer 3 firewall capabilities, and general JUNOS software robustness, it is clear that you can deploy a hardened Layer 2/Layer 3 network based strictly on EX platforms (and a RADIUS server, if desired).

Users who require deep packet inspection for real-time antivirus or intrusion detection and prevention, or who need stateful services such as NAT or IP Security, will need to augment their EX switches with other Juniper products that are designed for sophisticated IP services or security-related functions.

Chapter Review Questions

1. Which is true regarding MAC limiting?
    a. It is currently not supported because not learning all MAC addresses breaks bridging
    b. It can be based on an allowed list of MACs on a per-port basis
    c. It can be based on an allowed MAC number, per port

d. It is set on a VLAN basis, not on a port basis

e. Both B and C

2. Which is true about ARP inspection?

   a. It requires a DHCP snooping database

   b. It can operate without DHCP snooping

   c. It prevents IP address spoofing

   d. It is configured at the port rather than VLAN level

3. True or false: by default, all access links are considered trusted for DHCP.

4. Which of the following are configured on a VLAN basis?

   a. ARP inspection

   b. DHCP snooping

   c. IP Source Guard

   d. All of the above

5. Which of the following EAP methods are certificate-based?

   a. MAC

   b. EAP-MD5

   c. EAP-TLS

   d. All of the above

6. With regard to MAC authentication, which is true?

   a. Setting MAC-based authorization automatically disables EAP

   b. Local authentication is mandatory due to the lack of EAP on a non-responsive client

   c. Local authentication is always based on per-port databases

   d. Local authentication is global and applies to all 802.1X-1X-enabled interfaces unless constrained with the `interface` keyword

7. Which is the default supplicant mode?

   a. Single

   b. Single-secure

   c. Multiple

   d. None of the above

8. What is the difference between single and multiple supplicant modes?

   a. Single supports only one client at any time

   b. Multiple supports multiple clients but only when each has been authenticated

   c. Multiple supports multiple clients but requires only one to be authenticated

   d. Both A and B

9. What command clears EAP sessions on an interface?

   a. `clear eap interface <interface-name>`

   b. `clear dot1x interface <interface-name>`

   c. `clear interface dot1x <interface-name>`

   d. `restart dot1x-protocol`

10. What option allows you to move a supplicant to a particular VLAN when the RADIUS server is unreachable?

   a. `guest-vlan`

   b. `default-vlan`

   c. `native-vlan`

   d. `server-fail`

Chapter Review Answers

1. Answer: E. MAC limits on the EX are port-level settings, not VLAN-level settings. Failing to learn a MAC can result in flooding within the VLAN for that MAC, but this is not precluded by any specification, and you could choose to disable the port upon a violation if flooding is an issue. By limiting the number of MACs, you prevent MAC table overflow and the potential for flooding of previously learned addresses, which could pose a security or performance threat.

2. Answer: A. ARP inspection is based on the information contained in the DHCP snooping database. It prevents ARP spoofing, not IP address spoofing. The latter is still possible with a static ARP entry that would bypass the need for ARP. IP Source Guard prevents IP address spoofing.

3. Answer: False. Access links are untrusted for DHCP by default. Trunk interfaces are trusted by default. Use the `dhcp-trusted` setting at the `[edit ethernet-switch ing-options secure-access-port interface <interface-name>]` hierarchy to permit a DHCP server on an access link when you enable the DHCP snooping feature.

4. Answer: D. All of the features listed are set on a VLAN basis at the `[edit ethernet-switching-options secure-access-port vlan <vlan-name>]` hierarchy.

5. Answer: C. Of the options listed, only EAP-TLS is certificate-based. In fact, MAC authentication is not an EAP method at all!

6. Answer: D. MAC authorization can be local or via a RADIUS server and neither method disables EAP methods on the related interface. The former bypasses 802.1X but allows other clients to use it, while the latter permits EAP unless the `restrict` keyword is also added. When using local authentication, 802.1X is bypassed for matching MACs on any 802.1X-enabled interface, unless you choose to specify a list of allowed interfaces.

7. Answer: A. The default mode is single, which permits multiple MAC addresses once a single supplicant has been authorized.

8. Answer: B. Only B is correct. Single permits multiple clients but requires that only one authenticate. Multiple mode requires each MAC to authenticate to provide fine-grained control. The single-secure mode limits the interface to a single supplicant.

9. Answer: C. Only C is correct. Restarting the `dot1x` process clears sessions on all interfaces. Options A and B are not even real commands.

10. Answer: D. The `server-fail` options define how supplicants are handled when the server becomes unreachable. A guest VLAN is a different concept and is used for clients that fail to authenticate, but this implies that a RADIUS REJECT has been received, so it does not qualify as a non-responsive server. The concepts of default and native VLAN do not relate to access control.

IP Telephony

This chapter covers *IP telephony*, which involves using IP to replace the traditional private branch exchange (PBX) architecture, as well as extending IP all the way to the handset. This chapter does *not* cover Voice over IP (VoIP). This distinction may initially seem contradictory, so let's define these terms. Voice over IP is the process for digitizing your voice over a WAN connection in order to save costs. IP telephony is actually changing the phone system in order to leverage IP protocols to enhance productivity, usually in the form of IP phones and call managers. So, VoIP is actually a subset of IP telephony, and a full IP telephony system will use VoIP concepts and protocols.

In order to deploy an IP telephony system, you need to configure your routers, switches, phones, and call managers. This chapter focuses on the necessary switch configuration and discusses the process for making a phone call over the LAN, as opposed to a LAN-to-WAN phone call. The standards protocols for EX switches, as well as JUNOS software features, will also be examined.

The topics covered in this chapter include:

- Deployment scenarios
- Power over Ethernet (PoE)
- Link Layer Discovery Protocol (LLDP)
- Link Layer Discovery Protocol-Media Endpoint Discovery (LLDP-MED)
- Voice virtual LANs (VLANs)
- Case studies

Deployment Scenarios

An IP phone can be deployed in several ways. Figure 10-1 shows the most common scenario, in which the end station and phone attach to the same switch port. The phone acts as a simple switch to connect the end host to the switch. The advantage of this type of deployment is a high cost savings on switch ports. The disadvantage is possible degradation of the voice call due to the bursty nature of the data traffic coming from the end host. The problem compounds when multiple end hosts connect via the same IP phone.

In order to circumvent this issue, voice traffic should be prioritized over data traffic to ensure proper throughput, delay, and jitter. The most common way to accomplish this is to separate the voice traffic and data traffic into different broadcast domains or VLANs. You can do this by setting up a separate VLAN for the phone, called a *voice VLAN*, and configuring the phone with the proper voice VLAN. The end host normally sends untagged frames to the switch, which can reduce contention on the switch itself by giving voice a higher priority, but does not prevent a large burst from affecting the call.

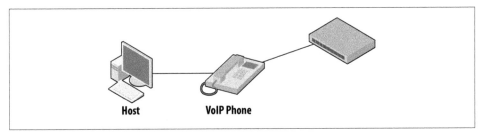

Figure 10-1. Typical IP telephony deployment

Figure 10-2 shows a deployment scenario that can eliminate any chance that a data burst will affect the call. This deployment allows the phone and the end host to each have their own physical port on the switch. In this case, the host and the phone are assigned into separate VLANs on the switch, which requires no configuration on the phone or host, as they are automatically in the correct domain due to the separate physical connection. For example, the switch can have an access VLAN of 100 configured for the phone, and an access VLAN of 200 for the host.

Because of cost reduction design approaches that turn up in many networks, this configuration is not common in enterprises today. Therefore, an alternative approach is to place the phone and host in a voice VLAN and separate the traffic based on a tagged or untagged frame. In other words, if the traffic is received from the phone as tagged, it is placed in the voice VLAN; untagged traffic is placed in the data VLAN. This may or may not require VLAN configuration of the IP phone, depending on the level of LLDP-MED support (discussed later in this chapter).

QoS or CoS?

It is recommended that you deploy class of service (CoS) in your network in order to achieve the proper quality of service (QoS) and prioritization for your voice calls. Even if traffic is separated on the access switch, there could still be congestion on the distribution, or on the core switch and router links.

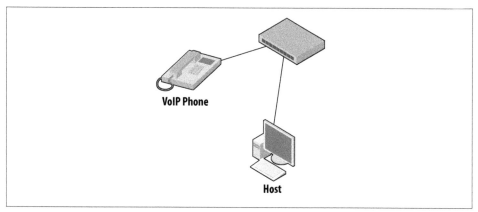

VoIP Phone

Host

Figure 10-2. Dedicated port deployment

 This chapter provides only a brief overview of CoS. For a detailed discussion of CoS, see Chapter 9 of *JUNOS Enterprise Routing*, by Doug Marschke and Harry Reynolds (O'Reilly).

It is very common to hear the terms *class of service* and *quality of service* used interchangeably. This chapter reserves the term *QoS* for individual network parameters, such as delay or loss probability, and uses *CoS* in a more granular fashion. CoS is the combined effect of applying specific QoS parameters to a packet stream, which should result in a service differentiation among the supported traffic classes in your network.

For example, consider going out to a club that has "VIP" or "table" service that is going to differentiate you from the normal clubgoers, who get all dressed up but then have to wait in line to get in—and actually go to the bar to get a drink. The club makes money by charging extra for VIP service; you're happier because the wait time to get into the club (delay) is decreased; and the table service allows you to consistently refill your expensive but bottomless drink (jitter). In the same way, QoS parameters help to achieve a certain CoS (VIP or not), and the bouncers make it their job to enforce these classes.

Here are the primary network QoS parameters:

Bandwidth
> Bandwidth is a measure of each link's information-carrying capacity, and is limited by the lesser bandwidth amount supported between two endpoints.

Delay
> Delay is a measure of the time taken to move a packet from one point to another. End-to-end delay is a cumulative function of serialization delays, propagation delays, and any queuing delays (buffering) that the packet may experience.

Delay variation (jitter)

Delay variation, often called jitter, is a measure of the variance in transfer delays between packets that make up a stream. Jitter is significant to real-time applications because the receiver must dimension its jitter buffer based on maximum jitter, which adds delays for all packets, and eventual loss when jitter values exceed buffer capacity.

Loss

Loss measures the percentage of packets not delivered. Loss can stem from transmission errors or can be due to discards stemming from congestion in packet-based networks, or from packets that are received out of order.

Loss pattern

The loss pattern defines the nature of a loss event as either bursty (short duration) or chronic, which is sometimes called a dribble error.

After deciding on QoS parameters, CoS is deployed on a switch using a variety of tools.

Figure 10-3 provides a big-picture view of the CoS processing stages associated with the EX Series switch.

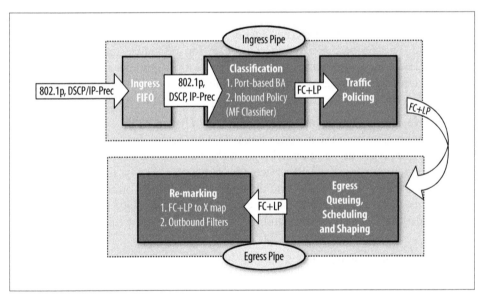

Figure 10-3. EX Series CoS processing stages

Our discussion begins with a packet arriving at the ingress interface. Here's a description of the operation and general capabilities of each CoS stage that a packet encounters as it makes its way from ingress to egress:

- Ingress CoS processing:

  Behavior aggregate classification
  > Packets arriving at the router are first subjected to the behavior aggregate (BA) classification stage. This stage sets the forwarding class (FC) and packet loss priority (PLP) using any of the supported BA classifier types, to include IP precedence, DiffServ Code Points (DSCPs), IEEE 802.1p, and so on.

  Multifield classification
  > The next processing stage is multifield classification. Here a firewall filter can be defined to match against numerous packet fields, incoming interfaces, and so on, to either set the FC and PLP or override the values set during previous BA classification.

  Ingress policing
  > When desired, a firewall policer can be applied to limit matching traffic, either by discard, by reclassification, or by marking excess with a loss priority of high (meaning in the event of congestion, a random early detection [RED] profile can be used to more aggressively drop PLP high traffic).

- Egress CoS processing:

  Rewrite marker
  > The rewrite marker stage allows you to alter one, or in some cases multiple, packet fields, as the packet is transmitted to downstream nodes. Normally you rewrite packet fields to accommodate downstream BA-based classification. Rewrite markers are indexed by protocol family and by FC—for example, writing a 001 pattern into the precedence field of all `family inet` packets that are classified as Best Effort.

  Queuing and scheduling
  > The queuing stage involves placing packet notifications into the corresponding FC queue, where they are serviced by a scheduler that factors priority and configured weight to determine when a packet should be dequeued from a given queue.

  RED/congestion control
  > The final CoS processing stage involves a *weighted RED drop decision*, based on protocol, loss priority, and average queue fill level. RED tends to operate at the head of the queue, and a RED decision is made against each packet selected for transmission at the scheduler stage.

Table 10-1 describes some of the CoS features in an EX Series switch. This chapter uses some of these features, but does not contain a full CoS discussion.

Table 10-1. EX Series CoS features

Feature	Description	Notes
BA classification	Classify packets based on DSCP, IP Precedence, or 802.1p	Can assign a port to be in trust or untrust mode
FCs	Map streams of packets to 1 of 16 FCs	Default of four: Best Effort, Assured Forwarding, Expedited Forwarding, or Network Control
Traffic policing	Limit transmission rate of a class of traffic	
Queuing and scheduling	Place a packet into a queue and transmit	Eight egress queues per port
		Strict priority queues or shaped deficit-weighted round-robin queues can be configured
Re-marking	Rewrite outgoing packet CoS values	802.1p or DSCP

Deployment Scenarios Summary

This section looked at the deployment scenarios and the QoS mechanisms that should be used in an IP telephony network. The most cost-effective model is to have data and voice in the same port, and in order to avoid packet loss due to bursty traffic CoS prioritization and scheduling are utilized. Next, let's take a look at some more tools that can be used in the IP telephony network, starting with Power over Ethernet.

Power over Ethernet

We discussed PoE capabilities for different EX Series platforms in Chapter 2. Recall that PoE is a standard defined in 802.3af that delivers power over standard copper Ethernet cable that is also delivering data. This technology allows various devices such as Wireless Access Points (WAPs), video cameras, and point-of-sale devices to be powered. Most importantly for this chapter, PoE supplies power for many IP phones.

PoE defines two type of devices: *power sourcing equipment* (PSE) and *powered devices* (PDs). For example, a PSE is a switch that provides power to a PD such as an IP phone.

The PoE standard defines different classes of attached devices that can require different levels of power. This allows for better utilization and distribution of power capacity. The classes are:

Class 0
    15.4 watts reserved
Class 1
    4 watts reserved
Class 2
    7 watts reserved
Class 3
    15.4 watts reserved

Class 4
> Reserved for future expansion

 Line loss can reduce the total amount of received power by up to 15% or 16%.

The default class is 0, and is used for any device that does not have a class defined. Also of interest is the fact that Class 0 and Class 3 both have 15.4 watts reserved. The primary difference is that only a device that supports 802.3af can be a Class 3 PD, whereas any device can be a Class 0 PD.

When configuring a PoE device, you may have a limited amount of total system power and want to distribute that power accordingly. This is often referred to as *power management* or a *power budget*. There are several ways to achieve power management:

- Statically define the amount of power on a given port.
- Dynamically have the system budget for the power actually consumed on the port.
- Allocate power on a port based on the class assignment (0 through 4).
- Allow PoE to run on only certain ports.

JUNOS Support for PoE

EX Series switch models provide either eight, 24, or 48 PoE ports while on AC power. You can extend the total number of PoE ports for an EX Series switch by inserting additional PoE cards if the power supply supports it.

There are three flavors of power supplies:

- 320-watt power supply, which supports eight PoE ports
- 600-watt power supply, which supports 24 PoE ports
- 930-watt power supply, which supports 48 PoE ports

However, if there is a switch that supports only eight PoE ports for the entire system, a higher-watt power supply will not increase the number of PoE ports. Also, if there is a mix of power supplies, the lowest watt value is always used by the system or in a Virtual Chassis (VC) by the individual Line Card (LC).

To verify the power supply and PoE ports, issue a show chassis hardware command. In this example, Ethanol has a 3,200-watt power supply and supports eight PoE ports:

```
lab@Ethanol> show chassis hardware
Hardware inventory:
Item            Version  Part number  Serial number   Description
Chassis                               BH0208188249    EX3200-24T
FPC 0           REV 07   750-021261   BH0208188249    EX3200-24T, 8 POE
```

```
    CPU                          BUILTIN      BUILTIN        FPC CPU
    PIC 0                        BUILTIN      BUILTIN        24x 10/100/1000 Base-T
    Power Supply 0    REV 02     740-020957   ATO508131072   PS 320W AC
    Fan Tray                                                 Fan Tray
```

Starting in JUNOS 9.2, you can configure power management as either *static* or *class-based*, as discussed in the previous section. In class-based power management, the maximum power is set by the class of the device. The default is static management, in which each interface is allocated 15.4 watts as maximum power; however, you can change this value with the maximum-power command:

```
lab@Ethanol# set poe interface ge-0/0/0 ?
Possible completions:
  <[Enter]>            Execute this command
+ apply-groups         Groups from which to inherit configuration data
+ apply-groups-except  Don't inherit configuration data from these groups
  disable             Disable PoE interface
  maximum-power        Maximum power (0..15.4 watts)
  priority             Priority options
> telemetries          Telemetries settings
  |                    Pipe through a command
[edit]
```

You can also set priority values for an interface of either high or low for cases where there is not enough power to supply all the ports. In this situation, a high-priority port will stay active over a lower-priority port that could be shut down. So, the most important devices—security devices, emergency phones, and so on—should receive high priority:

```
lab@Ethanol# set poe interface ge-0/0/0 priority ?
Possible completions:
  high                 High priority
  low                  Low priority
```

The switch contains a controller that regulates the power for all the PoE ports; you can view it with the show poe controller command. In this example, Ethanol has 130 watts of power to use, and is using only 3 watts:

```
lab@Ethanol> show poe controller

Controller  Maximum  Power          Guard band  Management
index       power    consumption
    0       130 W    3W             0W          Static
```

All PoE interfaces can be examined:

```
lab@Ethanol> show poe interface
Interface Admin status Oper status Max power Priority Power consumption  Class
  ge-0/0/0 Enabled     ON          15.4W     Low      3.6W               0
  ge-0/0/1 Enabled     OFF         15.4W     Low      0.0W               0
  ge-0/0/2 Enabled     OFF         15.4W     Low      0.0W               0
  ge-0/0/3 Enabled     OFF         15.4W     Low      0.0W               0
  ge-0/0/4 Enabled     OFF         15.4W     Low      0.0W               0
  ge-0/0/5 Enabled     OFF         15.4W     Low      0.0W               0
```

```
ge-0/0/6 Enabled      OFF        15.4W    Low    0.0W              0
ge-0/0/7 Enabled      OFF        15.4W    Low    0.0W              0
```

An individual PoE interface can be examined to view power consumption:

```
lab@Ethanol> show poe interface ge-0/0/0
PoE interface status:
PoE interface            : ge-0/0/0
Administrative status    : Enabled
Operational status       :   ON
Power limit on the interface : 15.4W
Priority                 :  Low
Power consumed           : 3.5W
Class of power device    :   0
```

PoE interfaces can be difficult to troubleshoot: either the device works or it doesn't. However, you can configure telemetry to show the history of an interface's power consumption. In the following code, Ethanol is configured to poll power data every minute, and to keep a log of entries for three hours, resulting in 180 total entries:

```
lab@Ethanol> show configuration poe
interface all {
    telemetries {
        interval 1;
        duration 3;
    }
}
```

In this case, a constant 3.5 watts is recorded every minute:

```
lab@Ethanol> show poe telemetries interface ge-0/0/0 all

Sl No    Timestamp                  Power    Voltage
  1      08-27-2008 17:51:25 UTC    3.5W     50.8V
  2      08-27-2008 17:50:25 UTC    3.5W     50.8V
  3      08-27-2008 17:49:25 UTC    3.5W     50.8V
  4      08-27-2008 17:48:25 UTC    3.5W     50.8V
  5      08-27-2008 17:47:25 UTC    3.6W     50.8V
  6      08-27-2008 17:46:25 UTC    3.5W     50.8V
  7      08-27-2008 17:45:25 UTC    3.6W     50.8V
  8      08-27-2008 17:44:25 UTC    3.5W     50.8V
  9      08-27-2008 17:43:25 UTC    3.5W     50.8V
 10      08-27-2008 17:42:25 UTC    3.5W     50.8V
 11      08-27-2008 17:41:25 UTC    3.5W     50.8V
```

PoE Summary

PoE is a fantastic tool that you can use for a variety of devices, including IP phones. The support of the end device will be an important driver if PoE is successful and useful in your network. However, with the port cost savings, it's wise to purchase phones with PoE support. Next, let's examine Link Layer Discovery Protocol, a concept that is probably very familiar to most people, though perhaps by a different name.

Link Layer Discovery Protocol

LLDP is a standards-based (IEEE 802.1AD) Layer 2 protocol that allows network devices to advertise their major capabilities on the same LAN segment. This protocol was modeled after proprietary protocols such as Cisco Discovery Protocol (CDP), Extreme Discovery Protocol (EDP), and the little-used Nortel Discovery Protocol (NDP). The information that is distributed is stored by its recipients in a standard Management Information Base (MIB) that can be captured by Simple Network Management Protocol (SNMP) for later analysis or topology creation.

LLDP is a one-way protocol, meaning that an LLDP agent can send and receive information, but can never solicit information. Also, LLDP operates over a Layer 2 or Layer 3 interface in untagged mode, which enables the protocol to build a topology regardless of any VLAN configuration.

LLDP sends its frames to a link local multicast address of 01-80-C2-00-00-OE, so frames are never relayed to another connected neighbor. Figure 10-4 shows the LLDP frame.

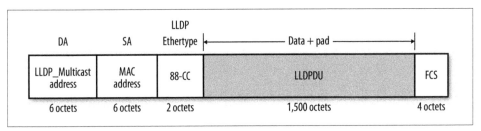

Figure 10-4. LLDP frame

The LLDP Protocol Data Unit (PDU) contains device capabilities in the form of *type length values* (TLVs), shown in Figure 10-5.

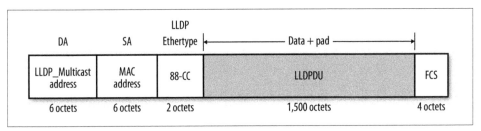

Figure 10-5. LLDP TLVs

Here are the four mandatory TLVs:

Chassis ID
> Identifies the chassis end station. This can be identified in several ways, depending on the chassis subtype, but the most common value is the Media Access Control (MAC) address of the transmitting station.

Port ID

Identifies the port component of the transmitting LLDP agent. As with the chassis, a port can be identified in several ways. The most common value is the name of the interface.

Time to Live (TTL)

Identifies the number of seconds that the recipient LLDP agent is to store the information. If this value is 0, it indicates that information associated with this system should be deleted.

End of LLDD-PDU

A 2-octet, all-zero TLV that marks the end of the TLV sequence.

Some optional TLVs may also be included in the LLDP PDU:

Port description

Advertises the port description, which is an alphanumeric string with a maximum of 256 characters.

System name

The name of the system, which is an alphanumeric string with a maximum of 256 characters. This is usually referred to as the *hostname*.

System description

Contains information about the software and current version.

System capabilities

Identifies the primary functions of the system and whether these primary functions are enabled. Table 10-2 shows the possible capabilities.

Management address

Contains the IPv4 management address of the system.

Organizational-specific TLVs

Contain specific information defined by an organization such as IEEE or IETF, or by a vendor. For example, VLAN IDs or maximum transmission units (MTUs) may be advertised.

Table 10-2. System capabilities

Bit	Capability	Reference
0	Other	N/A
1	Repeater	RFC 2108
2	Bridge	RFC 2674
3	WLAN access point	802.11
4	Router	RFC 1812
5	Telephone	RFC 2011
6	DOCSIS device	RFCs 2669 and 2670
7	End station only	RFC 2011

Bit	Capability	Reference
8–15	Reserved	N/A

LLDP sends periodic updates of these messages (the default is 30 seconds) to neighboring devices. Also, if any information needs to be updated on the device, a "triggered" update can be sent before the default transmit time. No authentication or acknowledgment is defined in LLDP messages. LLDP messages are held in the receiver's database for a time of:

msgTxInterval × msgTxHold

The default in JUNOS software is 30 × 4, or 120 seconds.

JUNOS LLDP

To configure LLDP in JUNOS, enable an interface under [edit protocols lldp]:

```
lab@Ethanol# set protocols lldp ?
Possible completions:
  advertisement-interval  Transmit interval for LLDP messages (5..32768 seconds)
+ apply-groups            Groups from which to inherit configuration data
+ apply-groups-except     Don't inherit configuration data from these groups
  disable                 Disable LLDP
  hold-multiplier         Hold timer interval for LLDP messages (2..10 seconds)
> interface               Interface configuration
  lldp-configuration-notification-interval  Time interval for LLDP notification
  ptopo-configuration-maximum-hold-time  Hold time for physical topology
connection entries
  ptopo-configuration-trap-interval  Interval for physical topology configuration
change trap
> traceoptions            Trace options for LLDP
  transmit-delay          Transmit delay time interval for LLDP messages (seconds)
```

The default configuration has all the interfaces enabled, as traceoptions can also be configured for troubleshooting:

```
lab@Ethanol> show configuration protocols lldp
traceoptions {
    file lldp.log;
    flag all;
}
interface all;
```

A variety of show lldp commands are available. The first command that you should run is the show lldp neighbors command. Ethanol has two neighbors in multiple interfaces in Brandy and Bourbon. Since the port info has to be unique per LLDP message, you see the option of the name of the interface or the SNMP index number for multiple interfaces between neighbors:

```
lab@Ethanol> show lldp neighbors
LocalInterface Chassis Id         Port info    System Name
ge-0/0/2.0     00:1f:12:30:96:80  ge-0/0/5.0   Brandy
```

```
ge-0/0/12.0    00:1f:12:30:96:80  142              Brandy
ge-0/0/8.0     00:1f:12:30:96:80  156              Brandy
ge-0/0/9.0     00:1f:12:30:96:80  157              Brandy
ge-0/0/3.0     00:1f:12:30:c4:80  ge-0/0/5.0       Bourbon
ge-0/0/12.0    00:1f:12:30:c4:80  143              Bourbon
```

You can view associated timers and supported TLVs with the `detail` switch. Notice some of the additional 802 and MED TLVs that are supported in this system:

```
lab@Ethanol> show lldp detail

LLDP                     : Enabled
Advertisement interval   : 30 seconds
Transmit delay           : 2 seconds
Hold timer               : 4 seconds
Notification interval    : 5 Second(s)
Config Trap Interval     : 60 seconds
Connection Hold timer    : 300 seconds

LLDP MED                 : Enabled
MED fast start count     : 3 Packets

Interface    LLDP       LLDP-MED    Neighbor count
all          Enabled    Enabled     6

Interface    Vlan-id    Vlan-name
ge-0/0/2.0   0          default
ge-0/0/3.0   0          default
ge-0/0/8.0   0          default
ge-0/0/9.0   0          default
ge-0/0/12.0  0          default

LLDP basic TLVs supported:
Chassis identifier, Port identifier, Port description, System name, System
description, System capabilities, Management address.

Supported LLDP 802 TLVs:
Power via MDI, Link aggregation, Maximum frame size, Port VLAN tag, Port
VLAN name.

Supported LLDP MED TLVs:
LLDP MED capabilities, Network policy, Endpoint location, Extended power
Via MDI.
```

To examine the local information that is placed in the TLVs, use the `local-information` switch:

```
lab@Ethanol> show lldp local-information

LLDP Local MIB details

Chassis ID   : 00:1f:12:30:98:40
System name  : Ethanol
System descr : Juniper Networks, Inc. ex3200-24t Espresso, version 9.2R1.9
               Build date: 2008-08-05 07:46:56 UTC
```

```
Interface name        Interface ID      Interface description
me0.0                 34                me0.0
ge-0/0/2.0            113               ge-0/0/2.0
ge-0/0/3.0            115               ge-0/0/3.0
ge-0/0/8.0            142               ge-0/0/8.0
ge-0/0/9.0            143               ge-0/0/9.0
ge-0/0/12.0           146               ge-0/0/12.0
```

All of this information is placed into MIBs that can be retrieved via SNMP. These MIBs include the local MIB, neighbors MIB, neighbors management MIB, and neighbors unknown MIB. Here's an example of the neighbors MIB shown from Ethanol:

```
lab@Ethanol> show lldp neighbors-mib
chasidsubtyp  portidsubtyp                          remote system description
capsup  capenab
4             7           Juniper Networks, Inc. ex3200-24t Espresso, version
9.2R1.9 Build date: 2008-08-05 07:46:56 UTC
4             7           Juniper Networks, Inc. ex3200-24t Espresso, version
9.2R1.9 Build date: 2008-08-05 07:46:56 UTC          40      8
4             7           Juniper Networks, Inc. ex3200-24t Espresso, version
9.2R1.9 Build date: 2008-08-05 07:46:56 UTC
4             7           Juniper Networks, Inc. ex3200-24t Espresso, version
9.2R1.9 Build date: 2008-08-05 07:46:56 UTC
4             7           Juniper Networks, Inc. ex3200-24t Espresso, version
9.2R1.9 Build date: 2008-08-05 07:46:56 UTC          40      8
4             7           Juniper Networks, Inc. ex3200-24t Espresso, version
9.2R1.9 Build date: 2008-08-05 07:46:56 UTC
```

Finally, LLDP statistics can be viewed:

```
lab@Ethanol> show lldp statistics
Interface   Received  Unknown TLVs  With Errors  Transmitted  Untransmitted
me0.0       0         0             0            197          1
ge-0/0/2.0 193        0             0            195          0
ge-0/0/3.0 194        0             0            195          0
ge-0/0/8.0 193        0             0            195          0
ge-0/0/9.0 193        0             0            195          0
ge-0/0/12.0 387       0             0            23155        0
```

 The command show lldp statistics seems to have a strange field of untransmitted frames. Due to the non-descriptive nature of this field, the field has been changed to Discarded in later code.

LLDP Summary

LLDP is the standards-based approach to proprietary protocols such as CDP. LLDP can be used for inventory, and to verify the connectivity and capabilities of LLDP neighbors. LLDP by itself is a broad protocol, independent from IP telephony. However, the next section discusses some extensions to LLDP that connect its functionality with IP telephony.

LLDP with Media Endpoint Discovery

With so many IP telephony pieces and vendors available, it's often difficult to ensure that each device is configured and operating properly. This is usually done with some type of manual configuration and discovery.

LLDP comes to the rescue with Media Endpoint Discovery, which is an LLDP extension defined in TIA-1057. TIA-1057 was specifically written in order to support interoperability and discovery between VoIP devices. These devices can include IP phones, gateways, IP media servers, and so on.

LLDP-MED provides the following benefits:

Network policy discovery
    Allows endpoints to advertise VLAN IDs and DSCP settings

PoE management
    Allows endpoints to advertise their PoE requirements

Inventory management discovery
    Stores all the relevant information (software, serial number, etc.) for inventory and reporting purposes

Location discovery
    Identifies the location of the IP phone on the switch port for E-911 services

Not all vendors support LLDP-MED. Examples of IP phones that support LLDP-MED functionality include:

- Avaya 9600 series with firmware 1.2.1
- Avaya 9600 series with firmware 2.6
- Cisco phones (7609G, 7911G, 7931G, 7941G, 7945G, 7961G, 7965G, 7970G, 7971G, and 7975G)
- Nortel phones (1110, 1150, 1210, 1220, and 1230)

LLDP-MED is essentially additional TLVs that are advertised in a standard LLDP message. So, all the LLDP features discussed previously are maintained. Essentially, new TLVs are defined; here are some common LLDP-MED TLVs:

LLDP-MED capabilities
    Define the capability of the IP telephony device and endpoint class. Class 1 is all LLDP devices, Class 2 includes devices with IP media capabilities, and Class 3 is end-user IP communications, such as a phone.

IEEE 802.3 MAC/PHY configuration/status
    Advertises the LAN speed and duplex.

Extended power via MDI
    Advertises the power requirements of the device.

Network policy

Contains VLAN IDs, Layer 2 priority values, or DSCP settings, as well as the application type (see Table 10-3).

LLDP-MED location identification

Contains information for emergency call services (emergency location identification number, or ELIN).

Inventory management

Contains information such as the hardware, firmware, and software version of the devices, as well as serial number, manufacturer name, asset ID, and model name.

Table 10-3. Application types for network policy TLV

Value	Value meaning	Description
0	Reserved	Reserved for future use
1	Voice	Used for dedicated IP telephony handsets
2	Voice signaling	Used for network topologies that require a different policy for voice signaling and media
3	Guest voice	A limited feature set for guest users
4	Guest voice signaling	Same as value 2 for the guest voice capability
5	Soft phone voice	For soft phone applications such as PCs
6	Video conferencing	Used for real-time dedicated video conferencing equipment
7	Streaming video	Used for broadcast or multicast video content distribution
8	Video signaling	Used for network topologies that require a different policy for video signaling and media

Finally, LLDP-MED works in conjunction with 802.1X, discussed in Chapter 9. When 802.1X is in effect, LLDP frames are advertised and processed only when the port is authenticated. If an IP phone and a PC are connected on the same switch port, authentication can happen separately. If only the IP phone or the PC is 802.1X-capable, use single-supplicant mode.

LLDP-MED and JUNOS

A EX Series switch will automatically advertise LLDP-MED values if they receive LLDP-MED TLVs. Initially, however, only LLDP TLVs are advertised, and the EX Series device will toggle when it receives LLDP-MED TLVs (Figure 10-6).

To enable LLDP-MED, configure interfaces under [edit protocols lldp-med]:

```
lab@Ethanol# show protocols lldp-med
interface all;
```

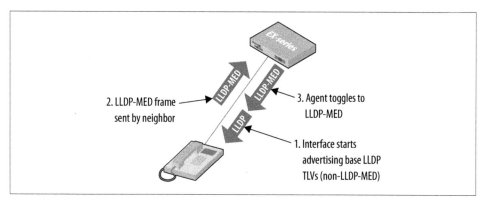

Figure 10-6. EX Series LLDP-MED behavior

To verify that LLDP-MED is enabled, use the `detail` switch:

```
lab@Ethanol> show lldp detail

LLDP                    : Enabled
Advertisement interval  : 30 seconds
Transmit delay          : 2 seconds
Hold timer              : 4 seconds
Notification interval   : 5 Second(s)
Config Trap Interval    : 60 seconds
Connection Hold timer   : 300 seconds

LLDP MED                : Enabled
MED fast start count    : 3 Packets

Interface      LLDP        LLDP-MED      Neighbor count
all            Enabled     Enabled       6

Interface    Vlan-id    Vlan-name
ge-0/0/2.0   0          default
ge-0/0/3.0   0          default
ge-0/0/8.0   0          default
ge-0/0/9.0   0          default
ge-0/0/12.0  0          default

LLDP basic TLVs supported:
Chassis identifier, Port identifier, Port description, System name, System
description, System capabilities, Management address.

Supported LLDP 802 TLVs:
Power via MDI, Link aggregation, Maximum frame size, Port VLAN tag, Port
VLAN name.

Supported LLDP MED TLVs:
LLDP MED capabilities, Network policy, Endpoint location, Extended power
Via MDI.
```

One of the few settings that currently can be configured is the frequency at which LLDP-MED advertisements are sent from the switch in the first second after it has detected an LLDP-MED device. This can be modified with the fast-start knob:

```
[edit protocols lldp-med]
lab@Ethanol# set ?
Possible completions:
+ apply-groups          Groups from which to inherit configuration data
+ apply-groups-except   Don't inherit configuration data from these groups
  disable               Disable LLDP
  fast-start            Discovery count for MED (1..10)
> interface             Interface configuration
```

You can also configure the location information that is advertised from the switch to the LLDP-MED device:

```
[edit protocols lldp-med]
lab@Ethanol# set interface all ?
Possible completions:
  <[Enter]>             Execute this command
+ apply-groups          Groups from which to inherit configuration data
+ apply-groups-except   Don't inherit configuration data from these groups
  disable               Disable LLDP
> location
  |                     Pipe through a command
```

The rest of the commands are the same as discussed in the LLDP section. Here is an example of a sample LLDP message being sent to an IP phone that was captured with the monitor traffic command:

```
07:18:17.250893 Out LLDP, length 245
        Chassis ID TLV (1), length 7
          Subtype MAC address (4): 0:1f:12:30:98:40
        Port ID TLV (2), length 4
          Subtype Local (7): 136
        Time to Live TLV (3), length 2: TTL 120s
        System Name TLV (5), length 7: Ethanol
        System Description TLV (6), length 96
          Juniper Networks, Inc. ex3200-24t Espresso, version 9.2R1.9 Build date:
2008-08-05 07:46:56 UTC
        Management Address TLV (8), length 24
          Management Address length 5, AFI IPv4 (1): 172.16.69.8
          Interface Index Interface Numbering (2): 34
          OID length 12\001\003\006\001\002\001\037\001\001\001\001"
        Port Description TLV (4), length 10: ge-0/0/0.0
        System Capabilities TLV (7), length 4
          System  Capabilities [Bridge, Router] (0x0014)
          Enabled Capabilities [Bridge] (0x0004)
        Organization specific TLV (127), length 9: OUI IEEE 802.3 Private
(0x00120f)
          Link aggregation Subtype (3)
            aggregation status [supported], aggregation port ID 0
        Organization specific TLV (127), length 6: OUI IEEE 802.3 Private
(0x00120f)
          Max frame size Subtype (4)
```

```
          MTU size 1514
        Organization specific TLV (127), length 6: OUI Ethernet bridged
(0x0080c2)
          0x0000:  0080 c201 0064
        Organization specific TLV (127), length 11: OUI Ethernet bridged
(0x0080c2)
          0x0000:  0080 c203 0000 0464 6174 61
        Organization specific TLV (127), length 12: OUI Ethernet bridged
(0x0080c2)
          0x0000:  0080 c203 0064 0576 6f69 6365
        Organization specific TLV (127), length 7: OUI ANSI/TIA
(0x0012bb)
          LLDP-MED Capabilities Subtype (1)
            Media capabilities [LLDP-MED capabilities, network policy, location
identification, extended power via MDI-PD] (0x0017)
            Device type [network connectivity] (0x04)
        Organization specific TLV (127), length 8: OUI ANSI/TIA
(0x0012bb)
          Network policy Subtype (2)
            Application type [voice] (0x01), Flags [Tagged]
            Vlan id 100, L2 priority 0, DSCP value 0 ^
        End TLV (0), length 0
```

LLDP-MED Summary

LLDP provides for a true plug-and-play voice solution. The capabilities of the IP telephony devices can be advertised to the switch, which can make dynamic adjustments. In turn, an IP phone can be automatically configured based on the information it receives over the switch port in the LLDP-MED TLVs. The final piece of the puzzle is to find an easy way to separate the voice and data traffic in order to provide reliable QoS for the duration of a phone call.

Voice VLAN

An important step in IP telephony concerns separating the data traffic from the voice traffic, especially in a deployment scenario such as the one depicted previously in Figure 10-1. A feature called the *voice VLAN* can be used for this purpose. The voice VLAN allows for tagged and untagged traffic to be accepted, and also allows you to associate each type of traffic with distinct and separate VLANs. This helps to prioritize voice traffic over data traffic and deploy CoS.

If LLDP-MED is used, the CoS implementation is greatly simplified, as the IP phones will be dynamically associated with the appropriate voice VLAN as well as 802.1p values based on the forwarding class settings. This essentially allows for plug-and-play functionality. If LLDP-MED is not used, the configuration is more manual in nature.

First, define the voice VLAN and associated VLAN (see Chapter 5 for information on VLAN configuration). This example also uses an FC called expedited-forwarding, so that value will be sent to the IP phone if there is LLDP-MED support:

```
ethernet-switching-options {
    voip {
        interface ge-0/0/0.0 {
            vlan voice;
            forwarding-class expedited-forwarding;
        }
    }
}

lab@Ethanol> show configuration vlans
voice {
    vlan-id 100;
}
```

 Prior to 9.3, the TLV for DSCP and LLDP-MED did not exist. This is tracked in PR 313953.

Next, apply the voice VLAN to the interface:

```
lab@Ethanol> show configuration interfaces ge-0/0/0
unit 0 {
    family ethernet-switching {
        vlan {
            members voice;
        }
    }
}
```

Then verify the configuration by examining the voice VLAN:

```
lab@Ethanol> show vlans voice detail
VLAN: voice, 802.1Q Tag: 100, Admin State: Enabled
Number of interfaces: 1 (Active = 1)
  Untagged interfaces: ge-0/0/0.0*
  Tagged interfaces: ge-0/0/0.0*
```

Case Studies

We will now examine a few quick case studies with a very simple topology, depicted in Figure 10-7. Several phones are connected off Cisco switch Schnapps, and due to port capacity issues, new phones are connected off Ethanol. The current call manager (Cisco Express Call Manager) is connected off Rum. The goal is to make a phone call from the IP phone at Ethanol to a phone located at Schnapps. It also should be noted that many of the IP phones will have a PC connected to it at a later time. We will look at two different scenarios off Ethanol. In the first scenario, LLDP-MED is not supported on the phone, and in the second it is supported.

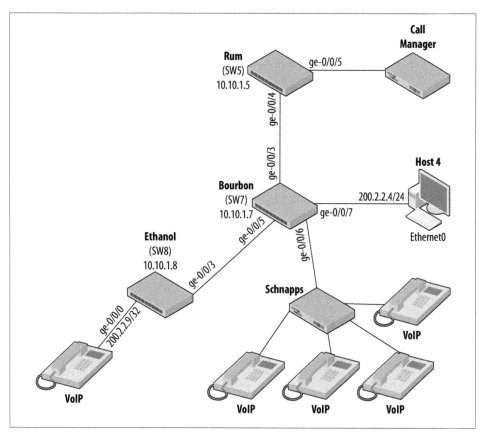

Figure 10-7. IP telephony topology

Without LLDP-MED Support

First, let's examine the current network and the VLANs that are being used. VLAN 100 is defined as the voice VLAN, in which all traffic for call setup and voice bearer traffic is placed. VLAN 200 is used for all other non-voice-related traffic, and is called the data VLAN. Rum is connected to the call manager for the IP telephony network. This is implemented on a Cisco 2811 router running IOS 12.4(15)T4 using Call Manager Express 4.2. The call manager acts as the controller for the IP telephony network, controlling the phone configuration, call routing, and call setup. In this network, it is also acting as a Dynamic Host Configuration Protocol (DHCP) server to avoid any static IP address configuration on the phone. The interface on Rum toward the call manager is configured as an access port and assigned to the voice VLAN:

```
lab@Rum> show configuration interfaces ge-0/0/5
unit 0 {
    family ethernet-switching {
        port-mode access;
```

```
            vlan {
                members voice;
            }
        }
    }
```

 For VLAN configurations and descriptions, please review Chapter 5.

Also, interface ge-0/0/4 toward Bourbon is configured as a trunk port to allow the voice and data VLAN to be transmitted:

```
lab@Rum> show vlans
Name          Tag    Interfaces
data          200
                     ge-0/0/4.0*
default
                     ge-0/0/0.0, ge-0/0/1.0, ge-0/0/2.0, ge-0/0/3.0,
                     ge-0/0/8.0
voice         100
                     ge-0/0/4.0*, ge-0/0/5.0*
```

Walking through the topology, we can see that Bourbon's VLANs are displayed. The interfaces toward Rum and toward Schnapps (ge-0/0/6) are also configured as trunk ports:

```
lab@Bourbon> show vlans
Name          Tag    Interfaces
data          200
                     ge-0/0/2.0, ge-0/0/3.0*, ge-0/0/5.0*, ge-0/0/6.0*,
                     ge-0/0/7.0
default
                     ge-0/0/1.0, ge-0/0/4.0, ge-0/0/12.0
voice         100
                     ge-0/0/3.0*, ge-0/0/5.0*, ge-0/0/6.0*
```

The interface toward Schnapps is configured as a trunk port because it will be receiving data from the IP phones on the voice VLANs, as well as data from attached PCs. Schnapps is a Cisco 3750 that supports PoE:

```
lab@Bourbon> show configuration interfaces ge-0/0/6
unit 0 {
    family ethernet-switching {
        port-mode trunk;
        vlan {
            members [ voice data ];
        }
    }
}
```

For completeness, the snippet of relevant interface configuration is also displayed for Schnapps. Notice that the Fast Ethernet connection toward an IP phone is using the same voice VLAN feature available in JUNOS:

```
interface FastEthernet1/0/1
 description Trunk Connection to Bourbon GE-0/0/5
 switchport trunk encapsulation dot1q
 switchport mode trunk
 spanning-tree portfast
!
interface FastEthernet1/0/2
 description Connection to VOIP-1
 switchport access vlan 200
 switchport voice vlan 100
 spanning-tree portfast
!
interface FastEthernet1/0/3
 switchport access vlan 200
 switchport mode access
 switchport voice vlan 100
 spanning-tree portfast
!
interface FastEthernet1/0/4
 switchport access vlan 200
 switchport mode access
 switchport voice vlan 100
 spanning-tree portfast
!
interface FastEthernet1/0/5
 switchport access vlan 200
 switchport mode access
 switchport voice vlan 100
 spanning-tree portfast
!
interface FastEthernet1/0/6
 switchport access vlan 200
 switchport mode access
 switchport voice vlan 100
 spanning-tree portfast
```

. . . .

Plug-and-play solution without LLDP-MED

A Cisco 7960 phone is plugged into Ethanol's ge-0/0/0 port. This phone does support PoE and powers up fine, but the issue is that it does not support LLDP-MED, and is auto-configured using a Cisco proprietary protocol. This protocol automatically configures the VLAN tag that the phone should use to transmit frames. If the phone does not receive this information, it will simply send untagged frames to the switch after a timeout period. Since JUNOS does not support this proprietary protocol, the phone must either be manually configured for the proper VLAN or configured as an access port in order to receive the untagged frames. The plug-and-play solution would be to have no static configuration configured on the phone, and the port configured as an access port in the voice VLAN:

```
lab@Ethanol> show configuration interfaces ge-0/0/0
unit 0 {
    family ethernet-switching {
        port-mode access;
        vlan {
            members voice;
        }
    }
}
```

Looking at the switching table on Ethanol, we can see the local phone's MAC address of 00:06:53:56:a8:f3, learned over the access port. Also shown are other IP phones that are being learned on the trunk port toward Bourbon:

```
lab@Ethanol> show ethernet-switching table
Ethernet-switching table: 8 entries, 6 learned
  VLAN            MAC address       Type       Age Interfaces
  voice           *                 Flood        - All-members
  voice           00:06:53:56:a8:f3 Learn        0 ge-0/0/0.0
  voice           00:0f:1f:26:6b:9a Learn        0 ge-0/0/3.0
  voice           00:19:30:a3:51:a8 Learn        0 ge-0/0/3.0
  voice           00:22:bd:ab:70:83 Learn        0 ge-0/0/3.0
  voice           00:30:94:c3:2a:7a Learn     3:47 ge-0/0/3.0
  data            *                 Flood        - All-members
  data            00:22:bd:ab:70:83 Learn        0 ge-0/0/3.0
```

This solution works fine if there is no PC or other device connected behind the IP phone. However, our design goal is to have a deployment such as the one in Figure 10-1, with a PC and IP phone on the same switch port. In the current configuration, if a PC were plugged into the phone attached to Ethanol, it would be mapped to the voice VLAN instead of the data VLAN. This could cause less than desirable results, such as connectivity issues and packet contention problems. The solution to this is to use the voice VLAN feature and tag the frames that are transmitted from the phone.

Voice VLAN and IP phone configuration

In order to solve the problem with the previous example, the voice VLAN feature should be implemented on Ethanol. This allows both untagged data traffic and tagged voice traffic to be received on a single switch port. The first step is to configure the IP phone for the proper VLAN. Without LLDP-MED, this will be a manual configuration on the phone. On the Cisco 7960 phone, this requires setting the Admin VLAN to 100 by using the menu and navigating to Networking Settings→Admin VLAN.

We configure the port toward the IP phone as an access port with the data VLAN, which allows untagged frames from locally attached PCs to be received on the switch port:

```
lab@Ethanol> show configuration interfaces ge-0/0/0
unit 0 {
    family ethernet-switching {
        port-mode access;
        vlan {
            members data;
        }
    }
}
```

Next, configure the voice VLAN under [edit ethernet-switching-options]. In this example, Ethanol's ge-0/0/0.0 interface will be assigned the voice VLAN with a tag of 100. This allows a VLAN tag of 100 to be received on the same interface as untagged frames. In addition, all VLAN 100 traffic will be classified into an FC called expedited-forwarding, and the proper DSCP bits will be sent to the phone. However, without LLDP-MED, this configuration is non-functional unless the phone accepts manual DSCP code point settings (see the sidebar "CoS for Voice Network" on page 584):

```
lab@Ethanol> show configuration ethernet-switching-options
voip {
    interface ge-0/0/0.0 {
        vlan 100;
        forwarding-class expedited-forwarding;
    }
}
```

View the VLANs on Ethanol to verify that both the data and voice VLANs are assigned to access port ge-0/0/0 and trunk port ge-0/0/3:

```
lab@Ethanol> show vlans
Name        Tag     Interfaces
data        200
                    ge-0/0/0.0*, ge-0/0/3.0*, ge-0/0/7.0
default
                    ge-0/0/2.0, ge-0/0/8.0, ge-0/0/9.0, ge-0/0/12.0
voice       100
                    ge-0/0/0.0*, ge-0/0/3.0*

lab@Ethanol> show vlans voice detail
VLAN: voice, 802.1Q Tag: 100, Admin State: Enabled
Number of interfaces: 2 (Active = 2)
  Tagged interfaces: ge-0/0/0.0*, ge-0/0/3.0*
```

The MAC table on Ethanol now shows that tagged and untagged frames are being received on ge-0/0/0:

```
lab@Ethanol> show ethernet-switching table
Ethernet-switching table: 9 entries, 7 learned
  VLAN            MAC address         Type       Age Interfaces
  voice           *                   Flood        - All-members
  voice           00:06:53:56:a8:f3   Learn        0 ge-0/0/0.0
  voice           00:0f:1f:26:6b:9a   Learn        0 ge-0/0/3.0
  voice           00:19:30:a3:51:a8   Learn        0 ge-0/0/3.0
  voice           00:22:bd:ab:70:83   Learn        0 ge-0/0/3.0
  voice           00:30:94:c3:2a:7a   Learn     3:47 ge-0/0/3.0
```

```
data                 *                    Flood       - All-members
data                 00:06:53:56:f8:b3    Learn    1:03 ge-0/0/0.0
data                 00:22:bd:ab:70:83    Learn       0 ge-0/0/3.0
```

A phone call is completed, and we see that MAC address 00:06:53:56:a8:f3 for the new IP phone is learned on all the switches:

```
lab@Bourbon> show ethernet-switching table
Ethernet-switching table: 11 entries, 9 learned
  VLAN            MAC address          Type        Age Interfaces
  data            *                    Flood       - All-members
  data            00:06:53:56:a8:f3    Learn    1:02 ge-0/0/5.0
  data            00:22:bd:ab:70:83    Learn       0 ge-0/0/6.0
  voice           *                    Flood       - All-members
  voice           00:03:e3:62:f4:7a    Learn       0 ge-0/0/6.0
  voice           00:06:53:56:a8:f3    Learn      25 ge-0/0/5.0
  voice           00:0f:1f:26:6b:9a    Learn       0 ge-0/0/6.0
  voice           00:19:30:a3:51:a8    Learn       0 ge-0/0/3.0
  voice           00:22:bd:ab:70:83    Learn       0 ge-0/0/6.0
  voice           00:30:94:c3:2a:7a    Learn       0 ge-0/0/6.0
  voice           00:30:94:c3:2f:3e    Learn       0 ge-0/0/6.0


lab@Rum> show ethernet-switching table
Ethernet-switching table: 11 entries, 9 learned
  VLAN            MAC address          Type        Age Interfaces
  voice           *                    Flood       - All-members
  voice           00:06:53:56:a8:f3    Learn       0 ge-0/0/4.0
  voice           00:0f:1f:26:6b:9a    Learn       0 ge-0/0/4.0
  voice           00:19:30:a3:51:a8    Learn       0 ge-0/0/5.0
  voice           00:22:bd:ab:70:83    Learn       0 ge-0/0/4.0
  voice           00:30:94:c3:2a:7a    Learn       0 ge-0/0/4.0
  voice           00:30:94:c3:2f:14    Learn       0 ge-0/0/4.0
  voice           00:30:94:c3:2f:3e    Learn       0 ge-0/0/4.0
  data            *                    Flood       - All-members
  data            00:06:53:56:a8:f3    Learn      29 ge-0/0/4.0
  data            00:22:bd:ab:70:83    Learn       0 ge-0/0/4.0
```

This is verified from the call manager log as the local phone 7502 calls phones 7503, 7504, and 7505:

```
Reports > Call History
ID Start Time.Originating Number Terminating Number Duration
1 18:29:11 EDT Sat Apr 18 2009 7502 7503 <Unknown>
2 18:29:15 EDT Sat Apr 18 2009 7502 7504 <Unknown>
3 18:29:18 EDT Sat Apr 18 2009 7502 7505 <Unknown>
```

CoS for Voice Network

It is recommended that CoS be deployed in your IP telephony network in order to prioritize voice traffic over data traffic. You do this by placing your voice traffic into a separate FC and then prioritizing that FC appropriately. If you are using the LLDP-MED feature, DSCP markings can be advertised to the phone during the voice VLAN configuration. This still requires a BA classifier or multifield classifier to map the packets coming from the phone to the correct FC.

First, create a classifier that maps bits to an FC:

```
class-of-service {
    classifiers {
        dscp test {
            import default;
            forwarding-class expedited-forwarding {
                loss-priority low code-points 101110;
                loss-priority high code-points
101111;
            }
        }
    }
```

Next, advertise those bit markings to the IP phone:

```
lab@Ethanol> show configuration ethernet-switching-options
voip {
    interface ge-0/0/0.0 {
        vlan 100;
        forwarding-class expedited-forwarding;
    }
}
```

Then apply the classifier to the incoming interface:

```
interfaces {
    ge-0/0/0 {
        unit 0 {
            classifiers {
                dscp test;
                ieee-802.1 test;
            }
```

However, if you are not using LLDP-MED, the DSCP settings will not be sent to the phone. Traffic can still be classified using either a BA classifier or a multifield classifier with a firewall filter. In this case, the phone must accept manual DSCP settings, or you must match on the default values. Most phones that use Skinny Call Control Protocol (SCCP) packets will be marked *CS3* or *AF31*, whereas voice Real-time Transport Protocol (RTP) packets will be marked *EF*. However, any standard firewall filter match criteria can work in a multifield (MF) classifier. The result should be that voice traffic is mapped to a separate FC, as seen in queue 5 from `Ethanol`:

```
lab@Ethanol> show interfaces queue ge-0/0/3
Physical interface: ge-0/0/3, Enabled, Physical link
is Up
  Interface index: 132, SNMP ifIndex: 114
Forwarding classes: 16 supported, 4 in use
Egress queues: 8 supported, 4 in use
Queue: 0, Forwarding classes: best-effort
  Queued:
    Packets            : Not Available
    Bytes              : Not Available
    Packets            :             13623
0 pps
    Bytes              :           2998056
0 bps
      Tail-dropped packets :               0
0 pps
Queue: 1, Forwarding classes: assured-forwarding
```

```
    Queued:
      Packets              : Not Available
      Bytes                : Not Available
      Packets              :                   0
  0 pps
      Bytes                :                   0
  0 bps
      Tail-dropped packets :                   0
  0 pps
  Queue: 5, Forwarding classes: expedited-forwarding
    Queued:
      Packets              : Not Available
      Bytes                : Not Available
      Packets              :                 512
  0 pps
      Bytes                :              109378
  0 bps
      Tail-dropped packets :                   0
  0 pps
  Queue: 7, Forwarding classes: network-control
    Queued:
      Packets              : Not Available
      Bytes                : Not Available
      Packets              :                   0
  0 pps
      Bytes                :                   0
  0 bps
      Tail-dropped packets :                   0
  0 pps
```

The following code snippet shows a very basic CoS configuration for the egress interface. This is displayed just as a syntax example:

```
lab@Ethanol> show configuration class-of-service
interfaces {
    ge-0/0/3 {
        scheduler-map VOIP;
    }
}
scheduler-maps {
    VOIP {
        forwarding-class best-effort scheduler be;
        forwarding-class expedited-forwarding
scheduler ef;
        forwarding-class network-control scheduler
nc;
    }
}
schedulers {
    ef {
        transmit-rate percent 10;
        priority low;
    }
    be {
        transmit-rate remainder;
        priority low;
    }
    nc {
        transmit-rate percent 5;
        priority low;
```

```
        }
    }
```

With LLDP-MED Support

For a true plug-and-play effect, it's best if the IP phone has LLDP-MED support. Here, the phone in the previous example is replaced by a Cisco G7960 with LLDP-MED support. Once the phone is plugged in, it is automatically recognized as an LLDP neighbor with LLDP-MED support:

```
lab@Ethanol> show lldp neighbors
LocalInterface Chassis Id       Port info    System Name
ge-0/0/3.0     00:1f:12:30:c4:80 ge-0/0/5.0  Bourbon
ge-0/0/0.0     0.0.0.0          SW PORT      SEP001193123586.cisco.com
```

You can view the details of the phone in the lldp neighbors interface command. This command includes important data such as capabilities, software versions, and model numbers:

```
lab@Ethanol> show lldp neighbors interface ge-0/0/0.0
LLDP Neighbor Information:
Index: 10 Time to live: 180 Time mark: Sun Sep  7 02:31:48 2008 Age: 35 secs
Local interface    : ge-0/0/0.0
Chassis type       : Network address
Chassis ID         : 0.0.0.0
Port type          : Locally assigned
Port ID            : 001193123586:P1
Port description   : SW PORT
System name        : SEP001193123586.cisco.com
System description : Cisco IP Phone CP-7970G,V, SCCP70.8-3-3SR2S

System capabilities
        Supported: Bridge Telephone
        Enabled  : Bridge Telephone
Media endpoint class: Class III Device
MED Firmware revision : 7970_64060118.bin
MED Software revision : SCCP70.8-3-3SR2Sn
MED Serial number     : INM08281G3T3SR2
MED Manufacturer name : Cisco Systems, Inc.
MED Model name        : CP-7970Gstem
MED Asset id : CP-7
```

The phone then receives the settings based on the following configuration:

```
lab@Ethanol> show configuration ethernet-switching-options
voip {
    interface ge-0/0/0.0 {
        vlan 100;
        forwarding-class expedited-forwarding;
    }
}
```

This configuration sends traffic via VLAN 100, using DSCP values that correspond to the EF class. This is seen in the following monitor traffic command (it should be noted that the switch was upgraded to 9.5R1.8 for this capture):

```
MD autoneg capability [10BASE-T hdx, 10BASE-T fdx, 100BASE-TX hdx, 100BASE-TX fdx,
Asym and Sym PAUSE for fdx, 1000BASE-{X LX SX CX} fdx, 1000BASE-T fdx] (0xa836)
        MAU type Unknown (0x0000)
    Organization specific TLV (127), length 7: OUI IEEE 802.3 Private
(0x00120f)
        Power via MDI Subtype (2)
        MDI power support [PSE, enabled], power pair signal, power class class2
    Organization specific TLV (127), length 9: OUI IEEE 802.3 Private
(0x00120f)
        Link aggregation Subtype (3)
        aggregation status [supported], aggregation port ID 0
    Organization specific TLV (127), length 6: OUI IEEE 802.3 Private
(0x00120f)
        Max frame size Subtype (4)
        MTU size 1514
    Organization specific TLV (127), length 6: OUI Ethernet bridged (0x0080c2)
        0x0000:  0080 c201 00c8
    Organization specific TLV (127), length 16: OUI Juniper (0x009069)
        0x0000:  0090 6901 4248 3032 3038 3138 3832 3439
    Organization specific TLV (127), length 11: OUI Ethernet bridged (0x0080c2)
        0x0000:  0080 c203 00c8 0464 6174 61
    Organization specific TLV (127), length 12: OUI Ethernet bridged (0x0080c2)
        0x0000:  0080 c203 0064 0576 6f69 6365
    Organization specific TLV (127), length 7: OUI ANSI/TIA (0x0012bb)
        LLDP-MED Capabilities Subtype (1)
        Media capabilities [LLDP-MED capabilities, network policy, location
identification, extended power via MDI-PD] (0x0017)
        Device type [network connectivity] (0x04)
    Organization specific TLV (127), length 7: OUI ANSI/TIA (0x0012bb)
        Extended power-via-MDI Subtype (4)
        Power type [PSE device], Power source [PSE - primary power source]
        Power priority [low] (0x03), Power 4.9 Watts
    Organization specific TLV (127), length 8: OUI ANSI/TIA (0x0012bb)
        Network policy Subtype (2)
        Application type [voice] (0x01), Flags [Tagged]
        Vlan id 100, L2 priority 4, DSCP value 46
    End TLV (0), length 0
```

Verify that the phone and call manager are communicating before the call by examining the keepalives to and from the phone:

```
CME#debug ephone keepalive
EPHONE keepalive debugging is enabled
CME#
Apr 23 03:24:11.149: ephone-5 If FastEthernet0/0 ETHERNET 200.2.2.8 via ARP
Apr 23 03:24:11.149: ephone-5[4][SEP003094C32A7A]:Keepalive socket[4]
SEP003094C32A7A
CME#
Apr 23 03:24:12.337: ephone-4 If FastEthernet0/0 ETHERNET 200.2.2.7 via ARP
Apr 23 03:24:12.337: ephone-4[5][SEP003094C32F14]:Keepalive socket[5]
SEP003094C32F14
```

```
CME#
CME#
Apr 23 03:24:29.429: ephone-3 If FastEthernet0/0 ETHERNET 200.2.2.12 via ARP
Apr 23 03:24:29.429: ephone-3[1][SEP003094C32F3E]:Keepalive socket[1]
SEP003094C32F3E
Apr 23 03:24:29.653: ephone-7 If FastEthernet0/0 ETHERNET 200.2.2.13 via ARP
Apr 23 03:24:29.653: ephone-7[3][SEP0003E362F47A]:Keepalive socket[3]
SEP0003E362F47A
CME#
Apr 23 03:24:41.145: ephone-5 If FastEthernet0/0 ETHERNET 200.2.2.8 via ARP
Apr 23 03:24:41.145: ephone-5[4][SEP003094C32A7A]:Keepalive socket[4]
SEP003094C32A7A
CME#
Apr 23 03:24:42.333: ephone-4 If FastEthernet0/0 ETHERNET 200.2.2.7 via ARP
Apr 23 03:24:42.333: ephone-4[5][SEP003094C32F14]:Keepalive socket[5] SEP00
```

A phone call is made (trust us on that), and the traffic is mapped to the EF class based on a BA classifier. In this case, we decided to use a DSCP and IEEE-802.1 classifier. Our phone used only the IEEE-802.1 settings when receiving both bit markings in LLDP-MED, but this may change based on phone model. Normally, we would just send the DSCP settings and use the proper DSCP classifier:

```
interfaces {
    ge-0/0/0 {
        unit 0 {
            classifiers {
                dscp test;
                ieee-802.1 test;
            }
        }
    }
}

lab@Ethanol> show interfaces queue ge-0/0/3
Physical interface: ge-0/0/3, Enabled, Physical link is Up
  Interface index: 132, SNMP ifIndex: 164
Forwarding classes: 16 supported, 4 in use
Egress queues: 8 supported, 4 in use
Queue: 0, Forwarding classes: best-effort
  Queued:
    Packets               : Not Available
    Bytes                 : Not Available
    Packets               :                454
    Bytes                 :              53163
    Tail-dropped packets  :                  0
Queue: 1, Forwarding classes: assured-forwarding
  Queued:
    Packets               : Not Available
    Bytes                 : Not Available
    Packets               :                  0
    Bytes                 :                  0
    Tail-dropped packets  :                  0
Queue: 5, Forwarding classes: expedited-forwarding
  Queued:
    Packets               : Not Available
    Bytes                 : Not Available
```

```
     Packets            :            2091
     Bytes              :          463790
     Tail-dropped packets :              0
Queue: 7, Forwarding classes: network-control
   Queued:
     Packets          : Not Available
     Bytes            : Not Available
     Packets            :             347
     Bytes              :           96989
     Tail-dropped packets :              0
```

Case Study Summary

We examined two case studies, one in which the IP phone supported LLDP-MED, and another in which LLDP-MED was supported on the end device. LLDP-MED provides true plug-and-play support, with no configuration needed on the IP phone. Without LLDP-MED, some manual configuration is required on the phone, but nothing too complex.

Conclusion

IP telephony can no longer be considered an immature technology—many offices around the world have IP phones on their desks. In fact, most people don't realize that their office utilizes IP telephony due to the improved voice quality, compression techniques, and ease of use. For years, IP telephony solutions were proprietary solutions, but the standards organizations have taken charge, creating features such as Power over Ethernet, LLDP-MED, and IP signaling protocols. With any luck, when you deploy your IP telephony network, all you will have to do is plug it in.

Chapter Review Questions

1. What is the most common IP telephony deployment scenario?

    a. IP phone plugged into a PC

    b. IP phone and PC plugged into different switch ports

    c. IP phone and PC plugged into the same switch port

    d. IP phone plugged into an IP gateway

2. What type of power management scheme allows for different wattage based on the type of device?

    a. Static

    b. Class

    c. Random

    d. Policy

3. True or false: every Ethernet port on an EX is PoE-capable by default.

4. In which layer does LLDP operate?

    a. Layer 2

    b. Layer 3

    c. Layer 4

    d. Layer 5

5. Which TLVs are valid for LLDP? (Choose two.)

    a. Port ID

    b. Router ID

    c. System ID

    d. Protocols supported

6. What is the primary goal of LLDP-MED?

    a. To adjust the BGP metric of IP phones

    b. To signal the existence of a phone to a call manager

    c. To discover the capabilities of an IP endpoint

    d. To provide a secure and encrypted neighbor discovery protocol

7. Which JUNOS feature allows the use of tagged and untagged frames on the same interface?

    a. Trunk mode

    b. Access mode

    c. Promiscuous mode

    d. Voice VLAN

8. When does an EX Series switch begin to send out LLDP-MED frames?

    a. As soon as LLDP-MED is configured

    b. When it receives an LLDP-MED TLV

    c. When the voice VLAN is configured

    d. When an RTP message is received

9. Where is the voice VLAN ID configured?

    a. `[edit vlans]`

    b. `[edit ethernet-switching-options]`

    c. `[edit interfaces]`

    d. `[edit voice]`

10. What are two ways to classify incoming packets in JUNOS software?

    a. BA classifiers

    b. Rewrite rule

c. Scheduler

d. Multifield classifier

e. IP classifier

11. How do you make a non-LLDP-MED phone work with an EX Series switch?

   a. You can't

   b. Configure the switch port as a trunk port

   c. Configure the switch port and an access port

   d. Plug the phone into a PC

12. Which command displays LLDP-MED information received from an IP phone?

   a. `show lldp-med detail`

   b. `show lldp neighbors interface <interface-name>`

   c. `show voice lldp`

   d. `show lldp-med database`

Chapter Review Answers

1. Answer: C. The most common and port-cost-efficient model is to plug the IP phone and PC into the same switch port. This is accomplished by using the data port on the IP phone and then plugging the IP phone into the switch port.

2. Answer: B. The class of device can determine the amount of power that can be assigned to a port.

3. Answer: C. Ah, a trick question. The number of PoE ports that can be supported depends on the card and power supply type.

4. Answer A. LLDP is a Layer 2, Data Link layer protocol.

5. Answer: A, C. These are two of the many TLVs that may be sent in an LLDP message.

6. Answer: C. LLDP-MED is an endpoint discovery protocol, which is used to learn the capabilities of an endpoint. This information could then be used to configure the device.

7. Answer: D. The voice VLAN feature allows for tagged and untagged frames to be received on the same port. This is usually reserved for use by an IP phone that is tagging packets and a PC that is sending untagged frames.

8. Answer: B. EX Series switches do not automatically advertise LLDP-MED information until they receive LLDP-MED information from an LLDP neighbor.

9. Answer: B. The voice VLAN is configured under [edit ethernet-switching-options] and not at the VLAN or interface level, like other VLAN configurations.

10. Answer: A, D. BA classifiers can classify packets based on 802.1p, DSCP, or IP precedence bits. A multifield classifier uses any field available in a firewall filter.

11. Answer: C. One option is to configure the switch as an access port to allow the phone to send untagged frames. This removes any configuration requirements on the phone itself; however, it does not provide data traffic separation if a PC is also attached to the switch port.

12. Answer: B. There are no specific LLDP-MED commands, so standard LLDP commands should be used to view the LLDP-MED information. This is seen most easily by viewing a particular interface.

High Availability

High Availability (HA) has been a hot topic ever since two devices were networked together over a piece of barbwire. What happens if the wire breaks? What is the backup plan, and how long will it take to solve a problem? That's the essence of High Availability: ensuring some level of reliability and redundancy. One of the difficult elements of HA is measuring the success of and understanding all the components involved. Is HA measured on a time factor, a packet loss, or a delay? The answer is all of the above, so when 99% reliability is advertised, you must look at the factors that go into the calculation of that measurement. To look at a non-network example, imagine you own a bar in San Francisco that makes top-notch margaritas.[*] You need to ensure that the drink tastes the same each time it is served. This requires the ingredients to be readily available and consistent. Some of the factors that could affect a margarita include:

- Type of tequila
- Number of fresh limes
- Amount of agave nectar

Many of these items can be quantitatively measured; we could say that if all the ingredients are measured and are available every day, 99% of the time the drink should taste the same. However, could any other factors affect the taste? The quality of the lime, for instance, or the quality of the agave nectar? Perhaps the type of food the person was eating at the same time could affect the taste. These variables could significantly lower that 99% value. The point is, just as you think you have a total grasp on HA, another wrench is thrown into the puzzle that may affect your measurements. The lesson learned: be aware of the factors that are measured when one vendor claims 99.99999% HA while another claims 99.99%. Is it really an apples-to-apples comparison?

Let's stagger out of the bar for a bit and look at the EX Series switch and its place in network HA. Remember, a switch is just one component of a network, and even if the switch locally fails to another link in subseconds, it does no good if the spanning tree

---

[*] Maybe one at 1824 Irving Street. Of course, this is all hypothetical.

takes 30 seconds for the coverage of traffic to that link. Therefore, always treat HA as a network measurement with many individual parts.

On an EX Series switch, HA can be achieved by offering redundant hardware, a JUNOS software feature, or new protocol features. We will examine all of these pieces in this chapter.

 HA is a very large topic that should really have more than a chapter—more like an entire book. Luckily, such a book exists; see *JUNOS High Availability (http://oreilly.com/catalog/9780596523046/)*, by James Sonderegger et al. (O'Reilly), for more information.

The topics covered in this chapter include:

- Hardware redundancy
- Bidirectional Forwarding Detection (BFD)
- Aggregated Ethernet (AE)
- Virtual Router Redundancy Protocol (VRRP)
- In-Service Software Upgrades (ISSU)

Hardware Redundancy

As we discussed in many of the chapters in this book, there are many redundant hardware components in EX3200 and EX4200–8200 series switches. See Table 11-1 for a quick summary.

Table 11-1. Redundant hardware components

Switch series	Redundant power	Fans	Redundant RE
3200	Yes (RPS)	No	No
4200	Yes	Yes	Yes (VC)
8200	Yes	Yes	Yes

The EX3200 is the smallest of the EX Series switches and, thus, does not require as much hardware redundancy. In the base model, it supports a single field-replaceable power supply unit (PSU) and a fan tray with a single blower. However, you can also use an optional remote power supply (RPS) to provide redundant power.

The EX4200 supports redundant load-sharing PSUs and a multiblower fan module, both of which can be hot-swapped in the field. As with the EX3200, the remote PSU option is also supported on the EX4200 for added redundancy.

The EX8200 series switches offer fully redundant routing engines (REs), switching fabric, power supplies, and cooling to maximize reliability and uptime.

The EX8208 provides two switch fabric/RE slots. The EX8208 uses a single fan tray that provides side-to-side airflow with redundant multispeed fans. It also contains a whopping six power supply bays, located on the front of the chassis base. Each AC power supply delivers 2,000 watts of power at high-line (12 A at 200–240 V) or 1,200 watts at low-line (15 A at 100–120 V). The actual number of power supplies required depends on the combination of Line Cards (LCs) installed and the desired level of redundancy. For example, 6,000 watts is required to support a chassis that is fully populated with sixty-four 10 Gigabit Ethernet (GE) ports, whereas 3,600 watts will support various 10 Gigabit Ethernet and Gigabit Ethernet LC combinations. At the time of this writing, a redundant RE chassis requires 1,400 watts of power and an eight-port 10 GE SFP + LC requires 880 watts. Consult the latest Juniper documentation for updated power consumption values.

The EX8216 also has two route engine slots on the front of the chassis and six power supply bays at the front base of the chassis. However, it also incorporates two fan trays (side-to-side airflow).

Routing Engine Failover

The 4200 and 8200 switches support redundant REs. The 8200 supports a redundant SRE in the same chassis, and the 4200 supports RE redundancy in a Virtual Chassis (VC) configuration.

Recall that in a VC, the master acts as the master RE, and the backup acts as the backup RE. The master RE, which is in the master of the VC, has full control over the VC configuration. The backup RE is in the backup of the VC configuration. Depending on the configuration, it stays in sync with the master RE in terms of protocol states, forwarding tables (FTs), and so forth. If the master becomes unavailable, the backup RE takes over the functions that the master RE performs.

As seen in this capture, the RE in slot 0 is the master RE for Vodkila:

```
lab@Vodkila> show chassis routing-engine
Routing Engine status:
  Slot 0:
    Current state                 Master
    Temperature                   43 degrees C / 109 degrees F
    DRAM                        1024 MB
    Memory utilization            19 percent
    CPU utilization:
      User                        11 percent
      Kernel                      21 percent
      Interrupt                    0 percent
      Idle                        67 percent
    Model                         EX4200-24T, 8 POE
    Serial ID                     BM0208269834
    Start time                    2005-01-24 10:05:37 UTC
    Uptime                        3 days, 10 hours, 32 minutes, 5 seconds
    Load averages:                1 minute   5 minute  15 minute
                                      0.30       0.25       0.17
```

```
Routing Engine status:
  Slot 1:
    Current state                     Backup
    Temperature                    40 degrees C / 104 degrees F
    DRAM                         1024 MB
    Memory utilization             21 percent
    CPU utilization:
      User                          9 percent
      Kernel                        2 percent
      Interrupt                     0 percent
      Idle                         88 percent
    Model                           EX4200-24T, 8 POE
    Serial ID                       BM0208269767
    Start time                      2005-01-24 10:05:38 UTC
    Uptime                          3 days, 10 hours, 32 minutes, 4 seconds
    Load averages:                  1 minute    5 minute  15 minute
                                        0.39        0.20        0.12
```

To connect to the backup RE, you can use either the Out of Band (OoB) management interface, if configured, or an internal channel. In this case, we are using the internal channel to connect to the backup RE:

```
lab@Vodkila> request session member 1
€
--- JUNOS 9.2R1.9 built 2008-08-05 07:25:22 UTC

lab@Vodkila-fpc1-BK>
```

Now is a good time to chat about what it really means to be the backup RE. By default, without any configuration, the backup RE is just waiting around to become the master in case of a failure. The backup does not run all processes and does not pull information from the Packet Forwarding Engine (PFE), so information such as PFE interfaces and chassis components is not available. In a failover scenario, the new RE would have to connect to the PFE, causing a reset and flush of all the memory tables that are used for forwarding:

```
lab@Vodkila-fpc1-BK> show chassis routing-engine

lab@Vodkila-fpc1-BK> show chassis hardware
Aborted! This command can only be used on the master routing engine.
lab@Vodkila> show interfaces terse
Interface               Admin Link Proto    Local           Remote
vcp-255/1/0             up    up
vcp-255/1/0.32768       up    up
vcp-0                   up    up
vcp-0.32768             up    up
vcp-1                   up    up
vcp-1.32768             up    up
bme0                    up    up
bme0.32768              up    up   inet     128.0.0.17/2
                                            128.0.0.32/2
                                   tnp      0x11
bme0.32770              down  up   eth-switch
dsc                     up    up
```

```
gre                       up    up
ipip                      up    up
lo0                       up    up
lo0.0                     up    up    inet      10.10.1.3         --> 0/0
lsi                       up    up
me0                       up    up
me0.0                     up    up    eth-switch
mtun                      up    up
pimd                      up    up
pime                      up    up
tap                       up    up
```

This causes default failover time to be much longer than usually desired, as the RE will have to connect to the PFE. This time increases if the RE has to restart any protocols, such as Open Shortest Path First (OSPF) or Rapid Spanning Tree Protocol (RSTP), that are running on the switch. We will discuss improvements to this later in this chapter. But first, let's start with a basic scenario.

Default Failover Layer 2

In our first scenario, we have Vodkila configured as a VC, and we are narrowing our focus between Scotch and Host1. Vodkila is in pure Layer 2 mode and Scotch is acting as the default gateway (200.2.2.254) for Host1 (Figure 11-1).

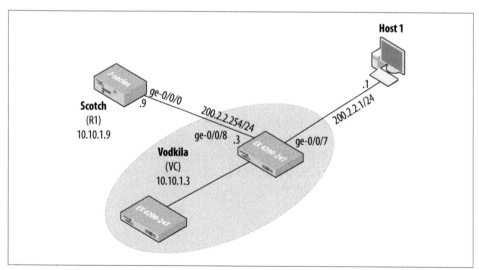

Figure 11-1. Layer 2 default RE switchover

Vodkila does not have any special configuration, so it's using the default parameters. This means if a failover of the primary RE occurs, the backup will have to connect to the PFE and re-create the tables. As a test, a ping is issued from Scotch to Host1. A failover is initiated on Vodkila using the request chassis RE master switch command. Here's the result:

```
lab@Scotch> ping 200.2.1.1 count 100
PING 200.2.1.1 (200.2.1.1): 56 data bytes
64 bytes from 200.2.1.1: icmp_seq=0 ttl=255 time=5.218 ms
64 bytes from 200.2.1.1: icmp_seq=1 ttl=255 time=39.622 ms
64 bytes from 200.2.1.1: icmp_seq=2 ttl=255 time=3.416 ms
64 bytes from 200.2.1.1: icmp_seq=3 ttl=255 time=3.426 ms
64 bytes from 200.2.1.1: icmp_seq=4 ttl=255 time=3.440 ms
64 bytes from 200.2.1.1: icmp_seq=5 ttl=255 time=3.440 ms
64 bytes from 200.2.1.1: icmp_seq=6 ttl=255 time=3.441 ms
64 bytes from 200.2.1.1: icmp_seq=7 ttl=255 time=10.293 ms
64 bytes from 200.2.1.1: icmp_seq=8 ttl=255 time=3.447 ms
64 bytes from 200.2.1.1: icmp_seq=9 ttl=255 time=3.419 ms
64 bytes from 200.2.1.1: icmp_seq=10 ttl=255 time=3.452 ms
64 bytes from 200.2.1.1: icmp_seq=11 ttl=255 time=32.833 ms
64 bytes from 200.2.1.1: icmp_seq=12 ttl=255 time=3.413 ms
64 bytes from 200.2.1.1: icmp_seq=13 ttl=255 time=4.952 ms
64 bytes from 200.2.1.1: icmp_seq=14 ttl=255 time=3.412 ms
36 bytes from 172.16.9.1: Destination Host Unreachable
Vr HL TOS  Len   ID Flg  off TTL Pro  cks        Src       Dst
 4  5  00 0054 4c68   0 0000  3f 01 b124 172.16.9.9  200.2.1.1

36 bytes from 172.16.9.1: Destination Host Unreachable
Vr HL TOS  Len   ID Flg  off TTL Pro  cks        Src       Dst
 4  5  00 0054 4c69   0 0000  3f 01 b123 172.16.9.9  200.2.1.1

64 bytes from 200.2.1.1: icmp_seq=25 ttl=255 time=26.386 ms
64 bytes from 200.2.1.1: icmp_seq=26 ttl=255 time=3.441 ms
64 bytes from 200.2.1.1: icmp_seq=27 ttl=255 time=3.445 ms
64 bytes from 200.2.1.1: icmp_seq=28 ttl=255 time=3.194 ms
64 bytes from 200.2.1.1: icmp_seq=29 ttl=255 time=3.441 ms
64 bytes from 200.2.1.1: icmp_seq=30 ttl=255 time=3.452 ms
64 bytes from 200.2.1.1: icmp_seq=31 ttl=255 time=39.320 ms
64 bytes from 200.2.1.1: icmp_seq=32 ttl=255 time=3.416 ms
64 bytes from 200.2.1.1: icmp_seq=33 ttl=255 time=3.435 ms
```

After packet 14, forwarding is not resumed until packet 25, with a loss of about 11 seconds. Two packets are received with a destination Host Unreachable, as Scotch has a default router to the "Internet" that is in use when the interface on Vodkila bounces. This failover is not the end of the world, but 11 seconds does not fall into the five 9s of uptime that we are trying to achieve.

Default Failover Layer 3

Let's change the scenario a bit and add some Layer 3 into the mix. In this case, Vodkila is the default gateway for Host1. Vodkila and Scotch are OSPF neighbors. Scotch can reach Host1 through an OSPF route advertised by Vodkila by virtue of a passive OSPF interface toward Host1 (see Figure 11-2):

```
lab@Scotch> show route 200.2.1.1

inet.0: 15 destinations, 16 routes (15 active, 0 holddown, 0 hidden)
+ = Active Route, - = Last Active, * = Both
```

```
200.2.1.0/24    *[OSPF/10] 00:20:28, metric 11
                > to 10.3.9.3 via ge-0/0/0.0
```

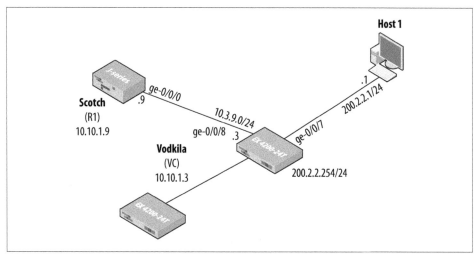

Figure 11-2. Layer 3 switchover

The same test is used as in the Layer 2 case, but Scotch issues a ping command to
Host1, while Vodkila issues a request chassis RE master acquire command:

```
lab@Scotch> ping 200.2.1.1 count 100
PING 200.2.1.1 (200.2.1.1): 56 data bytes
64 bytes from 200.2.1.1: icmp_seq=0 ttl=254 time=3.663 ms
64 bytes from 200.2.1.1: icmp_seq=1 ttl=254 time=3.472 ms
64 bytes from 200.2.1.1: icmp_seq=2 ttl=254 time=3.399 ms
64 bytes from 200.2.1.1: icmp_seq=3 ttl=254 time=4.179 ms
64 bytes from 200.2.1.1: icmp_seq=4 ttl=254 time=3.435 ms
64 bytes from 200.2.1.1: icmp_seq=5 ttl=254 time=5.212 ms
64 bytes from 200.2.1.1: icmp_seq=6 ttl=254 time=3.386 ms
64 bytes from 200.2.1.1: icmp_seq=7 ttl=254 time=3.430 ms
64 bytes from 200.2.1.1: icmp_seq=8 ttl=254 time=28.372 ms
64 bytes from 200.2.1.1: icmp_seq=9 ttl=254 time=3.431 ms
64 bytes from 200.2.1.1: icmp_seq=10 ttl=254 time=3.429 ms
64 bytes from 200.2.1.1: icmp_seq=11 ttl=254 time=5.213 ms
64 bytes from 200.2.1.1: icmp_seq=12 ttl=254 time=3.652 ms
64 bytes from 200.2.1.1: icmp_seq=13 ttl=254 time=3.429 ms
ping: sendto: No route to host
ping: sendto: No route to host
ping: sendto: No route to host
ping: sendto: No route to host
ping: sendto: No route to host
ping: sendto: No route to host
ping: sendto: No route to host
ping: sendto: No route to host
ping: sendto: No route to host
ping: sendto: No route to host
ping: sendto: No route to host
```

```
ping: sendto: No route to host
ping: sendto: No route to host
ping: sendto: No route to host
ping: sendto: No route to host
ping: sendto: No route to host
ping: sendto: No route to host
ping: sendto: No route to host
ping: sendto: No route to host
ping: sendto: No route to host
ping: sendto: No route to host
ping: sendto: No route to host
ping: sendto: No route to host
ping: sendto: No route to host
ping: sendto: No route to host
ping: sendto: No route to host
ping: sendto: No route to host
ping: sendto: No route to host
ping: sendto: No route to host
ping: sendto: No route to host
ping: sendto: No route to host
ping: sendto: No route to host
ping: sendto: No route to host
ping: sendto: No route to host
ping: sendto: No route to host
ping: sendto: No route to host
ping: sendto: No route to host
ping: sendto: No route to host
ping: sendto: No route to host
ping: sendto: No route to host
ping: sendto: No route to host
ping: sendto: No route to host
ping: sendto: No route to host
ping: sendto: No route to host
ping: sendto: No route to host
ping: sendto: No route to host
ping: sendto: No route to host
ping: sendto: No route to host
ping: sendto: No route to host
ping: sendto: No route to host
ping: sendto: No route to host
ping: sendto: No route to host
ping: sendto: No route to host
ping: sendto: No route to host
ping: sendto: No route to host
64 bytes from 200.2.1.1: icmp_seq=76 ttl=254 time=4.420 ms
64 bytes from 200.2.1.1: icmp_seq=77 ttl=254 time=4.158 ms
64 bytes from 200.2.1.1: icmp_seq=78 ttl=254 time=45.458 ms
64 bytes from 200.2.1.1: icmp_seq=79 ttl=254 time=3.429 ms
64 bytes from 200.2.1.1: icmp_seq=80 ttl=254 time=4.175 ms
64 bytes from 200.2.1.1: icmp_seq=81 ttl=254 time=4.167 ms
64 bytes from 200.2.1.1: icmp_seq=82 ttl=254 time=5.219 ms
64 bytes from 200.2.1.1: icmp_seq=83 ttl=254 time=4.168 ms
64 bytes from 200.2.1.1: icmp_seq=84 ttl=254 time=3.435 ms
64 bytes from 200.2.1.1: icmp_seq=85 ttl=254 time=4.431 ms
64 bytes from 200.2.1.1: icmp_seq=86 ttl=254 time=4.125 ms
```

```
64 bytes from 200.2.1.1: icmp_seq=87 ttl=254 time=3.423 ms
64 bytes from 200.2.1.1: icmp_seq=88 ttl=254 time=37.474 ms
64 bytes from 200.2.1.1: icmp_seq=89 ttl=254 time=4.155 ms
64 bytes from 200.2.1.1: icmp_seq=90 ttl=254 time=3.435 ms
64 bytes from 200.2.1.1: icmp_seq=91 ttl=254 time=3.430 ms
64 bytes from 200.2.1.1: icmp_seq=92 ttl=254 time=3.393 ms
64 bytes from 200.2.1.1: icmp_seq=93 ttl=254 time=3.403 ms
64 bytes from 200.2.1.1: icmp_seq=94 ttl=254 time=4.410 ms
64 bytes from 200.2.1.1: icmp_seq=95 ttl=254 time=4.158 ms
64 bytes from 200.2.1.1: icmp_seq=96 ttl=254 time=4.160 ms
64 bytes from 200.2.1.1: icmp_seq=97 ttl=254 time=3.421 ms
64 bytes from 200.2.1.1: icmp_seq=98 ttl=254 time=26.816 ms
64 bytes from 200.2.1.1: icmp_seq=99 ttl=254 time=3.427 ms

--- 200.2.1.1 ping statistics ---
100 packets transmitted, 38 packets received, 62% packet loss
round-trip min/avg/max/stddev = 3.386/7.105/45.458/9.731 ms
```

In this case, the failover time is much worse, with approximately 62 packets lost and about a full minute of time. When the failover occurs, the RE switches over in several seconds and is notified almost immediately. You can observe this by looking at the mastership log on Vodkila:

```
lab@vodkilla> show log mastership
Jan 27 20:37:23 *** mcontrol init V01 ***
Jan 27 21:09:26 mcontrol toggle sig from init ***
Jan 27 21:09:26 The local RE becomes the master
```

So, why the long failover? The No route to host output from the ping command is a hint. Essentially, the OSPF neighbor was lost, so the route to Host1 disappeared. Let's look at the first step in improving both scenarios.

Graceful Routing Engine Switchover

One of the problems in the default behavior is the fact that the PFE resets when an RE switchover occurs. A feature in JUNOS software called Graceful Routing Engine Switchover (GRES) removes the PFE reboot by synchronizing state between the two REs. It uses the internal control channel to synchronize kernel and forwarding information, as well as to send keepalives. These keepalives allow the backup RE to determine whether the master is present. If the backup RE does not receive a keepalive from the master RE after two seconds, it determines that the master RE has failed and acquires mastership.

When a switchover occurs, the PFE disconnects from the old master RE and connects to the new master RE without rebooting. Therefore, traffic should not be affected. The new master RE and PFE then synchronize to make sure all the information is up-to-date, as the FTs were not updated during the switchover.

To enable GRES, a simple command is issued on the switch, graceful switchover:

```
lab@Vodkila> show configuration chassis
redundancy {
```

```
        graceful-switchover;
}
```

When GRES is turned on, the command-line interface (CLI) prompt will change to indicate that the RE is either {master} or {backup}:

```
{master}
lab@Vodkila>
```

View the state of the synchronization between REs by issuing a show system switchover on the *backup* RE:

```
lab@Vodkila> show system switchover
Graceful switchover: On
Configuration database: Ready
Kernel database: Ready
Peer state: Steady State
```

The kernel and chassis states are synchronized between the two REs, which means that interface status and configuration can now be seen on the backup RE:

```
--- JUNOS 9.2R1.9 built 2008-08-05 07:25:22 UTC
{backup}
lab@Vodkila-fpc1-BK> show interfaces terse
Interface          Admin Link Proto  Local            Remote
vcp-255/1/0        up    up
vcp-255/1/0.32768  up    up
ge-0/0/0           up    up
ge-0/0/0.0         up    up   inet   10.3.5.1/24
ge-0/0/1           up    up
ge-0/0/2           up    up
ge-0/0/3           up    up
ge-0/0/3.0         up    up   inet   10.1.3.3/24
ge-0/0/4           up    down
ge-0/0/5           up    down
ge-0/0/6           up    down
ge-0/0/7           up    up
ge-0/0/7.0         up    up   inet   200.1.3.3/24
                                     200.2.1.50/24
                                     200.2.1.254/24
ge-0/0/8           up    up
ge-0/0/8.0         up    up   inet   10.3.9.3/24
ge-0/0/9           up    up
ge-0/0/9.0         up    up   inet   172.16.3.3/24
ge-0/0/10          up    down
ge-0/0/11          up    down
ge-0/0/12          up    down
ge-0/0/13          up    down
ge-0/0/14          up    down
ge-0/0/15          up    down
ge-0/0/16          up    down
ge-0/0/17          up    down
ge-0/0/18          up    down
ge-0/0/19          up    down
ge-0/0/20          up    down
ge-0/0/21          up    up
```

```
ge-0/0/22              up    down
ge-0/0/23              up    up
ge-0/1/1               up    down
ge-0/1/1.0             up    down eth-switch
ge-1/0/0               up    up
ge-1/0/0.0             up    up   inet     10.4.5.4/24
ge-1/0/1               up    up
ge-1/0/1.0             up    up   inet     10.4.6.4/24
ge-1/0/2               up    up
ge-1/0/2.0             up    up   inet     10.4.7.4/24
ge-1/0/3               up    down
ge-1/0/3.0             up    down inet     10.2.4.4/24
ge-1/0/4               up    down
ge-1/0/5               up    down
ge-1/0/6               up    down
ge-1/0/7               up    down
ge-1/0/8               up    down
ge-1/0/8.0             up    down inet     10.1.4.4/24
ge-1/0/9               up    down
ge-1/0/10              up    down
ge-1/0/11              up    down
ge-1/0/12              up    down
ge-1/0/13              up    down
ge-1/0/14              up    down
ge-1/0/15              up    down
ge-1/0/16              up    down
ge-1/0/17              up    down
ge-1/0/18              up    down
ge-1/0/19              up    down
ge-1/0/20              up    down
ge-1/0/21              up    up
ge-1/0/22              up    up
ge-1/0/23              up    down
ge-1/1/1               up    down
ge-2/0/0               up    down
ge-2/0/1               up    down
ge-2/0/2               up    down
ge-2/0/3               up    down
ge-2/0/4               up    down
ge-2/0/5               up    down
ge-2/0/6               up    down
ge-2/0/7               up    down
ge-2/0/8               up    down
ge-2/0/9               up    down
ge-2/0/10              up    down
ge-2/0/11              up    down
ge-2/0/12              up    down
ge-2/0/13              up    down
ge-2/0/14              up    down
ge-2/0/15              up    down
ge-2/0/16              up    down
ge-2/0/17              up    down
ge-2/0/18              up    down
ge-2/0/19              up    down
ge-2/0/20              up    down
```

```
ge-2/0/21            up    down
ge-2/0/22            up    up
ge-2/0/23            up    up
vcp-0               up    up
vcp-0.32768         up    up
vcp-1               up    up
vcp-1.32768         up    up
bme0                up    up
bme0.32768          up    up    inet     128.0.0.17/2
                                         128.0.0.32/2
                                tnp      0x11
bme0.32770          down  up    eth-switch
dsc                 up    up
gre                 up    up
ipip                up    up
lo0                 up    up
lo0.0               up    up    inet     10.10.1.3      --> 0/0
lsi                 up    up
me0                 up    up
me0.0               up    up    eth-switch
mtun                up    up
pimd                up    up
pime                up    up
tap                 up    up
vlan                up    up
```

 Enabling GRES also allows for the configuration of non-stop bridging. This feature additionally saves Layer 2 Control Protocol (L2CP) information by running the L2CP process (l2cpd) on the backup RE. This essentially maintains Spanning Tree Protocol (STP), RSTP, and Multiple Spanning Tree Protocol (MSTP). Currently, this is supported only on the MX Series, but future support on the EX Series is expected.

GRES with Layer 2

Returning to the previous scenario without GRES (Figure 11-2), Scotch is going to issue a ping command to Host1. With GRES turned on, the results are much better!

As before, the switchover command is issued on Vodkila:

```
lab@Vodkila> request chassis routing-engine master switch
Toggle mastership between routing engines ? [yes,no] (no) yes

lab@Vodkila>
```

A ping is issued from the switch during the switchover and no packet drops occur, as the PFE is never rebooted:

```
lab@Scotch> ping 200.2.1.1 count 50
PING 200.2.1.1 (200.2.1.1): 56 data bytes
64 bytes from 200.2.1.1: icmp_seq=0 ttl=255 time=10.226 ms
64 bytes from 200.2.1.1: icmp_seq=1 ttl=255 time=4.169 ms
64 bytes from 200.2.1.1: icmp_seq=2 ttl=255 time=3.429 ms
64 bytes from 200.2.1.1: icmp_seq=3 ttl=255 time=3.456 ms
```

```
64 bytes from 200.2.1.1: icmp_seq=4 ttl=255 time=3.443 ms
64 bytes from 200.2.1.1: icmp_seq=5 ttl=255 time=3.176 ms
64 bytes from 200.2.1.1: icmp_seq=6 ttl=255 time=3.430 ms
64 bytes from 200.2.1.1: icmp_seq=7 ttl=255 time=3.443 ms
64 bytes from 200.2.1.1: icmp_seq=8 ttl=255 time=4.174 ms
64 bytes from 200.2.1.1: icmp_seq=9 ttl=255 time=3.432 ms
64 bytes from 200.2.1.1: icmp_seq=10 ttl=255 time=3.177 ms
64 bytes from 200.2.1.1: icmp_seq=11 ttl=255 time=3.448 ms
64 bytes from 200.2.1.1: icmp_seq=12 ttl=255 time=3.430 ms
64 bytes from 200.2.1.1: icmp_seq=13 ttl=255 time=3.181 ms
64 bytes from 200.2.1.1: icmp_seq=14 ttl=255 time=3.435 ms
64 bytes from 200.2.1.1: icmp_seq=15 ttl=255 time=3.431 ms
64 bytes from 200.2.1.1: icmp_seq=16 ttl=255 time=3.434 ms
64 bytes from 200.2.1.1: icmp_seq=17 ttl=255 time=4.241 ms
64 bytes from 200.2.1.1: icmp_seq=18 ttl=255 time=3.291 ms
64 bytes from 200.2.1.1: icmp_seq=19 ttl=255 time=3.444 ms
64 bytes from 200.2.1.1: icmp_seq=20 ttl=255 time=4.173 ms
64 bytes from 200.2.1.1: icmp_seq=21 ttl=255 time=3.435 ms
64 bytes from 200.2.1.1: icmp_seq=22 ttl=255 time=5.204 ms
64 bytes from 200.2.1.1: icmp_seq=23 ttl=255 time=4.157 ms
64 bytes from 200.2.1.1: icmp_seq=24 ttl=255 time=4.954 ms
64 bytes from 200.2.1.1: icmp_seq=25 ttl=255 time=4.165 ms
64 bytes from 200.2.1.1: icmp_seq=26 ttl=255 time=5.215 ms
64 bytes from 200.2.1.1: icmp_seq=27 ttl=255 time=4.270 ms
64 bytes from 200.2.1.1: icmp_seq=28 ttl=255 time=74.606 ms
64 bytes from 200.2.1.1: icmp_seq=29 ttl=255 time=3.436 ms
64 bytes from 200.2.1.1: icmp_seq=30 ttl=255 time=3.430 ms
64 bytes from 200.2.1.1: icmp_seq=31 ttl=255 time=3.429 ms
64 bytes from 200.2.1.1: icmp_seq=32 ttl=255 time=3.432 ms
64 bytes from 200.2.1.1: icmp_seq=33 ttl=255 time=5.214 ms
64 bytes from 200.2.1.1: icmp_seq=34 ttl=255 time=3.433 ms
64 bytes from 200.2.1.1: icmp_seq=35 ttl=255 time=3.475 ms
64 bytes from 200.2.1.1: icmp_seq=36 ttl=255 time=3.427 ms
64 bytes from 200.2.1.1: icmp_seq=37 ttl=255 time=5.215 ms
64 bytes from 200.2.1.1: icmp_seq=38 ttl=255 time=40.354 ms
64 bytes from 200.2.1.1: icmp_seq=39 ttl=255 time=3.433 ms
64 bytes from 200.2.1.1: icmp_seq=40 ttl=255 time=5.208 ms
64 bytes from 200.2.1.1: icmp_seq=41 ttl=255 time=4.170 ms
64 bytes from 200.2.1.1: icmp_seq=42 ttl=255 time=3.433 ms
64 bytes from 200.2.1.1: icmp_seq=43 ttl=255 time=3.442 ms
64 bytes from 200.2.1.1: icmp_seq=44 ttl=255 time=4.186 ms
64 bytes from 200.2.1.1: icmp_seq=45 ttl=255 time=5.216 ms
64 bytes from 200.2.1.1: icmp_seq=46 ttl=255 time=3.429 ms
64 bytes from 200.2.1.1: icmp_seq=47 ttl=255 time=5.218 ms
64 bytes from 200.2.1.1: icmp_seq=48 ttl=255 time=32.536 ms
64 bytes from 200.2.1.1: icmp_seq=49 ttl=255 time=3.434 ms

--- 200.2.1.1 ping statistics ---
50 packets transmitted, 50 packets received, 0% packet loss
round-trip min/avg/max/stddev = 3.176/6.712/74.606/11.674 ms
```

GRES with Layer 3

The result is slightly different if we go back to the Layer 3 scenario, due to the fact that GRES syncs up the kernel and chassis information, but does not synchronize any information associated with the Routing Protocol Daemon (RPD). This process controls all routing protocols, so during a switchover, the process needs to be restarted. This means any dynamic routing protocols such as Routing Information Protocol (RIP), OSPF, or Border Gateway Protocol (BGP) will also be restarted.

Observe this behavior by once again manually initiating a switchover on Vodkila and issuing a ping command from Scotch:

```
lab@Scotch> ping 200.2.1.1 count 100
PING 200.2.1.1 (200.2.1.1): 56 data bytes
64 bytes from 200.2.1.1: icmp_seq=0 ttl=254 time=44.453 ms
64 bytes from 200.2.1.1: icmp_seq=1 ttl=254 time=3.423 ms
64 bytes from 200.2.1.1: icmp_seq=2 ttl=254 time=3.390 ms
64 bytes from 200.2.1.1: icmp_seq=3 ttl=254 time=3.168 ms
64 bytes from 200.2.1.1: icmp_seq=4 ttl=254 time=3.430 ms
64 bytes from 200.2.1.1: icmp_seq=5 ttl=254 time=4.413 ms
64 bytes from 200.2.1.1: icmp_seq=6 ttl=254 time=4.163 ms
64 bytes from 200.2.1.1: icmp_seq=7 ttl=254 time=5.217 ms
64 bytes from 200.2.1.1: icmp_seq=8 ttl=254 time=3.437 ms
64 bytes from 200.2.1.1: icmp_seq=9 ttl=254 time=4.163 ms
64 bytes from 200.2.1.1: icmp_seq=10 ttl=254 time=26.783 ms
64 bytes from 200.2.1.1: icmp_seq=11 ttl=254 time=5.213 ms
64 bytes from 200.2.1.1: icmp_seq=12 ttl=254 time=3.176 ms
64 bytes from 200.2.1.1: icmp_seq=13 ttl=254 time=3.427 ms
64 bytes from 200.2.1.1: icmp_seq=14 ttl=254 time=3.448 ms
64 bytes from 200.2.1.1: icmp_seq=15 ttl=254 time=5.971 ms
64 bytes from 200.2.1.1: icmp_seq=16 ttl=254 time=3.428 ms
64 bytes from 200.2.1.1: icmp_seq=17 ttl=254 time=3.399 ms
64 bytes from 200.2.1.1: icmp_seq=18 ttl=254 time=4.165 ms
64 bytes from 200.2.1.1: icmp_seq=19 ttl=254 time=3.407 ms
64 bytes from 200.2.1.1: icmp_seq=20 ttl=254 time=41.399 ms
64 bytes from 200.2.1.1: icmp_seq=21 ttl=254 time=5.978 ms
64 bytes from 200.2.1.1: icmp_seq=22 ttl=254 time=3.421 ms
64 bytes from 200.2.1.1: icmp_seq=23 ttl=254 time=4.430 ms
64 bytes from 200.2.1.1: icmp_seq=24 ttl=254 time=4.163 ms
64 bytes from 200.2.1.1: icmp_seq=25 ttl=254 time=3.392 ms
ping: sendto: No route to host
ping: sendto: No route to host
ping: sendto: No route to host
ping: sendto: No route to host
ping: sendto: No route to host
64 bytes from 200.2.1.1: icmp_seq=31 ttl=254 time=3.432 ms
64 bytes from 200.2.1.1: icmp_seq=32 ttl=254 time=3.424 ms
64 bytes from 200.2.1.1: icmp_seq=33 ttl=254 time=4.180 ms
64 bytes from 200.2.1.1: icmp_seq=34 ttl=254 time=4.963 ms
64 bytes from 200.2.1.1: icmp_seq=35 ttl=254 time=3.429 ms
64 bytes from 200.2.1.1: icmp_seq=36 ttl=254 time=4.162 ms
64 bytes from 200.2.1.1: icmp_seq=37 ttl=254 time=4.414 ms
64 bytes from 200.2.1.1: icmp_seq=38 ttl=254 time=3.397 ms
64 bytes from 200.2.1.1: icmp_seq=39 ttl=254 time=3.185 ms
```

```
64 bytes from 200.2.1.1: icmp_seq=40 ttl=254 time=26.110 ms
64 bytes from 200.2.1.1: icmp_seq=41 ttl=254 time=4.427 ms
64 bytes from 200.2.1.1: icmp_seq=42 ttl=254 time=3.433 ms
64 bytes from 200.2.1.1: icmp_seq=43 ttl=254 time=3.449 ms
64 bytes from 200.2.1.1: icmp_seq=44 ttl=254 time=3.402 ms
64 bytes from 200.2.1.1: icmp_seq=45 ttl=254 time=3.428 ms
64 bytes from 200.2.1.1: icmp_seq=46 ttl=254 time=4.157 ms
64 bytes from 200.2.1.1: icmp_seq=47 ttl=254 time=4.422 ms
64 bytes from 200.2.1.1: icmp_seq=48 ttl=254 time=5.211 ms
64 bytes from 200.2.1.1: icmp_seq=49 ttl=254 time=3.178 ms
64 bytes from 200.2.1.1: icmp_seq=50 ttl=254 time=68.221 ms
64 bytes from 200.2.1.1: icmp_seq=51 ttl=254 time=3.431 ms
64 bytes from 200.2.1.1: icmp_seq=52 ttl=254 time=3.452 ms
64 bytes from 200.2.1.1: icmp_seq=53 ttl=254 time=3.436 ms
64 bytes from 200.2.1.1: icmp_seq=54 ttl=254 time=3.396 ms
64 bytes from 200.2.1.1: icmp_seq=55 ttl=254 time=4.149 ms
64 bytes from 200.2.1.1: icmp_seq=56 ttl=254 time=3.407 ms
64 bytes from 200.2.1.1: icmp_seq=57 ttl=254 time=3.426 ms
64 bytes from 200.2.1.1: icmp_seq=58 ttl=254 time=3.172 ms
64 bytes from 200.2.1.1: icmp_seq=59 ttl=254 time=6.234 ms
64 bytes from 200.2.1.1: icmp_seq=60 ttl=254 time=32.687 ms
64 bytes from 200.2.1.1: icmp_seq=61 ttl=254 time=4.187 ms
64 bytes from 200.2.1.1: icmp_seq=62 ttl=254 time=3.446 ms
64 bytes from 200.2.1.1: icmp_seq=63 ttl=254 time=3.404 ms
64 bytes from 200.2.1.1: icmp_seq=64 ttl=254 time=4.163 ms
64 bytes from 200.2.1.1: icmp_seq=65 ttl=254 time=3.424 ms
64 bytes from 200.2.1.1: icmp_seq=66 ttl=254 time=3.430 ms
64 bytes from 200.2.1.1: icmp_seq=67 ttl=254 time=3.176 ms
64 bytes from 200.2.1.1: icmp_seq=68 ttl=254 time=5.211 ms
64 bytes from 200.2.1.1: icmp_seq=69 ttl=254 time=3.430 ms
64 bytes from 200.2.1.1: icmp_seq=70 ttl=254 time=26.575 ms
64 bytes from 200.2.1.1: icmp_seq=71 ttl=254 time=3.428 ms
64 bytes from 200.2.1.1: icmp_seq=72 ttl=254 time=3.390 ms
64 bytes from 200.2.1.1: icmp_seq=73 ttl=254 time=3.400 ms
64 bytes from 200.2.1.1: icmp_seq=74 ttl=254 time=3.615 ms
64 bytes from 200.2.1.1: icmp_seq=75 ttl=254 time=4.410 ms
64 bytes from 200.2.1.1: icmp_seq=76 ttl=254 time=5.209 ms
64 bytes from 200.2.1.1: icmp_seq=77 ttl=254 time=3.427 ms
64 bytes from 200.2.1.1: icmp_seq=78 ttl=254 time=3.433 ms
64 bytes from 200.2.1.1: icmp_seq=79 ttl=254 time=3.447 ms
64 bytes from 200.2.1.1: icmp_seq=80 ttl=254 time=40.982 ms
64 bytes from 200.2.1.1: icmp_seq=81 ttl=254 time=3.428 ms
64 bytes from 200.2.1.1: icmp_seq=82 ttl=254 time=3.429 ms
64 bytes from 200.2.1.1: icmp_seq=83 ttl=254 time=4.166 ms
64 bytes from 200.2.1.1: icmp_seq=84 ttl=254 time=3.413 ms
64 bytes from 200.2.1.1: icmp_seq=85 ttl=254 time=3.421 ms
64 bytes from 200.2.1.1: icmp_seq=86 ttl=254 time=3.172 ms
64 bytes from 200.2.1.1: icmp_seq=87 ttl=254 time=5.215 ms
64 bytes from 200.2.1.1: icmp_seq=88 ttl=254 time=4.164 ms
64 bytes from 200.2.1.1: icmp_seq=89 ttl=254 time=3.182 ms
64 bytes from 200.2.1.1: icmp_seq=90 ttl=254 time=33.141 ms
64 bytes from 200.2.1.1: icmp_seq=91 ttl=254 time=3.452 ms
64 bytes from 200.2.1.1: icmp_seq=92 ttl=254 time=3.169 ms
64 bytes from 200.2.1.1: icmp_seq=93 ttl=254 time=3.403 ms
64 bytes from 200.2.1.1: icmp_seq=94 ttl=254 time=3.415 ms
```

```
64 bytes from 200.2.1.1: icmp_seq=95 ttl=254 time=4.160 ms
64 bytes from 200.2.1.1: icmp_seq=96 ttl=254 time=4.417 ms
64 bytes from 200.2.1.1: icmp_seq=97 ttl=254 time=3.400 ms
64 bytes from 200.2.1.1: icmp_seq=98 ttl=254 time=3.429 ms
64 bytes from 200.2.1.1: icmp_seq=99 ttl=254 time=3.441 ms

--- 200.2.1.1 ping statistics ---
100 packets transmitted, 95 packets received, 5% packet loss
round-trip min/avg/max/stddev = 3.168/7.047/68.221/10.702 ms
```

The number of packets that were lost was reduced, but Scotch does lose packets when RPD is restarted and OSPF is reestablished.

Graceful Restart

With the advent of separate control and forwarding planes, and the need for a restart event of the control plane while keeping routing stable, *Graceful Restart* was created. The idea of Graceful Restart is that a restarting router can tell its neighbors about that event. The neighbors will, in turn, still use the restarting router for forwarding, as well as hide the restarting event from the rest of the network. The operation of the protocol during the restarting event varies, depending on the protocol. For instance, there are RFCs describing Graceful Restart for OSPF (RFC 3623) and for BGP (RFC 4724). Table 11-2 lists protocols and RFCs.

Table 11-2. Protocols and RFCs

Protocol	RFC	EX support (9.2R1.9)
OSPF	3623	Yes
BGP	4724	Yes
IS-IS	5306	Yes
OSPFv3	5187	No
RSVP	4558	No
LDP	3478	No
STP	N/A	Yes

In every protocol, the mechanism is the same but there are two types of devices: a restarting router and a helper. The restarting router (or switch) informs its neighbor that a restart will occur, which means that control messages will not be seen during a grace period, but the forwarding of packets will proceed as normal. The helper routers support this event by not informing their neighbors that a change has occurred during this grace period. If contact is reestablished before the grace period ends, the helper aids in synchronizing any data that may have changed during the switchover. If contact is not established and the grace time expires, the helper informs its neighbors of the event and no longer uses the restarting router for forwarding.

In JUNOS software, Graceful Restart is enabled globally under [edit routing-options], which turns on Graceful Restart for all supported protocols. The feature can be disabled if desired under the individual protocol configuration:

```
lab@Vodkila# set routing-options graceful-restart

{master}[edit]
lab@Vodkila# commit and-quit
```

With Graceful Restart turned on in Vodkila *and* Scotch, an RE switchover is initiated according to Figure 11-2, and no loss of packets occurs!

```
lab@Scotch> ping 200.2.1.1 count 100
PING 200.2.1.1 (200.2.1.1): 56 data bytes
64 bytes from 200.2.1.1: icmp_seq=0 ttl=254 time=3.933 ms
64 bytes from 200.2.1.1: icmp_seq=1 ttl=254 time=5.212 ms

--- 200.2.1.1 ping statistics ---
100 packets transmitted, 100 packets received, 0% packet loss
round-trip min/avg/max/stddev = 3.166/4.730/40.835/5.064 ms
```

OSPF Graceful Restart

Graceful Restart for OSPF uses one of the opaque link state advertisements (LSAs) in OSPF, number 9 with opaque type 3 and ID 0. This LSA has link local scope and is generated only during a restarting event. This LSA will contain information such as the grace timer, restarting reason, type of interface, and interface IP:

```
lab@Scotch> show ospf database

    OSPF link state database, Area 0.0.0.0
 Type      ID               Adv Rtr           Seq
Age Opt Cksum  Len
Router   10.10.1.3        10.10.1.3         0x80000017
125  0x22 0xb88b 168
Router  *10.10.1.9        10.10.1.9         0x8000000c
210  0x22 0xf08e  48
Network *10.3.9.9         10.10.1.9         0x80000005
725  0x22 0x635f  32

    OSPF Link-Local link state database, interface
ge-0/0/0.0
 Type      ID               Adv Rtr           Seq
Age Opt Cksum  Len
OpaqLoc  3.0.0.0          10.10.1.3         0x80000001
08  0x22 0xeb24  44

lab@Scotch> show ospf database detail  | find "OSPF
Link-Local"

    OSPF Link-Local link state database, interface
ge-0/0/0.0
 Type      ID               Adv Rtr           Seq
Age Opt Cksum  Len
OpaqLoc  3.0.0.0          10.10.1.3         0x80000001
14  0x22 0xeb24  44
  Link-Local grace LSA
```

```
        Grace 210
        Reason 1
        IPAddr 10.3.9.3
```

In this case, 10.10.1.3 is restarting on interface 10.3.9.2. The grace time is 210 seconds, and the device has been restarting for 14 seconds based on the age timer. The reason is type 1, which stands for software restart. Other options are software upgrade (type 2) and switch to a redundant route processor (type 3).

After the restarting event has occurred, the restarting router will flush the opaque LSAs and resynchronize the link state database (LSDB) with its neighbors to ensure that no information has changed or was lost during the switchover.

Non-Stop Routing

The problem with Graceful Restart is that a restarting event relies not only on the restarting router, but also on helpers. If the helpers do not support a Graceful Restart, packet loss will occur. This is especially problematic in a mixed-vendor environment, where some equipment could be configured as helpers and some not. A classic case is OSPF Graceful Restart on Juniper and Cisco. Cisco implemented a prestandard Graceful Restart mechanism that did not interoperate with Juniper. Since Graceful Restart was not supported in IOS, to alleviate the need for helper support, Juniper implemented a feature called NSR (Non-Stop Routing). This feature has all the same mechanisms as we previously discussed regarding GRES, except that it also runs the RPD on the backup RE, as well as synchronized Transmission Control Protocol (TCP) state information. As a result, during a failure, routing protocol state is maintained without the need for helper device support. In other words, the rest of the network has no idea that a restart event is happening, as it stays local to that device. To enable NSR, configure Non-Stop Routing under the [edit routing-options] hierarchy.

 At the time of this writing, NSR is not supported in EX devices. But this feature could be available by the time this book is published.

GRES, GR, NSR, Oh My!

We just threw a slew of acronyms at you, and you may be confused as to which features to use and the differences between them! The first step in any RE HA is to activate GRES in order to get kernel synchronization between two REs. Without this feature, the PFE resets after an RE change. Then, if the RPD needs to be involved in the process, either Graceful Restart or NSR needs to be configured in order to prevent protocols from advertising the restart event. Graceful Restart relies on neighbor devices, whereas NSR does not. NSR is the preferred method, if it is supported on your device.

VRRP

Anybody using a PC for Internet surfing, music downloads, gaming, or video downloads will use IP as the network protocol. The PC will have an IP address assigned, as well as a default gateway address to reach any destinations that are not on the local subnet.

If the default gateway is a single device and that device fails, the PC won't be able to reach destinations outside the local subnet. In other words, the infamous "the Internet is down" support call will occur. In a fault-tolerant network, it would be ideal to have a backup gateway device, without having to modify the configuration on any PCs, as well as being able to load-share with multiple PCs on the LAN.

Virtual Router Redundancy Protocol was created to eliminate single points of behavior that are inherent to static default routed networks. VRRP creates a logical grouping of multiple physical routers to a "virtual" router that will be used as the default gateway for end hosts. In modern networks, this device could physically be a router or any other device that supports VRRP, such as an EX Series switch. Regardless, VRRP allows the PC to always maintain the same gateway address even if the physical gateway has changed (Figure 11-3). The "routers" that are part of the same VRRP logical group will share this "virtual" IP address, as well as sharing a "virtual" Media Access Control (MAC) address. Essentially, VRRP describes an *election protocol* in order to maintain ownership of this virtual IP address and MAC address. One "router" in the VRRP group will be the master router, controlling the virtual IP address unless a failure occurs that results in a release of that ownership. Such a failure would cause another "router" to claim ownership of the virtual IP by issuing a VRRP message and a Gratuitous ARP to claim the virtual MAC address. Once a "router" becomes the master, it periodically advertises VRRP messages to indicate its overall health and reachability.

 Please consult Chapter 7 for more VRRP details.

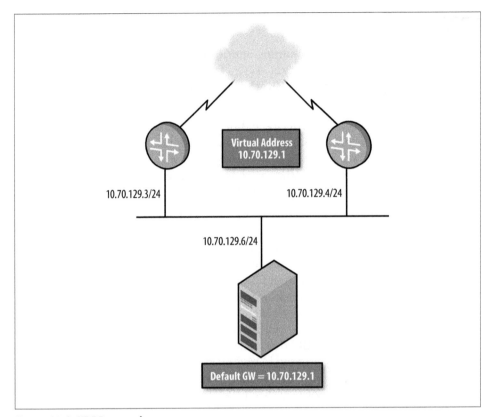

Figure 11-3. VRRP example

In-Service Software Upgrades

Another feature that is very useful is the idea of an In-Service Software Upgrade (ISSU). This is a software upgrade that allows you to move to a new JUNOS release with no disruption in the control plane, and little if any disruption in the forwarding plane. This feature is supported only on dual RE platforms, including the VC, and the REs must be running the same JUNOS software version before ISSU. Lastly, in order for this feature to work, GRES and NSR must also be enabled.

 As in the NSR warning, ISSU is currently not supported on EX Series devices, but the word is that it is planned.

Perform ISSU by issuing the following steps:

1. Enable GRES and non-stop active routing. Verify that the REs and protocols are synchronized.

2. Download the new software package from the Juniper Networks website and then copy the package to the router.

3. Issue the request system software `in-service-upgrade` command on the master RE.

During ISSU, both the master and backup will be upgraded to the new versions of the JUNOS software. Since this is not yet supported on the EX, we won't include additional details here.

Aggregated Ethernet

Aggregated Ethernet (AE) enables you to group multiple Ethernet interfaces to form a single physical link at the MAC layer, allowing you to increase your speed beyond what a single link can provide, as well as increasing redundancy. AE has had several names over the years. Officially, it is a standard in 802.1AX; however, many people still refer to it as 802.3ad because it appeared in section 41 of that standard. Other names that have been used by various vendors include *NIC teaming*, *port channel*, *port teaming*, *port trunking*, *link bundling*, and *EtherChannel*. In this chapter, however, we will refer to the link as AE.

 How does a standard get its name, such as 802.1AX or 802.3ad? Well, the 802 refers to an actual working group for the standard. The letters themselves have no meaning; letters are just incremented starting from a and going to z. So, in order: a, aa, ab, b, bb, and so on. The capitalization indicates a standalone document, whereas lowercase denotes a supplement to an existing document. At least, that was the idea when the naming convention was created.

Historically, link aggregation is configured statically, meaning that the local device bonds the physical interfaces together. This meant that there was no way to link path failures (if there was an intermediate device) or to know whether the other end of the link was also bundling the interface. For that reason, the industry-standard Link Aggregation Control Protocol (LACP) was created. Essentially, LACP sends messages to the remote end of the link to determine bundle parameters as well as link availability. This creates a "bundle" with multiple physical "members."

Although AE interfaces were originally created for redundancy, an added benefit may be the ability to balance traffic across each member link. The actual algorithm is vendor-independent, and could be platform-independent. Since load balancing is an outbound forwarding decision, in an EX Series switch load balancing is achieved by creating a hash on the Layer 3 and Layer 4 information in the packet. Every packet with the same hash will always take the same link, whereas a different hash may take another link. The key point is that a hash will always choose the same link.

You specify up to eight Ethernet ports when configuring a bundle. This interface can reside on any cards in the switch, as long as they have the same speed and are in full duplex (FD) mode. This includes a port on a VC with a VC having an additional limitation of 64 bundles, whereas an EX8200 series can support up to 127 bundles.

LACP in Action

LACP is not necessary in order to create an AE link. You can create the configuration statically on each link, essentially telling the system to bundle several links together and be on its way. However, there are some problems with static configuration:

- There is no keepalive mechanism to determine whether the entire path is up if there is an intermediate device between the bundle.
- There is no way to ensure that the port physically connects to the correct device (see Figure 11-3, shown previously).
- There is no way to ensure that the configuration is correct on both sides.

LACP solves this without any change to the actual Ethernet frame format used for transport across the bundle. Essentially, LACP periodically advertises the aggregation capabilities and state information across each member link. The partner (the other end of the link) compares its own configuration to what was received, and decides whether the link should be part of the bundle. Also, if a link was already determined to be part of the same bundle, and LACP messages are not received for a period of time, that link can be brought down.

These messages are sent on a link-by-link basis, asynchronously. No synchronization is necessary for the bundle.

On each aggregated link, a *partner* and an *actor* are created. The actor is the device that takes the action, and the partner is the device sending the message. An actor compares its local configuration with the received partner message. This messaging happens in both directions; as a result, a given link is always an actor or a partner—the name is relative to a given direction. This means, however, that an additional check can be made. In confusing terms, we can say that the partner's partner information is a view of the actor's current configuration. Thus, the partner's partner information can be compared with its own actor state to make sure the partner received the correct information from the actor. In simpler terms, the received state is sent back to the sender in order to verify that the sender information was received.

In order to make sure aggregated bundles are formed only with links that interconnect the same devices, some type of identifier must be used to identify the bundle. LACP uses a system ID to identify messages from a device, and only messages received from the partner with the same system ID will be considered for link aggregation.

The system ID is a 64-bit value, with 48 bits used for the MAC address and 16 bits used for a priority field. The priority field is used for network administrator control for dynamic changes to the bundle. The system with the lowest system identifier is allowed to make dynamic changes to its aggregated link capabilities. This prevents two partners from trying to make changes at the same time and causing instability. A port ID and priority are also given to each port; these determine the order in which a given link can be added to the bundle.

Also, the fact that a link terminates at the same partner system does not necessarily mean that it will be part of the link bundle. Each link is assigned a key value, and a link must have the same key in order to be eligible to be aggregated. This key is a 16-bit value assigned locally to a port, and links with different keys should never be aggregated together. Here are some reasons links between systems might have different key values:

- The link speeds are different (1 Gbps versus 10 Gbps).
- The physical limitations of the hardware vary; perhaps a link could be aggregated on only a single LC, and cannot span multiple LCs.
- A system administrator may configure a maximum number of links that can be part of the bundle, and that limit has already been reached.

Each system is in either *active* or *passive* mode. Active mode periodically sends LACP messages, and passive mode will not transmit messages until it receives a message from its partner. That way, if a switch connects to a device that does not support LACP, unnecessary transmissions are not performed. If at a later date that interface begins to support LACP, no reconfiguration is necessary.

As previously mentioned, LACP messages are sent at a periodic rate of every second (fast rate) or every 30 seconds (slow rate). The rate is determined by the partner; the fastest rate is the default.

LACP messages are always sent with address 01-80-C2-00-00-02, which is a link local address. Inside the LACP messages are type length values (TLVs) for the actor information, partner information, collector information, and terminator. The actor information is the information that has been sent to the other end of the link. The partner information is the actor's view of the received information. The collector information is for determining the timer to switch a flow from one link to another (more on this later), and the terminator TLV simply marks an end to the message. So, the partner/actor information transmitted in the TLV is:

- System priority
- System ID
- Key
- Port priority
- Port number
- State

The state information is 8 bits, and the bits are classified as shown in Table 11-3.

Table 11-3. Classification of 8-bit state information

Bit	Name	Definitions
0	Activity	0 = passive mode
		1 = active mode
1	Timeout	0 = long timeout (90 seconds)
		1 = short timeout (3 seconds)
2	Aggregation	0 = link cannot be aggregated
		1 = link can be aggregated
3	Synchronization	0 = port is not bound to aggregator
		1 = port is bound to aggregator
4	Collecting	0 = collector disabled
		1 = collector enabled
5	Distributing	0 = distributor disabled
		1 = distributor enabled
6	Defaulted	0 = device is using configured partner information
		1 = device is using received partner information
7	Expired	0 = LACP state is not expired
		1 = LACP state is expired

In an EX switch, you can view LACP messages by issuing a monitor traffic command
on the member link:

```
lab@Brandy> monitor traffic interface ge-0/0/3 layer2-headers extensive
15:47:46.338554 bpf_flags 0x80, Out
        Juniper PCAP Flags [Ext], PCAP Extension(s) total length 16
          Device Media Type Extension TLV #3, length 1, value: Ethernet (1)
          Logical Interface Encapsulation Extension TLV #6, length 1, value:
Ethernet (14)
          Device Interface Index Extension TLV #1, length 2, value: 35328
          Logical Interface Index Extension TLV #4, length 4, value: 68
        -----original packet-----
          2:1f:12:30:98:40 > 1:80:c2:0:0:2, ethertype Slow Protocols (0x8809),
length 124: LACPv1, length 110
        Actor Information TLV (0x01), length 20
          System 0:1f:12:30:98:40, System Priority 0, Key 1, Port 520, Port
Priority 127
          State Flags [Activity, Timeout, Aggregation, Synchronization,
Collecting, Distributing]
          0x0000: 001f 1230 9840 0001 007f 0208 3f00 0000
          0x000f: 0214
        Partner Information TLV (0x02), length 20
          System 0:1f:12:30:96:80, System Priority 0, Key 1, Port 520, Port
Priority 127
          State Flags [Activity, Timeout, Aggregation, Synchronization,
```

```
Collecting, Distributing]
        0x0000: 001f 1230 9680 0001 007f 0208 3f00 0000
        0x000f: 0310
    Collector Information TLV (0x03), length 16
     Max Delay 0
        0x0000: 0000 0000 0000 0000 0000 0000 0000
    Terminator TLV (0x00), length 0 (=52)
        0x0000: 0000 0000 0000 0000 0000 0000 0000 0000
        0x000f: 0000 0000 0000 0000 0000 0000 0000 0000
        0x001f: 0000 0000 0000 0000 0000 0000 0000 0000
        0x002f:
     packet exceeded snapshot
15:47:46.930512 bpf_flags 0x81,  In
    Juniper PCAP Flags [Ext, In], PCAP Extension(s) total length 16
        Device Media Type Extension TLV #3, length 1, value: Ethernet (1)
        Logical Interface Encapsulation Extension TLV #6, length 1, value:
Ethernet (14)
        Device Interface Index Extension TLV #1, length 2, value: 35328
        Logical Interface Index Extension TLV #4, length 4, value: 68
        -----original packet-----
        2:1f:12:30:96:80 > 1:80:c2:0:0:2, ethertype Slow Protocols (0x8809),
length 124: LACPv1, length 110
    Actor Information TLV (0x01), length 20
        System 0:1f:12:30:96:80, System Priority 0, Key 1, Port 520, Port
Priority 127
        State Flags [Activity, Timeout, Aggregation, Synchronization,
Collecting, Distributing]
        0x0000: 001f 1230 9680 0001 007f 0208 3f00 0000
        0x000f: 0214
    Partner Information TLV (0x02), length 20
        System 0:1f:12:30:98:40, System Priority 0, Key 1, Port 520, Port
Priority 127
        State Flags [Activity, Timeout, Aggregation, Synchronization,
Collecting, Distributing]
        0x0000: 001f 1230 9840 0001 007f 0208 3f00 0000
        0x000f: 0310
    Collector Information TLV (0x03), length 16
     Max Delay 0
        0x0000: 0000 0000 0000 0000 0000 0000 0000
    Terminator TLV (0x00), length 0 (=52)
        0x0000: 0000 0000 0000 0000 0000 0000 0000 0000
        0x000f: 0000 0000 0000 0000 0000 0000 0000 0000
        0x001f: 0000 0000 0000 0000 0000 0000 0000 0000
        0x002f:
     packet exceeded snapshot
```

You also can view messages by turning on traceoptions under [edit protocols lacp]:

```
lab@Brandy# show protocols
lacp {
    traceoptions {
        file lacp.trace;
        flag all;
    }
}
```

```
lab@Brandy# run show log lacp.trace

Aug 22 19:02:02 ge-0/0/9: incoming lacp pdu
Aug 22 19:02:02 Received LACP pdu - interface ge-0/0/9
Aug 22 19:02:02 lacp subtype 0x1 lacp version number 0x1
Aug 22 19:02:02 first tlv type 0x1 actor info len 0x14
Aug 22 19:02:02 actor sys priority 0x0 actor sys 00:1f:12:30:98:40
Aug 22 19:02:02 actor key 0x1 actor port priority 0x7f actor port 0x209 actor
state 0x3f
Aug 22 19:02:02 second tlv type 0x2 partner info len 0x14
Aug 22 19:02:02 partner sys priority 0x0 partner sys 00:1f:12:30:96:80
Aug 22 19:02:02 partner key 0x1 partner port priority 0x7f partner port 0x209
partner state 0x3f
Aug 22 19:02:02 third tlv type 0x3 collector info len 0x10 collector max del 0x0
fourth tlv type 0x0 terminator len 0
Aug 22 19:02:02 ge-0/0/9: lacp recv state transition from CURRENT to CURRENT
Aug 22 19:02:02 ge-0/0/9: lacp mux state transition from COLLECTING_DISTRIBUTING
to COLLECTING_DISTRIBUTING
Aug 22 19:02:02 lacp_periodic_transmission_machine: begin 0, lacp enable 1, port
1, actor state 0x3f, partner state 0x3f
Aug 22 19:02:02 ge-0/0/9: lacp xmit state transition from FAST_PERIODIC to
FAST_PERIODIC
Aug 22 19:02:03 ge-0/0/5: process timer expired on 0x1993600 in FAST_PERIODIC state
Aug 22 19:02:03 lacp_periodic_transmission_machine: begin 0, lacp enable 1, port
1, actor state 0x3f, partner state 0x3f
Aug 22 19:02:03 ge-0/0/5: lacp xmit state transition from FAST_PERIODIC to
PERIODIC_TX
Aug 22 19:02:03 ge-0/0/5: process_periodic_tx_state: need to transmit lacpdu
Aug 22 19:02:03 Transmitting LACP pdu - interface ge-0/0/5
Aug 22 19:02:03 lacp subtype 0x1 lacp version number 0x1
Aug 22 19:02:03 first tlv type 0x1 actor info len 0x14
Aug 22 19:02:03 actor sys priority 0x0 actor sys 00:1f:12:30:96:80
Aug 22 19:02:03 actor key 0x1 actor port priority 0x7f actor port 0x205 actor
state 0x3f
Aug 22 19:02:03 second tlv type 0x2 partner info len 0x14
Aug 22 19:02:03 partner sys priority 0x0 partner sys 00:1f:12:30:98:40
Aug 22 19:02:03 partner key 0x1 partner port priority 0x7f partner port 0x202
partner state 0x3f
Aug 22 19:02:03 third tlv type 0x3 collector info len 0x10 collector max del 0x0
fourth tlv type 0x0 terminator len 0
Aug 22 19:02:03 lacp_periodic_transmission_machine: begin 0, lacp enable 1, port
1, actor state 0x3f, partner state 0x3f
Aug 22 19:02:03 ge-0/0/5: lacp xmit state transition from PERIODIC_TX to
FAST_PERIODIC
Aug 22 19:02:03 ge-0/0/8: process timer expired on 0x1993800 in FAST_PERIODIC state
Aug 22 19:02:03 lacp_periodic_transmission_machine: begin 0, lacp enable 1, port
1, actor state 0x3f, partner state 0x3f
```

JUNOS Configuration

Configure a Link Aggregated Group (LAG) by specifying the link number as a physical
device and then associating a set of ports with the link. All the ports must have the same
speed and be in FD mode. JUNOS software for EX Series switches assigns a unique ID

---

and port priority to each port. The ID and priority are not configurable. In Figure 11-4, interfaces ge-0/0/3, ge-0/0/8, and ge-0/0/9 are all part of the bundle on AE0.

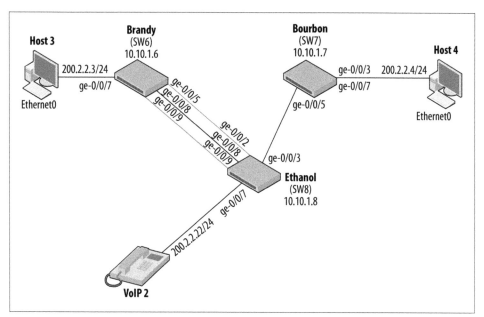

Figure 11-4. AE between Brandy and Ethanol

This AE interface is a Layer 2 interface providing virtual LAN (VLAN) trunking for two VLANs. The first step is to enable an AE interface on the switch at the [edit chassis] level. This enables a bundle to be formed and indicates the maximum number of bundles that are allowed (1–32). In this case, just one bundle is going to be configured:

```
[edit]
lab@Brandy# show chassis
aggregated-devices {
    ethernet {
        device-count 1;
    }
}
```

Next, associate the member link with a bundle (ae0) and configure the bundle interface itself. On the AE0 interface, LACP is configured, as the default operation is passive mode:

```
lab@Ethanol# show interfaces
ge-0/0/2 {
    ether-options {
        802.3ad ae0;
    }
}
ge-0/0/3 {
    unit 0 {
```

```
                family ethernet-switching {
                    port-mode trunk;
                    vlan {
                        members [ 20 10 ];
                    }
                }
            }
        }
    ge-0/0/8 {
        ether-options {
            802.3ad ae0;
        }
    }
    ge-0/0/9 {
        ether-options {
            802.3ad ae0;
        }
    }
    ge-0/0/12 {
        disable;
    }
    ge-0/1/0 {
        unit 0 {
            family ethernet-switching;
        }
    }
    xe-0/1/0 {
        unit 0 {
            family ethernet-switching;
        }
    }
    ge-0/1/1 {
        unit 0 {
            family ethernet-switching;
        }
    }
    xe-0/1/1 {
        unit 0 {
            family ethernet-switching;
        }
    }
    ge-0/1/2 {
        unit 0 {
            family ethernet-switching;
        }
    }
    ge-0/1/3 {
        unit 0 {
            family ethernet-switching;
        }
    }
    ae0 {
        unit 0 {
            family ethernet-switching {
                port-mode trunk;
```

```
                vlan {
                    members [ admin sales ];
                }
            }
        }
    }
    lo0 {
        unit 0 {
            family inet {
                address 10.10.1.8/32;
            }
        }
    }
    me0 {
        unit 0 {
            family inet {
                address 172.16.69.8/24;
            }
        }
    }
}
```

Verify that the interfaces are mapped to the correct bundle:

```
[edit]
lab@Brandy# run show interfaces terse
Interface               Admin Link Proto    Local            Remote
ge-0/0/0                up    up
ge-0/0/1                up    up
ge-0/0/1.0              up    up   eth-switch
ge-0/0/2                up    up
ge-0/0/2.0              up    up   eth-switch
ge-0/0/3                up    up
ge-0/0/3.0              up    up   eth-switch
ge-0/0/4                up    up
ge-0/0/4.0              up    up   eth-switch
ge-0/0/5                up    up
ge-0/0/5.0              up    up   aenet    --> ae0.0
ge-0/0/6                up    down
ge-0/0/7                up    up
ge-0/0/7.0              up    up   eth-switch
ge-0/0/8                up    up
ge-0/0/8.0              up    up   aenet    --> ae0.0
ge-0/0/9                up    up
ge-0/0/9.0              up    up   aenet    --> ae0.0
ge-0/0/10               up    down
ge-0/0/11               up    down
ge-0/0/12               up    up
ge-0/0/12.0             up    up   eth-switch
ge-0/0/13               up    down
ge-0/0/14               up    down
ge-0/0/15               up    down
ge-0/0/16               up    down
ge-0/0/17               up    down
ge-0/0/18               up    down
ge-0/0/19               up    down
ge-0/0/20               up    down
```

```
ge-0/0/21               up    down
ge-0/0/22               up    down
ge-0/0/23               up    down
ae0                     up    up
ae0.0                   up    up    eth-switch
bme0                    up    up
bme0.32768              up    up    inet    128.0.0.1/2
                                            128.0.0.16/2
                                            128.0.0.32/2
                                    tnp     0x10
dsc                     up    up
gre                     up    up
ipip                    up    up
loo                     up    up
loo.0                   up    up    inet    10.10.1.6      --> 0/0
lsi                     up    up
me0                     up    up
me0.0                   up    up    inet    172.16.69.6/24
mtun                    up    up
pimd                    up    up
pime                    up    up
tap                     up    up
vlan                    up    up
```

Next, verify that LACP is configured and working properly over each member link:

```
lab@Brandy# run show lacp interfaces
Aggregated interface: ae0
    LACP state:     Role   Exp   Def  Dist  Col  Syn  Aggr  Timeout  Activity
       ge-0/0/5    Actor    No    No   Yes  Yes  Yes   Yes     Fast    Active
       ge-0/0/5  Partner    No    No   Yes  Yes  Yes   Yes     Fast    Active
       ge-0/0/8    Actor    No    No   Yes  Yes  Yes   Yes     Fast    Active
       ge-0/0/8  Partner    No    No   Yes  Yes  Yes   Yes     Fast    Active
       ge-0/0/9    Actor    No    No   Yes  Yes  Yes   Yes     Fast    Active
       ge-0/0/9  Partner    No    No   Yes  Yes  Yes   Yes     Fast    Active
    LACP protocol:    Receive State     Transmit State          Mux State
       ge-0/0/5            Current      Fast periodic  Collecting distributing
       ge-0/0/8            Current      Fast periodic  Collecting distributing
       ge-0/0/9            Current      Fast periodic  Collecting distributing
```

 A link that is not eligible for part of the bundle but is locally configured for that bundle would be in the detached state:

```
lab@Brandy# run show lacp interfaces
Aggregated interface: ae0
    LACP state:        Role   Exp   Def  Dist  Col
  Syn  Aggr  Timeout  Activity
       ge-0/0/5       Actor    No    No   Yes  Yes
  Yes  Yes     Fast    Active
       ge-0/0/5     Partner    No    No   Yes  Yes
  Yes  Yes     Fast    Active
       ge-0/0/8       Actor    No    No   Yes  Yes
  Yes  Yes     Fast    Active
       ge-0/0/8     Partner    No    No   Yes  Yes
  Yes  Yes     Fast    Active
       ge-0/0/9       Actor    No    No   Yes  Yes
  Yes  Yes     Fast    Active
```

```
        ge-0/0/9      Partner   No   No  Yes  Yes
  Yes   Yes   Fast     Active
        ge-0/0/3      Actor     No  Yes   No   No
  No    Yes   Fast     Active
        ge-0/0/3      Partner   No  Yes   No   No
  No    Yes   Fast     Passive
        LACP protocol:   Receive State    Transmit State
  Mux State
        ge-0/0/5                Current    Fast periodic
  Collecting distributing
        ge-0/0/8                Current    Fast periodic
  Collecting distributing
        ge-0/0/9                Current    Fast periodic
  Collecting distributing   ge-0/0/3
  Defaulted     Fast periodic             Detached
```

Traffic should be seen transmitting over the bundle:

```
lab@Brandy> show interfaces ae0 extensive
Physical interface: ae0, Enabled, Physical link is Up
  Interface index: 128, SNMP ifIndex: 158, Generation: 131
  Link-level type: Ethernet, MTU: 1514, Speed: 3000mbps,
  MAC-REWRITE Error: None, Loopback: Disabled, Source filtering: Disabled,
  Flow control: Disabled, Minimum links needed: 1, Minimum bandwidth needed: 0
  Device flags   : Present Running
  Interface flags: SNMP-Traps Internal: 0x0
  Current address: 02:1f:12:30:96:80, Hardware address: 02:1f:12:30:96:80
  Last flapped   : Never
  Statistics last cleared: Never
  Traffic statistics:
   Input  bytes  :          21862038                0 bps
   Output bytes  :          49314673                0 bps
   Input  packets:            110918                0 pps
   Output packets:            542012                0 pps
  IPv6 transit statistics:
   Input  bytes  :                 0
   Output bytes  :                 0
   Input  packets:                 0
   Output packets:                 0
  Input errors:
   Errors: 0, Drops: 0, Framing errors: 0, Runts: 0, Giants: 0,
   Policed discards: 0, Resource errors: 0
  Output errors:
   Carrier transitions: 21, Errors: 0, Drops: 0, MTU errors: 0,
   Resource errors: 0

  Logical interface ae0.0 (Index 65) (SNMP ifIndex 159) (Generation 130)
    Flags: SNMP-Traps 0x0 Encapsulation: ENET2
    Statistics        Packets         pps        Bytes         bps
    Bundle:
        Input :              0           0            0            0
        Output:       94538569           0   5901432513            0
    Protocol eth-switch, MTU: 0, Generation: 142, Route table: 0
      Flags: Is-Primary, Trunk-Mode
```

Traffic should also be seen flowing over a member link, depending on the flows that are being sent:

```
lab@Brandy> show interfaces ge-0/0/5 extensive
Physical interface: ge-0/0/5, Enabled, Physical link is Up
  Interface index: 135, SNMP ifIndex: 118, Generation: 138
  Link-level type: Ethernet, MTU: 1514, Speed: 1000mbps,
  MAC-REWRITE Error: None, Loopback: Disabled, Source filtering: Disabled,
  Flow control: Disabled, Auto-negotiation: Enabled, Remote fault: Online
  Device flags   : Present Running
  Interface flags: SNMP-Traps Internal: 0x0
  Link flags     : None
  CoS queues     : 8 supported, 8 maximum usable queues
  Hold-times     : Up 0 ms, Down 0 ms
  Current address: 02:1f:12:30:96:80, Hardware address: 00:1f:12:30:96:86
  Last flapped   : 2008-08-22 18:22:22 UTC (00:02:21 ago)
  Statistics last cleared: Never
  Traffic statistics:
   Input  bytes  :             12442024                    0 bps
   Output bytes  :             39952569                  512 bps
   Input  packets:                56154                    0 pps
   Output packets:               487302                    1 pps
   IPv6 transit statistics:
    Input  bytes  :                    0
    Output bytes  :                    0
    Input  packets:                    0
    Output packets:                    0
  Input errors:
    Errors: 0, Drops: 0, Framing errors: 0, Runts: 0, Policed discards: 0,
    L3 incompletes: 0, L2 channel errors: 0, L2 mismatch timeouts: 0,
    FIFO errors: 0, Resource errors: 0
  Output errors:
    Carrier transitions: 7, Errors: 0, Drops: 0, Collisions: 0, Aged packets: 0,
    FIFO errors: 0, HS link CRC errors: 0, MTU errors: 0, Resource errors: 0
  Egress queues: 8 supported, 4 in use
  Queue counters:       Queued packets  Transmitted packets      Dropped packets
    0 best-effort                    0                54783                     0
    1 assured-forw                   0                    0                     0
    5 expedited-fo                   0                    0                     0
    7 network-cont                   0               432519                     0
  Active alarms  : None
  Active defects : None
  MAC statistics:                      Receive             Transmit
    Total octets                      12442024             39952569
    Total packets                        56154               487302
    Unicast packets                          0                    0
    Broadcast packets                        0                    0
    Multicast packets                    56154               487302
    CRC/Align errors                         0                    0
    FIFO errors                              0                    0
    MAC control frames                       0                    0
    MAC pause frames                         0                    0
    Oversized frames                         0
    Jabber frames                            0
    Fragment frames                          0
```

```
   VLAN tagged frames                      0
   Code violations                         0
 Filter statistics:
   Input packet count                      0
   Input packet rejects                    0
   Input DA rejects                        0
   Input SA rejects                        0
   Output packet count                             0
   Output packet pad count                         0
   Output packet error count                       0
   CAM destination filters: 0, CAM source filters: 0
 Autonegotiation information:
   Negotiation status: Complete
   Link partner:
       Link mode: Full-duplex, Flow control: None, Remote fault: OK,
       Link partner Speed: 1000 Mbps
   Local resolution:
       Flow control: None, Remote fault: Link OK
 Packet Forwarding Engine configuration:
   Destination slot: 0
   Direction : Output
   CoS transmit queue          Bandwidth         Buffer Priority    Limit
                       %        bps        %      usec
   0 best-effort      95     950000000     95      NA     low      none
   7 network-control   5      50000000      5      NA     low      none

 Logical interface ge-0/0/5.0 (Index 71) (SNMP ifIndex 119) (Generation 136)
   Flags: SNMP-Traps Encapsulation: ENET2
   Traffic statistics:
    Input  bytes  :                 0
    Output bytes  :           5834273
    Input  packets:                 0
    Output packets:             27391
    IPv6 transit statistics:
     Input  bytes  :                0
     Output bytes  :                0
     Input  packets:                0
     Output packets:                0
   Local statistics:
    Input  bytes  :                 0
    Output bytes  :           5834273
    Input  packets:                 0
    Output packets:             27391
   Transit statistics:
    Input  bytes  :                 0              0 bps
    Output bytes  :                 0              0 bps
    Input  packets:                 0              0 pps
    Output packets:                 0              0 pps
    IPv6 transit statistics:
     Input  bytes  :                0
     Output bytes  :                0
     Input  packets:                0
     Output packets:                0
   Protocol aenet, AE bundle: ae0.0, Generation: 148, Route table: 0
```

Source MAC Selection

The source MAC address of packets transmitted over the bundle must be the same across any member link. In JUNOS, the AE interface MAC addresses are chosen out of the possible public MAC addresses:

```
lab@Ethanol# run show chassis mac-addresses
    FPC 0   MAC address information:
        Public base address      00:1f:12:30:98:40
        Public count             64
```

This public MAC address is then assigned to the member links while the link is part of the bundle. This address is a reserved address that should never be forwarded:

```
lab@Ethanol# run show interfaces | match
02:1f:12:30:98:40
  Current address: 02:1f:12:30:98:40, Hardware
address: 00:1f:12:30:98:43
  Current address: 02:1f:12:30:98:40, Hardware
address: 00:1f:12:30:98:49
  Current address: 02:1f:12:30:98:40, Hardware
address: 00:1f:12:30:98:4a
  Current address: 02:1f:12:30:98:40, Hardware
address: 02:1f:12:30:98:40
```

Additional configuration options

By default, when there are multiple links in a bundle, all links are used for sending traffic and the bundle is considered in the Up state if at least one member link is up. To change this behavior, configure the minimum links command under the aggregated bundle. For example, setting the minimum links requirement to 2 means that at least two member links must be up, or the aggregated interface will be in the down state:

```
lab@Brandy# set interfaces ae0 aggregated-ether-options minimum-links ?
Possible completions:
  <minimum-links>      Minimum number of aggregated links (1..8)
[edit]
```

If you have only two links in a bundle, you can configure LACP link protection. This creates standby and backup links:

```
lab@Brandy# show interfaces ae0
aggregated-ether-options {
    link-protection;
    lacp {
        active;
    }
}
unit 0 {
    family ethernet-switching {
        port-mode trunk;
        vlan {
            members [ sales admin ];
        }
```

```
        }
    }

[edit]
lab@Brandy# show interfaces ge-0/0/8
ether-options {
    802.3ad {
        ae0;
        primary;
    }
}

[edit]
lab@Brandy# show interfaces ge-0/0/9
ether-options {
    802.3ad {
        ae0;
        backup;
    }
}
```

Load balancing over AE

When sending traffic over multiple member links, the draft specifies that traffic that is part of the same "conversation" should always be on the same link. This is done by creating a hash, based on the conversation and specifying that every packet with that hash takes the same link. The exact details of the conversation are not defined, but in JUNOS the hashing works as follows:

- Non-IP traffic hashes on the source and destination MAC addresses (SMAC and DMAC).
- IP traffic uses the SMAC and DMAC addresses and the IP address, and, if Layer 4 is present, port numbers.

These flows are not user-configurable at this time.

> The hash is used only for transit traffic over the AE link. CPU control packets are always sent out on the lowest-numbered link.

Bidirectional Forwarding Detection (BFD)

BFD is a protocol that was created to detect link failures very quickly. Essentially, it is a *very* fast hello protocol that gets around the slower protocol timers in OSPF, RIP, Intermediate System to Intermediate System (IS-IS), and so on. In a routed environment, this is especially useful when two Layer 3 devices are connected to a Layer 2 device in the middle (Figure 11-5). The problem is that the Router1 link could go down on Tequila, but Router2 would not detect the failure until the protocol timed out as a result of its local link remaining up. If OSPF was running, this could take up to 40 seconds with default timers. BFD was created to speed up this detection.

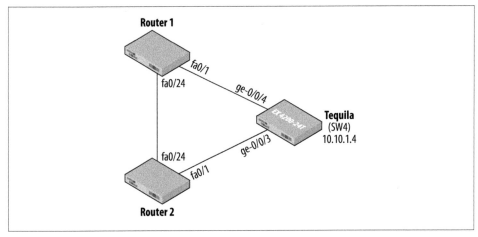

Figure 11-5. Layer 2 interconnection issue

The exact operation of BFD is beyond the scope of this book, but let's take a look at a basic description. BFD establishes a session between two endpoints over a particular link. If more than one link exists between two systems, multiple BFD sessions may be established to monitor each link. The session is established with a three-way hand-shake, and is torn down the same way.

BFD does not discover neighbors that are explicitly configured. It uses any underlying media that is available, and also piggybacks on protocols such as BGP, OSPF, or IS-IS. The idea is that BFD should be implemented in the PFE of the system for quicker detection and closer proximity. After all, this protocol is intended to protect the forwarding from one device to another. In a JUNOS EX Series switch, BFD is configured under the protocol values. Common protocols in which BFD is used include:

- OSPF
- IS-IS
- BGP
- RIP
- Static Routes
- Protocol Independent Multicast (PIM)

Here's a quick example. BFD is configured to ride on top of OSPF. The hello interval is configured for 200 milliseconds, with a total detect time of three times that value (600 ms):

```
lab@Rum# show protocols ospf
area 0.0.0.0 {
```

```
interface all {
    bfd-liveness-detection {
        version automatic;
        minimum-interval 200;
        transmit-interval {
            minimum-interval 200;
        }
    }
}
```

The 200 ms values are just an example, as many options can be set:

```
[edit]
lab@Rum# set protocols ospf area 0 interface all bfd-liveness-detection ?
Possible completions:
+ apply-groups            Groups from which to inherit configuration data
+ apply-groups-except     Don't inherit configuration data from these groups
> detection-time          Detection-time options
  minimum-interval        Minimum transmit and receive interval (milliseconds)
  minimum-receive-interval  Minimum receive interval (1..255000 milliseconds)
  multiplier              Detection time multiplier (1..255)
  no-adaptation           Disable adaptation
> transmit-interval       Transmit-interval options
  version                 BFD protocol version number
[edit]
```

Verify that the BFD session is up by issuing the run show bfd session extensive command:

```
[edit]
lab@Rum# run show bfd session extensive
                                    Detect   Transmit
Address                State  Interface  Time     Interval  Multiplier
10.5.7.7               Up     ge-0/0/4.0  0.600    0.200     3
  Client OSPF Area 0.0.0.0, TX interval 0.200, RX interval 0.200
  Session up time 00:00:13
  Local diagnostic NbrSignal, remote diagnostic None
  Remote state Up, version 1
  Min async interval 0.200, min slow interval 1.000
  Adaptive async TX interval 0.200, RX interval 0.200
  Local min TX interval 0.200, minimum RX interval 0.200, multiplier 3
  Remote min TX interval 0.200, min RX interval 0.200, multiplier 3
  Local discriminator 1, remote discriminator 1
  Echo mode disabled/inactive


1 session, 1 client
Cumulative transmit rate 10.0 pps, cumulative receive rate 10.0 pps
```

 For more information on BFD, please consult *JUNOS High Availability* (O'Reilly).

High Availability Summary

We examined several HA features in this chapter, including ISSU, AE, and BFD. These tools aid in HA goals. ISSU will allow you to upgrade the switch with minimal forwarding impact, while AE allows for high bandwidth and more reliability by implementing LACP. Lastly, we examined BFD as a mechanism to achieve fast hellos over links such as the very popular Ethernet that do not contain fast—or, at times, any—keepalive mechanism.

Conclusion

This chapter examined some of the pieces in High Availability on a single switch, to help improve your overall system availability. These strategies include hardware features such as redundant components, and software features such as ISSU and GRES. There are even protocols that increase availability, such as LACP for aggregated Ethernet and the ultra-fast hello mechanism of BFD. Remember that you are only as strong as your weakest link, and as you add more systems to your network, achieving High Availability becomes more difficult. As these systems are added, there is a greater probability of software bugs, component failures, and/or link failures. Keep these issues in mind as you are deciding to buy that new device for added redundancy!

Chapter Review Questions

1. Which two switches support the remote power supply option?
    a. EX3200
    b. EX4200
    c. EX8208
    d. EX8216
2. True or false: all EX Series switches have redundant routing engines.
3. Which feature must be enabled in order to avoid forwarding engine reset during a routing engine switchover?
    a. Hitless switchover
    b. Graceful Routing Engine Switchover
    c. VRRP
    d. BFD
4. How many routers can be in a given VRRP group?
    a. 16
    b. 8
    c. 6

d. Limitless

5. Which feature synchronizes routing state between two routing engines?

   a. Synchronized commits

   b. Graceful Restart

   c. Non-Stop Routing

   d. Mirror disk

6. True or false: Graceful Restart needs to be enabled only on the restarting switch.

7. True or false: you can have multiple backup routing engines in a Virtual Chassis.

8. Which end of the link transmits its LACP parameters?

   a. Actor

   b. Sender

   c. Partner

   d. Initiator

   e. Director

   f. Producer

9. Which JUNOS software CLI command allows the switch to respond only to initial LACP messages?

   a. `active`

   b. `listen`

   c. `partner`

   d. `passive`

10. Which JUNOS software command will show the state of the LACP messages?

    a. `show eth0`

    b. `show interfaces aggregated-ethernet`

    c. `show lacp interfaces`

    d. `show log lacp messages`

11. Which two protocols allow for the sending of BFD messages in JUNOS software?

    a. OSPF

    b. RSTP

    c. EIGRP

    d. RIP

12. Which configuration allows an aggregated Ethernet to be created on the chassis?

    a.
    ```
    aggregated-devices {
        ethernet {
            device-count 1;
        }
    }
    ```

b.
```
        ge-0/0/9 {
          ether-options {
              802.3ad ae0;
          }
        }
```

c.
```
        ae0 {
          unit 0 {
              family ethernet-switching
```

d.
```
        ae0 {
          ether-options {
              802.3ad ae0;
```

Chapter Review Answers

1. Answer: A, B. The 3200 and 4200 both support remote power supply options; however, support is not as common on the 4200, as it also has the option of a redundant power supply.

2. Answer: False. Only the 8200 series has redundant routing engines; however, the 4200 supports dual REs in a VC configuration.

3. Answer: B. GRES must be enabled to avoid a forwarding reset. This is not the default behavior of the switch, and must be configured.

4. Answer: D. VRRP can support multiple routers in a group, a single primary, and multiple backups. In most common deployments a single backup is used.

5. Answer: C. Non-Stop Routing copies routing tables, databases, and TCP state to the backup routing engine.

6. Answer: False. Graceful Restart has two types of routers: a restarting router and a helper router. The helper and restarting router must have Graceful Restart configured in order to understand the special messaging that will be exchanged.

7. Answer: False. Although each member switch may contain a routing engine, only a master and backup routing engine will be active at a given time. The other switch's routing engine will be in a cold start state.

8. Answer: C. The partner is the device that will actually send the messages, and the actor is the end of the link that acts on the message. So far, no one has added a producer or director to the script.

9. Answer: D. When in passive mode, the switch responds to LACP messages—the "Don't speak until I tell you to" method.

10. Answer: C. This is the only command that will show LACP message statistics and state.

11. Answer: A, D. OSPF and RIP currently support transport of BFD. Thankfully, EIGRP is not allowed in JUNOS, and spanning tree is not supported in BFD at the time of this writing.

12. Answer: A. The first step in creating an aggregated Ethernet bundle is to allow aggregated devices in the chassis. The device count will specify how many bundles are allowed on the system. Afterward, the actual interface parameters and bundle mappings are configured.

Glossary

Author's note: This is the same glossary we published in *JUNOS Enterprise Routing*, the companion volume to this book. We include it here for the reader focusing only on JUNOS switching and the EX platform.

3DES

Data Encryption Standard. Triple DES.

AAL

Asynchronous Transfer Mode (ATM) adaptation layer. A series of protocols enabling various types of traffic, including voice, data, image, and video, to run over an ATM network.

AAL5 mode

Asynchronous Transfer Mode (ATM) adaptation Layer 5. One of four ATM adaptation layers (AALs) recommended by the ITU-T. AAL5 is used predominantly for the transfer of classical IP over ATM. AAL5 is the least complex of the current AAL recommendations. It offers low-bandwidth overhead and simpler processing requirements in exchange for reduced bandwidth capacity and error-recovery capability. It is a Layer 2 circuit transport mode that allows you to send ATM cells between ATM2 IQ interfaces across a Layer 2 circuit-enabled network. You use Layer 2 circuit AAL5 transport mode to tunnel a stream of AAL5-encoded ATM segmentation and reassembly Protocol Data Units (SAR-PDUs) over a Multiprotocol Label Switching (MPLS) or IP backbone.

See also cell-relay mode, Layer 2 circuits, standard AAL5 mode, trunk mode.

ABR

Area border router. Router that belongs to more than one area. Used in Open Shortest Path First (OSPF).

See also OSPF.

access concentrator

Router that acts as a server in a Point-to-Point Protocol over Ethernet (PPPoE) session—for example, an E Series router.

accounting services

Method of collecting network data related to resource usage.

ACFC

Address and Control Field Compression. Enables routers to transmit packets without the two 1-byte address and control fields (0xff and 0x03) which are normal for Point-to-Point Protocol (PPP)-encapsulated packets, thus transmitting less data and conserving bandwidth. ACFC is defined in RFC 1661, "The Point-to-Point Protocol (PPP)."

See also PFC.

active route

Route chosen from all routes in the routing table to reach a destination. Active routes are installed into the forwarding table.

adaptive services

Set of services or applications that you can configure on an Adaptive Services PIC (ASP). The services and applications include stateful firewall, Network Address Translation (NAT), intrusion detection services (IDSs), Internet Protocol Security (IPSec), Layer 2 Tunneling Protocol (L2TP), and voice services.

See also tunneling protocol.

address match conditions

Use of an IP address as a match criterion in a routing policy or a firewall filter.

adjacency

Portion of the local routing information that pertains to the reachability of a single neighbor over a single circuit or interface.

Adjacency-RIB-In

Logical software table that contains Border Gateway Protocol (BGP) routes received from a specific neighbor.

Adjacency-RIB-Out

Logical software table that contains Border Gateway Protocol (BGP) routes to be sent to a specific neighbor.

ADM

Add/drop multiplexer. SONET functionality that allows lower-level signals to be dropped from a high-speed optical connection.

ADSL

Asymmetrical digital subscriber line. A technology that allows more data to be sent over existing copper telephone lines, using the public switched telephone network (PSTN). ADSL supports data rates from 1.5 Mbps to 9 Mbps when receiving data (downstream rate) and from 16 Kbps to 640 Kbps when sending data (upstream rate).

ADSL interface

Asymmetrical digital subscriber line interface. Physical WAN interface that connects a router to a digital subscriber line access multiplexer (DSLAM). An ADSL interface allocates line bandwidth asymmetrically. Downstream (provider-to-customer) data rates can be up to 8 Mbps for ADSL, 12 Mbps for ADSL2, and 25 Mbps for ADSL2+. Upstream (customer-to-provider) rates can be up to 800 Kbps for ADSL and 1

Mbps for ADSL2 and ADSL2+, depending on the implementation.

ADSL2 interface

ADSL interface that supports ITU-T Standard G.992.3 and ITU-T Standard G.992.4. ADSL2 allocates downstream (provider-to-customer) data rates of up to 12 Mbps and upstream (customer-to-provider) rates of up to 1 Mbps.

ADSL2+ interface

ADSL interface that supports ITU-T Standard G.992.5. ADSL2+ allocates downstream (provider-to-customer) data rates of up to 25 Mbps and upstream (customer-to-provider) rates of up to 1 Mbps.

AES

Advanced Encryption Standard. Defined in FIPS PUB 197. The AES algorithm uses keys of 128, 192, or 256 bits to encrypt and decrypt data in blocks of 128 bits.

aggregate route

Combination of groups of routes that have common addresses into a single entry in a routing table.

aggregated interface

Logical bundle of physical interfaces. The aggregated interface is managed as a single interface with one IP address. Network traffic is dynamically distributed across ports, so administration of data flowing across a given port is done automatically within the aggregated link. Using multiple ports in parallel provides redundancy and increases the link speed beyond the limits of any single port.

AH

Authentication header. A component of the IPSec protocol used to verify that the contents of a packet have not changed, and to validate the identity of the sender.

ALI

ATM line interface. Interface between Asynchronous Transfer Mode (ATM) and 3G systems.

See also **ATM**.

ANSI

American National Standards Institute. The U.S. representative to the International Organization for Standardization (ISO).

APN

Access point name. When mobile stations connect to IP networks over a wireless network, the Gateway GPRS Support Node (GGSN) uses the APN to distinguish among the connected IP networks (known as *APN networks*). In addition to identifying these connected networks, an APN is also a configured entity that hosts the wireless sessions, which are called Packet Data Protocol (PDP) contexts.

APQ

Alternate priority queuing. Dequeuing method that has a special queue, similar to strict-priority queuing (SPQ), which is visited only 50% of the time. The packets in the special queue still have a predictable latency, although the upper limit of the delay is higher than that with SPQ. Since the other configured queues share the remaining 50% of the service time, queue starvation is usually avoided.

See also **SPQ**.

APS

Automatic Protection Switching. Technology used by SONET add/drop multiplexers (ADMs) to protect against circuit faults between the ADM and a router and to protect against failing routers.

area

1. Routing subdomain that maintains detailed routing information about its own internal composition as well as routing information that allows it to reach other routing subdomains. In Intermediate System-to-Intermediate System Level 1 (IS-IS), an area corresponds to a Level 1 subdomain. 2. In IS-IS and Open Shortest Path First (OSPF), a set of contiguous networks and hosts within an Autonomous System

(AS) that have been administratively grouped together.

ARP

Address Resolution Protocol. Protocol used for mapping IPv4 addresses to Media Access Control (MAC) addresses.

See also **NDP**.

AS

Autonomous System. Set of routers under a single technical administration. Each AS normally uses a single Interior Gateway Protocol (IGP) and metrics to propagate routing information within the set of routers. Also called a *routing domain*.

ASBR

Autonomous System Boundary Router. In Open Shortest Path First (OSPF), a router that exchanges routing information with routers in other Autonomous Systems (ASs).

ASBR Summary LSA

OSPF link state advertisement (LSA) sent by an area border router (ABR) to advertise the router ID of an Autonomous System Boundary Router (ASBR) across an area boundary.

See also **ASBR**.

AS external link advertisement

OSPF link state advertisement (LSA) sent by Autonomous System Boundary Routers (ASBRs) to describe external routes that they have detected. These LSAs are flooded throughout the Autonomous System (AS) (except for stub areas).

ASIC

Application-specific integrated circuit. Specialized processors that perform specific functions on the router.

ASM

Adaptive Services Module. On a Juniper Networks M7i router, provides the same functionality as the Adaptive Services PIC (ASP).

ASM

Any Source Multicast. A network that supports both one-to-many and many-to-many communication models. An ASM network must determine all the sources of a group and deliver all of them to interested subscribers.

ASP

Adaptive Services PIC.

See also **adaptive services.**

AS path

In the Border Gateway Protocol (BGP), the route to a destination. The path consists of the Autonomous System (AS) numbers of all routers that a packet must go through to reach a destination.

ATM

Asynchronous Transfer Mode. A high-speed multiplexing and switching method utilizing fixed-length cells of 53 octets to support multiple types of traffic.

ATM-over-ADSL interface

Asynchronous Transfer Mode (ATM) interface used to send network traffic through a point-to-point connection to a DSL access multiplexer (DSLAM). ATM-over-ADSL interfaces are intended for asymmetrical digital subscriber line (ADSL) connections only, not for direct ATM connections.

atomic

Smallest possible operation. An atomic operation is performed either entirely or not at all. For example, if machine failure prevents a transaction from completing, the system is rolled back to the start of the transaction, with no changes taking place.

AUC

Authentication center. Part of the Home Location Register (HLR) in third-generation (3G) systems; performs computations to verify and authenticate a mobile phone user.

automatic policing

Policer that allows you to provide strict service guarantees for network traffic. Such guarantees are especially useful in the context of differentiated services for traffic-engineered label-switched paths (LSPs), providing better emulation for Asynchronous Transfer Mode (ATM) wires over a Multiprotocol Label Switching (MPLS) network.

auto-negotiation

Used by Ethernet devices to configure interfaces automatically. If interfaces support different speeds or different link modes (half duplex or full duplex), the devices attempt to settle on the lowest common denominator.

Autonomous System external link advertisement

OSPF link state advertisement (LSA) sent by Autonomous System Boundary Routers (ASBRs) to describe external routes that they have detected. These LSAs are flooded throughout the Autonomous System (AS) (except for stub areas).

Autonomous System path

In the Border Gateway Protocol (BGP), the route to a destination. The path consists of the Autonomous System (AS) numbers of all the routers a packet must pass through to reach a destination.

auto-RP

Method of electing and announcing the rendezvous point-to-group address mapping in a multicast network. JUNOS software supports this vendor-proprietary specification.

See also **RP.**

backbone area

In Open Shortest Path First (OSPF), an area that consists of all networks in area ID 0.0.0.0, their attached routers, and all area border routers (ABRs).

backbone router

Open Shortest Path First (OSPF) router with all operational interfaces within area 0.0.0.0.

backplane

See **midplane.**

backup designated router

Open Shortest Path First (OSPF) router on a broadcast segment that monitors the operation of the designated router and takes over its functions if the designated router fails.

BA classifier

Behavior aggregate classifier. A method of classification that operates on a packet as it enters the router. The packet header contents are examined, and this single field determines the class-of-service (CoS) settings applied to the packet.

See also multifield classifier.

bandwidth

Range of transmission frequencies a network can use, expressed as the difference between the highest and lowest frequencies of a transmission channel. In computer networks, greater bandwidth indicates a faster data transfer rate capacity.

bandwidth model

In Differentiated Services-aware traffic engineering, determines the value of the available bandwidth advertised by the Interior Gateway Protocols (IGPs).

bandwidth on demand

1. A technique to temporarily provide additional capacity on a link to handle bursts in data, videoconferencing, or other variable bit rate applications. Also called *flexible bandwidth allocation*. 2. On a Services Router, an Integrated Services Digital Network (ISDN) cost-control feature defining the bandwidth threshold that must be reached on links before a Services Router initiates additional ISDN data connections to provide more bandwidth.

B-channel

Bearer channel. A 64 Kbps channel used for voice or data transfer on an Integrated Services Digital Network (ISDN) interface.

See also D-channel.

BECN

Backward explicit congestion notification. In a Frame Relay network, a header bit transmitted by the destination device requesting that the source device send data more slowly. BECN minimizes the possibility that packets will be discarded when more packets arrive than can be handled.

See also FECN.

Bellman-Ford algorithm

Algorithm used in distance-vector routing protocols to determine the best path to all routes in the network.

BERT

Bit error rate test. A test that can be run on the following interfaces to determine whether they are operating properly: E1, E3, T1, T3, and channelized (DS3, OC3, OC12, and STM1) interfaces.

BFD

Bidirectional Forwarding Detection. A simple hello mechanism that detects failures in a network. Used with routing protocols to speed up failure detection.

BGP

Border Gateway Protocol. Exterior gateway protocol used to exchange routing information among routers in different Autonomous Systems (ASs).

bit field match conditions

Use of fields in the header of an IP packet as match criteria in a firewall filter.

bit rate

Number of bits transmitted per second.

BITS

Building Integrated Timing Source. Dedicated timing source that synchronizes all equipment in a particular building.

Blowfish

Unpatented, symmetric cryptographic method developed by Bruce Schneier and used in many commercial and freeware software applications. Blowfish uses variable-length keys of up to 448 bits.

BOOTP

Bootstrap protocol. A User Datagram Protocol (UDP)/IP-based protocol that allows a booting host to configure itself dynamically and without user supervision. BOOTP provides a means to notify a host of its assigned IP address, the IP address of a boot server host, and the name of a file to be loaded into memory and executed. Other configuration information, such as the local subnet mask, the local time offset, the addresses of default routers, and the addresses of various Internet servers, can also be communicated to a host using BOOTP.

bootstrap router

Single router in a multicast network responsible for distributing candidate rendezvous point (RP) information to all Physical Interface Module (PIM)-enabled routers.

BPDU

Bridge Protocol Data Unit. A Spanning Tree Protocol (STP) hello packet that is sent out at intervals to exchange information across bridges and detect loops in a network topology.

BRI

Basic Rate Interface. Integrated Services Digital Network (ISDN) interface intended for home and small enterprise applications. BRI consists of two 64 Kbps B-channels to carry voice or data, and one 16 Kbps D-channel for control and signaling.

See also B-channel, D-channel.

bridge

Device that uses the same communications protocol to connect and pass packets between two network segments. A bridge operates at Layer 2 of the Open Systems Interconnection (OSI) reference model.

broadcast

Operation of sending network traffic from one network node to all other network nodes.

BSC

Base station controller. Key network node in third-generation (3G) systems that supervises the functioning and control of multiple base transceiver stations.

BSS

Base station subsystem. Composed of the base transceiver station (BTS) and base station controller (BSC).

BSSGP

Base Station System GPRS Protocol. Processes routing and quality-of-service (QoS) information for the base station subsystem (BSS).

BTS

Base transceiver station. Mobile telephony equipment housed in cabinets and collocated with antennas. (Also known as a radio base station.)

buffers

Memory space for handling data in transit. Buffers compensate for differences in processing speed between network devices and handle bursts of data until they can be processed by slower devices.

bundle

1. Multiple physical links of the same type, such as multiple asynchronous lines, or physical links of different types, such as leased synchronous lines and dial-up asynchronous lines. 2. Collection of software that makes up a JUNOS software release.

bypass LSP

Carries traffic for a label-switched path (LSP) whose link-protected interface has failed. A bypass LSP uses a different interface and path to reach the same destination.

CA

Certificate authority. A trusted third-party organization that creates, enrolls, validates, and revokes digital certificates. The CA guarantees a user's identity and issues public and private keys for message encryption and decryption (coding and decoding).

CAC

Call admission control. In Differentiated Services-aware traffic engineering, checks for adequate bandwidth on the path before the label-switched path (LSP) is established. If the bandwidth is insufficient, the LSP is not established and an error is reported.

CAIDA

Cooperative Association for Internet Data Analysis. An association that provides tools and analyses promoting the engineering and maintenance of a robust, scalable Internet infrastructure. One tool, cflowd, allows you to collect an aggregate of sampled flows and send the aggregate to a specified host that runs the cflowd application available from CAIDA.

callback

Alternative feature to dial-in that enables a J Series services router to call back the caller from the remote end of a backup Integrated Services Digital Network (ISDN) connection. Instead of accepting a call from the remote end of the connection, the router rejects the call, waits a configured period of time, and calls a number configured on the router's dialer interface.

See also **dial-in**.

caller ID

Telephone number of the caller on the remote end of a backup Integrated Services Digital Network (ISDN) connection, used to dial in and to identify the caller. Multiple caller IDs can be configured on an ISDN dialer interface. During dial-in, the router matches the incoming call's caller ID against the caller IDs configured on its dialer interfaces. Each dialer interface accepts calls only from callers whose caller IDs are configured on it.

CAMEL

Customized Applications of Mobile Enhanced Logic. An ETSI standard for GSM networks that enhances the provision of Intelligent Network services.

candidate configuration

File maintained by the JUNOS software containing changes to the router's active configuration. This file becomes the active configuration when a user issues the `commit` command.

candidate RP advertisements

Information sent by routers in a multicast network when they are configured as a local rendezvous point (RP). This information is unicast to the bootstrap router for the multicast domain.

carrier-of-carriers VPN

Virtual private network (VPN) service supplied to a network service provider that is supplying either Internet service or VPN service to an end customer. For a carrier-of-carriers VPN, the customer's sites are configured within the same Autonomous System (AS).

CB

Control Board. On a T640 routing node, part of the host subsystem that provides control and monitoring functions for router components.

CBC

Cipher block chaining. A mode of encryption using 64 or 128 bits of fixed-length blocks in which each block of plain text is XORed with the previous cipher text block before being encrypted.

See also **XOR**.

CBR

Constant bit rate. For ATM1 and ATM2 intelligent queuing (IQ) interfaces, data that is serviced at a constant, repetitive rate. CBR is used for traffic that does not need to periodically burst to a higher rate, such as non-packetized voice and audio.

CCC

Circuit cross-connect. A JUNOS software feature that allows you to configure transparent connections between two circuits. A circuit can be a Frame Relay data-link connection identifier (DLCI), an Asynchronous

Transfer Mode (ATM) virtual channel, a Point-to-Point Protocol (PPP) interface, a Cisco High-Level Data Link Control (HDLC) interface, or a Multiprotocol Label Switching (MPLS) label-switched path (LSP).

CDMA

Code Division Multiple Access. Technology for digital transmission of radio signals between, for example, a mobile telephone and a base transceiver station (BTS).

CDMA2000

Radio transmission and backbone technology for the evolution to third-generation (3G) mobile networks.

CDR

Call Detail Record. A record containing data (such as origination, termination, length, and time of day) unique to a specific call.

CE device

Customer edge device. Router or switch in the customer's network that is connected to a service provider's provider edge (PE) router and participates in a Layer 3 virtual private network (VPN).

cell relay

Data transmission technology based on the use of small, fixed-size packets (cells) that can be processed and switched in hardware at high speeds. Cell relay is the basis for many high-speed network protocols, including Asynchronous Transfer Mode (ATM) and IEEE 802.6.

cell-relay mode

Layer 2 circuit transport mode that sends Asynchronous Transfer Mode (ATM) cells between ATM2 intelligent queuing (IQ) interfaces over a Multiprotocol Label Switching (MPLS) core network. You use Layer 2 circuit cell-relay transport mode to tunnel a stream of ATM cells over an MPLS or IP backbone.

See also **AAL5 mode, Layer 2 circuits, standard AAL5 mode, trunk mode.**

cell tax

Physical transmission capacity used by header information when sending data packets in an Asynchronous Transfer Mode (ATM) network. Each ATM cell uses a 5-byte header.

CFEB

Compact Forwarding Engine Board. In M7i and M10i routers, provides route lookup, filtering, and switching to the destination port.

cflowd

Application available from CAIDA that collects an aggregate of sampled flows and sends the aggregate to a specified host running the cflowd application.

CFM

Cubic feet per minute. Measure of air flow in volume per minute.

channel

Communication circuit linking two or more devices. A channel provides an input/output interface between a processor and a peripheral device, or between two systems. A single physical circuit can consist of one or many channels, or two systems carried on a physical wire or wireless medium. For example, the dedicated channel between a telephone and the central office (CO) is a twisted-pair copper wire.

See also **frequency-division multiplexed channel, time-division multiplexed channel.**

channel group

Combination of DS0 interfaces partitioned from a channelized interface into a single logical bundle.

channelized E1

A 2.048 Mbps interface that can be configured as a single clear-channel E1 interface or channelized into as many as 31 discrete DS0 interfaces. On most channelized E1 interfaces, time slots are numbered from 1 through 32, and time slot 1 is reserved for framing. On some legacy channelized E1 interfaces, time slots are numbered from 0 through 31, and time slot 0 is reserved for framing.

channelized interface

Interface that is a subdivision of a larger interface, minimizing the number of Physical Interface Cards (PICs) or Physical Interface Modules (PIMs) that an installation requires. On a channelized PIC or PIM, each port can be configured as a single clear channel or partitioned into multiple discrete T3, T1, E1, and DS0 interfaces, depending on the size of the channelized PIC or PIM.

channelized T1

A 1.544 Mbps interface that can be configured as a single clear-channel T1 interface or channelized into as many as 24 discrete DS0 interfaces. Time slots are numbered from 1 through 24.

CHAP

Challenge Handshake Authentication Protocol. A protocol that authenticates remote users. CHAP is a server-driven, three-step authentication mechanism that depends on a shared secret password that resides on both the server and the client.

chassisd

Chassis daemon. A JUNOS software process responsible for managing the interaction of the router's physical components.

CIDR

Classless Inter-Domain Routing. A method of specifying Internet addresses in which you explicitly specify the bits of the address to represent the network address instead of determining this information from the first octet of the address.

CIP

Connector Interface Panel. On an M160 router, the panel that contains connectors for the routing engines (REs), BITS interfaces, and alarm relay contacts.

CIR

Committed information rate. The CIR specifies the average rate at which packets are admitted to the network. As each packet enters the network, it is counted. Packets that do not exceed the CIR are marked green, which corresponds to low loss priority. Packets that exceed the CIR but are below the peak information rate (PIR) are marked yellow, which corresponds to medium loss priority.

See also trTCM, PIR.

Cisco-RP-Announce

Message advertised into a multicast network by a router configured as a local rendezvous point (RP) in an auto-RP network. A Cisco-RP-Announce message is advertised in Dense-mode Physical Interface Module (PIM) to the 224.0.1.39 multicast group address.

Cisco-RP-Discovery

Message advertised by the mapping agent in an auto-RP network. A Cisco-RP-Discovery message contains the rendezvous point (RP) to multicast group address assignments for the domain. It is advertised in Dense-mode Physical Interface Module (PIM) to the 224.0.1.40 multicast group address.

classification

In class of service (CoS), the examination of an incoming packet that associates the packet with a particular CoS servicing level. There are two kinds of classifiers: behavior aggregate (BA) and multifield.

See also BA classifier, multifield classifier.

classifier

Method of reading a sequence of bits in a packet header or label and determining how the packet should be forwarded internally and scheduled (queued) for output.

class type

In Differentiated Services-aware traffic engineering, a collection of traffic flows that are treated equally in a Differentiated Services domain. A class type maps to a queue and is much like a class-of-service (CoS) forwarding class in concept. It is also known as a *traffic class*.

clear channel

Interface configured on a channelized Physical Interface Card (PIC) or Physical Interface Module (PIM) that operates as a single channel, does not carry signaling, and uses the entire port bandwidth.

CLEC

(Pronounced "see-lek.") Competitive local exchange carrier. Company that competes with the already established local telecommunications business by providing its own network and switching.

CLEI

Common Language Equipment Identifier. Inventory code used to identify and track telecommunications equipment.

CLI

Command-line interface. Interface provided for configuring and monitoring the routing protocol software.

client peer

In a Border Gateway Protocol (BGP) route reflection, a member of a cluster that is not the route reflector.

See also **nonclient peer**.

CLNP

Connectionless Network Protocol. An ISO-developed protocol for Open Systems Interconnection (OSI) connectionless network service. CLNP is the OSI equivalent of IP.

CLNS

Connectionless Network Service. A Layer 3 protocol, similar to Internet Protocol version 4 (IPv4). CLNS uses network service access points (NSAPs) instead of the prefix addresses found in IPv4 to specify end systems and intermediate systems.

cluster

In the Border Gateway Protocol (BGP), a set of routers that have been grouped together. A cluster consists of one system that acts as a route reflector, along with any number of client peers. The client peers receive their route information only from the route reflector system. Routers in a cluster do not need to be fully meshed.

CO

Central office. The local telephone company building that houses circuit-switching equipment used for subscriber lines in a given area.

code-point alias

Name assigned to a pattern of code-point bits. This name is used, instead of the bit pattern, in the configuration of other class-of-service (CoS) components, such as classifiers, drop-profile maps, and rewrite rules.

command completion

Function of a router's command-line interface (CLI) that allows a user to enter only the first few characters in any command. Users access this function through the space bar or Tab key.

commit

JUNOS software command-line interface (CLI) configuration-mode command that saves changes made to a router configuration, verifies the syntax, applies the changes to the configuration currently running on the router, and identifies the resultant file as the current operational configuration.

commit script

Script that enforces custom configuration rules. A script runs each time a new candidate configuration is committed and inspects the configuration. If a configuration breaks your custom rules, the script can generate actions for the JUNOS software.

commit script macro

Sequence of commands that allows you to create custom configuration syntax to simplify the task of configuring a routing platform. By itself, your custom syntax has no operational impact on the routing platform. A corresponding commit script macro uses your custom syntax as input data for generating standard JUNOS configuration statements that execute your intended operation.

community

1. In the Border Gateway Protocol (BGP), a group of destinations that share a common property. Community information is included as one of the path attributes in BGP update messages. 2. In the Simple Network Management Protocol (SNMP), an authentication scheme that authorizes SNMP clients based on the source IP address of incoming SNMP packets, defines which Management Information Base (MIB) objects are available, and specifies the operations (read-only or read-write) allowed on those objects.

confederation

In the Border Gateway Protocol (BGP), a group of systems that appears to external Autonomous Systems (ASs) as a single AS.

configuration mode

JUNOS software mode that allows a user to alter the router's current configuration.

Connect

Border Gateway Protocol (BGP) neighbor state in which the local router has initiated the Transmission Control Protocol (TCP) session and is waiting for the remote peer to complete the TCP connection.

constrained path

In traffic engineering, a path determined using the CSPF algorithm. The ERO carried in the Resource Reservation Protocol (RSVP) packets contains the constrained path information.

See also **ERO**.

context node

Node that the Extensible Stylesheet Language for Transformations (XSLT) processor is currently examining. XSLT changes the context as it traverses the XML document's hierarchy.

See also **XSLT**.

context-sensitive help

Function of the router's command-line interface (CLI) that allows a user to request information on the JUNOS software hierarchy. You can access context-sensitive help in both operational and configuration modes.

contributing routes

Active IP routes in the routing table that share the same most-significant bits and are more specific than an aggregate or generated route.

control plane

Virtual network path used to set up, maintain, and terminate data plane connections.

See also **data plane**.

core

Central backbone of the network.

CoS

Class of service. Method of classifying traffic on a packet-by-packet basis using information in the type-of-service (ToS) byte to provide different service levels to different traffic.

cosd

Class-of-service (CoS) process that enables the routing platform to provide different levels of service to applications based on packet classifications.

CPE

Customer premises equipment. Telephone, modem, router, or other service provider equipment located at a customer site.

craft interface

Mechanisms used by a Communication Workers of America craftsperson to operate, administer, and maintain equipment or provision data communications. On a Juniper Networks router, the craft interface allows you to view status and troubleshooting information and perform system control functions.

CRL

Certificate revocation list. A list of digital certificates that have been invalidated, including the reasons for revocation and the names of the entities that issued them. A CRL prevents usage of digital certificates and signatures that have been compromised.

CRTP

Compressed Real-time Transport Protocol. Protocol that decreases the size of the IP, User Datagram Protocol (UDP), and Real-Time Transport Protocol (RTP) headers and works with reliable and fast point-to-point links for Voice over IP traffic. CRTP is defined in RFC 2508.

Crypto Accelerator Module

Processor card that speeds up certain cryptographic IP Security (IPSec) services on some J Series services routers. For the supported cryptographic algorithms, see the J Series documentation.

Crypto Officer

Superuser responsible for the proper operation of a router running JUNOS-FIPS software.

CSCP

Class Selector code point. Eight Differentiated Services code point (DSCP) values of the form $xxx000$ (where x can be 0 or 1). Defined in RFC 2474.

CSNP

Complete sequence number PDU. Packet that contains a complete list of all the label-switched paths (LSPs) in the Intermediate System-to-Intermediate System Level 1 (IS-IS) database.

CSP

Critical Security Parameter. On routers running JUNOS-FIPS software, a collection of cryptographic keys and passwords that must be protected at all times.

CSPF

Constrained Shortest Path First. A Multiprotocol Label Switching (MPLS) algorithm that has been modified to take into account specific restrictions when calculating the shortest path across the network.

CSU/DSU

Channel service unit/data service unit. A channel service unit connects a digital phone line to a multiplexer or other digital signal device. A data service unit connects a data terminating equipment (DTE) device to a digital phone line.

CVS

Concurrent Versions System. A widely used version control system for software development or data archives.

daemon

Background process that performs operations for the system software and hardware. Daemons normally start when the system software is booted, and run as long as the software is running. In the JUNOS software, daemons are also referred to as *processes*.

damping

Method of reducing the number of update messages sent between Border Gateway Protocol (BGP) peers, thereby reducing the load on these peers without adversely affecting the route convergence time for stable routes.

database description packet

Open Shortest Path First (OSPF) packet type used in the formation of an adjacency. The packet sends summary information about the local router's database to the neighboring router.

data-MDT

Data-driven multicast distribution tree (MDT) tunnel. A multicast tunnel created and deleted based on defined traffic loads and designed to ease loading on the default MDT tunnel.

data packet

Chunk of data transiting the router from the source to a destination.

data plane

Virtual network path used to distribute data between nodes.

See also control plane.

dcd

Device control process. A JUNOS software interface process (daemon).

DCE

Data circuit-terminating equipment. An RS-232C device, typically used for a modem or printer, or a network access and packet switching node.

D-channel

Delta channel. A circuit-switched channel that carries signaling and control for B-channels. In Basic Rate Interface (BRI) applications, it can also support customer packet data traffic at speeds up to 9.6 kbps.

See also B-channel, BRI.

DCU

Destination class usage. A means of tracking traffic originating from specific prefixes on the customer edge router and destined for specific prefixes on the provider core router, based on the IP source and destination addresses.

DE

Discard-eligible bit. In a Frame Relay network, a header bit notifying devices on the network that traffic can be dropped during congestion to ensure the delivery of higher-priority traffic.

deactivate

Method of modifying the router's active configuration. Portions of the hierarchy marked as inactive using this command are ignored during the router's commit process as though they were not configured at all.

dead interval

Amount of time that an Open Shortest Path First (OSPF) router maintains a neighbor relationship before declaring that neighbor as no longer operational. The JUNOS software uses a default value of 40 seconds for this timer.

dead peer detection

See DPD.

default address

Router address that is used as the source address on unnumbered interfaces.

default route

Route used to forward IP packets when a more specific route is not present in the routing table. Often represented as 0.0.0.0/0, the default route is sometimes referred to as the *route of last resort*.

demand circuit

Network segment whose cost varies with usage, according to a service level agreement (SLA) with a service provider. Demand circuits limit traffic based on either bandwidth (bits or packets transmitted) or access time.

See also multicast.

Dense mode

Method of forwarding multicast traffic to interested listeners. Dense mode forwarding assumes that most of the hosts on the network will receive the multicast data. Routers flood packets and prune unwanted traffic every three minutes.

DES

Data Encryption Standard. A method for encrypting information using a 56-bit key. Considered to be a legacy method and insecure for many applications.

See also 3DES.

designated router

In Open Shortest Path First (OSPF), a router selected by other routers that is responsible for sending link state advertisements (LSAs) that describe the network, thereby reducing the amount of network traffic and the size of the routers' topological databases.

destination prefix length

Number of bits of the network address used for the host portion of a Classless Inter-Domain Routing (CIDR) IP address.

DFC

Dynamic flow capture. Process of collecting packet flows that match a particular filter list to one or more content destinations using an on-demand control protocol that relays requests from one or more control sources.

DHCP

Dynamic Host Configuration Protocol. Allocates IP addresses dynamically so that they can be reused when no longer needed.

dial backup

Feature that reestablishes network connectivity through one or more backup Integrated Services Digital Network (ISDN) dialer interfaces after a primary interface fails. When the primary interface is reestablished, the ISDN interface is disconnected.

dialer filter

Stateless firewall filter that enables dial-on-demand routing backup when applied to a physical Integrated Services Digital Network (ISDN) interface and its dialer interface configured as a passive static route. The passive static route has a lower priority than dynamic routes. If all dynamic routes to an address are lost from the routing table and the router receives a packet for that address, the dialer interface initiates an ISDN backup connection and sends the packet over it.

See also dial-on-demand routing (DDR) backup, floating static route.

dialer interface (dl)

Logical interface for configuring dialing properties and the control interface for a backup Integrated Services Digital Network (ISDN) connection.

dialer profile

Set of characteristics configured for the Integrated Services Digital Network (ISDN) dialer interface. Dialer profiles allow the configuration of physical interfaces to be separated from the logical configuration of dialer interfaces required for ISDN connectivity. This feature also allows physical and logical interfaces to be bound together dynamically on a per-connection basis.

dialer watch

Dial-on-demand routing (DDR) backup feature that provides reliable connectivity without relying on a dialer filter to activate the Integrated Services Digital Network (ISDN) interface. The ISDN dialer interface monitors the existence of each route on a watch list. If all routes on the watch list are lost from the routing table, dialer watch initiates the ISDN interface for failover connectivity.

See also dial-on-demand routing (DDR) backup.

dial-in

Feature that enables J Series services routers to receive calls from the remote end of a backup Integrated Services Digital Network (ISDN) connection. The remote end of the ISDN call might be a service provider, a corporate central location, or a customer premises equipment (CPE) branch office. All incoming calls can be verified against caller IDs configured on the router's dialer interface.

See also callback.

dial-on-demand routing (DDR) backup

Feature that provides a J Series services router with full-time connectivity across an Integrated Services Digital Network (ISDN) line. When routes on a primary serial T1, E1, T3, E3, Fast Ethernet, or Point-to-Point Protocol over Ethernet (PPPoE) interface are lost, an ISDN dialer interface establishes a backup connection. To save connection time costs, the services router drops the ISDN connection after a configured period of inactivity. Services routers with ISDN interfaces support two types of DDR backup: on-demand routing with a dialer filter and with a dialer watch.

See also dialer filter, dialer watch.

Differentiated Services-aware traffic engineering

Type of constraint-based routing that can enforce different bandwidth constraints for different classes of traffic. It can also perform call admission control (CAC) on each traffic engineering class when a label-switched path (LSP) is established.

Differentiated Services domain

Routers in a network that have Differentiated Services enabled.

Diffie-Hellman

Method of key exchange across a nonsecure environment, such as the Internet. The Diffie-Hellman algorithm negotiates a session key without sending the key itself across the network by allowing each party to pick a partial key independently and send part of it to each other. Each side then calculates a common key value. This is a symmetrical method and keys are typically used for only a short time, then discarded and regenerated.

DiffServ

Differentiated Services (based on RFC 2474). DiffServ uses the type-of-service (ToS) byte to identify different packet flows on a packet-by-packet basis. DiffServ adds a Class Selector code point (CSCP) and a Differentiated Services code point (DSCP).

DiffServ-aware

Paradigm that gives different treatment to traffic based on the experimental (EXP) bits in the Multiprotocol Label Switching (MPLS) label header and allows you to provide multiple classes of service (CoS).

digital certificate

Electronic file based on private and public key technology that verifies the identity of the certificate's holder to protect data exchanged online. Digital certificates are issued by a certificate authority (CA).

Dijkstra algorithm

See SPF.

DIMM

Dual inline memory module. A 168-pin memory module that supports 64-bit data transfer.

direct routes

See interface routes.

disable

Method of modifying the router's active configuration. When portions of the

hierarchy are marked as disabled (mainly router interfaces), the router uses the configuration but ignores the disabled portions.

discard

JUNOS software syntax command used in a routing policy or a firewall filter. The command halts the logical processing of the policy or filter when a set of match conditions is met. The specific route or IP packet is dropped from the network silently. It can also be a next hop attribute assigned to a route in the routing table.

distance-vector

Method used in Bellman-Ford routing protocols to determine the best path to all routers in the network. Each router determines the distance (metric) to the destination and the vector (next hop) to follow.

Distributed Buffer Manager ASIC

Juniper Networks ASIC responsible for managing the router's packet storage memory.

DLCI

Data-link connection identifier. Identifier for a Frame Relay virtual connection (also called a *logical interface*).

DLSw

Data link switching. Method of tunneling IBM System Network Architecture (SNA) and NetBIOS traffic over an IP network. (The JUNOS software does not support NetBIOS.)

See also tunneling protocol.

DLSw circuit

Path formed by establishing data link control (DLC) connections between an end system and a local router configured for DLSw. Each DLSw circuit is identified by the circuit ID that includes the end system Media Access Control (MAC) address, local service access point (LSAP), and DLC port ID. Multiple DLSw circuits can operate over the same DLSw connection.

DLSw connection

Set of Transmission Control Protocol (TCP) connections between two data link switching (DLSw) peers that is established after the initial handshake and successful capabilities exchange.

DNS

Domain Name System. A system that stores information about hostnames and domain names. DNS provides an IP address for each hostname, and lists the email exchange servers accepting email addresses for each domain.

DoS

Denial of service. A system security breach in which network services become unavailable to users.

DPD

Dead peer detection. Protocol that recognizes the loss of the primary IPSec Internet Key Exchange (IKE) peer and establishes a secondary IPSec tunnel to a backup peer.

DRAM

Dynamic random access memory. Storage source on the router that can be accessed quickly by a process.

drop probability

Percentage value that expresses the likelihood that an individual packet will be dropped from the network.

See also drop profile.

drop profile

Mechanism of random early detection (RED) that defines parameters that allow packets to be dropped from the network. When you configure drop profiles, there are two important values: the queue fullness and the drop probability.

See also drop probability, queue fullness, RED.

DSAP

Destination service access point. Service access point (SAP) that identifies the destination for which a Logical Link Control Protocol Data Unit (LPDU) is intended.

DS0

Digital signal level 0. In T-carrier systems, a basic digital signaling rate of 64 Kbps. The DS0 rate forms the basis for the North American digital multiplex transmission hierarchy.

DS1

Digital signal level 1. In T-carrier systems, a digital signaling rate of 1.544 Mbps. A standard used in telecommunications to transmit voice and data among devices. Also known as T1.

See also T1.

DS3

Digital signal level 3. In T-carrier systems, a digital signaling rate of 44.736 Mbps. This level of carrier can transport 28 DS1-level signals and 672 DS0-level channels within its payload. Also known as T3.

See also T3.

DSCP

Differentiated Services code point or Diff-Serv code point. Values for a 6-bit field defined for IPv4 and IPv6 packet headers that can be used to enforce class-of-service (CoS) distinctions in routers.

DSU

Data service unit. A device used to connect data terminal equipment (DTE) to a digital phone line. DSU converts digital data from a router to voltages and encoding required by the phone line.

See also CSU/DSU.

DTCP

Dynamic Tasking Control Protocol. A means of communicating filter requests and acknowledgments between one or more clients and a monitoring platform, used in dynamic flow capture (DFC) and flow-tap configurations. The protocol is defined in Internet draft draft-cavuto-dtcp-00.txt.

DTD

Document type definition. Defines the elements and structure of an Extensible

Markup Language (XML) document or data set.

DTE

Data terminal equipment. An RS-232-C interface that a computer uses to exchange information with a serial device.

DVMRP

Distance Vector Multicast Routing Protocol. Distributed multicast routing protocol that dynamically generates IP multicast delivery trees using a technique called reverse-path multicasting (RPM) to forward multicast traffic to downstream interfaces.

DWDM

Dense wavelength-division multiplexing. Technology that enables data from different sources to be carried together on an optical fiber, with each signal carried on its own separate wavelength.

dynamic label-switched path

Multiprotocol Label Switching (MPLS) network path established by signaling protocols such as the Resource Reservation Protocol (RSVP) and Label Distribution Protocol (LDP).

E1

High-speed WAN digital communications protocol that operates at a rate of 2.048 Mbps.

E3

High-speed WAN digital communications protocol that operates at a rate of 34.368 Mbps and uses time-division multiplexing to carry 16 E1 circuits.

EAL3

Common Criteria Evaluation Assurance Level 3. Evaluation Assurance Level is an assurance and compliance requirement defined by Common Criteria. Higher levels have more stringent requirements.

EBGP

External BGP. A Border Gateway Protocol (BGP) configuration in which sessions are established between routers in different Autonomous Systems (ASs).

E-carrier

E stands for *European*. Standards that form part of the Synchronous Digital Hierarchy (SDH), in which groups of E1 circuits are bundled onto higher-capacity E3 links between telephone exchanges or countries. E-carrier standards are used just about everywhere in the world except North America and Japan, and are incompatible with the T-carrier standards.

ECC

Error checking and correction. The process of detecting errors during the transmission or storage of digital data and correcting them automatically. This usually involves sending or storing extra bits of data according to specified algorithms.

ECSA

Exchange Carriers Standards Association. A standards organization created after the divestiture of the Bell System to represent the interests of interexchange carriers.

edge router

In Multiprotocol Label Switching (MPLS), a router located at the beginning or end of a label-switching tunnel. An edge router at the beginning of a tunnel applies labels to new packets entering the tunnel. An edge router at the end of a tunnel removes labels from packets exiting the tunnel.

See also **MPLS**.

editor macros (Emacs)

Shortcut keystrokes used within the router's command-line interface (CLI). These macros move the cursor and delete characters based on the sequence you specify.

EGP

Exterior Gateway Protocol; an example is the Border Gateway Protocol (BGP).

egress router

In Multiprotocol Label Switching (MPLS), the last router in a label-switched path (LSP).

See also **ingress router**.

EIA

Electronic Industries Association. A U.S. trade group that represents manufacturers of electronic devices and sets standards and specifications.

EIA-530

Serial interface that employs the EIA-530 standard for the interconnection of data terminating equipment (DTE) and data circuit-terminating equipment (DCE).

EIR

Equipment identity register. A mobile network database that contains information about devices using the network.

embedded OS software

Software used by a Juniper Networks router to operate the physical router components.

EMI

Electromagnetic interference. Any electromagnetic disturbance that interrupts, obstructs, or otherwise degrades or limits the effective performance of electronics or electrical equipment.

end system

In Intermediate System-to-Intermediate System Level 1 (IS-IS), a network entity that sends and receives packets.

EPD

Early packet discard. For ATM2 interfaces only, a limit on the number of transmit packets that can be queued. Packets that exceed the limit are dropped.

See also **queue length**.

ERO

Explicit Route Object. An extension to the Resource Reservation Protocol (RSVP) that allows an RSVP PATH message to traverse an explicit sequence of routers that is independent of conventional shortest-path IP routing.

ESD

Electrostatic discharge. Stored static electricity that can damage electronic

equipment and impair electrical circuitry when released.

ES-IS

End System-to-Intermediate System. Protocol that resolves Layer 3 ISO network service access points (NSAPs) to Layer 2 addresses. ES-IS resolution is similar to the way the Address Resolution Protocol (ARP) resolves Layer 2 addresses for IPv4.

ESP

Encapsulating Security Payload. A protocol for securing packet flows for IPSec using encryption, data integrity checks, and sender authentication, which are added as a header to an IP packet. If an ESP packet is successfully decrypted, and no other party knows the secret key the peers share, the packet was not wiretapped in transit.

See also **AH**.

Established

Border Gateway Protocol (BGP) neighbor state that represents a fully functional BGP peering session.

Ethernet

Local area network (LAN) technology used for transporting information from one location to another, formalized in the IEEE standard 802.3. Ethernet uses either coaxial cable or twisted-pair cable. Transmission speeds for data transfer range from the original 10 Mbps, to Fast Ethernet at 100 Mbps, to Gigabit Ethernet at 1000 Mbps.

ETSI

European Telecommunications Standardization Institute. A nonprofit organization that produces voluntary telecommunications standards used throughout Europe.

eventd

Event policy process that performs configured actions in response to events on a routing platform that trigger system log messages.

exact

JUNOS software routing policy match type that represents only the route specified in a route filter.

exception packet

IP packet that is not processed by the normal packet flow through the Packet Forwarding Engine. Exception packets include local delivery information, expired Time to Live (TTL) packets, and packets with an IP option specified.

Exchange

Open Shortest Path First (OSPF) adjacency state in which two neighboring routers are actively sending database description packets to each other to exchange their database contents.

EXP bits

Experimental bits, also known as the class-of-service (CoS) bits, located in each Multiprotocol Label Switching (MPLS) label and used to encode the CoS value of a packet as it traverses a label-switched path (LSP).

export

Placing of routes from the routing table into a routing protocol.

ExStart

Open Shortest Path First (OSPF) adjacency state in which the neighboring routers negotiate to determine which router is in charge of the synchronization process.

Extensible Markup Language

See XML.

external metric

Cost included in a route when Open Shortest Path First (OSPF) exports route information from external Autonomous Systems (ASs). There are two types of external metrics: Type 1 and Type 2. Type 1 external metrics are equivalent to the link-state metric; that is, the cost of the route, used in the internal AS. Type 2 external metrics are greater than the cost of any path internal to the AS.

FA

Forwarding adjacency. Resource Reservation Protocol (RSVP) label-switched path (LSP) tunnel through which one or more other RSVP LSPs can be tunneled.

fabric schedulers

Identify a packet as high or low priority based on its forwarding class, and associate schedulers with the fabric priorities.

failover

Process by which a standby or secondary system component automatically takes over the functions of an active or primary component when the primary component fails or is temporarily shut down or removed for servicing. During failover, the system continues to perform normal operations with little or no interruption in service.

See also GRES.

Fast Ethernet

Term encompassing a number of Ethernet standards that carry traffic at the nominal rate of 100 Mbps, instead of the original Ethernet speed of 10 Mbps.

See also Ethernet, Gigabit Ethernet.

fast port

Fast Ethernet port on a J4300 services router, and either a Fast Ethernet port or DS3 port on a J6300 services router. Only enabled ports are counted. A two-port Fast Ethernet Physical Interface Module (PIM) with one enabled port counts as one fast port. The same PIM with both ports enabled counts as two fast ports.

fast reroute

Mechanism for automatically rerouting traffic on a label-switched path (LSP) if a node or link in an LSP fails, thus reducing the loss of packets traveling over the LSP.

FBF

Filter-based forwarding. A filter that classifies packets to determine their forwarding path within a router. FBF is used to redirect traffic for analysis.

FCS

Frame check sequence. A calculation that is added to a frame for error control. FCS is used in High-level Data Link Control (HDLC), Frame Relay, and other Data Link layer protocols.

FDDI

Fiber Distributed Data Interface. A set of ANSI protocols for sending digital data over fiber-optic cable. FDDI networks are token-passing networks, and support data rates of up to 100 Mbps (100 million bits). FDDI networks are typically used as backbones for WANs.

FEAC

Far-end alarm and control. A T3 signal used to send alarm or status information from the far-end terminal back to the near-end terminal, and to initiate T3 loopbacks at the far-end terminal from the near-end terminal.

FEB

Forwarding Engine Board. In M5 and M10 routers, provides route lookup, filtering, and switching to the destination port.

FEC

Forwarding equivalence class. Criterion used to forward a set of packets, with similar or identical characteristics, using the same Multiprotocol Label Switching (MPLS) label. Forwarding equivalence classes are defined in the base Label Distribution Protocol (LDP) specification and can be extended through the use of additional parameters. FECs are also represented in other LDPs.

FECN

Forward explicit congestion notification. In a Frame Relay network, a header bit transmitted by the source device requesting that the destination device slow down its requests for data. FECN and backward explicit congestion notification (BECN) minimize the possibility that packets will be discarded when more packets arrive than can be handled.

See also BECN.

FIFO

First in, first out. Scheduling method in which the first data packet stored in the queue is the first data packet removed from the queue. All JUNOS software interface queues operate in this mode by default.

filter

Process or device that screens packets based on certain characteristics, such as source address, destination address, or protocol, and forwards or discards packets that match the filter. Filters are used to control data packets or local packets.

See also packet.

FIPS

Federal Information Processing Standards. Defines, among other things, security levels for computer and networking equipment. FIPS is usually applied to military environments.

firewall

Security gateway positioned between two networks, usually between a trusted network and the Internet. A firewall ensures that all traffic that crosses it conforms to the organization's security policy. Firewalls track and control communications, deciding whether to pass, reject, discard, encrypt, or log them. Firewalls also can be used to secure sensitive portions of a local network.

firewall filter

See stateful firewall filter.

firmware

Instructions and data programmed directly into the circuitry of a hardware device for the purpose of controlling the device. Firmware is used for vital programs that must not be lost when the device is powered off.

first in, first out

See FIFO.

flap damping

See damping.

flapping

See route flapping.

flash drive

Non-volatile memory card in Juniper Networks M Series and T Series routing platforms used for storing a copy of the JUNOS software and the current and most recent router configurations. It also typically acts as the primary boot device.

Flexible PIC Concentrator

See FPC.

floating static route

Route with an administrative distance greater than the administrative distance of the dynamically learned versions of the same route. The static route is used only when the dynamic routes are no longer available. When a floating static route is configured on an interface with a dialer filter, the interface can be used for backup.

flood and prune

Method of forwarding multicast data packets in a Dense-mode network. Flooding and pruning occur every three minutes.

flow

Stream of routing information and packets that are handled by the Routing Engine (RE) and the Packet Forwarding Engine (PFE). The RE handles the flow of routing information between the routing protocols and the routing tables and between the routing tables and the forwarding tables, as well as the flow of local packets from the router physical interfaces to the RE. The PFE handles the flow of data packets into and out of the router's physical interfaces.

flow collection interface

Interface that combines multiple cflowd records into a compressed ASCII data file and exports the file to an FTP server for storage and analysis, allowing users to manipulate the output from traffic monitoring operations.

flow control action

JUNOS software syntax used in a routing policy or firewall filter. It alters the default logical processing of the policy or filter when a set of match conditions is met.

flow monitoring

Application that monitors the flow of traffic and enables lawful interception of packets transiting between two routers. Traffic flows can be passively monitored by an offline router or actively monitored by a router participating in the network.

flow-tap application

Application that uses Dynamic Tasking Control Protocol (DTCP) requests to intercept IPv4 packets in an active monitoring router and send a copy of packets that match filter criteria to one or more content destinations. Flow-tap configurations can be used in flexible trend analysis for detecting new security threats and lawfully intercepting data.

forwarding classes

Affect the forwarding, scheduling, and marking policies applied to packets as they transit a routing platform. The forwarding class plus the loss priority define the per-hop behavior. Also known as ordered aggregates in the IETF Differentiated Services architecture.

forwarding table

JUNOS software forwarding information base. The JUNOS routing protocol process installs active routes from its routing tables into the routing engine (RE) forwarding table. The kernel copies this forwarding table into the Packet Forwarding Engine (PFE), which determines which interface transmits the packets.

FPC

Flexible PIC Concentrator. An interface concentrator on which Physical Interface Cards (PICs) are mounted. An FPC is inserted into a slot in a Juniper Networks router.

See also PIC.

fractional E1

Interface that contains one or more of the 32 DS0 time slots that can be reserved from an E1 interface. (The first time slot is reserved for framing.)

fractional interface

Interface that contains one or more DS0 time slots reserved from an E1 or T1 interface. Fractional interfaces allow service providers to provision part of an E1 or T1 interface to one customer and the other part to another customer. The individual fractional interfaces connect to different destinations, and customers pay for only the bandwidth fraction used and not for the entire E1 or T1 interface.

Fractional interfaces can be configured on both channelized Physical Interface Cards (PICs) and Physical Interface Modules (PIMs) and unchannelized, regular E1 and T1 PICs and PIMs.

fractional T1

Interface that contains one or more of the 24 DS0 time slots that can be reserved from a T1 interface.

fragmentation

In the Transmission Control Protocol/Internet Protocol (TCP/IP), the process of breaking packets into the smallest maximum size packet data unit (PDU) supported by any of the underlying networks. In the OSI reference model, this process is known as segmentation. For JUNOS applications, split Layer 3 packets can then be encapsulated in Multilink Frame Relay (MLFR) or the Multilink Point-to-Point Protocol (MLPPP) for transport.

Frame Relay

Efficient replacement for the older X.25 protocol that does not require explicit acknowledgment of each frame of data. Frame Relay allows private networks to reduce costs by using shared facilities between the endpoint switches of a network managed by a Frame Relay service provider. Individual data-link connection identifiers (DLCIs) are assigned to ensure that each customer receives only its own traffic.

frequency-division multiplexed channel

Signals carried at different frequencies and transmitted over a single wire or wireless medium.

FRF

Frame Relay Forum. A technical committee that promotes Frame Relay by negotiating agreements and developing standards.

FRF.15

End-to-end Frame Relay Implementation Agreement. An implementation of Multilink Frame Relay (MLFR) using multiple virtual connections to aggregate logical bandwidth for end-to-end Frame Relay. Released by the Frame Relay Forum.

FRF.16

Multilink Frame Relay Implementation Agreement. An implementation of Multilink Frame Relay (MLFR) in which a single logical connection is provided by multiplexing multiple physical interfaces for user-to-network interface and network-to-network interface (UNI/NNI) connections. Released by the Frame Relay Forum.

FRU

Field Replaceable Unit. A router component that customers can replace onsite.

FTP

File Transfer Protocol. Application protocol that is part of the Transmission Control Protocol/Internet Protocol (TCP/IP) protocol stack. Used for transferring files among network nodes. FTP is defined in RFC 959.

Full

Open Shortest Path First (OSPF) adjacency state that represents a fully functional neighbor relationship.

fxp0

See management Ethernet interface.

fxp1

JUNOS software permanent interface used for communications between the routing engine (RE) and the Packet Forwarding

Engine (PFE). This interface is not present in all routers.

fxp2

JUNOS software permanent interface used for communications between the routing engine (RE) and the Packet Forwarding Engine (PFE). This interface is not present in all routers.

Garbage Collection Timer

Timer used in a distance-vector network that represents the time remaining before a route is removed from the routing table.

G-CDR

GGSN call detail record. Collection of charges in ASN.1 format that is eventually billed to a mobile station user.

generated route

Summary route that uses an IP address next hop to forward packets in an IP network. A generated route is functionally similar to an aggregated route.

GGSN

Gateway GPRS support node. A router that serves as a gateway between mobile networks and packet data networks.

Gigabit Ethernet

Term describing various technologies for implementing Ethernet networking at a nominal speed of one gigabit per second. Gigabit Ethernet is supported over both optical fiber and twisted-pair cable. Physical layer standards include 1000Base-T, 1 Gbps over CAT-5e copper cabling, and 1000Base-SX for short to medium distances over fiber.

See also **Ethernet, Fast Ethernet**.

GMPLS

Generalized Multiprotocol Label Switching. A protocol that extends the functionality of Multiprotocol Label Switching (MPLS) to include a wider range of label-switched path (LSP) options for a variety of network devices.

GPRS

General Packet Radio System. A packet-switched service that allows full mobility and wide-area coverage as information is sent and received across a mobile network.

Graceful Restart

Process that allows a router whose control plane is undergoing a restart to continue to forward traffic while recovering its state from neighboring routers. Without Graceful Restart, a control plane restart disrupts services provided by the router.

graceful switchover

JUNOS software feature that allows a change from the primary device, such as a routing engine (RE), to the backup device without interruption of packet forwarding.

gratuitous

ARP broadcast request for a router's own IP address to check whether that address is being used by another node. Primarily used to detect IP address duplication.

GRE

Generic Routing Encapsulation. A general tunneling protocol that can encapsulate many types of packets to enable data transmission through a tunnel. GRE is used with IP to create a virtual point-to-point link to routers at remote points in a network.

See also **tunneling protocol**.

GRES

Graceful Routing Engine Switchover. In a router that contains a master and a backup routing engine (RE), allows the backup RE to assume mastership automatically, with no disruption of packet forwarding.

group

Collection of related Border Gateway Protocol (BGP) peers.

group address

IP address used as the destination address in a multicast IP packet. The group address functionally represents the senders and interested receivers for a particular multicast data stream.

G.SHDSL

Symmetric high-speed digital subscriber line (SHDSL). Standard published in 2001 by the ITU-T with recommendation ITU G.991.2 G.SHDSL. G.SHDSL incorporates features of other DSL technologies such as asymmetrical DSL (ADSL).

See also SHDSL, ADSL.

GSM

Global System for Mobile Communications. A second-generation (2G) mobile wireless networking standard defined by ETSI that uses TDMA technology and operates in the 900 MHz radio band.

See also TDMA.

GTP

GPRS tunneling protocol. A protocol that transports IP packets between an SGSN and a GGSN.

See also tunneling protocol.

GTP-C

GGSN tunneling protocol, control. A protocol that allows an SGSN to establish packet data network access for a mobile station.

See also tunneling protocol.

GTP-U

GGSN tunneling protocol, user plane. A protocol that carries mobile station user data packets.

See also tunneling protocol.

hashing

Cryptographic technique applied over and over (iteratively) to a message of arbitrary length to produce a hash "message digest" or "signature" of fixed length that is appended to the message when it is sent. In security, used to validate that the contents of a message have not been altered in transit. The Secure Hash Algorithm (SHA-1) and Message Digest 5 (MD5) are commonly used hashes.

See also SHA-1, MD5.

HDLC

High-Level Data Link Control. An International Telecommunication Union (ITU) standard for a bit-oriented Data Link layer protocol on which most other bit-oriented protocols are based.

health monitor

JUNOS software extension to the RMON alarm system that provides predefined monitoring for filesystem, CPU, and memory usage. The health monitor also supports unknown or dynamic object instances such as JUNOS processes.

hello interval

Amount of time an Open Shortest Path First (OSPF) router continues to send a hello packet to each adjacent neighbor.

hello mechanism

Process used by a Resource Reservation Protocol (RSVP) router to enhance the detection of network outages in a Multiprotocol Label Switching (MPLS) network.

HLR

Home Location Register. Database containing information about a subscriber and the current location of a subscriber's mobile station.

HMAC

Hashed Message Authentication Code. A mechanism for message authentication that uses cryptographic hash functions. HMAC can be used with any iterative cryptographic hash function—for example, Message Digest 5 (MD5) or Secure Hash Algorithm (SHA-1)—in combination with a secret shared key. The cryptographic strength of HMAC depends on the properties of the underlying hash function. Defined in RFC 2104, "HMAC: Keyed-Hashing for Message Authentication."

hold down

Timer used by distance-vector protocols to prevent the propagation of incorrect routing knowledge to other routers in the network.

hold time

Maximum number of seconds allowed to elapse between successive keepalive or update messages that a Border Gateway Protocol (BGP) system receives from a peer.

host membership query

Internet Group Management Protocol (IGMP) packet sent by a router to determine whether interested receivers exist on a broadcast network for multicast traffic.

host membership report

Internet Group Management Protocol (IGMP) packet sent by an interested receiver for a particular multicast group address. Hosts send report messages when they first join a group or in response to a query packet from the local router.

host module

On an M160 router, provides the routing and system management functions of the router. Consists of the routing engine (RE) and Miscellaneous Control Subsystem (MCS).

host subsystem

On a T640 routing node, provides the routing and system management functions of the router. Consists of a routing engine (RE) and an adjacent Control Board (CB).

hot standby

In JUNOS, method used with link services intelligent queuing interfaces (LSQs) to enable rapid switchover between primary and secondary (backup) Physical Interface Cards (PICs).

See also **warm standby**.

HSCSD

High-Speed Circuit Switched Data. Circuit-switched wireless data transmission for mobile users, at data rates up to 38.4 Kbps.

HTTP

Hypertext Transfer Protocol. Method used to publish and receive information on the Web, such as text and graphics files.

HTTPS

Hypertext Transfer Protocol over Secure Sockets Layer. Similar to HTTP, with an added encryption layer that encrypts and decrypts user page requests and pages that are returned by a web server. Used for secure communication, such as payment transactions.

IANA

Internet Assigned Numbers Authority. A regulatory group that maintains all assigned and registered Internet numbers, such as IP and multicast addresses.

IBGP

Internal BGP. A Border Gateway Protocol (BGP) configuration in which sessions are established between routers in the same Autonomous System (AS).

ICMP

Internet Control Message Protocol. Used in router discovery, ICMP allows router advertisements that enable a host to discover addresses of operating routers on the subnet.

IDE

Integrated Drive Electronics. Type of hard disk on a routing engine (RE).

IDEA

International Data Encryption Algorithm. An algorithm that uses a 128-bit key and is one of the methods at the heart of Pretty Good Privacy (PGP). IDEA is patented by Ascom Tech AG and is popular in Europe.

Idle

Initial Border Gateway Protocol (BGP) neighbor state in which the local router refuses all incoming session requests.

IDS

Intrusion detection service. A service that inspects all inbound and outbound network activity and identifies suspicious patterns that may indicate a network or system attack from someone attempting to break into or compromise a system.

IEC

International Electrotechnical Commission.

See also ISO.

IEEE

Institute of Electrical and Electronics Engineers. An international professional society for electrical engineers.

IETF

Internet Engineering Task Force. An international community of network designers, operators, vendors, and researchers concerned with the evolution of Internet architecture and the smooth operation of the Internet.

I-frame

Information frame used to transfer data in sequentially numbered logical link control Protocol Data Units (LPDUs) between link stations.

IGMP

Internet Group Management Protocol. Used with multicast protocols to determine whether group members are present.

IGP

Interior Gateway Protocol, such as Intermediate System to Intermediate System Level 1 (IS-IS), Open Shortest Path First (OSPF), and the Routing Information Protocol (RIP).

IKE

Internet Key Exchange. Part of IPSec that provides ways to securely negotiate the shared private keys that the authentication header (AH) and Encapsulating Security Payload (ESP) portions of IPSec need to function properly. IKE employs Diffie-Hellman methods and is optional in IPSec (the shared keys can be entered manually at the endpoints).

ILMI

Integrated Local Management Interface. A specification developed by the ATM Forum that incorporates network management capabilities into the Asynchronous Transfer Mode (ATM) user-to-network interface (UNI) and provides bidirectional exchange of management information between UNI management entities (UMEs).

IMEI

International Mobile Station Equipment Identity. A unique code used to identify an individual mobile station to a GSM network.

import

Installation of routes from the routing protocols into a routing table.

IMSI

International Mobile Subscriber Identity. Information that identifies a particular subscriber to a GSM network.

IMT-2000

International Mobile Telecommunications 2000. Global standard for third-generation (3G) wireless communications, defined by a set of interdependent ITU recommendations. IMT-2000 provides a framework for worldwide wireless access by linking the diverse systems of terrestrial and satellite-based networks.

inet.0

Default JUNOS software routing table for IPv4 unicast routers.

inet.1

Default JUNOS software routing table for storing the multicast cache for active data streams in the network.

inet.2

Default JUNOS software routing table for storing unicast IPv4 routes specifically used to prevent forwarding loops in a multicast network.

inet.3

Default JUNOS software routing table for storing the egress IP address of a Multiprotocol Label Switching (MPLS) label-switched path.

inet.4

Default JUNOS software routing table for storing information generated by the Multicast Source Discovery Protocol (MSDP).

inet6.0

Default JUNOS software routing table for storing unicast IPv6 routes.

infinity metric

Metric value used in distance-vector protocols to represent an unusable route. For the Routing Information Protocol (RIP), the infinity metric is 16.

ingress router

In Multiprotocol Label Switching (MPLS), the first router in a label-switched path (LSP).

See also egress router.

Init

Open Shortest Path First (OSPF) adjacency state in which the local router has received a hello packet but bidirectional communication is not yet established.

insert

JUNOS software command that allows a user to reorder terms in a routing policy or a firewall filter, or to change the order of a policy chain.

instance.inetflow.0

Routing table that shows route flows through the Border Gateway Protocol (BGP).

inter-AS routing

Routing of packets among different Autonomous Systems (ASs).

See also EBGP.

intercluster reflection

In a Border Gateway Protocol (BGP) route reflection, the redistribution of routing information by a route reflector system to all nonclient peers (BGP peers not in the cluster).

See also route reflection.

interface cost

Value added to all received routes in a distance-vector network before they are placed into the routing table. The JUNOS software uses a cost of 1 for this value.

interface preservation

See link-state replication.

interface routes

Routes that are in the routing table because an interface has been configured with an IP address. Also called *direct routes*.

intermediate system

In Intermediate System-to-Intermediate System Level 1 (IS-IS), the network entity that sends and receives packets and can also route packets.

Internet Processor ASIC

Juniper Networks ASIC responsible for using the forwarding table to make routing decisions within the Packet Forwarding Engine (PFE). The Internet Processor ASIC also implements firewall filters.

interprovider VPN

Virtual private network (VPN) that provides connectivity between separate Autonomous Systems (ASs) with separate border edge routers. It is used by VPN customers who have connections to several different Internet service providers (ISPs), or different connections to the same ISP in different geographic regions, each of which has a different AS.

intra-AS routing

Routing of packets within a single Autonomous System (AS).

See also IBGP.

I/O Manager ASIC

Juniper Networks ASIC responsible for segmenting data packets into 64-byte J-cells and for queuing resultant cells before transmission.

IP

Internet Protocol. The protocol used for sending data from one point to another on the Internet.

IPCP

IP Control Protocol. Protocol that establishes and configures IP over the Point-to-Point Protocol (PPP).

IPSec

IP Security. A standard way to add security to Internet communications. The secure aspects of IPSec are usually implemented in three parts: the authentication header (AH), the Encapsulating Security Payload (ESP), and the Internet Key Exchange (IKE).

IQ

Intelligent queuing. M Series and T Series routing platform interfaces that offer granular quality-of-service (QoS) capabilities; extensive statistics on packets and bytes that are transmitted, received, or dropped; and embedded diagnostic tools.

IRDP

ICMP Router Discovery Protocol. A protocol that enables a host to determine the address of a router that it can use as a default gateway.

ISAKMP

Internet Security Association and Key Management Protocol. A protocol that allows the receiver of a message to obtain a public key and use digital certificates to authenticate the sender's identity. ISAKMP is key-exchange-independent; that is, it supports many different key exchanges.

See also IKE, Oakley.

ISDN

Integrated Services Digital Network. A set of digital communications standards that enable the transmission of information over existing twisted-pair telephone lines at higher speeds than standard analog telephone service. An ISDN interface provides multiple B-channels (bearer channels) for data

and one D-channel for control and signaling information.

See also B-channel, D-channel.

IS-IS

Intermediate System-to-Intermediate System. A link-state, interior gateway routing protocol for IP networks that also uses the Shortest Path First (SPF) algorithm to determine routes.

ISO

International Organization for Standardization. A worldwide federation of standards bodies that promotes international standardization and publishes international agreements as International Standards.

ISP

Internet service provider. Company that provides access to the Internet and related services.

ITU-T

International Telecommunication Union Telecommunication Standardization (formerly known as the CCITT). Group supported by the United Nations that makes recommendations and coordinates the development of telecommunications standards for the entire world.

ITU-T Rec. G.992.1

International standard that defines the asymmetrical digital subscriber line (ADSL). Annex A defines how ADSL works over twisted-pair copper (POTS) lines. Annex B defines how ADSL works over Integrated Services Digital Network (ISDN) lines.

jbase

JUNOS software package containing updates to the kernel.

jbundle

JUNOS software package containing all possible software package files.

J-cell

A 64-byte data unit used within the Packet Forwarding Engine (PFE). All IP packets processed by a Juniper Networks router are segmented into J-cells.

jdocs

JUNOS software package containing the documentation set.

jitter

Small random variation introduced into the value of a timer to prevent multiple timer expirations from becoming synchronized. In real-time applications such as Voice over IP and video, variation in the rate at which packets in a stream are received that can cause quality degradation.

jkernel

JUNOS software package containing the basic components of the software.

Join message

Physical Interface Module (PIM) message sent hop by hop upstream toward a multicast source or the rendezvous point (RP) of the domain. It requests that multicast traffic be sent downstream to the router originating the message.

jpfe

JUNOS software package containing the embedded OS software for operating the Packet Forwarding Engine (PFE).

jroute

JUNOS software package containing the software used by the routing engine (RE).

J-Web

Graphical web browser interface to the JUNOS Internet software on routing platforms. With the J-Web interface, you can monitor, configure, diagnose, and manage the routing platform from a PC or laptop that has Hypertext Transfer Protocol (HTTP) or HTTP over Secure Sockets Layer (HTTPS) enabled.

keepalive message

Message sent between network devices to inform each other that they are still active.

kernel

Basic software component of the JUNOS software. The kernel operates the various processes used to control the router's operations.

kernel forwarding table

See forwarding table.

kmd

Key management process that provides IP-Sec authentication services for encryption of Physical Interface Cards (PICs).

L2TP

Layer 2 Tunneling Protocol. A procedure for secure communication of data across a Layer 2 network that enables users to establish Point-to-Point Protocol (PPP) sessions between tunnel endpoints. L2TP uses profiles for individual user and group access to ensure secure communication that is as transparent as possible to both end users and applications.

See also tunneling protocol.

label

In Multiprotocol Label Switching (MPLS), a 20-bit unsigned integer from 0 through 1,048,575, used to identify a packet traveling along a label-switched path (LSP).

Label Distribution Protocol

See LDP.

label object

Resource Reservation Protocol (RSVP) message object that contains the label value allocated to the next downstream router.

label pop operation

Function performed by a Multiprotocol Label Switching (MPLS) router in which the top label in a label stack is removed from the data packet.

label push operation

Function performed by a Multiprotocol Label Switching (MPLS) router in which a new label is added to the top of the data packet.

label request object

Resource Reservation Protocol (RSVP) message object that requests each router along the path of a label-switched path (LSP) to allocate a label for forwarding.

label swap operation

Function performed by a Multiprotocol Label Switching (MPLS) router in which the top label in a label stack is replaced with a new label before the data packet is forwarded to the next hop router.

label values

A 20-bit field in a Multiprotocol Label Switching (MPLS) header used by routers to forward data traffic along an MPLS label-switched path (LSP).

LAN PHY

Local Area Network Physical Layer Device. A physical layer device that allows 10 Gigabit Ethernet wide area links to use existing Ethernet applications.

See also PHY, WAN PHY.

Layer 2 circuits

Collection of transport modes that accept a stream of Asynchronous Transfer Mode (ATM) cells, convert them to an encapsulated Layer 2 format, and then tunnel them over a Multiprotocol Label Switching (MPLS) or IP backbone, where a similarly configured routing platform segments these packets back into a stream of ATM cells, to be forwarded to the virtual circuit configured for the far-end routing platform. Layer 2 circuits are designed to transport Layer 2 frames between provider edge (PE) routing platforms across a Label Distribution Protocol (LDP)-signaled MPLS backbone.

See also AAL5 mode, cell-relay mode, standard AAL5 mode, trunk mode.

Layer 2 VPN

Provides a private network service among a set of customer sites using a service provider's existing Multiprotocol Label Switching (MPLS) and IP network. A customer's data is separated from other data using software rather than hardware. In a Layer 2 VPN, the Layer 3 routing of customer traffic occurs within the customer's network.

Layer 3 VPN

Provides a private network service among a set of customer sites using a service provider's existing Multiprotocol Label Switching (MPLS) and IP network. A customer's routes and data are separated from other routes and data using software rather than hardware. In a Layer 3 VPN, the Layer 3 routing of customer traffic occurs within the service provider's network.

LCC

Line Card Chassis. Term used by the JUNOS command-line interface (CLI) to refer to a T640 routing node in a routing matrix.

LCP

Link Control Protocol. A traffic controller used to establish, configure, and test data-link connections for the Point-to-Point Protocol (PPP).

LDAP

Lightweight Directory Access Protocol. Software protocol used for locating resources on a public or private network.

LDP

Label Distribution Protocol. A protocol for distributing labels in non-traffic-engineered applications. LDP allows routers to establish label-switched paths (LSPs) through a network by mapping network-layer routing information directly to Data Link layer switched paths.

leaf node

Terminating node of a multicast distribution tree. A router that is a leaf node only has receivers and does not forward multicast packets to other routers.

LFI

Link fragmentation and interleaving. A method that reduces excessive delays by fragmenting long packets into smaller packets and interleaving them with real-time frames. For example, short delay-sensitive packets, such as packetized voice, can race ahead of larger delay-insensitive packets, such as common data packets.

liblicense

Library that includes messages generated for routines for software license management.

libpcap

Implementation of the pcap application programming interface. libpcap is used by a program to capture packets traveling over a network.

See also pcap.

limited operational environment

Term used to describe the restrictions placed on FIPS-certified equipment.

See also FIPS.

line loopback

Method of troubleshooting a problem with physical transmission media in which a transmission device in the network sends the data signal back to the originating router.

link

Communication path between two neighbors. A link is up when communication is possible between the two endpoints.

link protection

Method of establishing bypass label-switched paths (LSPs) to ensure that traffic going over a specific interface to a neighboring router can continue to reach the router if that interface fails. The bypass LSP uses a different interface and path to reach the same destination.

link services intelligent queuing interfaces

See LSQ.

link-state acknowledgment

Open Shortest Path First (OSPF) data packet used to inform a neighbor that a link-state update packet has been successfully received.

link-state database

All routing knowledge in a link-state network is contained in this database. Each router runs the Shortest Path First (SPF) algorithm against this database to locate the best network path to each destination in the network.

link-state PDU

Packet that contains information about the state of adjacencies to neighboring systems.

link-state replication

Addition to the SONET Automatic Protection Switching (APS) functionality that helps promote redundancy of the link Physical Interface Cards (PICs) used in LSQ configurations. If the active SONET PIC fails, links from the standby PIC are used without causing a link renegotiation. Also called *interface preservation*.

link-state request list

List generated by an Open Shortest Path First (OSPF) router during the exchange of database information while forming an adjacency. Advertised information by a neighbor that the local router does not contain is placed in this list.

link-state request packet

Open Shortest Path First (OSPF) data packet used by a router to request database information from a neighboring router.

link-state update

Open Shortest Path First (OSPF) data packet that contains one of multiple link state advertisements (LSAs). It is used to advertise routing knowledge into the network.

LLC

Logical Link Control. Data Link layer protocol used on a LAN. LLC1 provides connectionless data transfer, and LLC2 provides connection-oriented data transfer.

LLC frame

Unit of data that contains specific information about the LLC layer and identifies line protocols associated with the layer.

See also LLC.

LMI

Local management interface. Enhancements to the basic Frame Relay specifications, providing support for the following:

- A keepalive mechanism that verifies the flow of data
- A multicast mechanism that provides a network server with a local data-link connection identifier (DLCI) and multicast DLCI
- In Frame Relay networks, global addressing that gives DLCIs global instead of local significance
- A status mechanism that provides a switch with ongoing status reports on known DLCIs

LMP

Link Management Protocol. Part of GMPLS, a protocol used to define a forwarding adjacency between peers and to maintain and allocate resources on the traffic engineering links.

load balancing

Process that installs all next hop destinations for an active route in the forwarding table. You can use load balancing across multiple paths between routers. The behavior of load balancing depends on the version of the Internet Processor ASIC in the router. Also called *per-packet load balancing*.

loading

Open Shortest Path First (OSPF) adjacency state in which the local router sends link-state request packets to its neighbor and waits for the appropriate link-state updates from that neighbor.

local packet

Chunk of data destined for or sent by the routing engine (RE).

local preference

Optional Border Gateway Protocol (BGP) path attribute carried in internal BGP update packets that indicate the degree of preference for an external route.

local RIB

Logical software table that contains Border Gateway Protocol (BGP) routes used by the local router to forward data packets.

local significance

Concept used in a Multiprotocol Label Switching (MPLS) network where the label values are unique only between two neighbor routers.

logical interface

On a physical interface, the configuration of one or more units that include all addressing, protocol information, and other logical interface properties that enable the physical interface to function.

logical operator

Characters used in a firewall filter to represent a Boolean AND or OR operation.

logical router

Logical routing device that is partitioned from an M Series or T Series routing platform. Each logical router independently performs a subset of the tasks performed by the main router and has a unique routing table, interfaces, policies, and routing instances.

longer

JUNOS software routing policy match type that represents all routes more specific than the given subnet, but not the given subnet itself. It is similar to a mathematical greater-than operation.

loopback interface (lo0)

Interface that is always available because it is independent of any physical interfaces. When configured with an address, the loopback interface is the default address for the routing platform and any unnumbered interfaces.

See also **unnumbered interface**.

loose hop

In the context of traffic engineering, a path that can use any router or any number of other intermediate (transit) points to reach the next address in the path. (Definition from RFC 791, modified to fit LSPs.)

loss-priority map

Maps the loss priority of incoming packets based on code point values.

lower-speed IQ interfaces

E1, NxDS0, and T1 interfaces configured on an intelligent queuing (IQ) Physical Interface Card (PIC).

LPDU

LLC protocol Data Unit. LLC frame on a data link switching (DLSw) network.

See also LLC frame.

LSA

Link state advertisement. Open Shortest Path First (OSPF) data structure that is advertised in a link-state update packet. Each LSA uniquely describes a portion of the OSPF network.

LSI

Label-switched interface. A logical interface supported by the JUNOS software that provides virtual private network (VPN) services (such as VPLS and Layer 3 VPNs) normally provided by a Tunnel Services PIC.

LSP

1. Label-switched path. Sequence of routers that cooperatively perform Multiprotocol Label Switching (MPLS) operations for a packet stream. The first router in an LSP is called the ingress router, and the last router in the path is called the egress router. An LSP is a point-to-point, half-duplex connection from the ingress router to the egress router. (The ingress and egress routers cannot be the same router.) 2. See link-state PDU.

LSQ

Link services intelligent queuing interfaces. Interfaces configured on the Adaptive Services PIC (ASP) or Adaptive Services Module (ASM) that support Multilink Point-to-Point Protocol (MLPPP) and Multilink Frame Relay (MLFR) traffic and also fully support JUNOS class-of-service (CoS) components.

LSR

Label-switching router. A router on which Multiprotocol Label Switching (MPLS) is enabled and that can process label-switched packets.

MAC

Media Access Control. In the OSI seven-layer networking model defined by the IEEE, MAC is the lower sublayer of the Data Link layer. The MAC sublayer governs protocol access to the physical network medium. By using the MAC addresses that are assigned to all ports on a router, multiple devices on the same physical link can uniquely identify one another at the Data Link layer.

See also MAC address.

MAC address

Serial number permanently stored in a device adapter to uniquely identify the device.

See also MAC.

MAM

Maximum allocation bandwidth constraints model. In Differentiated Services-aware traffic engineering, a constraint model that divides the available bandwidth among the different classes. Sharing of bandwidth among the class types is not allowed.

management Ethernet interface

Permanent interface that provides an Out-of-Band method, such as Secure Shell (SSH) and Telnet, to connect to the routing platform. The Simple Network Management Protocol (SNMP) can use the management interface to gather statistics from the routing platform. Called *fxp0* on some routing platforms.

See also permanent interface.

mapping agent

Router used in an auto-RP multicast network to select the rendezvous point (RP) for all multicast group addresses. The RP is then advertised to all other routers in the domain.

martian address

Network address about which all information is ignored.

martian route

Network routes about which all information is ignored. The JUNOS software does not allow martian routes in the inet.0 routing table.

MAS

Mobile network access subsystem. A GSN application subsystem that contains the access server.

master

Router in control of the Open Shortest Path First (OSPF) database exchange during an adjacency formation.

match

Logical concept used in a routing policy or firewall filter. A match denotes the criteria used to find a route or IP packet before an action is performed.

match type

JUNOS software syntax used in a route filter to better describe the routes that should match the policy term.

MBGP

Multiprotocol Border Gateway Protocol. An extension to the Border Gateway Protocol (BGP) that allows you to connect multicast topologies within and between BGP Autonomous Systems (ASs).

MBone

Multicast Backbone. An interconnected set of subnetworks and routers that support the delivery of IP multicast traffic. The MBone is a virtual network that is layered on top of sections of the physical Internet.

MCS

Miscellaneous Control Subsystem. On the M40e and M160 routers, provides control and monitoring functions for router components and SONET clocking for the router.

MD5

Message Digest 5. A one-way hashing algorithm that produces a 128-bit hash used for generating message authentication signatures. MD5 is used in authentication header (AH) and Encapsulating Security Payload (ESP).

See also hashing, SHA-1.

MDRR

Modified deficit round robin. A method for selecting queues to be serviced.

See also queue.

MDT

Multicast distribution tree. The path between the sender (host) and the multicast group (receiver or listener).

mean time between failures

See MTBF.

MED

Multiple exit discriminator. An optional Border Gateway Protocol (BGP) path attribute consisting of a metric value that is used to determine the exit point to a destination when all other factors determining the exit point are equal.

mesh

Network topology in which devices are organized in a manageable, segmented manner with many, often redundant, interconnections between network nodes.

message aggregation

Extension to the Resource Reservation Protocol (RSVP) specification that allows neighboring routers to bundle up to 30 RSVP messages into a single protocol packet.

mgd

Management daemon. JUNOS software process responsible for managing all user access to the router.

MIB

Management Information Base. Definition of an object that can be managed by the Simple Network Management Protocol (SNMP).

midplane

Physically separates front and rear cavities inside the chassis, distributes power from the power supplies, and transfers packets and signals between router components, which plug into it.

MLD

Multicast listener discovery. A protocol that manages the membership of hosts and routers in multicast groups. IPv6 multicast routers use MLD to learn, for each of their attached physical networks, which groups have interested listeners.

MLFR

Multilink Frame Relay. Logically ties together individual circuits, creating a bundle. The logical equivalent of the Multilink Point-to-Point Protocol (MLPPP), MLFR is used for Frame Relay traffic instead of Point-to-Point Protocol (PPP) traffic. FRF.15 and FRF.16 are two implementations of MLFR.

MLPPP

Multilink Point-to-Point Protocol. Enables you to bundle multiple Point-to-Point Protocol (PPP) links into a single logical link between two network devices to provide an aggregate amount of bandwidth. The technique is often called *bonding* or *link aggregation*. Defined in RFC 1990.

See also **PPP**.

MMF

Multimode fiber. Optical fiber supporting the propagation of multiple frequencies of light. MMF is used for relatively short distances because the modes tend to disperse over longer lengths (called *modal*

dispersion). For longer distances, single-mode fiber (sometimes called *monomode*) is used.

See also **single-mode fiber**.

mobile station

Mobile device, such as a cellular phone or a mobile personal digital assistant (PDA).

mobile transport subsystem

See **MTS**.

MPLS

Multiprotocol Label Switching. Mechanism for engineering network traffic patterns that functions by assigning to network packets short labels that describe how to forward them through the network. Also called *label switching*.

See also **traffic engineering**.

MPLS EXP classifier

Class-of-service (CoS) behavior classifier for classifying packets based on the Multiprotocol Label Switching (MPLS) experimental bit.

See also **EXP bits**.

MPS

Mobile point-to-point control subsystem. A GGSN application subsystem that controls all functionality associated with a particular connection.

MRRU

Maximum received reconstructed unit. Similar to the maximum transmission unit (MTU), but is specific to link services interfaces.

See also **MTU**.

MSA

Multisource Agreement. The definition of a fiber-optic transceiver module that conforms to the 10 Gigabit Ethernet standard.

See also **XENPAK module**.

MSC

Mobile Switching Center. Provides origination and termination functions to calls from a mobile station user.

MSDP

Multicast Source Discovery Protocol. A protocol used to connect multicast routing domains to allow the domains to discover multicast sources from other domains. It typically runs on the same router as the Physical Interface Module (PIM) Sparse mode rendezvous point (RP).

MSISDN

Mobile Station Integrated Services Digital Network number. A number that callers use to reach a mobile services subscriber.

MTBF

Mean time between failures. Measure of hardware component reliability.

MTS

Mobile transport subsystem. A GSN application subsystem that implements all the protocols used by the GSN.

MTU

Maximum transmission unit. Limit on the data size for a network.

multicast

Operation of sending network traffic from one network node to multiple network nodes.

multicast-scope number

Number used for configuring the multicast scope. Configuring a scope number constrains the scope of a multicast session. The number value can be any hexadecimal number from 0 through F. The multicast-scope value is a number from 0 through 15, or a specified keyword with an associated prefix range. For example, link-local (value = 2), corresponding prefix 224.0.0.0/24.

multiclass LSP

In Differentiated Services-aware traffic engineering, a multiclass label-switched path (LSP) functions like a standard LSP, but also allows you to reserve bandwidth for multiple class types. The experimental (EXP) bits of the Multiprotocol Label Switching (MPLS) header are used to distinguish between class types.

multiclass MLPPP

Enables multiple classes of service while using the Multilink Point-to-Point Protocol (MLPPP). Defined in RFC 2686, "The Multi-Class Extension to Multi-Link PPP."

multifield classifier

Method for classifying traffic flows. Unlike a behavior aggregate (BA) classifier, a multifield classifier examines multiple fields in the packet to apply class-of-service (CoS) settings. Examples of fields that a multifield classifier examines include the source and destination addresses of the packet, as well as the source and destination port numbers of the packet.

See also BA classifier, classification.

multihoming

Network topology that uses multiple connections between customer and provider devices to provide redundancy.

MVS

Mobile visitor register subsystem.

named path

JUNOS software syntax that specifies a portion of or the entire network path that should be used as a constraint in signaling a Multiprotocol Label Switching (MPLS) label-switched path (LSP).

NAPT

Network Address Port Translation. A method that translates the addresses and transport identifiers of many private hosts into a few external addresses and transport identifiers to make efficient use of globally registered IP addresses. NAPT extends the level of translation beyond that of basic Network Address Translation (NAT).

See also NAT.

NAT

Network Address Translation. A method of concealing a set of host addresses on a private network behind a pool of public addresses. It can be used as a security measure to protect the host addresses from direct targeting in network attacks.

NCP

Network Control Protocol. A traffic controller used to establish and configure different network layer protocols for the Point-to-Point Protocol (PPP).

NDP

Neighbor Discovery Protocol. Protocol used by IPv6 nodes on the same link to discover each other's presence, determine each other's Link layer addresses, find routers, and maintain reachability information about the paths to active neighbors. NDP is defined in RFC 2461 and is equivalent to the Address Resolution Protocol (ARP) used with IPv4.

See also ARP.

neighbor

Adjacent system reachable by traversing a single subnetwork. An immediately adjacent router. Also called a *peer*.

NET

Network entity title. Network address defined by the ISO network architecture and used in CLNS-based networks.

NetBIOS

Network basic input/output system. An application programming interface used by programs on a LAN. NetBIOS provides a uniform set of commands for requesting the lower-level services required to manage names, conduct sessions, and send datagrams between nodes on a network.

network interface

Interface, such as an Ethernet or SONET/ SDH interface, that primarily provides traffic connectivity.

See also PIC, services interface.

network link advertisement

Open Shortest Path First (OSPF) link state advertisement (LSA) flooded throughout a single area by designated routers to describe all routers attached to the network.

network LSA

Open Shortest Path First (OSPF) link state advertisement (LSA) sent by the designated router on a broadcast or NBMA segment. It advertises the subnet associated with the designated router's segment.

network summary LSA

Open Shortest Path First (OSPF) link state advertisement (LSA) sent by an area border router (ABR) to advertise internal OSPF routing knowledge across an area boundary.

See also ABR.

NIC

Network Information Center. Internet authority responsible for assigning Internet-related numbers, such as IP addresses and Autonomous System (AS) numbers.

See also IANA.

NIST

National Institute of Standards and Technology. A nonregulatory U.S. federal agency whose mission is to develop and promote measurement, standards, and technology.

NLRI

Network layer reachability information. Information carried in Border Gateway Protocol (BGP) packets and used by the Multiprotocol Border Gateway Protocol (MBGP).

nonclient peer

In a Border Gateway Protocol (BGP) route reflection, a BGP peer that is not a member of a cluster.

See also client peer.

notification cell

JUNOS software data structure generated by the Distribution Buffer Manager ASIC that represents the header contents of an IP packet. The Internet Processor ASIC uses the notification cell to perform a forwarding table lookup.

Notification message

A Border Gateway Protocol (BGP) message that informs a neighbor about an error

condition, and then in some cases terminates the BGP peering session.

not-so-stubby area
See NSSA.

NSAP
Network service access point. Connection
to a network that is identified by a network
address.

n-selector
Last byte of a nonclient peer address.

NSR
Nonstop routing. A high-availability feature
that allows a routing platform with redundant routing engines (REs) to preserve routing information on the backup RE and
switch over from the primary RE to the
backup RE without alerting peer nodes that
a change has occurred. NSR uses the Graceful RE Switchover (GRES) infrastructure to
preserve interface, kernel, and routing
information.

NSSA
Not-so-stubby area. In Open Shortest Path
First (OSPF), a type of stub area in which
external routes can be flooded.

NTP
Network Time Protocol. A protocol used to
synchronize computer clock times on a
network.

Null Register message
Physical Interface Module (PIM) message
sent by the first hop router to the rendezvous
point (RP). The message informs the RP that
the local source is still actively sending multicast packets into the network.

See also RP.

numeric range match conditions
Use of numeric values (protocol and port
numbers) in the header of an IP packet to
match criteria in a firewall filter.

Oakley
Key determination protocol based on the
Diffie-Hellman algorithm that provides
added security, including authentication.

Oakley was the key-exchange algorithm
mandated for use with the initial version of
ISAKMP, although other algorithms can be
used. Oakley describes a series of key exchanges called *modes* and details the services
provided by each; for example, Perfect Forward Secrecy for keys, identity protection,
and authentication.

See also ISAKMP.

OAM
Operation, Administration, and Maintenance. An ATM Forum specification for
monitoring Asynchronous Transfer Mode
(ATM) virtual connections. OAM performs
standard loopback, fault detection and notification, and remote defect identification
for each connection, verifying that the connection is up and the router is operational.

OC
Optical carrier. In SONET, the OC level
indicates the transmission rate of digital
signals on optical fiber.

OC3
SONET line with a transmission speed of
155.52 Mbps (payload of 150.336 Mbps)
using fiber-optic cables. For SDH interfaces,
OC3 is also known as STM1.

OC12
SONET line with a transmission speed of
622 Mbps using fiber-optic cables.

Open message
Border Gateway Protocol (BGP) message
that allows two neighbors to negotiate the
parameters of the peering session.

OpenConfirm
Border Gateway Protocol (BGP) neighbor
state that shows that a valid Open message
was received from the remote peer.

OpenSent
Border Gateway Protocol (BGP) neighbor
state that shows that an Open message was
sent to the remote peer and the local router
is waiting for an Open message to be
returned.

operational mode

JUNOS software mode that allows a user to view statistics and information about the router's current operating status.

op script

Operational script. Extensible Stylesheet Language for Transformations (XSLT) script written to automate network troubleshooting and network management. Op scripts can perform any function available through JUNOScript remote procedure calls (RPCs).

origin

In the Border Gateway Protocol (BGP), an attribute that describes the source of the route.

orlonger

JUNOS software routing policy match type that represents all routes more specific than the given subnet, including the given subnet itself. It is similar to a mathematical greater-than-or-equal-to operation.

OSI

Open Systems Interconnection. Standard reference model for how messages are transmitted between two points on a network.

OSPF

Open Shortest Path First. A link-state Interior Gateway Protocol (IGP) that makes routing decisions based on the Shortest Path First (SPF) algorithm (also referred to as the Dijkstra algorithm).

OSPF hello packet

Message sent by each Open Shortest Path First (OSPF) router to each adjacent router. It is used to establish and maintain the router's neighbor relationships.

overlay network

Network design in which a logical Layer 3 topology (IP subnets) is operating over a logical Layer 2 topology (Asynchronous Transfer Mode permanent virtual circuits [ATM PVCs]). Layers in the network do not have knowledge of each other, and each layer requires separate management and operation.

oversubscription

Method that allows provisioning of more bandwidth than the line rate of the physical interface.

P2MP LSP

See point-to-multipoint LSP.

package

Collection of files that make up a JUNOS software component.

packet

Fundamental unit of information (message or fragment of a message) carried in a packet-switched network; for example, the Internet.

See also PSN.

packet aging

Occurs when packets in the output buffer are overwritten by newly arriving packets. This happens because the available buffer size is greater than the available transmission bandwidth.

packet capture

1. Packet sampling method in which entire IPv4 packets flowing through a router are captured for analysis. Packets are captured in the routing engine (RE) and stored as libpcap-formatted files on the router. Packet capture files can be opened and analyzed offline with packet analyzers such as tcpdump and Ethereal. 2. J-Web packet sampling method for quickly analyzing router control traffic destined for or originating from the RE. You can either decode and view the captured packets in the J-Web interface as they are captured, or save the packets to a file and analyze them offline with packet analyzers such as Ethereal. J-Web packet capture does not capture transient traffic.

See also traffic sampling.

Packet Forwarding Engine

Portion of the router that processes packets by forwarding them between input and output interfaces.

packet or cell switching

Transmission of packets from many sources over a switched network.

PADI

PPPoE Active Discovery Initiation packet. A Point-to-Point Protocol over Ethernet (PPPoE) initiation packet that is broadcast by the client to start the discovery process.

PADO

PPPoE Active Discovery Offer packet. A Point-to-Point Protocol over Ethernet (PPPoE) offer packet that is sent to the client by one or more access concentrators in reply to a PPPoE Active Discovery Initiation (PADI) packet.

PADR

PPPoE Active Discovery Request packet. A Point-to-Point Protocol over Ethernet (PPPoE) packet sent by the client to one selected access concentrator to request a session.

PADS

PPPoE Active Discovery Session Confirmation packet. A Point-to-Point Protocol over Ethernet (PPPoE) packet sent by the selected access concentrator to confirm the session.

PADT

PPPoE Active Discovery Termination packet. A Point-to-Point Protocol over Ethernet (PPPoE) packet sent by either the client or the access concentrator to terminate a session.

passive flow monitoring

Technique to intercept and observe specified data network traffic by using a routing platform such as a monitoring station that is not participating in the network.

path attribute

Information about a Border Gateway Protocol (BGP) route, such as the route origin, Autonomous System (AS) path, and next hop router.

PathErr message

Resource Reservation Protocol (RSVP) message indicating that an error has occurred along an established label-switched path (LSP). The message is advertised upstream toward the ingress router and does not remove any RSVP soft state from the network.

PathTear message

Resource Reservation Protocol (RSVP) message indicating that the established label-switched path (LSP) and its associated soft state should be removed by the network. The message is advertised downstream hop by hop toward the egress router.

pcap

Software library for packet capturing.

See also libpcap.

PC Card

(Previously known as a PCMCIA Card.) The removable storage media that ships with each router that contains a copy of the JUNOS software. The PC Card is based on standards published by the Personal Computer Memory Card International Association (PCMCIA).

PCI

Peripheral Component Interconnect. Standard, high-speed bus for connecting computer peripherals. Used on the routing engine (RE).

PCI Express

Peripheral Component Interconnect Express. Next-generation, higher-bandwidth bus for connecting computer peripherals. A PCI Express bus uses point-to-point bus topology with a shared switch rather than the shared bus topology of a standard PCI bus. The shared switch on a PCI Express bus provides centralized traffic routing and management and can prioritize traffic. On some J Series services routers, PCI Express slots are backward compatible with PCI and can accept Physical Interface Modules

(PIMs) intended for either PCI Express or PCI slots.

PCMCIA

Personal Computer Memory Card International Association. Industry group that promotes standards for credit-card-size memory and I/O devices.

PDH

Plesiochronous Digital Hierarchy. Developed to carry digitized voice more efficiently. Evolved into the North American, European, and Japanese Digital Hierarchies, in which only a discrete set of fixed rates is available; namely, NxDS0 (DS0 is a 64 kbps rate).

PDP

Packet data protocol. Network protocol, such as IP, used by packet data networks connected to a GPRS network.

PDU

Protocol Data Unit. A packet of data passed across a network. The term refers to a specific layer of the OSI seven-layer model and a specific protocol.

PEC

Policing equivalence classes. In traffic policing, a set of packets that are treated the same way by the packet classifier.

peer

Immediately adjacent router with which a protocol relationship has been established. Also called a *neighbor*.

peering

Practice of exchanging Internet traffic with directly connected peers according to commercial and contractual agreements.

PEM

1. Privacy Enhanced Mail. A technique for securely exchanging electronic mail over a public medium. 2. Power Entry Module. Distributes DC power within the router chassis. Supported on M40e, M160, M320, and T Series routing platforms.

penultimate router

Last transit router before the egress router in a Multiprotocol Label Switching (MPLS) label-switched path (LSP).

permanent interface

Interface that is always present in the routing platform.

See also **management Ethernet interface**, **transient interface**.

persistent change

Commit-script-generated configuration change that is copied to the candidate configuration. Persistent changes remain in the candidate configuration unless you explicitly delete them.

See also **transient change**.

PE router

Provider edge router. A router in the service provider's network that is connected to a customer edge (CE) device and participates in a virtual private network (VPN).

PFC

Protocol Field Compression. Normally, Point-to-Point Protocol (PPP)-encapsulated packets are transmitted with a 2-byte protocol field. For example, IPv4 packets are transmitted with the protocol field set to 0x0021, and Multiprotocol Label Switching (MPLS) packets are transmitted with the protocol field set to 0x0281. For all protocols with identifiers from 0x0000 through 0x00ff, PFC enables routers to compress the protocol field to one byte, as defined in RFC 1661, "The Point-to-Point Protocol (PPP)." PFC allows you to conserve bandwidth by transmitting less data.

See also **ACFC**.

PFS

Perfect Forward Secrecy protocol. A protocol derived from an encryption system that changes encryption keys often and ensures that no two sets of keys have any relationship to each other. If one set of keys is compromised, only communications using those keys are at risk. An example of a system that uses PFS is Diffie-Hellman.

PGM

Pragmatic General Multicast. A protocol layer that can be used between the IP layer and the multicast application on sources, receivers, and routers to add reliability, scalability, and efficiency to multicast networks.

PGP

Pretty Good Privacy. A strong cryptographic technique invented by Philip Zimmerman in 1991.

PHP

Penultimate hop popping. A mechanism used in a Multiprotocol Label Switching (MPLS) network that allows the transit router before the egress router to perform a label pop operation and forward the remaining data (often an IPv4 packet) to the egress router.

PHY

1. Special electronic integrated circuit or functional block of a circuit that performs encoding and decoding between a pure digital domain (on-off) and a modulation in the analog domain. 2. Open Systems Interconnection (OSI) physical layer. Layer 1 of the OSI model that defines the physical link between devices.

See also LAN PHY, WAN PHY.

physical interface

Port on a Physical Interface Card (PIC) or Physical Interface Module (PIM).

Physical Interface Module

A network interface card installed in a J Series services router to provide physical connections to a LAN or WAN. PIMs can be fixed or removable and interchangeable. The PIM receives incoming packets from the network and transmits outgoing packets to the network. Each PIM is equipped with a dedicated network processor that forwards incoming data packets to and receives outgoing data packets from the routing engine (RE). During this process, the PIM performs

framing and line-speed signaling for its medium type—for example, E1, serial, Fast Ethernet, or Integrated Services Digital Network (ISDN).

PIC

Physical Interface Card. A network interface-specific card that can be installed on a Flexible PIC Concentrator (FPC) in the router.

PIC I/O Manager

Juniper Networks ASIC responsible for receiving and transmitting information on the physical media. It performs media-specific tasks within the Packet Forwarding Engine (PFE).

PIR

Peak information rate. The PIR must be equal to or greater than the committed information rate (CIR), and both must be configured to be greater than 0. Packets that exceed the PIR are marked red, which corresponds to high loss priority.

See also CIR, trTCM.

PKI

Public key infrastructure. A hierarchy of trust that enables users of a public network to securely and privately exchange data through the use of public and private cryptographic key pairs that are obtained and shared with peers through a trusted authority.

PLMN

Public Land Mobile Network. A telecommunications network for mobile stations.

PLP

Packet loss priority. Used to determine the random early detection (RED) drop profile when a packet is queued. You can set it by configuring a classifier or policer. The system supports two PLP designations: low and high.

PLP bit

Packet loss priority bit. Used to identify packets that have experienced congestion or are from a transmission that exceeded a service provider's customer service license agreement. This bit can be used as part of a router's congestion control mechanism and can be set by the interface or by a filter.

PLR

Point of local repair. The ingress router of a backup tunnel or a detour label-switched path (LSP).

point-to-multipoint connection

Unidirectional connection in which a single source system transmits data to multiple destination end systems. Point-to-multipoint is one of two fundamental connection types.

See also **point-to-point connection**.

point-to-multipoint LSP

Resource Reservation Protocol (RSVP)-signaled label-switched path (LSP) with a single source and multiple destinations.

point-to-point connection

Unidirectional or bidirectional connection between two end systems. Point-to-point is one of two fundamental connection types.

See also **point-to-multipoint connection**.

poison reverse

Method used in distance-vector networks to avoid routing loops. Each router advertises routes back to the neighbor it received them from with an infinity metric assigned.

policer

Filter that limits traffic of a certain class to a specified bandwidth or burst size. Packets exceeding the policer limits are discarded, or assigned to a different forwarding class, a different loss priority, or both.

policing

Method of applying rate limits on bandwidth and burst size for traffic on a particular interface.

policy chain

Application of multiple routing policies in a single location. The policies are evaluated in a predefined manner and are always

followed by the default policy for the specific application location.

pop

Removal of the last label, by a router, from a packet as it exits a Multiprotocol Label Switching (MPLS) domain.

port mirroring

Method in which a copy of an IPv4 packet is sent from the routing platform to an external host address or a packet analyzer for analysis.

PPP

Point-to-Point Protocol. A Link layer protocol that provides multiprotocol encapsulation. PPP is used for Link layer and Network layer configuration. Provides a standard method for transporting multiprotocol datagrams over point-to-point links. Defined in RFC 1661.

pppd

Point-to-Point Protocol daemon that processes packets that use the Point-to-Point Protocol (PPP).

PPPoE

Point-to-Point Protocol over Ethernet. Network protocol that encapsulates Point-to-Point Protocol (PPP) frames in Ethernet frames and connects multiple hosts over a simple bridging access device to a remote access concentrator.

PPPoE over ATM

Point-to-Point Protocol over Ethernet frames in Asynchronous Transfer Mode. Network protocol that encapsulates Point-to-Point Protocol over Ethernet (PPPoE) frames in Asynchronous Transfer Mode (ATM) frames for digital subscriber line (DSL) transmission, and connects multiple hosts over a simple bridging access device to a remote access concentrator.

precedence bits

First three bits in the type-of-service (ToS) byte. On a Juniper Networks router, these bits are used to sort or classify individual packets as they arrive at an interface. The classification determines the queue to which the packet is directed upon transmission.

preference

Desirability of a route to become the active route. A route with a lower preference value is more likely to become the active route. The preference is an arbitrary value from 0 through 255 that the routing protocol process uses to rank routes received from different protocols, interfaces, or remote systems.

preferred address

On an interface, the default local address used for packets sourced by the local router to destinations on the subnet.

prefix-length-range

JUNOS software routing policy match type representing all routes that share the same most-significant bits. The prefix length of the route must also lie between the two supplied lengths in the route filter.

primary address

On an interface, the address used by default as the local address for broadcast and multicast packets sourced locally and sent out the interface.

primary contributing route

Contributing route with the numerically smallest prefix and smallest JUNOS software preference value. This route is the default next hop used for a generated route.

primary interface

Router interface that packets go out on when no interface name is specified and when the destination address does not specify a particular outgoing interface.

promiscuous mode

Used with Asynchronous Transfer Mode (ATM) CCC Cell Relay encapsulation, enables mapping of all incoming cells from an interface port or from a virtual path (VP) to a single label-switched path (LSP) without restricting the VCI number.

protocol address

Logical Layer 3 address assigned to an interface within the JUNOS software.

protocol families

Grouping of logical properties within an interface configuration; for example, the inet, inet4, and Multiprotocol Label Switching (MPLS) families.

Protocol Independent Multicast (PIM)

A protocol-independent multicast routing protocol. PIM Dense mode is a flood-and-prune protocol. PIM Sparse mode routes to multicast groups that use join messages to receive traffic. PIM Sparse-Dense mode allows some multicast groups to be dense groups (flood and prune) and some groups to be sparse groups (join and leave).

protocol preference

A 32-bit value assigned to all routes placed into the routing table. The protocol preference is used as a tiebreaker when multiple exact routes are placed into the table by different protocols.

provider router

Router in the service provider's network that is not connected to a customer edge (CE) device.

Prune message

Physical Interface Module (PIM) message sent upstream to a multicast source or the rendezvous point (RP) of the domain. The message requests that multicast traffic stop being transmitted to the router originating the message.

PSN

Packet-switched network. Network in which messages or fragments of messages (packets) are sent to their destinations through the most expedient route, as determined by a routing algorithm. Packet switching optimizes bandwidth in a network and minimizes latency.

PSNP

Partial sequence number PDU. A packet that contains only a partial list of the label-switched paths (LSPs) in the Intermediate System-to-Intermediate System Level 1 (IS-IS) link-state database.

public key infrastructure

See PKI.

push

Addition of a label or stack of labels, by a router, to a packet as it enters a Multiprotocol Label Switching (MPLS) domain.

PVC

Permanent virtual circuit. A software-defined logical connection in a network.

See also SVC.

QoS

Quality of service. Performance, such as transmission rates and error rates, of a communications channel or system.

quad-wide

Type of Physical Interface Card (PIC) that combines the PIC and Flexible PIC Concentrator (FPC) within a single FPC slot.

qualified next hop

Next hop for a static route that allows a second next hop for the same static route to have different metric and preference properties from the original next hop.

querier router

Physical Interface Module (PIM) router on a broadcast subnet responsible for generating Internet Group Management Protocol (IGMP) query messages for the segment.

queue

First-in, first-out (FIFO) number of packets waiting to be forwarded over a router interface. You can configure the minimum and maximum sizes of the packet queue, queue admission policies, and other parameters to manage the flow of packets through the router.

queue fullness

For random early detection (RED), the memory used to store packets expressed as a percentage of the total memory allocated for that specific queue.

See also **drop profile**.

queue length

For ATM1 interfaces only, a limit on the number of transmit packets that can be queued. Packets that exceed the limit are dropped.

See also EPD.

queuing

In routing, the arrangement of packets waiting to be forwarded. Packets are organized into queues according to their priority, time of arrival, or other characteristics, and are processed one at a time. After a packet is sent to the outgoing interface on a router, it is queued for transmission on the physical media. The amount of time a packet is queued on the router is determined by the availability of the outgoing physical media, bandwidth, and amount of traffic using the interface.

RA

Registration authority. A trusted third-party organization that acts on behalf of a certificate authority (CA) to verify the identity of a digital certificate user.

radio frequency interference

See RFI.

RADIUS

Remote Authentication Dial-In User Service. An authentication method for validating users who attempt to access the router using Telnet.

RBOC

(Pronounced "are-bock.") Regional Bell operating company. Regional telephone companies formed as a result of the divestiture of the Bell System.

RC2, RC4, RC5

RSA codes. A family of proprietary (RSA Data Security, Inc.) encryption schemes often used in web browsers and servers. These codes use variable-length keys up to 2,048 bits.

RDBMS

Relational database management system. A system that presents data in a tabular form with a means of manipulating the tabular data with relational operators.

RDM

Russian-dolls bandwidth allocation model. An allocation model that makes efficient use of bandwidth by allowing the class types to share bandwidth. RDM is defined in the Internet draft draft-ietf-tewg-diff-te-russian-03.txt, "Russian Dolls Bandwidth Constraints Model for Diff-Serv-aware MPLS Traffic Engineering."

receive

Next hop for a static route that allows all matching packets to be sent to the routing engine (RE) for processing.

recursive lookup

Method of consulting the routing table to locate the actual physical next hop for a route when the supplied next hop is not directly connected.

RED

Random early detection. Gradual drop profile for a given class that is used for congestion avoidance. RED tries to anticipate incipient congestion by dropping a small percentage of packets from the head of the queue to ensure that a queue never actually becomes congested.

refresh reduction

In the Resource Reservation Protocol (RSVP), an extension that addresses the problems of scaling, reliability, and latency when Refresh messages are used to cover message loss.

Register message

Physical Interface Module (PIM) message unicast by the first hop router to the rendezvous point (RP) that contains the multicast packets from the source encapsulated within its data field.

Register Stop message

Physical Interface Module (PIM) message sent by the rendezvous point (RP) to the first hop router to halt the sending of encapsulated multicast packets.

registration authority

See RA.

reject

Next hop for a configured route that drops all matching packets from the network and returns an Internet Control Message Protocol (ICMP) message to the source IP address. Also used as an action in a routing policy or firewall filter.

rename

JUNOS software command that allows a user to change the name of a routing policy, firewall filter, or any other variable character string defined in the router configuration.

Request message

Routing Information Protocol (RIP) message used by a router to ask for all or part of the routing table from a neighbor.

resolve

Next hop for a static route that allows the router to perform a recursive lookup to locate the physical next hop for the route.

Response message

Routing Information Protocol (RIP) message used to advertise routing information into a network.

result cell

JUNOS software data structure generated by the Internet Processor ASIC after performing a forwarding table lookup.

ResvConf message

Resource Reservation Protocol (RSVP) message that allows the egress router to receive an explicit confirmation message from a neighbor that its Resv message was received.

ResvErr message

Resource Reservation Protocol (RSVP) message indicating that an error has occurred along an established label-switched path

(LSP). The message is advertised downstream toward the egress router and it does not remove any RSVP soft state from the network.

ResvTear message

Resource Reservation Protocol (RSVP) message indicating that the established label-switched path (LSP) and its associated soft state should be removed by the network. The message is advertised upstream toward the ingress router.

revert timer

For SONET Automatic Protection Switching (APS), a timer that specifies the amount of time (in seconds) to wait after the working circuit has become functional before making the working circuit active again.

rewrite rules

Set the appropriate class-of-service (CoS) bits in an outgoing packet. This allows the next downstream router to classify the packet into the appropriate service group.

RFC

Request for Comments. Internet standard specifications published by the Internet Engineering Task Force (IETF).

RFI

Radio frequency interface. Interference from high-frequency electromagnetic waves emanating from electronic devices.

RIB

Routing information base. A logical data structure used by the Border Gateway Protocol (BGP) to store routing information.

See also **routing table**.

RID

Router ID. An IP address used by a router to uniquely identify itself to a routing protocol. This address may not be equal to a configured interface address.

RIP

Routing Information Protocol. Used in IPv4 networks, a distance-vector interior gateway protocol that makes routing decisions based on hop count.

RIPng

Routing Information Protocol next generation. Used in IPv6 networks, a distance-vector interior gateway protocol that makes routing decisions based on hop count.

RMON

Remote monitoring. A standard Management Information Base (MIB) that defines current and historical Media Access Control (MAC)-layer statistics and control objects, allowing you to capture real-time information across the entire network. This allows you to detect, isolate, diagnose, and report potential and actual network problems.

RNC

Radio network controller. Manages the radio part of the network in UMTS.

route distinguisher

A 6-byte value identifying a virtual private network (VPN) that is prefixed to an IPv4 address to create a unique IPv4 address. The new address is part of the VPN IPv4 address family, which is a Border Gateway Protocol (BGP) address family added as an extension to BGP. It allows you to configure private addresses within the VPN by preventing overlap with the private addresses in other VPNs.

route filter

JUNOS software syntax used in a routing policy to match an individual route or a group of routes.

route flapping

Condition of network instability whereby a route is announced and withdrawn repeatedly, often as a result of an intermittently failing link.

route identifier

IP address of the router from which a Border Gateway Protocol (BGP), Interior Gateway Protocol (IGP), or Open Shortest Path First (OSPF) packet originated.

route redistribution

Method of placing learned routes from one protocol into another protocol operating on the same router. The JUNOS software accomplishes this with a routing policy.

route reflection

In the Border Gateway Protocol (BGP), the configuration of a group of routers into a cluster in which one system acts as a route reflector, redistributing routes from outside the cluster to all routers in the cluster. Routers in a cluster do not need to be fully meshed.

router ID

See RID.

router-link advertisement

Open Shortest Path First (OSPF) link state advertisement (LSA) flooded throughout a single area by all routers to describe the state and cost of the router's links to the area.

router LSA

Open Shortest Path First (OSPF) link state advertisement (LSA) sent by each router in the network. It describes the local router's connected subnets and their metric values.

router priority

Numerical value assigned to an Open Shortest Path First (OPSF) or Intermediate System-to-Intermediate System Level 1 (IS-IS) interface that is used as the first criterion in electing the designated router or designated intermediate system, respectively.

routing engine

Portion of the router that handles all routing protocol processes, as well as other software processes that control the router's interfaces, some of the chassis components, system management, and user access to the router.

routing instance

Collection of routing tables, interfaces, and routing protocol parameters. The set of interfaces is contained in the routing tables, and the routing protocol parameters control the information in the routing tables.

routing matrix

Terabit routing system interconnecting up to four T640 routing nodes and a TX Matrix platform to deliver up to 2.56 terabits per second (Tbps) of subscriber switching capacity.

routing table

Common database of routes learned from one or more routing protocols. All routes are maintained by the JUNOS routing protocol process.

RP

Rendezvous point. For Physical Interface Module (PIM) Sparse mode, a core router acting as the root of the distribution tree in a shared tree.

RPC

Remote procedure call. A type of protocol that allows a computer program running on one computer to cause a function on another computer to be executed without explicitly coding the details for this interaction.

rpd

JUNOS software routing protocol process (daemon). A user-level background process responsible for starting, managing, and stopping the routing protocols on a Juniper Networks router.

RPF

Reverse path forwarding. An algorithm that checks the unicast routing table to determine whether there is a shortest path back to the source address of the incoming multicast packet. Unicast RPF helps to determine the source of denial-of-service (DoS) attacks and rejects packets from unexpected source addresses.

RPM

1. Reverse-path multicasting. Routing algorithm used by the Distance Vector Multicast Routing Protocol (DVMRP) to forward multicast traffic. 2. Real-time Performance Monitoring. A tool for creating active probes to track and monitor traffic.

RRO

Record route object. A Resource Reservation Protocol (RSVP) message object that notes the IP address of each router along the path of a label-switched path (LSP).

RSVP

Resource Reservation Protocol. A signaling protocol that establishes a session between two routers to transport a specific traffic flow.

RSVP Path message

Resource Reservation Protocol (RSVP) message sent by the ingress router downstream toward the egress router. It begins the establishment of a soft state database for a particular label-switched path (LSP).

RSVP Resv message

Resource Reservation Protocol (RSVP) message sent by the egress router upstream toward the ingress router. It completes the establishment of the soft state database for a particular label-switched path (LSP).

RSVP signaled LSP

Label-switched path (LSP) that is dynamically established using Resource Reservation Protocol (RSVP) Path and Resv messages.

RSVP-TE

RSVP traffic engineering; Resource Reservation Protocol (RSVP) with traffic engineering extensions as defined by RFC 3209. These extensions allow RSVP to establish label-switched paths (LSPs) in Multiprotocol Label Switching (MPLS) networks.

See also MPLS, RSVP.

RTP

Real-time Transport Protocol. An Internet protocol that provides mechanisms for the transmission of real-time data, such as audio, video, or voice, over IP networks. Compressed RTP is used for Voice over IP traffic.

RTVBR

Real-time variable bit rate. For ATM2 intelligent queuing (IQ) interfaces, data that is serviced at a higher priority rate than other VBR data. RTVBR is suitable for carrying packetized video and audio. RTVBR provides better congestion control and latency guarantees than non-real-time VBR.

SA

Security association. An IPSec term that describes an agreement between two parties about what rules to use for authentication and encryption algorithms, key exchange mechanisms, and secure communications.

sampling

Method whereby the sampling key based on the IPv4 header is sent to the routing engine (RE). There, the key is placed in a file, or cflowd packets based on the key are sent to a cflowd server.

SAP

1. Session Announcement Protocol. Used with multicast protocols to handle session conference announcements. 2. Service access point. Device that identifies routing protocols and provides the connection between the network interface card and the rest of the network.

SAR

Segmentation and reassembly. Buffering used with Asynchronous Transfer Mode (ATM).

SCB

System Control Board. On an M40 router, the part of the Packet Forwarding Engine (PFE) that performs route lookups, monitors system components, and controls Flexible PIC Concentrator (FPC) resets.

SCC

Switch-card chassis. Term used by the JUNOS command-line interface (CLI) to refer to the TX Matrix platform in a routing matrix.

SCEP

Simple Certificate Enrollment Protocol. A protocol for digital certificates that supports certificate authority (CA) and registration authority (RA) public key distribution, certificate enrollment, certificate revocation, certificate queries, and certificate revocation list (CRL) queries.

SCG

SONET Clock Generator. On a T640 routing node, provides the Stratum 3 clock signal for the SONET/SDH interfaces. Also provides external clock inputs.

scheduler maps

In class of service (CoS), associate schedulers with forwarding classes.

See also schedulers, forwarding classes.

schedulers

Define the priority, bandwidth, delay buffer size, rate control status, and random early detection (RED) drop profiles to be applied to a particular forwarding class for packet transmission.

See also scheduler maps.

scheduling

Method of determining which type of packet or queue is transmitted before another. An individual router interface can have multiple queues assigned to store packets. The router then determines which queue to service based on a particular method of scheduling. This process often involves a determination of which type of packet should be transmitted before another. For example, first in, first out (FIFO).

See also FIFO.

SCP

Secure copy. Means of securely transferring computer files between a local and remote host or between two remote hosts, using the Secure Shell (SSH) protocol.

SCU

Source class usage. A means of tracking traffic originating from specific prefixes on the provider core router and destined for specific prefixes on the customer edge router, based on the IP source and destination addresses.

SDH

Synchronous Digital Hierarchy. A CCITT variation of the SONET standard.

SDP

Session Description Protocol. Used with multicast protocols to handle session conference announcements.

SDRAM

Synchronous dynamic random access memory. An electronic standard in which the inputs and outputs of SDRAM data are synchronized to an externally supplied clock, allowing for extremely fast consecutive read and write capacity.

SDX software

Service Deployment System software. A customizable Juniper Networks product with which service providers can rapidly deploy IP services—such as video on demand (VoD), IP television, stateful firewalls, Layer 3 virtual private networks (VPNs), and bandwidth on demand (BoD)—to hundreds of thousands of subscribers over a variety of broadband access technologies.

services interface

Interface that provides specific capabilities for manipulating traffic before it is delivered to its destination; for example, the adaptive services interface and the tunnel services interface.

See also network interface.

session attribute object

Resource Reservation Protocol (RSVP) message object used to control the priority, preemption, affinity class, and local rerouting of the label-switched path (LSP).

SFM

Switching and Forwarding Module. On an M160 router, a component of the Packet Forwarding Engine (PFE) that provides route lookup, filtering, and switching to Flexible PIC Concentrators (FPCs).

SFP

Small Form-factor Pluggable transceiver. A transceiver that provides support for optical or copper cables. SFPs are hot-insertable and hot-removable.

See also XFP.

SGSN

Serving GPRS Support Node. Device in the mobile network that requests PDP contexts with a GGSN.

SHA-1

Secure Hash Algorithm 1. A secure hash algorithm standard defined in FIPS PUB 180-1 (SHA-1). Developed by the National Institute of Standards and Technology (NIST), SHA-1 (which effectively replaces SHA-0) produces a 160-bit hash for message authentication. Longer-hash variants include SHA-224, SHA-256, SHA-384, and SHA-512 (sometimes grouped under the name "SHA-2"). SHA-1 is more secure than Message Digest 5 (MD5).

See also hashing, MD5.

sham link

Unnumbered point-to-point intra-area link advertised by a type 1 link state advertisement (LSA).

shaping rate

In class of service (CoS), controls the maximum rate of traffic transmitted on an interface.

See also traffic shaping.

shared scheduling and shaping

Allocation of separate pools of shared resources to subsets of logical interfaces belonging to the same physical port.

shared tree

Multicast forwarding tree established from the rendezvous point (RP) to the last hop router for a particular group address.

SHDSL

Symmetric high-speed digital subscriber line. A standardized multirate symmetric DSL that transports rate-adaptive symmetrical data across a single copper pair at data rates from 192 Kbps to 2.3 Mbps, or from 384 Kbps to 4.6 Mbps over two pairs, covering applications served by HDSL, SDSL, T1, E1, and services beyond E1. SHDSL conforms to the following recommendations: ITU G.991.2 G.SHDSL, ETSI TS 101-524 SDSL, and the ANSI T1E1.4/2001-174 G.SHDSL.

See also G.SHDSL.

shim header

Location of the Multiprotocol Label Switching (MPLS) header in a data packet. The JUNOS software always places (shims) the header between the existing Layer 2 and Layer 3 headers.

Shortest Path First

See SPF.

shortest-path tree

See SPT.

SIB

Switch Interface Board. On a T640 routing node, provides the switching function to the destination Packet Forwarding Engine (PFE).

signaled path

In traffic engineering, an explicit path; that is, a path determined using Resource Reservation Protocol (RSVP) signaling. The ERO carried in the packets contains the explicit path information.

Simple Network Management Protocol

See SNMP.

simplex interface

Interface that treats packets it receives from itself as the result of a software loopback process. The interface does not consider these packets when determining whether the interface is functional.

single-mode fiber

Optical fiber designed for transmission of a single ray or mode of light as a carrier and used for long-distance signal transmission. For short distances, multimode fiber is used.

See also MMF.

SIP

Session Initiation Protocol. An Adaptive Services application protocol option used for setting up sessions between endpoints on the Internet. Examples include telephony, fax, videoconferencing, file exchange, and person-to-person sessions.

SNA

System Network Architecture. IBM proprietary networking architecture consisting of a protocol stack that is used primarily in banks and other financial transaction networks.

SNMP

Simple Network Management Protocol. A protocol governing network management and the monitoring of network devices and their functions.

soft state

In Resource Reservation Protocol (RSVP), controls state in hosts and routers that expires if not refreshed within a specified amount of time.

SONET

Synchronous Optical Network. A high-speed (up to 2.5 Gbps) synchronous network specification developed by Bellcore and designed to run on optical fiber. STS1 is the basic building block of SONET. Approved as an international standard in 1988.

See also **SDH**.

source-based tree

Multicast forwarding tree established from the source of traffic to all interested receivers for a particular group address. It is often used in a Dense-mode forwarding environment.

Sparse mode

Method of operating a multicast domain where sources of traffic and interested receivers meet at a central rendezvous point (RP). A Sparse-mode network assumes that there are very few receivers for each group address.

SPF

Shortest Path First. An algorithm used by Intermediate System-to-Intermediate System Level 1 (IS-IS) and Open Shortest Path First (OSPF) to make routing decisions based on the state of network links. Also called the *Dijkstra algorithm.*

SPI

Security Parameter Index. In IPSec, a numeric identifier used with the destination address and security protocol to identify a security association (SA). When Internet Key Exchange (IKE) is used to establish an SA, the SPI is randomly derived. When manual configuration is used for an SA, the SPI must be entered as a parameter.

SPID

Service Profile Identifier. Used only in Basic Rate Interface (BRI) implementations of the Integrated Services Digital Network (ISDN). The SPID specifies the services available on the service provider switch and defines the feature set ordered when the ISDN service is provisioned.

split horizon

Method used in distance-vector networks to avoid routing loops. Each router does not advertise routes back to the neighbor from which it received them.

SPQ

Strict-priority queuing. A dequeuing method that provides a special queue that is serviced until it is empty. The traffic sent to this queue tends to maintain a lower latency and more consistent latency numbers than traffic sent to other queues.

See also **APQ**.

SPT

Shortest-path tree. An algorithm that builds a network topology that attempts to minimize the path from one router (the root) to other routers in a routing area.

SQL

> Structured Query Language. International standard language used to create, modify, and select data from relational databases.

src port

> Transmission Control Protocol (TCP) or User Datagram Protocol (UDP) port for the source IP address in a packet.

SS7

> Signaling System 7. A protocol used in telecommunications for delivering calls and services.

SSAP

> Source service access point. Device that identifies the origin of an LPDU on a data link switching (DLSw) network.

SSB

> System and Switch Board. On an M20 router, a Packet Forwarding Engine (PFE) component that performs route lookups and component monitoring and monitors Flexible PIC Concentrator (FPC) operation.

SSH

> Secure Shell. A protocol that uses strong authentication and encryption for remote access across a nonsecure network. SSH provides remote login, remote program execution, file copy, and other functions. In a Unix environment, SSH is intended as a secure replacement for rlogin, rsh, and rcp.

SSH/TLS

> Secure Shell with Transport Layer Security. A combination of two standard methods used to secure communications over the Internet. TLS is the name of a standard protocol based on SSL 3.0 and is defined in RFC 2246. In combination, SSH/TLS is also known as SSHv2 and uses FIPS-restricted cipher sets in a FIPS environment.

SSL

> Secure Sockets Layer. A protocol that encrypts security information using public-private key technology, which requires a paired private key and authentication certificate, before transmitting data across a network.

SSM

> Source-specific multicast. A service that allows a client to receive multicast traffic directly from the source. Typically, SSM uses a subset of the Physical Interface Module (PIM) Sparse-mode functionality along with a subset of IGMPv3 to create a shortest-path tree (SPT) between the client and the source, but it builds the SPT without the help of a rendezvous point (RP).

SSP

> Switch-to-Switch Protocol. Protocol implemented between two data link switching (DLSw) routers that establishes connections, locates resources, forwards data, and handles error recovery and flow control.

SSRAM

> Synchronous static random access memory. Used for storing routing tables, packet pointers, and other data such as route lookups, policer counters, and other statistics to which the microprocessor needs quick access.

standard AAL5 mode

> Transport mode that allows multiple applications to tunnel the Protocol Data Units of their Layer 2 protocols over an Asynchronous Transfer Mode (ATM) virtual circuit. You use this transport mode to tunnel IP packets over an ATM backbone.
>
> See also AAL5 mode, cell-relay mode, Layer 2 circuits, trunk mode.

starvation

> Problem that occurs when lower-priority traffic, such as data and protocol packets, is locked out (starved) because a higher-priority queue uses all of the available transmission bandwidth.

stateful firewall filter

> Type of firewall filter that evaluates the context of connections, permits or denies traffic based on the context, and updates this information dynamically. Context includes IP

source and destination addresses, port numbers, Transmission Control Protocol (TCP) sequencing information, and TCP connection flags. The context established in the first packet of a TCP session must match the context contained in all subsequent packets if a session is to remain active.

See also stateless firewall filter.

stateful firewall recovery

Recovery strategy that preserves parameters concerning the history of connections, sessions, or application status before failure.

See also stateless firewall recovery.

stateless firewall filter

Type of firewall filter that statically evaluates the contents of packets transiting the router and packets originating from or destined for the routing engine (RE). Packets are accepted, rejected, forwarded, or discarded and collected, logged, sampled, or subjected to classification according to a wide variety of packet characteristics. Sometimes called access control lists (ACLs) or simply firewall filters, stateless firewall filters protect the processes and resources owned by the RE. A stateless firewall filter can evaluate every packet, including fragmented packets. In contrast to a stateful firewall filter, a stateless firewall filter does not maintain information about connection states.

See also stateful firewall filter.

stateless firewall recovery

Recovery strategy that does not attempt to preserve the history of connections, sessions, or application status before failure.

See also stateful firewall recovery.

static LSP

See static path.

static path

In the context of traffic engineering, a static route that requires hop-by-hop manual configuration. No signaling is used to create or maintain the path. Also called a *static LSP*.

static route

Explicitly configured route that is entered into the routing table. Static routes have precedence over routes chosen by dynamic routing protocols.

static RP

One of three methods of learning the rendezvous point (RP) to group address mapping in a multicast network. Each router in the domain must be configured with the required RP information.

S/T interface

System reference point/terminal reference point interface. A four-pair connection between the Integrated Services Digital Network (ISDN) provider service and the customer terminal equipment.

STM

Synchronous transport module. CCITT specification for SONET at 155.52 Mbps.

strict

In the context of traffic engineering, a route that must go directly to the next address in the path. (Definition from RFC 791, modified to fit LSPs.)

strict hop

Routers in a Multiprotocol Label Switching (MPLS) named path that must be directly connected to the previous router in the configured path.

STS

Synchronous transport signal. Synchronous transport signal level 1 is the basic building block signal of SONET, operating at 51.84 Mbps. Faster SONET rates are defined as STS-n, where n is an integer by which the basic rate of 51.84 Mbps is multiplied.

See also SONET.

stub area

In Open Shortest Path First (OSPF), an area through which, or into which, Autonomous System (AS) external advertisements are not flooded.

STU-C

Symmetric high-speed digital subscriber line (SHDSL) transceiver unit—central office. Equipment at the telephone company central office that provides SHDSL connections to remote user terminals.

STU-R

Symmetric high-speed digital subscriber line (SHDSL) transceiver unit—remote. Equipment at the customer premises that provides SHDSL connections to remote user terminals.

sub-LSP

Part of a point-to-multipoint label-switched path (LSP). A sub-LSP carries traffic from the main LSP to one of the egress Provider Edge (PE) routers. Each point-to-multipoint LSP has multiple sub-LSPs.

See also point-to-multipoint LSP.

subnet mask

Number of bits of the network address used for the host portion of a Class A, Class B, or Class C IP address.

subrate value

Value that reduces the maximum allowable peak rate by limiting the High-level Data Link Control (HDLC)-encapsulated payload. The subrate value must exactly match that of the remote channel service unit (CSU).

summary link advertisement

Open Shortest Path First (OSPF) link-state advertisement (LSA) flooded throughout the advertisement's associated areas by area border routers (ABRs) to describe the routes that they know about in other areas.

SVC

Switched virtual connection. A dynamically established, software-defined logical connection that stays up as long as data is being transmitted. When transmission is complete, the software tears down the SVC.

See also PVC.

sysid

System identifier. Portion of the ISO non-client peer. The system ID can be any six bytes that are unique throughout a domain.

syslog

System log. A method for storing messages to a file for troubleshooting or record-keeping. It can also be used as an action within a firewall filter to store information to the messages file.

T1

Basic physical layer protocol used by the Digital Signal level 1 (DS1) multiplexing method in North America. A T1 interface operates at a bit rate of 1.544 Mbps and can support 24 DS0 channels.

T3

Physical layer protocol used by the Digital Signal level 3 (DS3) multiplexing method in North America. A T3 interface operates at a bit rate of 44.736 Mbps.

TACACS+

Terminal Access Controller Access Control System Plus. Authentication method for validating users who attempt to access the router using Telnet.

tail drop

Queue management algorithm for dropping packets from the input end (tail) of the queue when the length of the queue exceeds a configured threshold.

See also RED.

T-carrier

Generic designator for any of several digitally multiplexed telecommunications carrier systems originally developed by Bell Labs and used in North America and Japan.

TCM

Tricolor marking. Traffic policing mechanism that extends the functionality of class-of-service (CoS) traffic policing by providing three levels of drop precedence (loss priority or PLP) instead of two. There are two types of TCM: single-rate and two-rate. The

JUNOS software currently supports two-rate TCM only.

See also trTCM.

TCP

Transmission Control Protocol. Works in conjunction with IP to send data over the Internet. Divides a message into packets and tracks the packets from point of origin to destination.

tcpdump

Unix packet monitoring utility used by the JUNOS software to view information about packets sent or received by the routing engine (RE).

TCP port 179

Well-known port number used by the Border Gateway Protocol (BGP) to establish a peering session with a neighbor.

TDMA

Time-Division Multiplex Access. A type of multiplexing in which two or more channels of information are transmitted over the same link, where the channels take turns to use the link. Each link is allocated a different time interval ("slot" or "slice") for the transmission of each channel. For the receiver to distinguish one channel from the other, some kind of periodic synchronizing signal or distinguishing identifier is required.

See also GSM.

TEI

Terminal Endpoint Identifier. A terminal endpoint can be any Integrated Services Digital Network (ISDN)-capable device attached to an ISDN network. The TEI is a number between 0 and 127, where 0 through 63 are used for static TEI assignment, 64 through 126 are used for dynamic assignment, and 127 is used for group assignment.

terminating action

Action in a routing policy or firewall filter that halts the logical software processing of a policy or filter.

terms

Used in a routing policy or firewall filter to segment the policy or filter into small match and action pairs.

through

JUNOS software routing policy match type representing all routes that fall between the two supplied prefixes in the route filter.

Time-Division Multiplex Access

See TDMA.

time-division multiplexed channel

Channel derived from a given frequency and transmitted over a single wire or wireless medium. The channel is preassigned a time slot whether or not there is data to transmit.

timeout timer

Used in a distance-vector protocol to ensure that the current route is still usable for forwarding traffic.

TNP

Trivial Network Protocol. A Juniper Networks proprietary protocol automatically configured on an internal interface by the JUNOS software. TNP is used to communicate between the routing engine (RE) and components of the Packet Forwarding Engine (PFE), and is critical to the operation of the router.

token-bucket algorithm

Used in a rate-policing application to enforce an average bandwidth while allowing bursts of traffic up to a configured maximum value.

ToS

Type of service. The method of handling traffic using information extracted from the fields in the ToS byte to differentiate packet flows.

totally stubby area

Open Shortest Path First (OSPF) area type that prevents Type 3, 4, and 5 link state advertisements (LSAs) from entering the non-backbone area.

traffic engineering

Process of selecting the paths chosen by data traffic to balance the traffic load on the various links, routers, and switches in the network. (Definition from *http://www.ietf.org/ internet-drafts/draft-ietf-mpls-framework -04.txt.*)

See also MPLS.

traffic engineering class

In Differentiated Services-aware traffic engineering, a paired class type and priority.

traffic engineering class map

In Differentiated Services-aware traffic engineering, a map among the class types, priorities, and traffic engineering classes. The traffic engineering class mapping must be consistent across the Differentiated Services domain.

traffic policing

Examines traffic flows and discards or marks packets that exceed service-level agreements (SLAs).

traffic sampling

Method used to capture individual packet information of traffic flow at a specified time period. The sampled traffic information is placed in a file and stored on a server for various types of analysis.

See also packet capture.

traffic shaping

Reduces the potential for network congestion by placing packets in a queue with a shaper at the head of the queue. Traffic shaping tools regulate the rate and volume of traffic admitted to the network.

See also shaping rate.

transient change

Commit-script-generated configuration change that is loaded into the checkout configuration, but not into the candidate configuration. Transient changes are not saved in the configuration if the associated commit script is deleted or deactivated.

See also persistent change.

transient interface

Interface that can be configured on a routing platform depending on your network needs. Unlike a permanent interface that is required for router operation, a transient interface can be disabled or removed without affecting the basic operation of the router.

See also FPC, PIC, permanent interface.

transit area

In Open Shortest Path First (OSPF), an area used to pass traffic from one adjacent area to the backbone or to another area if the backbone is more than two hops away from an area.

transit router

In Multiprotocol Label Switching (MPLS), any intermediate router in the label-switched path (LSP) between the ingress router and the egress router.

transport mode

IPSec mode of operation in which the data payload is encrypted, but the original IP header is left untouched. The IP addresses of the source or destination can be modified if the packet is intercepted. Because of its construction, transport mode can be used only when the communication endpoint and cryptographic endpoint are the same. Virtual private network (VPN) gateways that provide encryption and decryption services for protected hosts cannot use transport mode for protected VPN communications.

See also tunnel mode.

transport plane

See data plane.

trap

Reports significant events occurring on a network device, most often errors or failures. Simple Network Management Protocol (SNMP) traps are defined in either standard or enterprise-specific Management Information Bases (MIBs).

triggered updates

Used in a distance-vector protocol to reduce the time for the network to converge. When a router has a topology change, it immediately sends the information to its neighbors instead of waiting for a timer to expire.

trTCM

Two-rate TCM polices traffic according to the color classification (loss priority) of each packet. Traffic policing is based on two rates: the committed information rate (CIR) and the peak information rate (PIR). Two-rate TCM is defined in RFC 2698, "A Two Rate Three Color Marker."

See also CIR, PIR.

trunk mode

Layer 2 circuit cell-relay transport mode that allows you to send Asynchronous Transfer Mode (ATM) cells between ATM2 intelligent queuing (IQ) interfaces over a Multiprotocol Label Switching (MPLS) core network. You use Layer 2 circuit trunk mode (as opposed to standard Layer 2 circuit cell-relay mode) to transport ATM cells over an MPLS core network that is implemented between other vendors' switches or routers. The multiple connections associated with a trunk increase bandwidth and provide failover redundancy.

See also AAL5 mode, cell-relay mode, Layer 2 circuits, standard AAL5 mode.

Tspec object

Resource Reservation Protocol (RSVP) message object that contains information such as the bandwidth request of the label-switched path (LSP) as well as the minimum and maximum packets supported.

tunnel

Private, secure path through an otherwise public network.

tunnel endpoint

Last node of a tunnel where the tunnel-related headers are removed from the packet, which is then passed on to the destination network.

tunneling protocol

Network protocol that encapsulates one protocol or session inside another. When protocol A is encapsulated within protocol B, A treats B as though it were a Data Link layer. Tunneling can be used to transport a network protocol through a network that would not otherwise support it. Tunneling can also be used to provide various types of virtual private network (VPN) functionality such as private addressing.

tunnel mode

IPSec mode of operation in which the entire IP packet, including the header, is encrypted and authenticated and a new virtual private network (VPN) header is added, protecting the entire original packet. This mode can be used by both VPN clients and VPN gateways, and protects communications that come from or go to non-IPSec systems.

See also transport mode.

tunnel services interface

Provides the capability of a Tunnel Services PIC on an Adaptive Services PIC (ASP).

See also Tunnel Services PIC.

Tunnel Services PIC

Physical Interface Card (PIC) that allows the router to perform the encapsulation and de-encapsulation of IP datagrams. The Tunnel Services PIC supports IP-IP, Generic Routing Encapsulation (GRE), and Physical Interface Module (PIM) register encapsulation and de-encapsulation. When the Tunnel Services PIC is installed, the router can be a PIM rendezvous point (RP) or a PIM first hop router for a source that is directly connected to the router.

TX Matrix platform

Routing platform that provides the centralized switching fabric of the routing matrix.

UDP

User Datagram Protocol. In Transmission Control Protocol/Internet Protocol (TCP/IP), a connectionless transport layer protocol that exchanges datagrams without acknowledgments or guaranteed delivery, requiring that error processing and retransmission be handled by other protocols.

U interface

User reference point interface. A single-pair connection between the local Integrated Services Digital Network (ISDN) provider and the customer premises equipment.

UME

UNI management entity. The code residing in the Asynchronous Transfer Mode (ATM) devices at each end of a UNI (user-to-network interface) circuit that functions as a Simple Network Management Protocol (SNMP) agent, maintaining network and connection information specified in a Management Information Base (MIB).

UMTS

Universal mobile telecommunications system. Provides third-generation (3G), packet-based transmission of text, digitized voice, video, and multimedia, at data rates up to 2 Mbps.

UNI

User-to-network interface. ATM Forum specification that defines an interoperability standard for the interface between a router or an Asynchronous Transfer Mode (ATM) switch located in a private network and the ATM switches located within the public carrier networks. Also used to describe similar connections in Frame Relay networks.

unicast

Operation of sending network traffic from one network node to another individual network node.

unit

JUNOS software syntax that represents the logical properties of an interface.

unnumbered interface

Logical interface that is configured without an IP address.

Update message

Border Gateway Protocol (BGP) message that advertises path attributes and routing knowledge to an established neighbor.

update timer

Used in a distance-vector protocol to advertise routes to a neighbor on a regular basis.

UPS

Uninterruptible power supply. A device that sits between a power supply and a router or other device and prevents power-source events, such as outages and surges, from affecting or damaging the device.

upto

JUNOS software routing policy match type representing all routes that share the same most-significant bits and whose prefix length is smaller than the supplied subnet in the route filter.

UTC

Coordinated Universal Time. Historically referred to as Greenwich Mean Time (GMT), a high-precision atomic time standard that tracks Universal Time (UT) and is the basis for legal civil time all over the world. Time zones around the world are expressed as positive and negative offsets from UTC.

UTRAN

UMTS Terrestrial Radio Access Network. The WCDMA radio network in UMTS.

VBR

Variable bit rate. For ATM1 and ATM2 intelligent queuing (IQ) interfaces, data that is serviced at a varied rate within defined limits. VBR traffic adds the ability to statistically oversubscribe user traffic.

VC

Virtual circuit. A software-defined logical connection between two network devices that is not a dedicated connection but acts as though it is. It can be either permanent (PVC) or switched (SVC). VCs are used in Asynchronous Transfer Mode (ATM), Frame Relay, and X.25. In EX-specific context, VC stands for "Virtual Chassis," which refers to the interconnection of up to 10 ERX 4200s to form a single logical entity.

See also VPI, VCI, PVC, SVC.

VCI

1. Vapor corrosion inhibitor. Small cylinder packed with the router that prevents corrosion of the chassis and components during shipment. 2. Virtual circuit identifier. A 16-bit field in the header of an Asynchronous Transfer Mode (ATM) cell that indicates the particular virtual circuit the cell takes through a virtual path. Also called a *logical interface*.

See also VPI.

virtual channel

Enables queuing, packet scheduling, and accounting rules to be applied to one or more logical interfaces.

See also virtual channel group.

virtual channel group

Combines virtual channels into a group and then applies the group to one or more logical interfaces.

See also virtual channel.

virtual circuit

Represents a logical connection between two Layer 2 devices in a network.

virtual link

In Open Shortest Path First (OSPF), a link created between two routers that are part of the backbone but are not physically contiguous.

virtual loopback tunnel interface

See VT.

virtual path

Combination of multiple virtual circuits between two devices in an Asynchronous Transfer Mode (ATM) network.

VLAN

Virtual LAN. A logical group of network devices that appear to be on the same LAN, regardless of their physical location. VLANs

are configured with management software, and are extremely flexible because they are based on logical, rather than physical, connections.

VLAN-tagged frame

Tagged frame whose tag header carries both virtual LAN (VLAN) identification and priority information.

VPI

Virtual path identifier. An 8-bit field in the header of an Asynchronous Transfer Mode (ATM) cell that indicates the virtual path the cell takes.

See also VCI.

VPLS

Virtual private LAN service. An Ethernet-based multipoint-to-multipoint Layer 2 virtual private network (VPN) service used for interconnecting multiple Ethernet LANs across a Multiprotocol Label Switching (MPLS) backbone. VPLS is specified in the IETF draft "Virtual Private LAN Service."

VPN

Virtual private network. A private data network that uses a public Transmission Control Protocol/Internet Protocol (TCP/IP) network, typically the Internet, while maintaining privacy with a tunneling protocol, encryption, and security procedures.

See also tunneling protocol.

VRF instance

Virtual private network (VPN) routing and forwarding instance. A Virtual Route and Forwarding (VRF) instance for a Layer 3 VPN implementation consists of one or more routing tables, a derived forwarding table, a set of interfaces that use the forwarding table, and a set of policies and routing protocols that determine what goes into the forwarding table.

VRF table

Routing instance table that stores Virtual Route and Forwarding (VRF) routing information.

See also VRF instance.

VRRP

Virtual Router Redundancy Protocol. On Fast Ethernet and Gigabit Ethernet interfaces, allows you to configure virtual default routers.

VT

Virtual loopback tunnel interface. VT interface that loops packets back to the Packet Forwarding Engine (PFE) for further processing, such as looking up a route in a Virtual Route and Forwarding (VRF) routing table or looking up an Ethernet Media Access Control (MAC) address. A virtual loopback tunnel interface can be associated with a variety of Multiprotocol Label Switching (MPLS) and virtual private network (VPN)-related applications, including VRF routing instances, VPLS routing instances, and point-to-multipoint label-switched paths (LSPs).

warm standby

Method that enables one backup Adaptive Services PIC (ASP) to support multiple active ASPs, without providing guaranteed recovery times.

WAN PHY

Wide Area Network Physical Layer Device. A physical layer device that allows 10 Gigabit Ethernet wide-area links to use fiber-optic cables and other devices intended for SONET/SDH.

See also LAN PHY, PHY.

WAP

Wireless Application Protocol. A standard protocol that enables mobile users to access the Internet in a limited fashion if WAP is supported and enabled on the mobile device, server, and wireless network. WAP users can send and receive email and access websites in text format only (WAP does not support graphics).

WCDMA

Wideband Code Division Multiple Access. Radio interface technology used in most third-generation (3G) systems.

WDM

Wavelength-division multiplexing. Technique for transmitting a mix of voice, data, and video over various wavelengths (colors) of light.

WINS

Windows Internet Name Service. A Windows name resolution service for network basic input/output system (NetBIOS) names. WINS is used by hosts running NetBIOS over TCP/IP (NetBT) to register NetBIOS names and resolve NetBIOS names to IP addresses.

WRR

Weighted round-robin. Scheme used to decide the queue from which the next packet should be transmitted.

XENPAK

Standard that defines a type of pluggable fiber-optic transceiver module that is compatible with the 10 Gigabit Ethernet (10 GE) standard.

XENPAK module

10 Gigabit Ethernet fiber-optic transceiver. XENPAK modules are hot-insertable and hot-removable.

See also MSA.

XENPAK Multisource Agreement
See MSA.

XENPAK-SR 10Base-SR XENPAK

Media type that supports a link length of 26 meters on standard Fiber Distributed Data Interface (FDDI)-grade multimode fiber (MMF). Up to 300-meter link lengths are possible with 2000 MHz/km MMF (OM3).

XENPAK-ZR 10GBase-ZR XENPAK

Media type used for long-reach, single-mode (80–120 km) 10 Gigabit Ethernet metro applications.

XFP

10 Gigabit Small Form-factor Pluggable transceiver. A transceiver that provides support for fiber-optic cables. XFPs are hot-insertable and hot-removable.

XML

See also SFP.

XML

Extensible Markup Language. Language used for defining a set of markers, called tags, which define the function and hierarchical relationships of the parts of a document or data set.

XML schema

Definition of the elements and structure of one or more Extensible Markup Language (XML) documents. Similar to a document type definition (DTD), but with additional information and written in XML.

XOR

Exclusive or. A logical operator (exclusive disjunction) in which the operation yields the result of true when one, and only one, of its operands is true.

XPath

Standard used in Extensible Stylesheet Language for Transformations (XSLT) to specify and locate elements in the input document's Extensible Markup Language (XML) hierarchy. XPath is fully described in the World Wide Web Consortium (W3C) specification at *http://w3c.org/TR/xpath*.

XSLT

Extensible Stylesheet Language for Transformations. A standard for processing Extensible Markup Language (XML) data developed by the World Wide Web Consortium (W3C). XSLT performs XML-to-XML transformations, turning an input XML hierarchy into an output XML hierarchy. The XSLT specification is on the W3C website at *http://www.w3c.org/TR/xslt*.

zeroize

Process of removing all sensitive information, such as cryptographic keys and user passwords, from a router running JUNOS-FIPS.

Index

We'd like to hear your suggestions for improving our indexes. Send email to *index@oreilly.com*.

primary network parameters, 561
question mark (?), 99, 129
queuing, 563
quotation marks ("), 85

R

RADIUS (Remote Authentication Dial-In User
    Service)
    accounting information, 537
    authentication support, 532–534, 550–555
    security considerations, 509
    server configuration, 538, 540
    setting secondary configuration, 125
    user authentication case study, 128–130
random early detection (RED), 563
Rapid Spanning Tree Protocol (see RSTP)
read-only class, 127
readvertisement, 426, 465
Real-time Transport Protocol (RTP), 585
recovering passwords, 183–187
recursion, 417
RED (random early detection), 563
redistribute command, 392
redundancy
    default failover Layer 2, 599–600
    default failover Layer 3, 600–603
    EX hardware overview, 50, 59, 596
    GR support, 610
    GRES support, 603–610
    NSR support, 612
    routing engine failover, 597–599
    VC configuration options, 216
    VRRP support, 404
Redundant Trunk Group (see RTG)
regular expressions
    CLI support, 102
    common operators, 129
reject (filter action), 480
reject next hop, 412
remote access
    setting initial configuration, 122
    setting secondary configuration, 134–138
Remote Authentication Dial-In User Service
    (see RADIUS)
remote bridging, 38
Remote login protocol, 134
remote power supply (RPS), 49
rename function, 104
rendezvous point (RP), 144

repeaters
    5 4 3 rule, 37
    bus in a box, 37
    functionality, 37
request chassis routing-engine command, 256
request chassis routing-engine master switch
    command, 606
request command, 79
request session member command, 238
request system command, 224
request system configuration rescue delete
    command, 182
request system software command, 236
request system storage cleanup command, 169
request virtual-chassis command, 222
request virtual-chassis recycle command, 237
request virtual-chassis renumber command,
    237, 238
request virtual-chassis vc-port command, 204
REs (routing engines)
    control plane considerations, 56
    EX hardware overview, 50, 53
    failover considerations, 597–599
    intersystem packet flows, 214
    permanent interfaces, 143
    PFE support, 57
    VC configuration options, 221
rescue configuration, 182
Resource Reservation Protocol (see RSVP)
restart command, 73, 79, 234
restart ethernet-switching command, 305
rewriting markers, 563
Reynolds, Harry, 47, 126, 385, 430, 453, 508,
    560
RFC 1058, 422
RFC 1388, 422
RFC 1812, 569
RFC 2011, 569
RFC 2108, 569
RFC 2131, 522
RFC 2453, 422
RFC 2464, 269
RFC 2669, 569
RFC 2670, 569
RFC 2674, 569
RFC 2866, 537
RFC 3164, 170
RFC 3478, 610
RFC 3576, 536

About the Authors

Harry Reynolds has more than 29 years of experience in the networking industry, with the last 19 years focused on LANs and LAN interconnection. He is CCIE #4977 and JNCIE #3 certified, and holds various other industry and teaching certifications. Harry was a contributing author of *Juniper Network Complete Reference* (McGraw-Hill), wrote the JNCIE and JNCIP Study Guides (for Sybex Books), and coauthored *JUNOS Enterprise Routing* (*http://oreilly.com/catalog/9780596514426/*) with Doug Marschke (O'Reilly). Prior to joining Juniper, Harry served time in the U.S. Navy as an avionics technician, worked for equipment manufacturer Micom Systems, and spent much time developing and presenting hands-on technical training curricula targeted to both enterprise and service provider needs. Harry has presented classes for organizations such as American Institute, American Research Group, Hill Associates, and Data Training Resources. Harry is currently employed by Juniper Networks as a senior test engineer in the JUNOS software Core Protocols group. Harry has also functioned at Juniper as a consulting engineer on an aerospace routing contract, and as a senior education services engineer working on courseware and certification offerings.

Doug Marschke is an engineering graduate from the University of Michigan and is currently the chief technology officer at Proteus Networks. He is JNCIE-ER #3, JNCIE-M #41, and JNCIS-FW certified. He has been heavily involved in the Juniper certification exams from the start, contributing to the test writing, and is the coauthor of the current JNCIE Enterprise Exam. Doug also coauthored *JUNOS Enterprise Routing* with Harry Reynolds. Doug currently spends his time working with governments, service providers, and enterprises to optimize their IP networks for better performance, cost, and reliability. He also flies around the world and back, sharing his knowledge in a variety of training classes and seminars with topics ranging from troubleshooting to design to certification preparation. If Doug is not on the road, you can find him at his bar in San Francisco, the Taco Shop at Underdogs, ready to serve you a drink.

Colophon

The animal on the cover of *JUNOS Enterprise Switching* is a gorgeted bird of paradise (*Astrapia nigra*), also known as an Arfak Astrapia. The male of this species is black with iridescent purple, green, and gold plumage, whereas the female is predominantly brown. One of the larger birds of paradise, it measures 30 inches with a long, broad tail. It is native to Indonesia and resides in the Arfak Mountains in West Papua, where it subsists on a diet of tropical fruits. Protected by many laws as well as by its geographical isolation, the Arfak Astrapia is not considered a threatened species.

All birds of paradise are members of the family *Paradisaeidae*, and are found only in tropical forests in Indonesia, Papua New Guinea, and parts of Australia. They are famous for their bright plumage and long, elaborate feathers. Perhaps the best-known species is the greater bird of paradise, *Paradisaea apoda*. Native traders in the 18th century sold these birds for decoration, usually removing their wings and feet, which led to the mistaken belief in Europe that the birds were limbless and held permanently aloft by their plumes. This is why Linnaeus named the species *apoda*, "without feet."

The hunting of birds of paradise has occurred for thousands of years, and was heaviest in the late 19th and early 20th centuries when the plumes were used in the millinery trade. Although these birds are now protected, some hunting is still permitted to the tribal societies of New Guinea, many of which use the plumes ceremonially.

The cover image is from Cassell's *Natural History*. The cover font is Adobe ITC Garamond. The text font is Linotype Birka; the heading font is Adobe Myriad Condensed; and the code font is LucasFont's TheSansMonoCondensed.